Blind Equalization and System Identification

Chong-Yung Chi, Chih-Chun Feng,
Chii-Horng Chen and Ching-Yung Chen

Blind Equalization and System Identification

Batch Processing Algorithms, Performance and Applications

With 112 Figures

Springer

Chong-Yung Chi, PhD
Department of Electrical Engineering and Institute of Communications Engineering
National Tsing Hua University
Hsinchu
Taiwan 30013
R.O.C.

Chih-Chun Feng, PhD
Computer and Communications Research Laboratories
Industrial Technology Research Institute
Chutung
Hsinchu
Taiwan 310
R.O.C.

Chii-Horng Chen, PhD
MediaTek Inc.
Science-based Industrial Park
Hsinchu
Taiwan 300
R.O.C.

Ching-Yung Chen, PhD
Computer and Communications Research Laboratories
Industrial Technology Research Institute
Chutung
Hsinchu
Taiwan 310
R.O.C.

British Library Cataloguing in Publication Data
Blind equalization and system identification: batch processing algorithms, performance and applications
1. Signal processing - Digital techniques 2. Discrete-time systems 3. System identification
I. Chi, Chong-Yung
621.3'822
ISBN-10: 1846280222

Library of Congress Control Number: 2005936396

ISBN 1-84628-022-2 ISBN 1-84628-218-7 (eBook) Printed on acid-free paper
ISBN 978-1-84628-022-1

9 8 7 6 5 4 3 2 1

Springer Science+Business Media
springer.com

To my wife, Yi-Teh Lee, and my mother, Yun-Shiow Chen
— Chong-Yung Chi

To my wife, Shu-Hua Wu, and my son, Qian Feng
— Chih-Chun Feng

To my wife, Amy Chang, and my son, Louis Chen
— Chii-Horng Chen

To my wife, Shu-Lin Chiu, and my parents, Lian-Sheng Chen and Li-Hua Liu
— Ching-Yung Chen

Preface

Discrete-time signal processing has appeared to be one of the major momenta to the advances of science and engineering over the past several decades because of the rapid progress of digital and mixed-signal integrated circuits in processing speed, functionality, and cost effectiveness. In many science and engineering applications, the signals of interest are distorted by an unknown physical system (channel) and only a set of discrete-time measurements with system (channel) induced distortion is available. Because both the source signals and measurements are basically random in nature, discrete-time statistical signal processing has played a central role in extracting the source signals of interest and estimating the system characteristics since the 1980s. When the system is known or can be accurately estimated at extra cost, estimation of the source signals via a statistical optimum filter (such as the Wiener filter) is usually straightforward. A typical example is that in digital communications; training or pilot signals are often contained regularly or periodically in transmitted signals to facilitate channel estimation at the receiving end at the expense of bandwidth. Other than digital communications, such an arrangement usually cannot apply to other fields such as seismic exploration, speech analysis, ultrasonic nondestructive evaluation, texture image analysis and classification, etc. These therefore necessitate exploration of blind equalization and system identification, which have been challenging research areas for a long time and continue to be so.

Thus far, there have been developed a great many blind equalization and system identification algorithms, from one-dimensional (1-D) to two-dimensional (2-D) signals, and from single-input single-output (SISO) to multiple-input multiple-output (MIMO) systems. Some of them are closely related but with different perspectives, and may share certain common properties and characteristics proven from their performance analysis. We have studied the two problems for more than ten years and felt that a unified treatment of blind equalization and system identification for 1-D and 2-D (real or complex) signals as well as those for SISO and MIMO (real or complex) systems, from theory, performance analysis, simulation, to implementation and applications, is instrumental in building up sufficient background and learning state-of-the-art works. This book is devoted to such a unified treatment. It is designed as a textbook of graduate-level courses in Discrete-time Random Processes, Statistical Signal Processing, and Blind Equalization and System Identification. It is also suitable for researchers and practicing professionals who are working in the areas of digital communications, statistical signal processing, source separation, speech processing, image processing, seismic exploration, sonar, radar, and so on.

This book comprises eight chapters. Chapter 1 provides an overview of blind equalization and system identification. Chapters 2 and 3 review some requisite background mathematics and statistical signal processing for ease of grasping the material presented in the following chapters. The reader who is familiar with the background may skip these two chapters. Chapter 4 introduces the (1-D) SISO blind equalization algorithms as well as some of their applications, while Chapters 5 and 6 concentrate on the (1-D) MIMO case and Chapters 7 and 8 on the 2-D (SISO) case. Chapter 4 is thought of as a prerequisite for ease of understanding Chapters 5 through 8, while Chapters 5 and 6 can be read independently of Chapters 7 and 8, and vice versa. We have tried our best to make the treatment uniform to all the eight chapters. Some homework problems and computer assignments are included at the end of each chapter (Chapters 2 through 8) so that one can fully understand the material of each chapter through these exercises. The solution manual of homework problems can be obtained by contacting the publisher.

We would like to thank Yu-Hang Lin for his assistance in portraying many figures in Chapters 2 through 4 and his endeavors to optimize the book editing as well as drawings, and Chun-Hsien Peng for providing some computer code for use in some of simulation examples. We also thank Dr Wing-Kin Ma for his valuable comments and suggestions, and Tsung-Han Chan and some of the graduate students of the first author for their proofreading of this book.

Hsinchu, Taiwan
July 2005

Chong-Yung Chi, National Tsing Hua University[1]
Chih-Chun Feng, CCL/ITRI[2]
Chii-Horng Chen, Infineon-ADMtek Co., Ltd.
Ching-Yung Chen, CCL/ITRI[2]

[1] Institute of Communications Engineering, Department of Electrical Engineering
[2] Computer & Communications Research Laboratories, Industrial Technology Research Institute

Contents

1

Introduction

Equalization or *deconvolution* is essentially a signal processing procedure to restore a set of source signals which were distorted by an unknown linear (or nonlinear) system, whereas *system identification* is a signal processing procedure to identify and estimate the unknown linear (or nonlinear) system. The two problems arise in a variety of engineering and science areas such as digital communications, speech signal processing, image signal processing, biomedical signal processing, seismic exploration, ultrasonic nondestructive evaluation, underwater acoustics, radio astronomy, sonar, radar, and so on. As the source signals are known *a priori*, the design of system identification algorithms will be straightforward and effective. Certainly, the design of equalization algorithms will also be straightforward and effective when the system is completely known in advance. When both the source signals and the system are unknown, the equalization and system identification problems are far from trivial. Obviously, the two problems are closely related to each other, and therefore similar design philosophies may frequently apply to the design of both equalization algorithms and system identification algorithms. This book will provide an in-depth introduction to the design, analyses, and applications of equalization algorithms and system identification algorithms with only a given set of discrete-time measurements.

Problems and Approaches

Signals of interest, denoted as $\mathbf{u}[n]$ (a $K \times 1$ vector), may not be measured by a set of M sensors directly, but measurements (or data), denoted as $\mathbf{y}[n]$ (an $M \times 1$ vector), may be related to $\mathbf{u}[n]$ by

$$\mathbf{y}[n] = \sum_{k=-\infty}^{\infty} \mathbf{H}[k]\mathbf{u}[n-k] + \mathbf{w}[n] = \sum_{k=-\infty}^{\infty} \mathbf{H}[n-k]\mathbf{u}[k] + \mathbf{w}[n] \tag{1.1}$$

as shown in Fig. 1.1, where $\mathbf{H}[n]$ (an $M \times K$ matrix) is called the signal distorting system, and $\mathbf{w}[n]$ (an $M \times 1$ vector) is the sensor noise plus other unknown effects of the physical system. For instance, $\mathbf{u}[n]$ includes K users' symbol sequences and $\mathbf{H}[n]$ stands for the channel impulse response matrix in wireless communications that basically accounts for multiple access interference and intersymbol interference. The K-input M-output model above is also referred to as a multiple-input multiple-output (MIMO) *convolutional model* or an MIMO linear time-invariant (LTI) system, including special cases of a single-input multiple-output (SIMO) system for $K = 1$ and $M > 1$ and a single-input single-output (SISO) system for $M = K = 1$. Two interesting cases about the MIMO system $\mathbf{H}[n]$ are as follows.

- As $\mathbf{H}[n] = \mathbf{0}$ for all $n \neq 0$ (i.e. an instantaneous or memoryless MIMO system), measurements $\mathbf{y}[n]$ are actually a mixture of multiple sources $\mathbf{u}[n]$. Extracting $\mathbf{u}[n]$ from measurements $\mathbf{y}[n]$ for this case is the widely known source separation problem or independent component analysis problem as all the source signals are independent, which can be found in such applications as telecommunications, speech and acoustic signal processing, and biomedical pattern analysis.
- For a two-dimensional (2-D) SISO system, the independent variable n in measurements $\mathbf{y}[n]$ that usually represents the time index, must be replaced by a 1×2 row vector (n_1, n_2) that represents the spatial index in general. The corresponding system is referred to as a linear shift-invariant (LSI) system and has also been popularly used for modeling textures in image processing.

The equalization problem is to find an optimum $K \times M$ equalizer or deconvolution filter $\mathbf{V}[n]$ such that the equalizer outputs (also called the equalized signals or deconvolved signals)

$$\widehat{\mathbf{u}}[n] = \sum_{k=-\infty}^{\infty} \mathbf{V}[k]\mathbf{y}[n-k] \tag{1.2}$$

approximate $\mathbf{u}[n]$ well. Approaches to equalization algorithm design can be divided into two categories. One is the direct approach for which the optimum equalizer is directly obtained by using the data and all the available prior information without involving estimation of the system $\mathbf{H}[n]$. The other approach, the indirect approach, generally consists of the following steps: (i) estimation, if needed, of the system $\mathbf{H}[n]$ by means of system identification algorithms, (ii) estimation, if needed, of other related parameters such as autocorrelations of the data, and (iii) design of an equalizer with the estimated system and parameters for the retrieval of the source signals. Similarly, approaches to system identification can also be divided into direct and indirect approaches, except that equalization is performed prior to system estimation for the indirect approach.

For the direct approach, the equalizer coefficients are obtained by means of equalization algorithms, which can be divided into the following three types

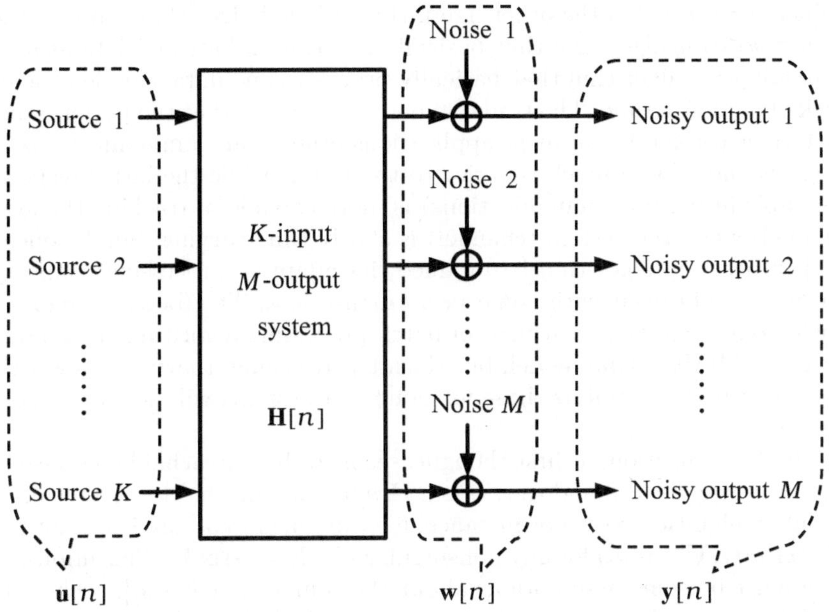

Fig. 1.1 Data model of interest

according to the available amount of training or pilot signals (terms used in communications).

- Nonblind equalization algorithms. This type of algorithm designs the so-called *nonblind equalizer* by fully exploiting prior information about the system and/or the available training or pilot signals. For example, in digital communications, the prior information about the source signals is commonly provided via training or pilot sequences sent from the transmitting end at the expense of system resources (e.g. bandwidth).
- Blind equalization algorithms. This type of algorithm designs the so-called *blind equalizer* with only the data. Training or pilot sequences are therefore not provided for communication systems, giving the benefit of resource savings. The equalizer is usually obtained according to some statistical criteria of the data.
- Semi-blind equalization algorithms. This type of algorithm designs the so-called *semi-blind equalizer* as a weighted mixture of the designs of the non-blind equalization and blind equalization when the amount of training or pilot source signals is not sufficient for obtaining a non-blind equalizer with acceptable performance.

Though non-blind equalizers have been widely used in digital communications, training or pilot signals are usually not available in other science and engineering areas. On the other hand, the designed algorithm can be either a *batch processing* algorithm that basically processes a batch of data at a time, or an adaptive algorithm that basically processes one data sample or a small block of data at a time. The trade-off between them is that the performance of the former (preferable in most applications other than communications) definitely is and can be much superior to the latter, while the latter (especially preferable in digital communications) is more capable of tracking the system (channel) when the system (channel) is slowly time varying. Surely, once the batch processing algorithm is obtained, its adaptive processing counterpart can also be obtained with some performance loss. This book therefore will provide only a unified treatment of batch processing algorithms based on the preceding MIMO signal model, but the adaptive counterparts for slowly time-varying systems as well as those for nonlinear systems will not be covered in this book.

Blind equalization, at first thought, seems to be unreachable because both the source signals $\mathbf{u}[n]$ and the system $\mathbf{H}[n]$ are unknown. Moreover, some potential ambiguities exist. For instance, measurements $\mathbf{y}[n]$ are invariant to the pair $(\mathbf{H}[n]\mathbf{U}, \mathbf{U}^{-1}\mathbf{u}[n])$ for any nonsingular $K \times K$ matrix $\mathbf{U}$. This implies that there must be some assumptions about the source signals $\mathbf{u}[n]$, such as temporally correlated or independent and spatially independent, and Gaussian or nonGaussian sources, and some conditions about the system $\mathbf{H}[n]$, such as full rank, $M \geq K$, a certain parametric form, and nonnegativity for all the coefficients of the system. These are essential in designing the physical system for data measuring or sensing so that the unknown source signals $\mathbf{u}[n]$ can be extracted uniquely to some degree from measurements $\mathbf{y}[n]$. Definitely, the more the source features and system characteristics are taken into account, the more accurate the extracted source signals in general. In fact, a great many algorithms have been successful in blind equalization by exploitation of the properties of the source signals such as their statistical properties, constellation properties, etc. Representatives exploiting the statistical properties include maximum-likelihood (ML) algorithms, second-order statistics (SOS) based algorithms, higher-order (≥ 3) statistics (HOS) based algorithms, and second-order cyclostationary statistics (SOCS) based algorithms. The ML algorithms derive the blind equalizer according to a presumed probability density function of the data, while the SOS, HOS and SOCS based algorithms design the blind equalizer using the SOS, HOS and SOCS of the data, respectively, as the names indicate. Among these algorithms, this book will deal with only the SOS and HOS blind equalization algorithms, with particular emphasis on an SOS based equalization algorithm, namely the *linear prediction approach*, and two HOS based equalization algorithms, namely the *maximum normalized cumulant (MNC) equalization algorithm* and the *super-exponential (SE) equalization algorithm.*

As for blind system identification, a parametric model for the system of interest is usually used in the design of system identification algorithms because estimation of the system therefore becomes a parameter estimation problem that often leads to mathematically tractable solutions with predictable performance. Widely used parametric models include the well-known autoregressive (AR) model, the moving-average (MA) model and the autoregressive moving-average (ARMA) model. Nevertheless, the stability issue is always a concern in system estimation and equalization. Recently, a Fourier series based model (FSBM), which is a parametric frequency-domain model ensuring stability for any LTI system, has also been proposed for use in statistical signal processing including blind equalization and system identification. The model for the equalizer or deconvolution filter considered in this book is either an MA model or an FSBM model so that the stability issue is avoided.

Historical Perspective

Linear prediction was specifically utilized for engineering applications by Norbert Wiener in the 1940s [1–3]. Since then the linear prediction approach has been widely used in spectrum analysis, blind deconvolution, blind system identification, and so forth. Because SOS (autocorrelations or power spectra) of the data are blind to the phase of the unknown system, it cannot be applied to equalization of nonminimum-phase systems. On the other hand, since the mid-1980s, the problem of SISO blind equalization and system identification has been tackled using HOS owing to the fact that HOS (cumulants or polyspectra) contain not only system magnitude information, but also system phase information.

Regarding HOS based approaches, the inverse filter criteria for equalization algorithm are due to Wiggins [4] and Donoho [5] for the real case, and due to Ulrych and Walker [6] and Shalvi and Weinstein [7] for the complex case. The application of the inverse filter criteria to deconvolution was originally for seismic exploration, in which the associated criterion was termed the minimum entropy deconvolution criterion based on the visual appearance of the deconvolved signal. However, to reflect the physical meaning of the inverse filter criteria, they are collectively renamed the MNC criterion in this book and are introduced in a unified manner including fundamental theory, performance analyses, and implementation.

The SE equalization algorithm under the SISO framework was proposed by Shalvi and Weinstein [8] in the 1990s with application to baud-spaced equalization in digital communications. Other blind equalization and system identification algorithms such as polyspectra based algorithms, constant modulus (CM) algorithm, etc., have been reported in detail in [9–15] and references therein.

In 1991, Tong, Xu and Kailath [16] proposed the blind identifiability and equalizability of SIMO LTI systems using only the SOS of system outputs. Their work led to a number of SOS based SIMO blind system estimation and equalization algorithms such as the subspace and least-squares (LS) estimation approaches reported in [17–21]. Unfortunately, these approaches cannot be extended to the MIMO case as the system inputs are temporally white.

Instead, under the assumption that system inputs are temporally colored with different power spectra, the identifiability of an MIMO FIR system using SOS of the system outputs has been proven by Hua and Tugnait [22]; meanwhile, some SOS based blind system identification and equalization methods have been reported [23–27].

On the other hand, blind equalization of SIMO systems using HOS was first investigated by Treichler and Larimore [28] as an extension of the CM algorithm. Further extensions of the CM algorithm to a general MIMO case were then reported in [29, 30]. In the meantime, the inverse filter criteria using third-order and fourth-order cumulants were first extended to the MIMO case by Tugnait [31] and then generalized by Chi and Chen [32] using all higher-order cumulants; the SE algorithm was first extended to the MIMO case by Martone [33] assuming that all the system inputs have the same cumulant, which was then generalized with the assumption relaxed by Yeung and Yau [34] and Inouye and Tanabe [35].

Two-dimensional blind deconvolution and system identification have been intensive research topics and therefore a lot of work has been reported since 1980, such as the 2-D linear prediction approach [36–38], LS solution based methods [39–41], and ML methods [40,42,43]. Kashyap *et al.* [39–41] estimate 2-D AR parameters by the LS solution of a set of linear equations formed from autocorrelations (SOS) of the data. Assuming that the data are Gaussian, ML estimators have been used to estimate AR parameters [40, 42] and ARMA parameters [43]. In light of the fact that SOS are blind to the system phase, only the magnitude information of 2-D LSI systems can be extracted by the above SOS based approaches.

The drawback (blind to the system phase) inherent in SOS has motivated the research of HOS based 2-D blind equalization and system identification algorithms over the last decade. Estimation of AR parameters by the LS solution of a set of linear equations formed from higher-order cumulants of the data have been reported [44,45]. Hall and Giannakis [46] proposed two 2-D inverse filter criteria for estimating AR parameters. As the AR parameters are obtained, the MA parameters are estimated either by a closed-form solution [45, 46] or by cumulant matching [46] using cumulants of the residual signal obtained by removing the AR part from the data. Other methods for jointly estimating AR and MA parameters are also reported, such as an inverse filter criteria based method [47] and a polyspectral matching method [48]. The estimated AR parameters have been used for texture synthesis [46–48] and classification [46]. Recently, Chen *et al.* [49] identified the 2-D system through estimation of the amplitude and phase parameters of 2-D FSBM and then the estimated amplitude parameters were used for texture image classification.

Both blind equalization and system identification have continued to be active research areas, and their rapid advances can be verified by new theory, analysis, computational complexity reduction, efficient real-time implementation, and more successful applications. Whatever is covered, this book cannot be comprehensive but merely provides the reader with a unified introduction to the concepts, philosophies and performance analyses of some blind equalization and system identification algorithms that we believe are essential, effective and efficient so that the reader can save a lot of time learning the necessary background and can proceed to the advanced research and development in these areas.

Book Organization

This book is organized as follows. Chapter 2 briefly reviews some mathematical background needed in this book, including linear algebra, mathematical analysis, optimization theory, and the well-known LS method. Through this review, most notation to be used in the subsequent chapters is introduced. In addition to the application to blind equalization and system identification, optimization theory, which is developed in terms of the complex-valued framework, is also applicable in other areas.

Chapter 3 deals with several fundamental topics of statistical signal processing, including discrete-time signals and systems, random variables, random processes, and estimation theory.

Chapter 4 introduces some widely used SISO blind equalization algorithms, including the linear prediction approach, the MNC equalization algorithm and the SE equalization algorithm, along with their analyses and improvements. In algorithm improvements, a hybrid framework of MNC and SE equalization algorithms, referred to as the hybrid MNC equalization algorithm, is thought of as one of the best blind equalization algorithms based on performance, convergence speed and computational load. This chapter also includes some simulation examples for testing these algorithms, as well as the application of these algorithms to seismic exploration, speech signal processing and baud-spaced equalization in digital communications. The material in this chapter covers almost all the essential concepts and design philosophies of blind equalization or deconvolution algorithms in terms of the SISO framework, and thus continues to be the foundation of blind equalization and system identification for both the MIMO and 2-D cases.

Chapter 5 introduces some widely used MIMO blind equalization algorithms based on either SOS or HOS of MIMO system outputs. A subspace approach and a linear prediction approach using SOS are introduced for the SIMO case, and a matrix pencil method is introduced for the MIMO case. For HOS based approaches, the MNC and SE equalization algorithms are introduced for an MIMO system with temporally independently and identically distributed inputs followed by their properties and relations, which lead to an improved MNC equalization algorithm, namely the MIMO hybrid MNC equalization algorithm. By making use of the MIMO hybrid MNC equalization algorithm and a greatest common divisor (GCD) computation algorithm, an equalization-GCD equalization algorithm is introduced for an MIMO system with temporally colored inputs. All the SIMO and MIMO blind equalization algorithms introduced are tested through simulation for performance evaluation and comparison, and verification of their analytical properties and relations.

Chapter 6 introduces some applications of the MIMO hybrid MNC equalization algorithm in the areas of signal processing and digital communications. In particular, for the SIMO case it introduces straightforward applications of fractionally spaced equalization and blind maximum ratio combining, as well as advanced applications in blind system identification (BSI) and multiple time delay estimation. On the other hand, applications of the MIMO hybrid MNC equalization algorithm for the MIMO case include blind beamforming for source separation and multiuser detection in wireless communications. In each of the applications introduced, a discrete-time SIMO or MIMO system model must be established prior to use of the MIMO hybrid MNC equalization algorithm, in addition to certain constraints, structures, and considerations on the system or the equalizer.

Chapter 7 begins with a review of 2-D deterministic signals, systems and linear random processes (random fields), and then provides an introduction to some 2-D deconvolution algorithms that, we believe, are effective in such applications as image restoration, image model identification, texture synthesis, texture image classification, and so forth. The 2-D deconvolution algorithms introduced include the 2-D linear prediction approach, 2-D MNC and SE deconvolution algorithms, and 2-D hybrid MNC deconvolution algorithm. Some simulation results are also provided to demonstrate the performance of these 2-D deconvolution algorithms.

Chapter 8 introduces a 2-D BSI algorithm, which is an iterative fast Fourier transform based nonparametric algorithm using the 2-D hybrid MNC deconvolution algorithm. Application of this 2-D BSI algorithm to texture synthesis is also introduced. This chapter also introduces a 2-D FSBM based parametric BSI algorithm that includes an amplitude estimator using SOS and two phase estimators using HOS, with application to texture image classification.

References

1. N. Wiener, *Extrapolation, Interpolation and Smoothing of Stationary Time Series with Engineering Applications.* Massachusetts: MIT Press, 1949.
2. J. Makhoul, "Linear prediction: A tutorial review," *Proc. IEEE*, vol. 63, no. 4, pp. 561–580, Apr. 1975.
3. S. Haykin, *Modern Filters.* New York: Macmillan, 1989.
4. R. A. Wiggins, "Minimum entropy deconvolution," *Geoexploration*, vol. 16, pp. 21–35, 1978.
5. D. L. Donoho, On Minimum Entropy Deconvolution, in *Applied Time Series Analysis II*, D.F. Findly, ed. New York: Academic Press, 1981.
6. T. J. Ulrych and C. Walker, "Analytic minimum entropy deconvolution," *Geophysics*, vol. 47, no. 9, pp. 1295–1302, Sept. 1982.
7. O. Shalvi and E. Weinstein, "New criteria for blind deconvolution of nonminimum phase systems (channels)," *IEEE Trans. Information Theory*, vol. 36, no. 2, pp. 312–321, Mar. 1990.

8. O. Shalvi and E. Weinstein, "Super-exponential methods for blind deconvolution," *IEEE Trans. Informtion Theory*, vol. 39, no. 2, pp. 504–519, Mar. 1993.
9. J. M. Mendel, "Tutorial on higher-order statistics (spectra) in signal processing and system theory: Theoretical results and some applications," *Proc. IEEE*, vol. 79, no. 3, pp. 278–305, Mar. 1991.
10. C. L. Nikias and A. P. Petropulu, *Higher-order Spectra Analysis: A Nonlinear Signal Processing Framework*. New Jersey: Prentice-Hall, 1993.
11. S. Haykin, ed., *Blind Deconvolution*. New Jersey: Prentice-Hall, 1994.
12. C. R. Johnson, Jr., P. Schniter, T. J. Endres, J. D. Behm, D. R. Brown, and R. A. Casas, "Blind equalization using the constant modulus criterion: A review," *Proc. IEEE*, vol. 86, no. 10, pp. 1927–1950, Oct. 1998.
13. J. K. Tugnait, L. Tong, and Z. Ding, "Single-user channel estimation and equalization," *IEEE Signal Processing Magazine*, vol. 17, no. 3, pp. 17–28, May 2000.
14. G. B. Giannakis, Y. Hua, P. Stoica, and L. Tong, ed., *Signal Processing Advances in Wireless and Mobile Communications*. Upper Saddle River: Prentice-Hall, 2001.
15. Z. Ding and Y. Li, ed., *Blind Equalization and Identification*. New York: Marcel Dekker, 2001.
16. L. Tong, G. Xu, and T. Kailath, "A new approach to blind identification and equalization of multipath channels," in *Record of the 25th Asilomar Conference on Signals, Systems and Computers*, Pacific Grove, California, Nov. 4–6, 1991, pp. 856–860.
17. L. Tong, G. Xu and T. Kailath, "Blind identification and equalization based on second-order statistics: A time domain approach," *IEEE Trans. Information Theory*, vol. 40, no. 2, pp. 340–349, Mar. 1994.
18. E. Moulines, P. Duhamel, J.-F. Cardoso, and S. Mayrargue, "Subspace methods for the blind identification of multichannel FIR filters," *IEEE Trans. Signal Processing*, vol. 43, no. 2, pp. 516–525, Feb. 1995.
19. K. Abed-Meraim, E. Moulines, and P. Loubaton, "Prediction error methods for second-order blind identification," *IEEE Trans. Signal Processing*, vol. 45, no. 3, pp. 694–705, Mar. 1997.
20. L. Tong and S. Perreau, "Multichannel blind identification: From subspace to maximum likelihood methods," *Proc. IEEE*, vol. 86, no. 10, pp. 1951–1968, Oct. 1998.
21. C. B. Papadias and D. T. M. Slock, "Fractionally spaced equalization of linear polyphase channels and related blind techniques based on multichannel linear prediction," *IEEE Trans. Signal Processing*, vol. 47, no. 3, pp. 641–654, Mar. 1999.
22. Y. Hua and J. K. Tugnait, "Blind identifiability of FIR-MIMO systems with colored input using second-order statistics," *IEEE Signal Processing Letters*, vol. 7, no. 12, pp. 348–350, Dec. 2000.
23. A. Gorokhov and P. Loubaton, "Subspace-based techniques for blind separation of convolutive mixtures with temporally correlated sources," *IEEE Trans. Circuits and Systems I*, vol. 44, no. 9, pp. 813–820, Sept. 1997.
24. K. Abed-Meraim and Y. Hua, "Blind identification of multi-input multi-output system using minimum noise subspace," *IEEE Trans. Signal Processing*, vol. 45, no. 1, pp. 254–258, Jan. 1997.
25. C. T. Ma, Z. Ding, and S. F. Yau, "A two-stage algorithm for MIMO blind deconvolution of nonstationary colored signals," *IEEE Trans. Signal Processing*, vol. 48, no. 4, pp. 1187–1192, Apr. 2000.

26. Y. Hua, S. An, and Y. Xiang, "Blind identification and equalization of FIR MIMO channels by BIDS," in *Proc. IEEE International Conference on Acoustics, Speech, and Signal Processing*, Salt Lake, Utah, May 7–11, 2001, pp. 2157–2160.
27. S. An and Y. Hua, "Blind signal separation and blind system identification of irreducible MIMO channels," in *Proc. Sixth IEEE International Symposium on Signal Processing and its Applications*, Kuala Lumpur, Malaysia, Aug. 13–16, 2001, pp. 276–279.
28. J. R. Treichler and M. G. Larimore, "New processing techniques based on the constant modulus adaptive algorithm," *IEEE Trans. Acoustics, Speech, and Signal Processing*, vol. 33, no. 2, pp. 420–431, Apr. 1985.
29. J. K. Tugnait, "Blind spatio-temporal equalization and impulse response estimation for MIMO channels using a Godard cost function," *IEEE Trans. Signal Processing*, vol. 45, no. 1, pp. 268–271, Jan. 1997.
30. Y. Li and K. J. R. Liu, "Adaptive blind source separation and equalization for multiple-input/multiple-output systems," *IEEE Trans. Information Theory*, vol. 44, no. 7, pp. 2864–2876, Nov. 1998.
31. J. K. Tugnait, "Identification and deconvolution of multichannel linear non-Gaussian processes using higher-order statistics and inverse filter criteria," *IEEE Trans. Signal Processing*, vol. 45, no. 3, pp. 658–672, Mar. 1997.
32. C.-Y. Chi and C.-H. Chen, "Cumulant based inverse filter criteria for MIMO blind deconvolution: Properties, algorithms, and application to DS/CDMA systems in multipath," *IEEE Trans. Signal Processing*, vol. 49, no. 7, pp. 1282–1299, July 2001.
33. M. Martone, "Non-Gaussian multivariate adaptive AR estimation using the super exponential algorithm," *IEEE Trans. Signal Processing*, vol. 44, no. 10, pp. 2640–2644, Oct. 1996.
34. K. L. Yeung and S. F. Yau, "A cumulant-based super-exponential algorithm for blind deconvolution of multi-input multi-output systems," *Signal Processing*, vol. 67, no. 2, pp. 141–162, 1998.
35. Y. Inouye and K. Tanebe, "Super-exponential algorithms for multichannel blind deconvolution," *IEEE Trans. Signal Processing*, vol. 48, no. 3, pp. 881–888, Mar. 2000.
36. D. E. Dudgeon and R. M. Mersereau, *Multidimensional Digital Signal Processing*. New Jersey: Prentice-Hall, 1984.
37. S. M. Kay, *Modern Spectral Estimation: Theory and Application.* New Jersey: Prentice-Hall, 1988.
38. S. R. Parker and A. H. Kayran, "Lattice parameter autoregressive modeling of two-dimensional fields-Part I: The quarter-plane case," *IEEE Trans. Acoustics, Speech and Signal Processing*, vol. 32, no. 4, pp. 872–885, Aug. 1984.
39. R. L. Kashyap, R. Chellappa, and A. Khotanzad, "Texture classification using features derived from random field models," *Pattern Recognition Letters*, vol. 1, no. 1, pp. 43–50, 1982.
40. R. L. Kashyap and R. Chellappa, "Estimation and choice of neighbors in spatial-interaction models of images," *IEEE Trans. Information Theory*, vol. 29, no. 1, pp. 58–72, Jan. 1983.
41. R. Chellappa and S. Chatterjee, "Classification of texture using Gaussian Markov random fields," *IEEE Trans. Acoustics, Speech and Signal Processing*, vol. 33, no. 4, pp. 959–963, Apr. 1985.

42. G. Sharma and R. Chellappa, "Two-dimensional spectrum estimation using noncausal autoregressive models," *IEEE Trans. Information Theory*, vol. 32, no. 2, pp. 268–275, Feb. 1986.
43. A. M. Tekalp, H. Kaufman, and J. E. Woods, "Identification of image and blur parameters for the restoration of noncausal blurs," *IEEE Trans. Acoustics, Speech and Signal Processing*, vol. 34, no. 4, pp. 963–972, Aug. 1986.
44. S. Bhattacharya, N. C. Ray, and S. Sinha, "2-D signal modelling and reconstruction using third-order cumulants," *Signal Processing*, vol. 62, pp. 61–72, 1997.
45. A. Swami, G. B. Giannakis, and J. M. Mendel, "Linear modeling of multidimensional non-Gaussian processes using cumulants," *Multidimensional Systems and Signal Processing*, vol. 1, pp. 11–37, 1990.
46. T. E. Hall and G. B. Giannakis, "Image modeling using inverse filtering criteria with application to textures," *IEEE Trans. Image Processing*, vol. 5, no. 6, pp. 938–949, June 1996.
47. J. K. Tugnait, "Estimation of linear parametric models of nonGaussian discrete random fields with application to texture synthesis," *IEEE Trans. Image Processing*, vol. 3, no. 2, pp. 109–127, Feb. 1994.
48. T. E. Hall and G. B. Giannakis, "Bispectral analysis and model validation of texture images," *IEEE Trans. Image Processing*, vol. 4, no. 7, pp. 996–1009, July 1995.
49. C.-H. Chen, C.-Y. Chi, and C.-Y. Chen, "2-D Fourier series based model for nonminimum-phase linear shift-invariant systems and texture image classification," *IEEE Trans. Signal Processing*, vol. 50, no. 4, pp. 945–955, Apr. 2002.

2

Mathematical Background

In this chapter, we briefly review some mathematical background needed in this book, including linear algebra, mathematical analysis, and optimization theory. Through this review, most notation to be used in subsequent chapters is introduced. We then present the well-known least-squares method as an application of linear algebra and optimization theory.

2.1 Linear Algebra

We start with a review of vectors, vector spaces and matrices, and then introduce two powerful tools for matrix decomposition, namely eigendecomposition and singular value decomposition. The usefulness of matrix decomposition will become evident in the remaining parts of this book.

2.1.1 Vectors and Vector Spaces

Vectors

In this book, vectors are denoted by bold lowercase letters. For example, we denote an $N \times 1$ vector

$$\mathbf{x} = \begin{pmatrix} x_1 \\ x_2 \\ \vdots \\ x_N \end{pmatrix} = (x_1, x_2, ..., x_N)^T$$

where x_n is a real or complex scalar representing the nth entry (component) of $\mathbf{x}$ and the superscript 'T' represents vector transposition. The complex-conjugate transposition, or *Hermitian*, of $\mathbf{x}$ is given by

$$\mathbf{x}^H = (\mathbf{x}^T)^* = (x_1^*, x_2^*, ..., x_N^*)$$

where the superscript 'H' denotes the Hermitian operator and the superscript '$*$' complex conjugation.

Let $\mathbf{x} = (x_1, x_2, ..., x_N)^T$ and $\mathbf{y} = (y_1, y_2, ..., y_N)^T$ be two $N \times 1$ vectors. The *inner product* of $\mathbf{x}$ and $\mathbf{y}$ is defined as

$$\langle \mathbf{x}, \mathbf{y} \rangle = \sum_{n=1}^{N} x_n y_n^* = \mathbf{y}^H \mathbf{x}, \tag{2.1}$$

which is also referred to as the *Euclidean inner product* of $\mathbf{x}$ and $\mathbf{y}$. The *length*, or *norm*, of the vector $\mathbf{x}$ is defined as

$$||\mathbf{x}|| = \left(\sum_{n=1}^{N} |x_n|^2 \right)^{1/2} = \sqrt{\mathbf{x}^H \mathbf{x}}, \tag{2.2}$$

which is also referred to as the *Euclidean norm* of $\mathbf{x}$. Other types of norms will be defined later and, for convenience, we will always use the definition (2.2) for the norm of $\mathbf{x}$ unless specified otherwise. A vector whose norm equals unity is called a *unit vector*. Furthermore, the *geometrical relationship* between two vectors $\mathbf{x}$ and $\mathbf{y}$ is given as follows: [1, p. 15]

$$\cos\phi = \begin{cases} \dfrac{|\langle \mathbf{x}, \mathbf{y} \rangle|}{||\mathbf{x}|| \cdot ||\mathbf{y}||} = \dfrac{|\mathbf{y}^H \mathbf{x}|}{||\mathbf{x}|| \cdot ||\mathbf{y}||}, & 0 \le \phi \le \pi/2 \text{ (complex)} \\ \dfrac{\langle \mathbf{x}, \mathbf{y} \rangle}{||\mathbf{x}|| \cdot ||\mathbf{y}||} = \dfrac{\mathbf{y}^T \mathbf{x}}{||\mathbf{x}|| \cdot ||\mathbf{y}||}, & 0 \le \phi \le \pi \text{ (real)} \end{cases} \tag{2.3}$$

where ϕ is the angle between $\mathbf{x}$ and $\mathbf{y}$. As depicted in Fig. 2.1, the relationship can be interpreted by viewing the inner product $\langle \mathbf{x}, \mathbf{y}/||\mathbf{y}|| \rangle$ as the projection of $\mathbf{x}$ onto the unit vector $\mathbf{y}/||\mathbf{y}||$. With the geometrical interpretation, $\mathbf{x}$ and $\mathbf{y}$ are said to be *orthogonal* if $\mathbf{x}^H \mathbf{y} = \mathbf{y}^H \mathbf{x} = 0$. Furthermore, if $\mathbf{x}$ and $\mathbf{y}$ are orthogonal and have the unit norm, then they are said to be *orthonormal*.

The geometrical relationship given by (2.3) is closely related to the following *Cauchy–Schwartz inequality* or *Schwartz inequality*.[1]

Theorem 2.1 (Cauchy–Schwartz Inequality). *Let* $\mathbf{x} = (x_1, x_2, ..., x_N)^T$ *and* $\mathbf{y} = (y_1, y_2, ..., y_N)^T$ *be real or complex nonzero vectors. Then*

$$|\mathbf{y}^H \mathbf{x}| \le ||\mathbf{x}|| \cdot ||\mathbf{y}|| \tag{2.4}$$

and the equality holds if and only if $\mathbf{x} = \alpha \mathbf{y}$ *where* $\alpha \neq 0$ *is an arbitrary real or complex scalar.*

The Cauchy–Schwartz inequality further leads to the following inequality:

[1] The Russians also refer to the Cauchy–Schwartz inequality as the *Cauchy–Schwartz–Buniakowsky inequality* [2].

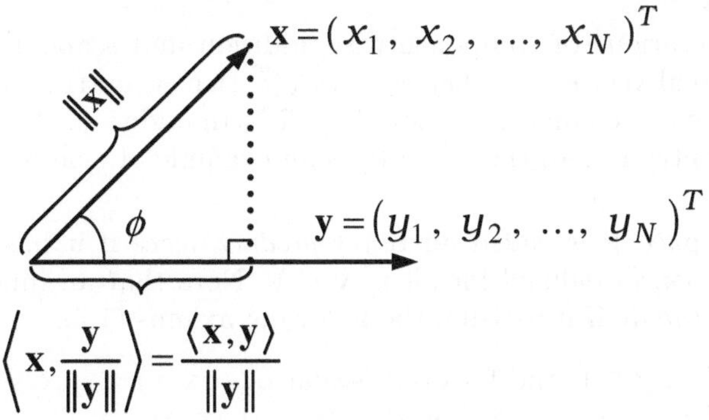

Fig. 2.1 The geometrical relationship between two vectors **x** and **y**

Theorem 2.2 (Triangle Inequality). *Let* $\mathbf{x} = (x_1, x_2, ..., x_N)^T$ *and* $\mathbf{y} = (y_1, y_2, ..., y_N)^T$ *be real or complex nonzero vectors. Then*

$$\|\mathbf{x} + \mathbf{y}\| \leq \|\mathbf{x}\| + \|\mathbf{y}\|. \tag{2.5}$$

The proofs of the two theorems are left as exercises (Problems 2.1 and 2.2).

Vector Spaces

A *vector space* is a non-empty set of elements along with several rules for the operations of addition and scalar multiplication of elements. The elements can be vectors, sequences, functions, etc., and are also referred to as *vectors* without confusion. Let $\mathcal{V}$ denote a vector space and the vectors (elements) in $\mathcal{V}$ be also denoted by bold lowercase letters. Then for each pair of vectors $\mathbf{x}$ and $\mathbf{y}$ in $\mathcal{V}$ there is a unique vector $\mathbf{x}+\mathbf{y}$ in $\mathcal{V}$ (the operation of addition) and for each scalar α there is a unique vector $\alpha\mathbf{x}$ in $\mathcal{V}$ (the operation of scalar multiplication). Furthermore, the operations of addition and scalar multiplication must satisfy the following axioms [3–5].

(VS1) For all $\mathbf{x}$, $\mathbf{y} \in \mathcal{V}$, $\mathbf{x} + \mathbf{y} = \mathbf{y} + \mathbf{x}$.
(VS2) For all $\mathbf{x}$, $\mathbf{y}$, $\mathbf{z} \in \mathcal{V}$, $(\mathbf{x} + \mathbf{y}) + \mathbf{z} = \mathbf{x} + (\mathbf{y} + \mathbf{z})$.
(VS3) For all $\mathbf{x} \in \mathcal{V}$, there exists a zero vector $\mathbf{0} \in \mathcal{V}$ such that $\mathbf{x} + \mathbf{0} = \mathbf{x}$.
(VS4) For each $\mathbf{x} \in \mathcal{V}$, there exists a vector $\mathbf{y} \in \mathcal{V}$ such that $\mathbf{x} + \mathbf{y} = \mathbf{0}$.
(VS5) For all $\mathbf{x}$, $\mathbf{y} \in \mathcal{V}$ and for every scalar α, $\alpha(\mathbf{x} + \mathbf{y}) = \alpha\mathbf{x} + \alpha\mathbf{y}$.
(VS6) For all $\mathbf{x} \in \mathcal{V}$ and for all scalars α and β, $(\alpha + \beta)\mathbf{x} = \alpha\mathbf{x} + \beta\mathbf{x}$.
(VS7) For all $\mathbf{x} \in \mathcal{V}$ and for all scalars α and β, $(\alpha\beta)\mathbf{x} = \alpha(\beta\mathbf{x})$.
(VS8) For all $\mathbf{x} \in \mathcal{V}$, there exists a scalar 1 such that $1 \cdot \mathbf{x} = \mathbf{x}$.

A subset of a vector space $\mathcal{V}$, denoted by $\mathcal{W}$, is called a *subspace* of $\mathcal{V}$ if $\mathcal{W}$ itself is a vector space under the operations of addition and scalar multiplication defined on $\mathcal{V}$. An example is as follows.

Example 2.3
Under the operations of componentwise addition and scalar multiplication, the set of all real vectors $\mathbf{x} = (x_1, x_2, ..., x_N)^T$ (whose entries are real) forms a real vector space, commonly denoted by $\mathcal{R}^N$. In addition, the set of all real $\mathbf{x}$ whose nth entry is zero (i.e. $x_n = 0$) is an example of a subspace of $\mathcal{R}^N$. □

A vector space $\mathcal{V}$ is called an *inner product space* if it has a legitimate inner product $\langle \mathbf{x}, \mathbf{y} \rangle$ defined for all $\mathbf{x}$, $\mathbf{y} \in \mathcal{V}$. Note that an inner product is said to be *legitimate* if it satisfies the following axioms [3, 5].

(IPS1) For all $\mathbf{x}$, $\mathbf{y} \in \mathcal{V}$ and for every scalar α, $\langle \alpha\mathbf{x}, \mathbf{y} \rangle = \alpha \langle \mathbf{x}, \mathbf{y} \rangle$.
(IPS2) For all $\mathbf{x}$, $\mathbf{y}$, $\mathbf{z} \in \mathcal{V}$, $\langle \mathbf{x}+\mathbf{y}, \mathbf{z} \rangle = \langle \mathbf{x}, \mathbf{z} \rangle + \langle \mathbf{y}, \mathbf{z} \rangle$.
(IPS3) For all $\mathbf{x} \in \mathcal{V}$, $\langle \mathbf{x}, \mathbf{x} \rangle \geq 0$, and $\langle \mathbf{x}, \mathbf{x} \rangle = 0$ if and only if $\mathbf{x} = \mathbf{0}$.
(IPS4) For all $\mathbf{x}$, $\mathbf{y} \in \mathcal{V}$, $\langle \mathbf{x}, \mathbf{y} \rangle = (\langle \mathbf{y}, \mathbf{x} \rangle)^*$.

Similarly, a vector space $\mathcal{V}$ is called a *normed vector space* if it has a legitimate norm $\|\mathbf{x}\|$ defined for all $\mathbf{x} \in \mathcal{V}$. A norm is said to be *legitimate* if it satisfies the following axioms [3, 5].

(NVS1) For all $\mathbf{x} \in \mathcal{V}$ and for every scalar α, $\|\alpha\mathbf{x}\| = |\alpha| \cdot \|\mathbf{x}\|$.
(NVS2) For all $\mathbf{x}$, $\mathbf{y} \in \mathcal{V}$, $\|\mathbf{x}+\mathbf{y}\| \leq \|\mathbf{x}\| + \|\mathbf{y}\|$.
(NVS3) For all $\mathbf{x} \in \mathcal{V}$, $\|\mathbf{x}\| \geq 0$, and $\|\mathbf{x}\| = 0$ if and only if $\mathbf{x} = \mathbf{0}$.

It is important to note [5, pp. 14–15] that a legitimate inner product for a vector space $\mathcal{V}$ always induces a legitimate norm for $\mathcal{V}$ via the relation

$$\|\mathbf{x}\| = \sqrt{\langle \mathbf{x}, \mathbf{x} \rangle} \quad \text{for all } \mathbf{x} \in \mathcal{V}.$$

Such a norm is referred to as an *induced norm.* An example is as follows.

Example 2.4 (Euclidean Space)
It can be easily shown that for the real vector space $\mathcal{R}^N$ (see Example 2.3), the Euclidean inner product defined as (2.1) is legitimate and induces the Euclidean norm defined as (2.2). Accordingly, $\mathcal{R}^N$ along with the Euclidean inner product is an inner product space, while $\mathcal{R}^N$ along with the Euclidean norm is a normed vector space. The former is known as the *Euclidean space* [4].

□

Let $\mathbf{q}_1$, $\mathbf{q}_2$, ..., $\mathbf{q}_N$ be the vectors in a vector space $\mathcal{V}$. Then they are said to *span* the subspace $\mathcal{W}$ if $\mathcal{W}$ consists of all linear combinations of $\mathbf{q}_1$, $\mathbf{q}_2$, ..., $\mathbf{q}_N$. Specifically, every vector $\mathbf{w}$ in $\mathcal{W}$ can be expressed as

$$\mathbf{w} = \alpha_1\mathbf{q}_1 + \alpha_2\mathbf{q}_2 + \cdots + \alpha_N\mathbf{q}_N$$

where α_k are scalars. For vectors $\mathbf{q}_1$, $\mathbf{q}_2$, ..., $\mathbf{q}_N$ in $\mathcal{V}$, one can determine their linear interdependence via the following equation:

$$c_1\mathbf{q}_1 + c_2\mathbf{q}_2 + \cdots + c_N\mathbf{q}_N = \mathbf{0}$$

where c_k are scalars and $\mathbf{0}$ is a zero vector defined by (VS3). If this equation holds true only when $c_1 = c_2 = \cdots = c_N = 0$, then $\mathbf{q}_1$, $\mathbf{q}_2$, ..., $\mathbf{q}_N$ are said to be *linearly independent*; otherwise, they are *linearly dependent*. If $\mathbf{q}_1$, $\mathbf{q}_2$, ..., $\mathbf{q}_N$ are linearly independent and span the vector space $\mathcal{V}$, they are called a *basis* for $\mathcal{V}$. A vector space $\mathcal{V}$ is said to be *finite-dimensional* if the number of linearly independent vectors in its basis is finite; otherwise, it is said to be *infinite-dimensional*.

A set S in an inner product space $\mathcal{V}$ is called an *orthogonal set* if every pair of vectors $\mathbf{q}_k, \mathbf{q}_m \in S$ is orthogonal, i.e. $\langle \mathbf{q}_k, \mathbf{q}_m \rangle = 0$ for $k \neq m$. Furthermore, if every vector $\mathbf{q}_k \in S$ has the unit norm, i.e. $\|\mathbf{q}_k\| = 1$, then the orthogonal set S is said to be *orthonormal*. In other words, an orthonormal set does not contain the zero vector $\mathbf{0}$. A basis for an inner product space $\mathcal{V}$ is said to be an *orthonormal basis* if it is an orthonormal set. For example, the set

$$\{\boldsymbol{\eta}_1 = (1,0,0,...,0)^T,\ \boldsymbol{\eta}_2 = (0,1,0,...,0)^T,\ ...,\ \boldsymbol{\eta}_N = (0,0,...,0,1)^T\} \quad (2.6)$$

is an orthonormal basis, referred to as the *standard basis*, for the Euclidean space $\mathcal{R}^N$ (see Example 2.4) where $\boldsymbol{\eta}_k$ denotes a unit vector whose kth entry equals unity and the remaining entries equal zero. Note that any basis can be transformed into an orthonormal basis via the process of *Gram–Schmidt orthogonalization* [1–3, 5].

2.1.2 Matrices

In this book, matrices are denoted by bold uppercase letters. For example,

$$\mathbf{A} = \begin{pmatrix} a_{11} & a_{12} & \cdots & a_{1K} \\ a_{21} & a_{22} & \cdots & a_{2K} \\ \vdots & \vdots & \ddots & \vdots \\ a_{M1} & a_{M2} & \cdots & a_{MK} \end{pmatrix} \quad (2.7)$$

denotes an $M \times K$ matrix whose (m,k)th entry (component) is a_{mk}, a real or complex scalar. We also use the shorthand representation

$$[\mathbf{A}]_{m,k} = a_{mk}$$

to specify the matrix $\mathbf{A}$. The *transposition* of $\mathbf{A}$ is

$$[\mathbf{A}^T]_{m,k} = [\mathbf{A}]_{k,m} = a_{km} \quad (2.8)$$

and $(\mathbf{A}^T)^T = \mathbf{A}$ where the superscript 'T' stands for matrix transposition. The complex-conjugate transposition, or *Hermitian*, of $\mathbf{A}$ is

$$[\mathbf{A}^H]_{m,k} = [\mathbf{A}^*]_{k,m} = a_{km}^* \quad (2.9)$$

and $(\mathbf{A}^H)^H = \mathbf{A}$ where the superscript 'H' stands for the Hermitian operation. The matrix $\mathbf{A}$ is said to be *square* if $M = K$. It is further said to be *symmetric* if $\mathbf{A}^T = \mathbf{A}$ for $\mathbf{A}$ real, and *Hermitian* if $\mathbf{A}^H = \mathbf{A}$ for $\mathbf{A}$ complex. Note that $\mathbf{A}^H = \mathbf{A}^T$ as $\mathbf{A}$ is real. For matrices $\mathbf{A}$ and $\mathbf{B}$, $(\mathbf{AB})^T = \mathbf{B}^T\mathbf{A}^T$, $(\mathbf{AB})^H = \mathbf{B}^H\mathbf{A}^H$, $(\mathbf{A}+\mathbf{B})^T = \mathbf{A}^T + \mathbf{B}^T$, and $(\mathbf{A}+\mathbf{B})^H = \mathbf{A}^H + \mathbf{B}^H$.

Let us further represent the $M \times K$ matrix $\mathbf{A}$ given in (2.7) by

$$\mathbf{A} = (\mathbf{a}_1, \mathbf{a}_2, ..., \mathbf{a}_K) = \begin{pmatrix} \mathbf{b}_1^T \\ \mathbf{b}_2^T \\ \vdots \\ \mathbf{b}_M^T \end{pmatrix}$$

where $\mathbf{a}_k = (a_{1k}, a_{2k}, ..., a_{Mk})^T$, $k = 1, 2, ..., K$, are called the *column vectors* of $\mathbf{A}$ and $\mathbf{b}_m^T = (a_{m1}, a_{m2}, ..., a_{mK})$, $m = 1, 2, ..., M$, the *row vectors* of $\mathbf{A}$. The subspace spanned by the column vectors is called the *column space* of $\mathbf{A}$, while the subspace spanned by the row vectors is called the *row space* of $\mathbf{A}$. The number of linearly independent column vectors of $\mathbf{A}$ is equal to the number of linearly independent row vectors of $\mathbf{A}$, that is defined as the *rank* of $\mathbf{A}$, denoted by $\text{rank}\{\mathbf{A}\}$. Note that $\text{rank}\{\mathbf{A}\} = \text{rank}\{\mathbf{A}^H\} \le \min\{M, K\}$ and $\text{rank}\{\mathbf{A}^H\mathbf{A}\} = \text{rank}\{\mathbf{A}\mathbf{A}^H\} = \text{rank}\{\mathbf{A}\}$. When $\text{rank}\{\mathbf{A}\} = \min\{M, K\}$, the matrix $\mathbf{A}$ is said to be of *full rank*; otherwise, it is *rank deficient*.

The *inverse* of an $M \times M$ square matrix $\mathbf{A}$ is also an $M \times M$ square matrix, denoted by $\mathbf{A}^{-1}$, which satisfies

$$\mathbf{A}\mathbf{A}^{-1} = \mathbf{A}^{-1}\mathbf{A} = \mathbf{I} \tag{2.10}$$

where

$$\mathbf{I} = \begin{pmatrix} 1 & 0 & \cdots & 0 \\ 0 & 1 & \cdots & 0 \\ \vdots & \vdots & \ddots & \vdots \\ 0 & 0 & \cdots & 1 \end{pmatrix} \tag{2.11}$$

is the $M \times M$ *identity matrix*. If $\mathbf{A}$ is of full rank, then $\mathbf{A}^{-1}$ exists and $\mathbf{A}$ is said to be *invertible* or *nonsingular*. On the other hand, if $\mathbf{A}$ is rank deficient, then it does not have an inverse and is accordingly said to be *noninvertible* or *singular*. For nonsingular matrices $\mathbf{A}$ and $\mathbf{B}$, $(\mathbf{A}^T)^{-1} = (\mathbf{A}^{-1})^T$, $(\mathbf{A}^H)^{-1} = (\mathbf{A}^{-1})^H$, and $(\mathbf{AB})^{-1} = \mathbf{B}^{-1}\mathbf{A}^{-1}$.

Consider an $M \times M$ square matrix $\mathbf{A}$ with $[\mathbf{A}]_{m,k} = a_{mk}$. The *determinant* of $\mathbf{A}$ is commonly denoted by $\det\{\mathbf{A}\}$ or $|\mathbf{A}|$. For $M = 1$, the matrix $\mathbf{A}$ reduces to a scalar a_{11} and its determinant is defined as $\det\{a_{11}\} = a_{11}$. For $M \ge 2$, the determinant $\det\{\mathbf{A}\}$ can be defined in terms of the determinants of the associated $(M-1) \times (M-1)$ matrices as follows:

$$\det\{\mathbf{A}\} = \sum_{m=1}^{M} (-1)^{m+k} \cdot a_{mk} \cdot \det\{\mathbf{A}_{mk}\} \quad \text{for any } k \in \{1, 2.., M\}$$
$$= \sum_{k=1}^{M} (-1)^{m+k} \cdot a_{mk} \cdot \det\{\mathbf{A}_{mk}\} \quad \text{for any } m \in \{1, 2.., M\} \qquad (2.12)$$

where $\mathbf{A}_{mk}$ is an $(M-1) \times (M-1)$ matrix obtained by deleting the mth row and kth column of $\mathbf{A}$. For example, if $M = 2$, $\det\{\mathbf{A}\}$ is given by

$$\det\left\{\begin{pmatrix} a_{11} & a_{12} \\ a_{21} & a_{22} \end{pmatrix}\right\} = (-1)^{1+1} \cdot a_{11} \cdot \det\{a_{22}\} + (-1)^{2+1} \cdot a_{21} \cdot \det\{a_{12}\}$$
$$= a_{11}a_{22} - a_{21}a_{12}.$$

Note that $\det\{\mathbf{A}^T\} = \det\{\mathbf{A}\}$, $\det\{\mathbf{A}^H\} = [\det\{\mathbf{A}\}]^*$, and $\det\{\alpha\mathbf{A}\} = \alpha^M \cdot \det\{\mathbf{A}\}$ for a scalar α. For square matrices $\mathbf{A}$ and $\mathbf{B}$, $\det\{\mathbf{AB}\} = \det\{\mathbf{A}\}\det\{\mathbf{B}\}$. If $\mathbf{A}$ is nonsingular, then $\det\{\mathbf{A}\} \neq 0$ and $\det\{\mathbf{A}^{-1}\} = 1/\det\{\mathbf{A}\}$. On the other hand, the *trace* of $\mathbf{A}$, denoted by $\text{tr}\{\mathbf{A}\}$, is defined as

$$\text{tr}\{\mathbf{A}\} = \sum_{m=1}^{M} a_{mm}, \qquad (2.13)$$

i.e. the sum of the diagonal elements of $\mathbf{A}$. As $M = 1$, the matrix $\mathbf{A}$ reduces to a scalar a_{11} and its trace $\text{tr}\{a_{11}\} = a_{11}$. If $\mathbf{A}$ is an $M \times K$ matrix and $\mathbf{B}$ is a $K \times M$ matrix, then $\text{tr}\{\mathbf{AB}\} = \text{tr}\{\mathbf{BA}\}$. As a special case, for column vectors $\mathbf{x}$ and $\mathbf{y}$, the trace $\text{tr}\{\mathbf{xy}^H\} = \text{tr}\{\mathbf{y}^H\mathbf{x}\} = \mathbf{y}^H\mathbf{x}$.

Let $\mathbf{A}$ be an $M \times M$ Hermitian matrix and $\mathbf{x}$ be an $M \times 1$ vector, then the quadratic function

$$Q(\mathbf{x}) \triangleq \mathbf{x}^H\mathbf{A}\mathbf{x} \qquad (2.14)$$

is called the *Hermitian form* of $\mathbf{A}$. The Hermitian matrix $\mathbf{A}$ is said to be *positive semidefinite* or *nonnegative definite* if $Q(\mathbf{x}) \geq 0$ for all $\mathbf{x} \neq \mathbf{0}$, and is said to be *positive definite* if $Q(\mathbf{x}) > 0$ for all $\mathbf{x} \neq \mathbf{0}$. In the same fashion, $\mathbf{A}$ is *negative semidefinite* or *nonpositive definite* if $Q(\mathbf{x}) \leq 0$ for all $\mathbf{x} \neq \mathbf{0}$, and *negative definite* if $Q(\mathbf{x}) < 0$ for all $\mathbf{x} \neq \mathbf{0}$.

An *eigenvector* of an $M \times M$ square matrix $\mathbf{A}$ is an $M \times 1$ nonzero vector, denoted by $\mathbf{q}$, which satisfies

$$\mathbf{Aq} = \lambda\mathbf{q} \qquad (2.15)$$

where λ is a scalar.[2] The scalar λ is an *eigenvalue* of $\mathbf{A}$ corresponding to the eigenvector $\mathbf{q}$. One can see from (2.15) that for any nonzero constant α,

---

[2] More precisely, the vector $\mathbf{q}$ is a *right eigenvector* of $\mathbf{A}$ if $\mathbf{Aq} = \lambda\mathbf{q}$, and a *left eigenvector* of $\mathbf{A}$ if $\mathbf{q}^H\mathbf{A} = \lambda\mathbf{q}^H$. In this book, "eigenvector" implies "right eigenvector" [6].

$\mathbf{A}(\alpha\mathbf{q}) = \lambda(\alpha\mathbf{q})$. This implies that any scaled version of $\mathbf{q}$ is also an eigenvector of $\mathbf{A}$ corresponding to the same eigenvalue λ. Eigenvectors which are orthogonal (i.e. $\mathbf{q}_m^H\mathbf{q}_n = 0$ for eigenvectors $\mathbf{q}_m$ and $\mathbf{q}_n$) and have the unit norm are referred to as *orthonormal eigenvectors.*

Special Forms of Matrices

A complex square matrix $\mathbf{U}$ is called a *unitary matrix* if it satisfies

$$\mathbf{U}\mathbf{U}^H = \mathbf{U}^H\mathbf{U} = \mathbf{I}, \tag{2.16}$$

i.e. $\mathbf{U}^H = \mathbf{U}^{-1}$ and $|\det\{\mathbf{U}\}| = 1$. Similarly, a real square matrix $\mathbf{V}$ is called an *orthogonal matrix* if it satisfies

$$\mathbf{V}\mathbf{V}^T = \mathbf{V}^T\mathbf{V} = \mathbf{I}, \tag{2.17}$$

i.e. $\mathbf{V}^T = \mathbf{V}^{-1}$ and $\det\{\mathbf{V}\} = 1$. Obviously, the identity matrix $\mathbf{I}$ is an orthogonal matrix.

A *diagonal matrix* is an $M \times M$ square matrix defined as

$$\mathbf{D} = \text{diag}\{d_1, d_2, ..., d_M\} = \begin{pmatrix} d_1 & 0 & \cdots & 0 \\ 0 & d_2 & \cdots & 0 \\ \vdots & \vdots & \ddots & \vdots \\ 0 & 0 & \cdots & d_M \end{pmatrix}. \tag{2.18}$$

If the diagonal matrix $\mathbf{D}$ is nonsingular, i.e. $\det\{\mathbf{D}\} = d_1 d_2 \cdots d_M \neq 0$, then its inverse

$$\mathbf{D}^{-1} = \text{diag}\left\{\frac{1}{d_1}, \frac{1}{d_2}, ..., \frac{1}{d_M}\right\}. \tag{2.19}$$

An *upper triangular matrix* is an $M \times M$ square matrix defined as

$$\mathbf{U} = \begin{pmatrix} u_{11} & u_{12} & \cdots & u_{1M} \\ 0 & u_{22} & \cdots & u_{2M} \\ \vdots & \vdots & \ddots & \vdots \\ 0 & 0 & \cdots & u_{MM} \end{pmatrix}, \tag{2.20}$$

and a *lower triangular matrix* defined as

$$\mathbf{L} = \begin{pmatrix} l_{11} & 0 & \cdots & 0 \\ l_{21} & l_{22} & \cdots & 0 \\ \vdots & \vdots & \ddots & \vdots \\ l_{M1} & l_{M2} & \cdots & l_{MM} \end{pmatrix}. \tag{2.21}$$

From (2.12), it follows that $\det\{\mathbf{U}\} = u_{11}u_{22}\cdots u_{MM}$ and $\det\{\mathbf{L}\} = l_{11}l_{22}\cdots l_{MM}$.

A *Toeplitz matrix* is an $M \times M$ square matrix defined as

$$\mathbf{R} = \begin{pmatrix} r_0 & r_1 & \cdots & r_{M-2} & r_{M-1} \\ r_{-1} & r_0 & \ddots & & r_{M-2} \\ \vdots & \ddots & \ddots & \ddots & \vdots \\ r_{-M+2} & & \ddots & r_0 & r_1 \\ r_{-M+1} & r_{-M+2} & \cdots & r_{-1} & r_0 \end{pmatrix}, \tag{2.22}$$

i.e. the entries on each of the diagonals are equal. Note that a Toeplitz matrix can be completely specified by its first column and first row.

A matrix $\mathbf{A}$ is called a 2×2 *partitioned matrix* if it can be expressed as

$$\mathbf{A} = \begin{pmatrix} \mathbf{A}_{11} & \mathbf{A}_{12} \\ \mathbf{A}_{21} & \mathbf{A}_{22} \end{pmatrix} \tag{2.23}$$

where $\mathbf{A}_{11}$, $\mathbf{A}_{12}$, $\mathbf{A}_{21}$, and $\mathbf{A}_{22}$ are the submatrices of $\mathbf{A}$. Manipulations of the submatrices for partitioned matrices are similar to those of the entries for general matrices. In particular, the Hermitian of $\mathbf{A}$ can be written as

$$\mathbf{A}^H = \begin{pmatrix} \mathbf{A}_{11}^H & \mathbf{A}_{21}^H \\ \mathbf{A}_{12}^H & \mathbf{A}_{22}^H \end{pmatrix}. \tag{2.24}$$

Furthermore, if $\mathbf{B}$ is also a 2×2 partitioned matrix given by

$$\mathbf{B} = \begin{pmatrix} \mathbf{B}_{11} & \mathbf{B}_{12} \\ \mathbf{B}_{21} & \mathbf{B}_{22} \end{pmatrix},$$

then

$$\mathbf{AB} = \begin{pmatrix} \mathbf{A}_{11}\mathbf{B}_{11} + \mathbf{A}_{12}\mathbf{B}_{21} & \mathbf{A}_{11}\mathbf{B}_{12} + \mathbf{A}_{12}\mathbf{B}_{22} \\ \mathbf{A}_{21}\mathbf{B}_{11} + \mathbf{A}_{22}\mathbf{B}_{21} & \mathbf{A}_{21}\mathbf{B}_{12} + \mathbf{A}_{22}\mathbf{B}_{22} \end{pmatrix} \tag{2.25}$$

where $\mathbf{B}_{11}$, $\mathbf{B}_{12}$, $\mathbf{B}_{21}$, and $\mathbf{B}_{22}$ are the submatrices with suitable sizes for the submatrix multiplications in $\mathbf{AB}$.

Matrix Formulas and Properties

The following theorem provides a useful formula for the derivation of matrix inverse [7, 8].

Theorem 2.5 (Matrix Inversion Lemma). *Let* $\mathbf{R}$ *be a nonsingular* $M \times M$ *matrix given by*

$$\mathbf{R} = \mathbf{A} + \mathbf{BCD} \tag{2.26}$$

where $\mathbf{A}$ *is a nonsingular* $M \times M$ *matrix,* $\mathbf{B}$ *is an* $M \times K$ *matrix,* $\mathbf{C}$ *is a nonsingular* $K \times K$ *matrix, and* $\mathbf{D}$ *is a* $K \times M$ *matrix. Then the inverse of* $\mathbf{R}$ *can be expressed as*

$$\mathbf{R}^{-1} = \mathbf{A}^{-1} - \mathbf{A}^{-1}\mathbf{B}(\mathbf{C}^{-1} + \mathbf{D}\mathbf{A}^{-1}\mathbf{B})^{-1}\mathbf{D}\mathbf{A}^{-1}. \tag{2.27}$$

A special case of the matrix inversion lemma is given as follows [7].[3]

Corollary 2.6 (Woodbury's Identity). *Let* $\mathbf{R}$ *be a nonsingular* $M \times M$ *matrix given by*

$$\mathbf{R} = \mathbf{A} + \alpha \mathbf{u}\mathbf{u}^H \tag{2.28}$$

where $\mathbf{A}$ *is a nonsingular* $M \times M$ *matrix,* $\mathbf{u}$ *is an* $M \times 1$ *vector, and* α *is a scalar. Then the inverse of* $\mathbf{R}$ *can be expressed as*

$$\mathbf{R}^{-1} = \mathbf{A}^{-1} - \frac{\alpha \mathbf{A}^{-1}\mathbf{u}\mathbf{u}^H\mathbf{A}^{-1}}{1 + \alpha \mathbf{u}^H \mathbf{A}^{-1}\mathbf{u}}. \tag{2.29}$$

The proof of Theorem 2.5 is left as an exercise (Problem 2.3), while Corollary 2.6 can be proved simply by substituting $\mathbf{B} = \mathbf{u}$, $\mathbf{C} = \alpha$ and $\mathbf{D} = \mathbf{u}^H$ into (2.27).

Moreover, two theorems regarding partitioned matrices are stated as follows [7, p. 572], [9, pp. 166–168], and the proofs are left as exercises (Problems 2.4 and 2.5).

Theorem 2.7. *Let* $\mathbf{A}$ *be a square matrix given as the partitioned form of (2.23). Then the determinant of* $\mathbf{A}$ *can be expressed as*

$$\det\{\mathbf{A}\} = \det\{\mathbf{A}_{11}\} \cdot \det\{\mathbf{A}_{22} - \mathbf{A}_{21}\mathbf{A}_{11}^{-1}\mathbf{A}_{12}\} \tag{2.30}$$

provided that $\mathbf{A}_{11}$ *is a nonsingular square matrix, or equivalently*

$$\det\{\mathbf{A}\} = \det\{\mathbf{A}_{22}\} \cdot \det\{\mathbf{A}_{11} - \mathbf{A}_{12}\mathbf{A}_{22}^{-1}\mathbf{A}_{21}\} \tag{2.31}$$

provided that $\mathbf{A}_{22}$ *is a nonsingular square matrix.*

Theorem 2.8. *Let* $\mathbf{A}$ *be a nonsingular square matrix given as the partitioned form of (2.23) where* $\mathbf{A}_{11}$ *and* $\mathbf{A}_{22}$ *are also nonsingular square matrices. Then the inverse of* $\mathbf{A}$ *can be expressed as*

---

[3] For ease of later use, we give a slightly generalized statement of Woodbury's identity by including a scalar α. As $\alpha = 1$, it reduces to the normal statement of Woodbury's identity.

$$\mathbf{A}^{-1} = \begin{pmatrix} \mathbf{B}_{11} & \mathbf{B}_{12} \\ \mathbf{B}_{21} & \mathbf{B}_{22} \end{pmatrix} \tag{2.32}$$

where

$$\begin{aligned}
\mathbf{B}_{11} &= (\mathbf{A}_{11} - \mathbf{A}_{12}\mathbf{A}_{22}^{-1}\mathbf{A}_{21})^{-1} \\
\mathbf{B}_{12} &= -(\mathbf{A}_{11} - \mathbf{A}_{12}\mathbf{A}_{22}^{-1}\mathbf{A}_{21})^{-1}\mathbf{A}_{12}\mathbf{A}_{22}^{-1} \\
\mathbf{B}_{21} &= -(\mathbf{A}_{22} - \mathbf{A}_{21}\mathbf{A}_{11}^{-1}\mathbf{A}_{12})^{-1}\mathbf{A}_{21}\mathbf{A}_{11}^{-1} \\
\mathbf{B}_{22} &= (\mathbf{A}_{22} - \mathbf{A}_{21}\mathbf{A}_{11}^{-1}\mathbf{A}_{12})^{-1}.
\end{aligned}$$

In the following, we summarize several matrix properties and leave the proofs as exercises (Problems 2.6, 2.7 and 2.8).

Property 2.9. *A positive definite matrix is nonsingular.*

Property 2.10. *The eigenvalues of a Hermitian matrix are all real.*

Property 2.11. *The eigenvalues of a positive definite (positive semidefinite) matrix are all real positive (nonnegative).*

Property 2.12. *The inverse of a positive definite matrix is also positive definite.*

Property 2.13. *For any matrix* $\mathbf{A}$*, both* $\mathbf{A}^H\mathbf{A}$ *and* $\mathbf{A}\mathbf{A}^H$ *are positive semidefinite.*

Property 2.14. *The eigenvectors of a Hermitian matrix corresponding to distinct eigenvalues are orthogonal.*

Although Property 2.14 is for the case of distinct eigenvalues, one can always find a complete set of orthogonal eigenvectors, or equivalently, orthonormal eigenvectors for any Hermitian matrix, no matter whether its eigenvalues are distinct or not [2, p. 297].

As a consequence, if $\mathbf{A}$ is a positive definite matrix, then its inverse $\mathbf{A}^{-1}$ exists (by Property 2.9) and is also positive definite (by Property 2.12). Furthermore, the eigenvalues of both matrices $\mathbf{A}$ and $\mathbf{A}^{-1}$ are all real positive (by Property 2.11).

2.1.3 Matrix Decomposition

Among the available tools of matrix decomposition, two representatives, *eigendecomposition* and *singular value decomposition (SVD)*, to be presented are of importance in the area of statistical signal processing. In particular, the eigendecomposition is useful in developing subspace based algorithms, while the SVD is powerful in solving least-squares problems as well as in determining the numerical rank of a real or complex matrix in the presence of roundoff errors (due to finite precision of computing machines).

Eigendecomposition

According to the foregoing discussion (the paragraph following Property 2.14), we can always find a complete set of M orthonormal eigenvectors for an $M \times M$ Hermitian matrix $\mathbf{A}$. As such, let $\mathbf{u}_1$, $\mathbf{u}_2$, ..., $\mathbf{u}_M$ be the M orthonormal eigenvectors of $\mathbf{A}$ corresponding to the eigenvalues λ_1, λ_2, ..., λ_M. Then, by definition,

$$\mathbf{A}(\mathbf{u}_1, \mathbf{u}_2, ..., \mathbf{u}_M) = (\lambda_1 \mathbf{u}_1, \lambda_2 \mathbf{u}_2, ..., \lambda_M \mathbf{u}_M)$$

or

$$\mathbf{AU} = \mathbf{U\Lambda} \tag{2.33}$$

where $\mathbf{\Lambda} = \text{diag}\{\lambda_1, \lambda_2, ..., \lambda_M\}$ is an $M \times M$ diagonal matrix and $\mathbf{U} = (\mathbf{u}_1, \mathbf{u}_2, ..., \mathbf{u}_M)$ is an $M \times M$ unitary matrix since $\mathbf{u}_1$, $\mathbf{u}_2$, ..., $\mathbf{u}_M$ are orthonormal. From (2.33), it follows that

$$\mathbf{A} = \mathbf{U\Lambda U}^H = \sum_{m=1}^{M} \lambda_m \mathbf{u}_m \mathbf{u}_m^H. \tag{2.34}$$

Equation (2.34) is called the *eigendecomposition* or the *spectral decomposition* of $\mathbf{A}$. Moreover, when $\mathbf{A}$ is nonsingular, (2.34) leads to

$$\mathbf{A}^{-1} = \mathbf{U}^{-H}\mathbf{\Lambda}^{-1}\mathbf{U}^{-1} = \mathbf{U\Lambda}^{-1}\mathbf{U}^H = \sum_{m=1}^{M} \frac{1}{\lambda_m} \mathbf{u}_m \mathbf{u}_m^H. \tag{2.35}$$

Singular Value Decomposition

The SVD is stated in the following theorem and, for clarity, is illustrated in Fig. 2.2. The theorem is called the *SVD theorem*, or the *Autonne–Eckart–Young theorem* in recognition of the originators [10].[4]

Theorem 2.15 (SVD Theorem). *Let $\mathbf{A}$ be an $M \times K$ real or complex matrix with* $\text{rank}\{\mathbf{A}\} = r$. *Then there exist an $M \times M$ unitary matrix*

$$\mathbf{U} = (\mathbf{u}_1, \mathbf{u}_2, ..., \mathbf{u}_M) \tag{2.36}$$

and a $K \times K$ unitary matrix

$$\mathbf{V} = (\mathbf{v}_1, \mathbf{v}_2, ..., \mathbf{v}_K) \tag{2.37}$$

such that the matrix $\mathbf{A}$ can be decomposed as

---

[4] The SVD was established for real square matrices by Beltrami and Jordan in the 1870s, for complex square matrices by Autonne in 1902, and for general rectangular matrices by Eckart and Young in 1939 [10].

$$\mathbf{A} = \mathbf{U}\boldsymbol{\Sigma}\mathbf{V}^H = \sum_{m=1}^{r} \lambda_m \mathbf{u}_m \mathbf{v}_m^H \tag{2.38}$$

where $\mathbf{u}_i$ *are* $M \times 1$ *vectors,* $\mathbf{v}_i$ *are* $K \times 1$ *vectors, and*

$$\boldsymbol{\Sigma} = \begin{pmatrix} \boldsymbol{\Lambda} & \mathbf{0} \\ \mathbf{0} & \mathbf{0} \end{pmatrix} \tag{2.39}$$

is an $M \times K$ *matrix. The matrix* $\boldsymbol{\Lambda} = \mathrm{diag}\{\lambda_1, \lambda_2, ..., \lambda_r\}$ *is an* $r \times r$ *diagonal matrix where* λ_i *are real and* $\lambda_1 \geq \lambda_2 \geq \cdots \geq \lambda_r > 0$.

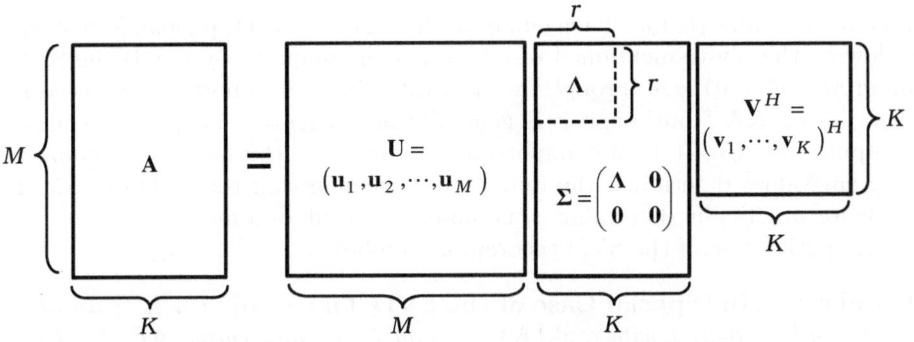

Fig. 2.2 Illustration of the SVD for an $M \times K$ matrix $\mathbf{A}$ with $M > K > r = \mathrm{rank}\{\mathbf{A}\}$ where $\boldsymbol{\Lambda} = \mathrm{diag}\{\lambda_1, \lambda_2, ..., \lambda_r\}$

As shown in Appendix 2A, the SVD theorem can be proved by either of the two approaches, *Approach I* and *Approach II*, where Approach I starts from the matrix $\mathbf{A}^H\mathbf{A}$ and Approach II from the matrix $\mathbf{A}\mathbf{A}^H$. Some important results regarding both approaches are summarized as follows.

- Results from Approach I. The nonnegative real numbers λ_1, λ_2, ..., λ_K are identical to the positive square roots of the eigenvalues of the $K \times K$ matrix $\mathbf{A}^H\mathbf{A}$ and the column vectors $\mathbf{v}_1$, $\mathbf{v}_2$, ..., $\mathbf{v}_K$ of $\mathbf{V}$ are the corresponding orthonormal eigenvectors. The positive real numbers λ_1, λ_2, ..., λ_r, together with $\lambda_{r+1} = \cdots = \lambda_K = 0$ (since $\mathrm{rank}\{\mathbf{A}\} = r$), are called the *singular values* of $\mathbf{A}$, while the vectors $\mathbf{v}_1$, $\mathbf{v}_2$, ..., $\mathbf{v}_K$ are called the *right singular vectors* of $\mathbf{A}$. With λ_m and $\mathbf{v}_m$ computed from $\mathbf{A}^H\mathbf{A}$, the column vectors $\mathbf{u}_1$, $\mathbf{u}_2$, ..., $\mathbf{u}_r$ in $\mathbf{U}$ are accordingly determined via (see (2.204))

$$(\mathbf{u}_1, \mathbf{u}_2, ..., \mathbf{u}_r) = \left(\frac{\mathbf{A}\mathbf{v}_1}{\lambda_1}, \frac{\mathbf{A}\mathbf{v}_2}{\lambda_2}, ..., \frac{\mathbf{A}\mathbf{v}_r}{\lambda_r} \right), \tag{2.40}$$

  while the remaining column vectors $\mathbf{u}_{r+1}$, $\mathbf{u}_{r+2}$, ..., $\mathbf{u}_M$ (allowing some choices) are chosen such that $\mathbf{U}$ is unitary. The vectors $\mathbf{u}_1$, $\mathbf{u}_2$, ..., $\mathbf{u}_M$ are called the *left singular vectors* of $\mathbf{A}$.

- Results from Approach II. The singular values $\lambda_1, \lambda_2, ..., \lambda_M$ of $\mathbf{A}$ are identical to the positive square roots of the eigenvalues of the $M \times M$ matrix $\mathbf{A}\mathbf{A}^H$ and the left singular vectors $\mathbf{u}_1, \mathbf{u}_2, ..., \mathbf{u}_M$ are the corresponding orthonormal eigenvectors. With λ_m and $\mathbf{u}_m$ computed from $\mathbf{A}\mathbf{A}^H$, the right singular vectors $\mathbf{v}_1, \mathbf{v}_2, ..., \mathbf{v}_r$ are accordingly determined via (see (2.213))

$$(\mathbf{v}_1, \mathbf{v}_2, ..., \mathbf{v}_r) = \left(\frac{\mathbf{A}^H\mathbf{u}_1}{\lambda_1}, \frac{\mathbf{A}^H\mathbf{u}_2}{\lambda_2}, ..., \frac{\mathbf{A}^H\mathbf{u}_r}{\lambda_r} \right), \tag{2.41}$$

while the remaining right singular vectors $\mathbf{v}_{r+1}, \mathbf{v}_{r+2}, ..., \mathbf{v}_K$ are chosen such that $\mathbf{V}$ is unitary.

As a result, a matrix may have numerous forms of SVD [11, p. 309]. Moreover, following the above-mentioned results, one can compute (by hand) the SVD of an $M \times K$ matrix $\mathbf{A}$ through the eigenvalues and orthonormal eigenvectors of $\mathbf{A}^H\mathbf{A}$ or $\mathbf{A}\mathbf{A}^H$, although it is generally not suggested for finite-precision computation [10]. It is also important to note that the number of nonzero singular values determines the rank of $\mathbf{A}$, revealing that the SVD provides a basis for practically determining the numerical rank of a matrix.

A special case of the SVD theorem is as follows.

Corollary 2.16 (Special Case of the SVD Theorem). *Let* $\mathbf{A}$ *be an* $M \times M$ *Hermitian matrix with* $\text{rank}\{\mathbf{A}\} = r$ *and* $\mathbf{A}$ *is nonnegative definite. Then the matrix* $\mathbf{A}$ *can be decomposed as*

$$\mathbf{A} = \mathbf{U}\mathbf{\Sigma}\mathbf{U}^H = \sum_{m=1}^{r} \lambda_m \mathbf{u}_m \mathbf{u}_m^H \tag{2.42}$$

where $\mathbf{\Sigma} = \text{diag}\{\lambda_1, ..., \lambda_r, \lambda_{r+1}, ..., \lambda_M\}$ *is an* $M \times M$ *diagonal matrix and* $\mathbf{U} = (\mathbf{u}_1, \mathbf{u}_2, ..., \mathbf{u}_M)$ *is an* $M \times M$ *unitary matrix. The singular values* $\lambda_1 \geq \cdots \geq \lambda_r > \lambda_{r+1} = \cdots = \lambda_M = 0$ *are the eigenvalues of* $\mathbf{A}$ *and the singular vectors* $\mathbf{u}_1, \mathbf{u}_2, ..., \mathbf{u}_M$ *are the corresponding orthonormal eigenvectors.*

The proof is left as an exercise (Problem 2.10). Comparing (2.42) with (2.34) reveals that for a Hermitian matrix $\mathbf{A}$, the SVD of $\mathbf{A}$ is equivalent to the eigendecomposition of $\mathbf{A}$.

2.2 Mathematical Analysis

This section briefly reviews the fundamentals of mathematical analysis, including sequences, series, Hilbert spaces, vector spaces of sequences and functions, and pays attention to the topic of Fourier series. Some of these topics need the background of functions, provided in Appendix 2B.

2.2.1 Sequences

A *sequence* is regarded as a list of real or complex numbers in a definite order:

$$a_m, a_{m+1}, ..., a_{n-1}, a_n$$

where a_k, $k = m, m+1, ..., n$, are called the *terms* of the sequence. The sequence is denoted by $\{a_k\}_{k=m}^{n}$ or, briefly, $\{a_k\}$. One should not confuse a sequence $\{a_k\}_{k=m}^{n}$ with a set $\{a_k, k = m, m+1, ..., n\}$; the order of a_k is meaningless for the latter. Moreover, a sequence $\{a_k\}$ is said to be an *infinite sequence* if it has infinitely many terms. A natural concern about a one-sided infinite sequence, $\{a_k\}_{k=1}^{\infty}$, is whether it converges or not, that is the topic to be dealt with next.

Sequences of Numbers

A real or complex sequence $\{a_k\}_{k=1}^{\infty}$ is said to *converge* to a real or complex number a if

$$\lim_{k\to\infty} a_k = a, \tag{2.43}$$

i.e. for every real number $\varepsilon > 0$ there exists an integer N such that

$$|a_k - a| < \varepsilon \quad \text{for all } k \geq N \tag{2.44}$$

where N is, in general, dependent on ε. If $\{a_k\}$ does not converge, it is called a *divergent sequence* [12]. A sequence $\{a_k\}$ is said to be *bounded* if $|a_k| \leq A$ for all k where A is a positive constant. A real sequence $\{a_k\}_{k=1}^{\infty}$ is said to be *increasing* (*decreasing*) or, briefly, *monotonic* if $a_k \leq a_{k+1}$ ($a_k \geq a_{k+1}$) for all k, and is said to be *strictly increasing* (*strictly decreasing*) if $a_k < a_{k+1}$ ($a_k > a_{k+1}$) for all k. A theorem regarding monotonic sequences is as follows [13, p. 61].

Theorem 2.17. *If $\{a_k\}_{k=1}^{\infty}$ is a monotonic and bounded real sequence, then $\{a_k\}_{k=1}^{\infty}$ converges.*

The proof is left as an exercise (Problem 2.11).

From a sequence $\{a_k\}_{k=1}^{\infty}$, one can obtain another sequence, denoted by $\{\sigma_n\}_{n=1}^{\infty}$, composed of the arithmetic mean

$$\sigma_n = \frac{a_1 + a_2 + \cdots + a_n}{n}. \tag{2.45}$$

The arithmetic mean σ_n is also referred to as the *nth Cesàro mean* of the sequence $\{a_n\}$ [14]. A related theorem is stated as follows [15, p. 138].

Theorem 2.18. *If a real or complex sequence $\{a_k\}_{k=1}^{\infty}$ is bounded and converges to a real or complex number a, then the sequence of arithmetic mean $\{\sigma_n\}_{n=1}^{\infty}$ also converges to the number a where σ_n is defined as (2.45).*

The proof, again, is left as an exercise (Problem 2.12). When the sequence of arithmetic means $\{\sigma_n\}$ converges to a, we say that the original sequence $\{a_k\}$ is *Cesàro summable* to a. Since the average operation in (2.45) may smooth out occasional fluctuations in $\{a_k\}$, it is expected that $\{\sigma_n\}$, in general, tends to converge even if $\{a_k\}$ is divergent. An example is given as follows.

Example 2.19
Consider that $a_k = (-1)^k$. The sequence $\{a_k\}_{k=1}^{\infty}$ is bounded by 1, but it diverges since $a_k = 1$ for k even and $a_k = -1$ for k odd. On the other hand, the arithmetic mean $\sigma_n = 0$ for n even and $\sigma_n = -1/n$ for n odd. This indicates that $\lim_{n\to\infty} \sigma_n = 0$, namely, $\{a_k\}$ is Cesàro summable to zero. □

Sequences of Functions

Now consider a sequence of real or complex functions, $\{a_k(x)\}_{k=1}^{\infty}$. Since $a_k(x)$ is a function of x, the convergence of $\{a_k(x)\}_{k=1}^{\infty}$ may further depend on the value of x.

The sequence $\{a_k(x)\}_{k=1}^{\infty}$ is said to *converge pointwise* to a real or complex function $a(x)$ on an interval $[x_{\mathrm{L}}, x_{\mathrm{U}}]$ if

$$\lim_{k\to\infty} a_k(x) = a(x) \quad \text{for every point } x \in [x_{\mathrm{L}}, x_{\mathrm{U}}], \tag{2.46}$$

i.e. for every real number $\varepsilon > 0$ and every point $x \in [x_{\mathrm{L}}, x_{\mathrm{U}}]$ there exists an integer N such that

$$|a_k(x) - a(x)| < \varepsilon \quad \text{for all } k \geq N \tag{2.47}$$

where N may depend on ε and x. When the integer N is independent of x, $\{a_k(x)\}_{k=1}^{\infty}$ is said to *converge uniformly* to $a(x)$ on the interval $[x_{\mathrm{L}}, x_{\mathrm{U}}]$. In other words, a uniformly convergent sequence $\{a_k(x)\}$ exhibits similar local behaviors of convergence for all $x \in [x_{\mathrm{L}}, x_{\mathrm{U}}]$, as illustrated in Fig. 2.3.

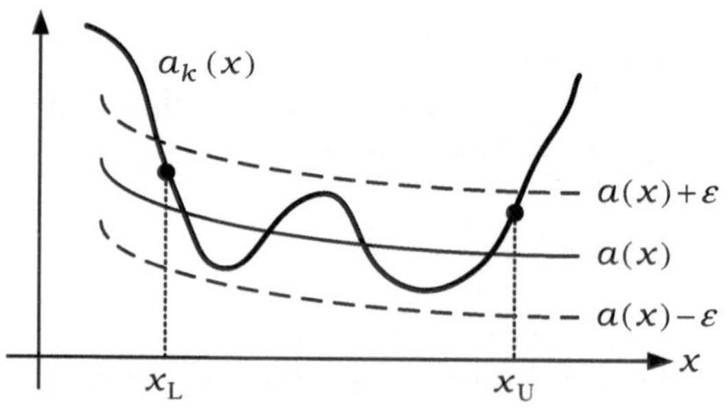

Fig. 2.3 A uniformly convergent sequence $\{a_k(x)\}_{k=1}^{\infty}$ on an interval $[x_{\mathrm{L}}, x_{\mathrm{U}}]$

It is important to emphasize that even if every $a_k(x)$ is a continuous function, a pointwise convergent sequence $\{a_k(x)\}_{k=1}^{\infty}$ may still converge to a discontinuous function $a(x)$. The following example demonstrates this fact [13, p. 320], [16, p. 171].

Example 2.20
As shown in Fig. 2.4, the sequence $\{x^k\}_{k=1}^{\infty}$ converges pointwise to the function $a(x)$ on $[0, 1]$ where

$$a(x) = \begin{cases} 0, & 0 \leq x < 1, \\ 1, & x = 1. \end{cases}$$

That is, $a(x)$ has a discontinuity at $x = 1$, although every function x^k is continuous on $[0, 1]$.

□

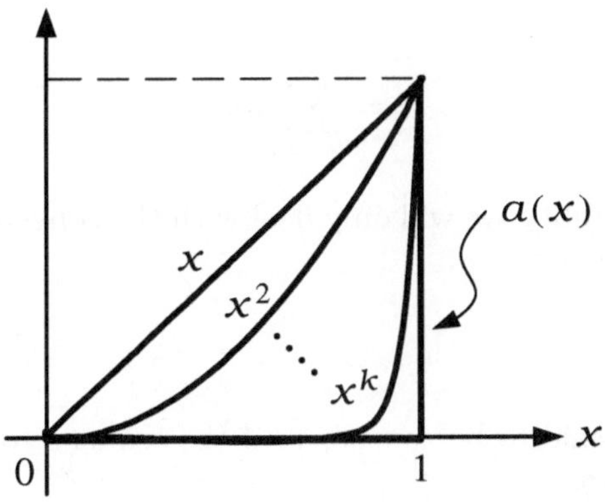

Fig. 2.4 Pointwise convergence of the sequence $\{x^k\}_{k=1}^{\infty}$ to a discontinuous function $a(x)$ on $[0, 1]$

Unlike pointwise convergence, uniform convergence ensures continuity, as the following theorem states [16, p. 174].

Theorem 2.21. *If the sequence* $\{a_k(x)\}_{k=1}^{\infty}$ *converges uniformly to a function* $a(x)$ *on an interval* $[x_L, x_U]$ *where every* $a_k(x)$ *is continuous on* $[x_L, x_U]$*, then the function* $a(x)$ *must be continuous on* $[x_L, x_U]$.

We leave the proof as an exercise (Problem 2.13). Theorem 2.21 implies that if every $a_k(x)$ is continuous but $a(x)$ is discontinuous, then it is not possible for the sequence $\{a_k(x)\}$ to converge uniformly to $a(x)$.

Example 2.22
Consider, again, the pointwise convergent sequence $\{x^k\}_{k=1}^{\infty}$ in Example 2.20. According to Theorem 2.21, it is clear that $\{x^k\}_{k=1}^{\infty}$ is not uniformly convergent on $[0,1]$ since $a(x)$ has a discontinuity at $x=1$.

□

2.2.2 Series

Closely related to a real or complex sequence $\{a_k\}_{k=m}^{n}$, a *series* is defined as $\sum_{k=m}^{n} a_k$. The series $\sum_{k=1}^{\infty} a_k$ is called a *one-sided infinite sequence* with the *nth partial sum* defined as

$$s_n = \sum_{k=1}^{n} a_k, \tag{2.48}$$

while the series $\sum_{k=-\infty}^{\infty} a_k$ is called a *two-sided infinite sequence* with the *nth partial sum* defined as

$$s_n = \sum_{k=-n}^{n} a_k. \tag{2.49}$$

Without loss of generality, we will only deal with the convergence of one-sided infinite series for brevity.

Series of Numbers

A series $\sum_{k=1}^{\infty} a_k$ is said to be *convergent* if the sequence of its partial sums, $\{s_n\}_{n=1}^{\infty}$, converges to

$$s \triangleq \sum_{k=1}^{\infty} a_k, \tag{2.50}$$

i.e. $\lim_{n\to\infty} s_n = s$ where s is called the *sum* or *value* of the series. From (2.48) and (2.50), it follows that if $\sum_{k=1}^{\infty} a_k$ is convergent, then the following condition always holds:

$$\lim_{n\to\infty} a_n = \lim_{n\to\infty} (s_n - s_{n-1}) = s - s = 0. \tag{2.51}$$

Moreover, a series $\sum_{k=1}^{\infty} a_k$ is said to be *absolutely convergent* if the series $\sum_{k=1}^{\infty} |a_k|$ is convergent.

Tests for Divergence and Convergence

In using a series $\sum_{k=1}^{\infty} a_k$, it is important to know whether $\sum_{k=1}^{\infty} a_k$ converges or diverges. The condition given by (2.51) suggests a test as follows.

Theorem 2.23 (Divergence Test). *Suppose $\sum_{k=1}^{\infty} a_k$ is a real or complex series to be tested. If the condition given by (2.51) is not satisfied, then the series $\sum_{k=1}^{\infty} a_k$ is divergent.*

Since the condition given by (2.51) is only necessary, but not sufficient, for convergence, it cannot be used for convergence testing. An example using the divergence test is as follows.

Example 2.24 (Geometric Series)
The partial sum of the geometric series $\sum_{k=1}^{\infty} ar^k$ can be expressed as

$$s_n = \sum_{k=1}^{n} ar^k = ar\frac{1-r^n}{1-r}.$$

If $|r| < 1$, then the geometric series converges with

$$\lim_{n\to\infty} s_n = \frac{ar}{1-r}.$$

On the other hand, if $|r| \geq 1$, then $\lim_{n\to\infty} ar^n \neq 0$ which does not satisfy the condition given by (2.51), and thus the geometric series diverges. □

The following test is useful for testing the convergence of a real series.

Theorem 2.25 (Integral Test). *Suppose $\sum_{k=1}^{\infty} a_k$ is a real series to be tested where $a_k \geq 0$ for all k. Find a continuous, positive, and decreasing function $f(x)$ on $[1,\infty)$ such that $f(k) = a_k$.*

- *If $\int_1^{\infty} f(x)dx$ is finite, then the series $\sum_{k=1}^{\infty} a_k$ is convergent.*
- *If $\int_1^{\infty} f(x)dx$ is infinite, then the series $\sum_{k=1}^{\infty} a_k$ is divergent.*

The proof is left as an exercise (Problem 2.14). An example using the integral test is as follows.

Example 2.26
To test the convergence of the real series $\sum_{k=1}^{\infty} 1/k^2$, let $f(x) = 1/x^2$. It is clear that $f(x)$ is continuous, positive, and decreasing on $[1,\infty)$ and $f(k) = 1/k^2$. By the integral test, $\int_1^{\infty} f(x)dx = 1$ implies that $\sum_{k=1}^{\infty} 1/k^2$ converges. □

Series of Functions

Now consider a series $\sum_{k=1}^{\infty} a_k(x)$ whose nth partial sum is given by

$$s_n(x) = \sum_{k=1}^{n} a_k(x) \tag{2.52}$$

where $a_k(x)$ is a real or complex function of a real independent variable x. The series $\sum_{k=1}^{\infty} a_k(x)$ is said to *converge pointwise* to a real or complex function $s(x)$ if the sequence of its partial sums, $\{s_n(x)\}_{n=1}^{\infty}$, converges pointwise to $s(x)$, and is said to *converge uniformly* to $s(x)$ if the sequence $\{s_n(x)\}_{n=1}^{\infty}$ converges uniformly to $s(x)$.

According to the above-mentioned definitions, the convergence theory for sequences of functions can similarly apply to series of functions. In particular, uniform convergence of series also implies pointwise convergence of series, but the converse may not be true. Moreover, a pointwise convergent series $\sum_{k=1}^{\infty} a_k(x)$ may converge to a discontinuous function $s(x)$, even if every function $a_k(x)$ is continuous. And the following theorem directly follows from Theorem 2.21.

Theorem 2.27. *If the series $\sum_{k=1}^{\infty} a_k(x)$ converges uniformly to a function $s(x)$ on an interval $[x_L, x_U]$ where every $a_k(x)$ is continuous on $[x_L, x_U]$, then the function $s(x)$ is also continuous on $[x_L, x_U]$.*

As a remark, let us emphasize that there is no connection between uniform convergence and absolute convergence [4, p. 765].

Test for Uniform Convergence

The following test is most commonly used for testing the uniform convergence of series.

Theorem 2.28 (Weierstrass M-Test). *Suppose $\sum_{k=1}^{\infty} a_k(x)$ is a real or complex series to be tested on an interval $[x_L, x_U]$. If there exists a convergent series $\sum_{k=1}^{\infty} M_k$ such that each term $M_k \geq |a_k(x)|$ for all $x \in [x_L, x_U]$, then the series $\sum_{k=1}^{\infty} a_k(x)$ is uniformly and absolutely convergent on $[x_L, x_U]$.*

Since the proof is lengthy and can be found, for instance, in [13], it is omitted here. An example using the Weierstrass M-test is provided as follows.

Example 2.29

Suppose $\sum_{k=1}^{\infty} e^{jkx}/k^2$ is the series to be tested on $[-\pi, \pi)$. Because $\left|e^{jkx}/k^2\right| \leq 1/k^2$ for all $x \in [-\pi, \pi)$ and $\sum_{k=1}^{\infty} 1/k^2$ converges (see Example 2.26), by the Weierstrass M-test, the series $\sum_{k=1}^{\infty} e^{jkx}/k^2$ is uniformly and absolutely convergent on $[-\pi, \pi)$.

□

2.2.3 Hilbert Spaces, Sequence Spaces and Function Spaces

Hilbert Spaces

Consider a sequence of real or complex vectors, denoted by $\{\mathbf{a}_n\}_{n=1}^{\infty}$, in a normed vector space $\mathcal{V}$. The sequence $\{\mathbf{a}_n\}_{n=1}^{\infty}$ is said to *converge in the norm* or, briefly, *converge* to a real or complex vector $\mathbf{a} \in \mathcal{V}$ if

$$\lim_{n\to\infty} \|\mathbf{a} - \mathbf{a}_n\| = 0. \tag{2.53}$$

Convergence in the norm is also often referred to as *convergence in the mean.* A sequence $\{\mathbf{a}_n\}_{n=1}^{\infty}$ in $\mathcal{V}$ is called a *Cauchy sequence* if for every real number $\varepsilon > 0$ there exists an integer N such that

$$\|\mathbf{a}_n - \mathbf{a}_m\| < \varepsilon \quad \text{for all } n > m \geq N. \tag{2.54}$$

Regarding Cauchy sequences, we have the following related theorem, whose proof is left as an exercise (Problem 2.15).

Theorem 2.30. *Every convergent sequence in a norm vector space $\mathcal{V}$ is a Cauchy sequence.*

The converse of Theorem 2.30, however, may be true for some norm vector spaces. If every Cauchy sequence in a norm vector space $\mathcal{V}$ converges to a vector in $\mathcal{V}$, then the normed vector space $\mathcal{V}$ is said to be *complete.* A complete normed vector space is also referred to as a *Banach space* [17].

Definition 2.31 (Hilbert Space). *A vector space $\mathcal{V}$ along with a legitimate norm and a legitimate inner product is said to be a Hilbert space if the normed vector space (i.e. $\mathcal{V}$ along with the legitimate norm) is complete and the inner product can induce the norm.*

As an example, the vector space $\mathcal{R}^N$ (see Example 2.3) along with the Euclidean norm and the Euclidean inner product is an N-dimensional Hilbert space [5,14,18], which we also refer to as the N-dimensional Euclidean space $\mathcal{R}^N$ for convenience.

Sequence Spaces

Consider a real or complex sequence $\{a_n\}_{n=1}^{\infty}$ which is bounded and satisfies

$$\left(\sum_{n=1}^{\infty} |a_n|^p\right)^{1/p} < \infty, \quad \text{for } 1 \leq p < \infty. \tag{2.55}$$

Let $\mathcal{V}$ be the set composed of all such sequences. Then, under the operations of componentwise addition and scalar multiplication of sequences, the set $\mathcal{V}$ can easily be shown to be a vector space (satisfying the axioms (VS1) through

(VS8)). The vector space $\mathcal{V}$ is a sequence space, commonly referred to as the *ℓ^p space* or, briefly, ℓ^p [5, 13, 17, 18].

For notational simplicity, let $\mathbf{a} = (a_1, a_2, ..., a_n, ...)^T$ denote a vector corresponding to $\{a_n\}_{n=1}^{\infty} \in \ell^p$. The inner product of sequences $\{a_n\}_{n=1}^{\infty}$ and $\{b_n\}_{n=1}^{\infty} \in \ell^p$ is defined as

$$\langle \mathbf{a}, \mathbf{b} \rangle = \sum_{n=1}^{\infty} a_n b_n^*, \tag{2.56}$$

while the *ℓ^p norm* of $\{a_n\}_{n=1}^{\infty} \in \ell^p$ is defined as

$$\|\mathbf{a}\|_p = \begin{cases} \left(\sum_{n=1}^{\infty} |a_n|^p \right)^{1/p}, & \text{for } 1 \leq p < \infty \\ \sup_{n=1,2,...} \{|a_n|\}, & \text{for } p = \infty \end{cases} \tag{2.57}$$

where the notation 'sup' stands for the *least upper bound* or the *supremum* of a set of real numbers.[5] From (2.56) and (2.57), it follows that only the ℓ^2 norm (i.e. $p = 2$) can be induced from (2.56). Furthermore, the ℓ^2 space along with the inner product defined as (2.56) and the ℓ^2 norm is known as an infinite-dimensional Hilbert space [14, p. 75]. As such, in what follows, the ℓ^2 space always refers to this Hilbert space for convenience.

Moreover, for ease of later use, we restate the Cauchy–Schwartz inequality in terms of two-sided sequences as follows.

Theorem 2.32 (Cauchy–Schwartz Inequality). *Suppose $\{a_n\}_{n=-\infty}^{\infty}$ and $\{b_n\}_{n=-\infty}^{\infty}$ are real or complex nonzero sequences with $\sum_{n=-\infty}^{\infty} |a_n|^2 < \infty$ and $\sum_{n=-\infty}^{\infty} |b_n|^2 < \infty$. Then*

$$\left| \sum_{n=-\infty}^{\infty} a_n b_n^* \right| \leq \left(\sum_{n=-\infty}^{\infty} |a_n|^2 \right)^{1/2} \left(\sum_{n=-\infty}^{\infty} |b_n|^2 \right)^{1/2} \tag{2.58}$$

and the equality holds if and only if $a_n = \alpha b_n$ for all n where $\alpha \neq 0$ is an arbitrary real or complex scalar.

Also with regard to two-sided sequences, the following inequality is useful in development of blind equalization algorithms [19, 20].

Theorem 2.33. *Suppose $\{a_n\}_{n=-\infty}^{\infty}$ is a real or complex nonzero sequence with $\sum_{n=-\infty}^{\infty} |a_n|^s < \infty$ where s is an integer and $1 \leq s < \infty$. Then*

---

[5] One should not confuse "supremum" with "maximum." A set which is bounded above has a supremum, but may not have a maximum (the largest element of the set) [12, p. 16]. For instance, the set $\{1 - (1/n), n = 1 \sim \infty\}$ has a supremum equal to one, but does not have any maximum.

$$\left(\sum_{n=-\infty}^{\infty}|a_n|^l\right)^{1/l} \leq \left(\sum_{n=-\infty}^{\infty}|a_n|^s\right)^{1/s} \tag{2.59}$$

and the equality holds if and only if a_n *has only one nonzero term where* l *is an integer and* $l > s$*.*

See Appendix 2C for the proof.

Function Spaces

Consider a real or complex functions $f(x)$ which is bounded and satisfies

$$\left(\int_{x_\mathrm{L}}^{x_\mathrm{U}}|f(x)|^p dx\right)^{1/p}, \quad \text{for } 1 \leq p < \infty. \tag{2.60}$$

Then the set of all such functions forms a function space (a vector space) under the operations of pointwise addition and scalar multiplication of functions. The function space is commonly referred to as the $\mathcal{L}^p[x_\mathrm{L}, x_\mathrm{U}]$ space or, briefly, $\mathcal{L}^p[x_\mathrm{L}, x_\mathrm{U}]$ [5, 13, 17, 18].

Define the inner product of functions $f(x)$ and $g(x) \in \mathcal{L}^p[x_\mathrm{L}, x_\mathrm{U}]$ as

$$\langle f, g\rangle = \int_{x_\mathrm{L}}^{x_\mathrm{U}} f(x)g(x)^* dx \tag{2.61}$$

and the $\mathcal{L}^p$ *norm* of $f(x) \in \mathcal{L}^p[x_\mathrm{L}, x_\mathrm{U}]$ as

$$\|f\|_p = \left(\int_{x_\mathrm{L}}^{x_\mathrm{U}}|f(x)|^p dx\right)^{1/p}, \quad \text{for } 1 \leq p < \infty. \tag{2.62}$$

Only the $\mathcal{L}^2$ norm ($p = 2$) can be induced from (2.61). More importantly, due to the operation of integration in (2.62), $\|f\|_2 = 0$ merely implies that $f(x) = 0$ *almost everywhere* on $[x_\mathrm{L}, x_\mathrm{U}]$, that is, $f(x)$ may not be identically zero on a set of points on which the integration is "negligible."[6] From this, it follows that the inner product defined as (2.61) does not satisfy the axiom (IPS3) and the $\mathcal{L}^2$ norm does not satisfy the axiom (NVS3). To get around this difficulty, we adopt the following convention: $\|f\|_2 = 0$ implies that $f(x)$ is a zero function, i.e. $f(x) = 0$ for all $x \in [x_\mathrm{L}, x_\mathrm{U}]$. With this convention, the $\mathcal{L}^2[x_\mathrm{L}, x_\mathrm{U}]$ space along with the inner product defined as (2.61) and the $\mathcal{L}^2$ norm is also known as an infinite-dimensional Hilbert space [18, p. 193]. In what follows, the $\mathcal{L}^2[x_\mathrm{L}, x_\mathrm{U}]$ space always refers to this Hilbert space for convenience.

Moreover, the Cauchy–Schwartz inequality described in Theorem 2.1 is applicable to the $\mathcal{L}^2[x_\mathrm{L}, x_\mathrm{U}]$ space, that is restated here in terms of functions with the above convention.

---

[6] The set of points on which integration is "negligible" is called a set of *measure zero* [13, 14].

Theorem 2.34 (Cauchy–Schwartz Inequality). *Suppose* $f(x)$ *and* $g(x)$ *are real or complex nonzero functions on* $[x_{\rm L}, x_{\rm U}]$ *with* $\int_{x_{\rm L}}^{x_{\rm U}} |f(x)|^2 dx < \infty$ *and* $\int_{x_{\rm L}}^{x_{\rm U}} |g(x)|^2 dx < \infty$. *Then*

$$\left| \int_{x_{\rm L}}^{x_{\rm U}} f(x) g(x)^* dx \right| \leq \left\{ \int_{x_{\rm L}}^{x_{\rm U}} |f(x)|^2 dx \right\}^{1/2} \left\{ \int_{x_{\rm L}}^{x_{\rm U}} |g(x)|^2 dx \right\}^{1/2} \tag{2.63}$$

and the equality holds if and only if $f(x) = \alpha g(x)$ *for all* $x \in [x_{\rm L}, x_{\rm U}]$ *where* $\alpha \neq 0$ *is an arbitrary real or complex scalar.*

Approximations in Function Spaces

Let us emphasize that any function in $\mathcal{L}^2[x_{\rm L}, x_{\rm U}]$ is actually viewed as a vector in the vector space. As such, convergence for a sequence of functions in $\mathcal{L}^2[x_{\rm L}, x_{\rm U}]$ means convergence in the norm for a sequence of vectors, that is closely related to the problem of minimum mean-square-error (MMSE) approximation in $\mathcal{L}^2[x_{\rm L}, x_{\rm U}]$ as revealed below.

Let $\{\phi_1(x), \phi_2(x), ..., \phi_n(x)\}$ be a set of real or complex orthogonal functions in $\mathcal{L}^2[x_{\rm L}, x_{\rm U}]$ where

$$\int_{x_{\rm L}}^{x_{\rm U}} \phi_k(x) \phi_m^*(x) dx = \begin{cases} E_\phi, & k = m, \\ 0, & k \neq m. \end{cases} \tag{2.64}$$

Given a real or complex function $f(x) \in \mathcal{L}^2[x_{\rm L}, x_{\rm U}]$, let us consider the problem of approximating the nth partial sum

$$s_n(x) = \sum_{k=-n}^{n} \theta_k \phi_k(x) \tag{2.65}$$

to the function $f(x)$ in the MMSE sense, i.e. finding the optimal parameters $\theta_{-n}, \theta_{-n+1}, ..., \theta_n$ such that the following mean-square-error (MSE) is minimum:

$$J_{\rm MSE}(\theta_k) = \int_{x_{\rm L}}^{x_{\rm U}} |f(x) - s_n(x)|^2 \, dx. \tag{2.66}$$

By substituting (2.65) into (2.66) and using (2.64), we obtain

$$
\begin{aligned}
J_{\mathrm{MSE}}(\theta_k) &= \int_{x_\mathrm{L}}^{x_\mathrm{U}} |f(x)|^2 dx + E_\phi \sum_{k=-n}^{n} |\theta_k|^2 \\
&\quad - \sum_{k=-n}^{n} \left[\theta_k^* \int_{x_\mathrm{L}}^{x_\mathrm{U}} f(x)\phi_k^*(x)dx + \theta_k \int_{x_\mathrm{L}}^{x_\mathrm{U}} f^*(x)\phi_k(x)dx \right] \\
&= \int_{x_\mathrm{L}}^{x_\mathrm{U}} |f(x)|^2 dx + E_\phi \sum_{k=-n}^{n} \left| \theta_k - \frac{1}{E_\phi} \int_{x_\mathrm{L}}^{x_\mathrm{U}} f(x)\phi_k^*(x)dx \right|^2 \\
&\quad - E_\phi \sum_{k=-n}^{n} \left| \frac{1}{E_\phi} \int_{x_\mathrm{L}}^{x_\mathrm{U}} f(x)\phi_k^*(x)dx \right|^2 .
\end{aligned}
$$

This implies that the optimal θ_k, denoted by $\widehat{\theta}_k$, is given by

$$
\widehat{\theta}_k = \frac{1}{E_\phi} \int_{x_\mathrm{L}}^{x_\mathrm{U}} f(x)\phi_k^*(x)dx \quad \text{for } k = -n, -n+1, ..., n, \tag{2.67}
$$

and the corresponding minimum value of $J_{\mathrm{MSE}}(\theta_k)$ is given by

$$
\min\{J_{\mathrm{MSE}}(\theta_k)\} = \int_{x_\mathrm{L}}^{x_\mathrm{U}} |f(x)|^2 dx - E_\phi \sum_{k=-n}^{n} |\widehat{\theta}_k|^2. \tag{2.68}
$$

Since (2.68) holds for any n and $J_{\mathrm{MSE}}(\theta_k) \geq 0$ (see (2.66)), letting $n \to \infty$ leads to the following inequality.

Theorem 2.35 (Bessel's Inequality). *Suppose $\{\phi_1(x), \phi_2(x), ..., \phi_n(x)\}$ is a set of real or complex orthogonal functions in $\mathcal{L}^2[x_\mathrm{L}, x_\mathrm{U}]$. If $f(x)$ is a real or complex function in $\mathcal{L}^2[x_\mathrm{L}, x_\mathrm{U}]$, then optimal approximation of the series $\sum_{k=-\infty}^{\infty} \theta_k \phi_k(x)$ to $f(x)$ in the MMSE sense gives*

$$
\sum_{k=-\infty}^{\infty} |\widehat{\theta}_k|^2 \leq \frac{1}{E_\phi} \int_{x_\mathrm{L}}^{x_\mathrm{U}} |f(x)|^2 dx < \infty \tag{2.69}
$$

where $\widehat{\theta}_k$ is the optimal θ_k and $E_\phi = \int_{x_\mathrm{L}}^{x_\mathrm{U}} |\phi_k(x)|^2 dx$.

From (2.66) and (2.62), it follows that when the sequence of functions $\{s_n(x)\}_{n=1}^{\infty}$ converges in the norm to $f(x) \in \mathcal{L}^2[x_\mathrm{L}, x_\mathrm{U}]$,

$$
\lim_{n\to\infty} \|f - s_n\|_2 = \lim_{n\to\infty} \sqrt{J_{\mathrm{MSE}}(\widehat{\theta}_k)} = 0. \tag{2.70}
$$

Correspondingly, Bessel's inequality (2.69) becomes the equality

$$
\sum_{k=-\infty}^{\infty} |\widehat{\theta}_k|^2 = \frac{1}{E_\phi} \int_{x_\mathrm{L}}^{x_\mathrm{U}} |f(x)|^2 dx, \tag{2.71}
$$

which is known as *Parseval's equality* or *Parseval's relation.* Owing to (2.70), convergence in the norm for the $\mathcal{L}^2[x_\mathrm{L}, x_\mathrm{U}]$ space is also referred to as *convergence in the mean-square (MS) sense.*

2.2.4 Fourier Series

Fourier series are of great importance in developing the theory of mathematical analysis, and have widespread applications in the areas of science and engineering such as signal representation and analysis in signal processing.

Consider that $f(x)$ is a periodic function with period 2π. When $f(x)$ is real, the *Fourier series* of $f(x)$ is given by

$$f(x) = \frac{a_0}{2} + \sum_{k=1}^{\infty}(a_k \cos kx + b_k \sin kx) \tag{2.72}$$

where a_k and b_k are given by

$$a_k = \frac{1}{\pi}\int_{-\pi}^{\pi} f(x)\cos(kx)dx, \quad k = 0, 1, 2, ... \tag{2.73}$$

$$b_k = \frac{1}{\pi}\int_{-\pi}^{\pi} f(x)\sin(kx)dx, \quad k = 1, 2, ... \tag{2.74}$$

The real numbers a_k and b_k are called the *Fourier coefficients* of $f(x)$. Note that $\{1, \cos kx, \sin kx, k = 1 \sim \infty\}$ is a set of orthogonal functions satisfying (2.64) ($E_\phi = \pi$). From (2.73) and (2.74), one can see that if $f(x)$ is odd, then $a_k = 0$ for all k; whereas if $f(x)$ is even, $b_k = 0$ for all k. On the other hand, when $f(x)$ is complex, the Fourier series of $f(x)$ is given by

$$f(x) = \sum_{k=-\infty}^{\infty} c_k e^{jkx} \tag{2.75}$$

where c_k, a Fourier coefficient of $f(x)$, is a complex number given by

$$c_k = \frac{1}{2\pi}\int_{-\pi}^{\pi} f(x)e^{-jkx}dx. \tag{2.76}$$

Note that $\{e^{jkx}, k = -\infty \sim \infty\}$ is also a set of orthogonal functions satisfying (2.64) ($E_\phi = 2\pi$).

Next, let us discuss the existence of Fourier series. In particular, we are concerned with the sufficient conditions under which the Fourier series given by (2.75) converges.

Local Behavior of Convergence

With the nth partial sum defined as

$$s_n(x) = \sum_{k=-n}^{n} c_k e^{jkx}, \tag{2.77}$$

the convergence problem of the Fourier series given by (2.75) is the same as that of the sequence $\{s_n(x)\}_{n=1}^{\infty}$.

Pointwise Convergence

It was believed, for a long time, that if the periodic function $f(x)$ is continuous, then the Fourier series would converge to $f(x)$ for all $x \in [-\pi, \pi)$ (i.e. pointwise convergence). Actually, there do exist continuous periodic functions whose Fourier series diverge at a given point or even everywhere; see [14, pp. 83–87] for an example of such functions. This implies that pointwise convergence requires some additional conditions on $f(x)$ as follows [18].

Theorem 2.36 (Pointwise Convergence Theorem). *Suppose $f(x)$ is a real or complex periodic function of period 2π. Then, under the conditions that (i) $f(x)$ is piecewise continuous on $[-\pi, \pi)$ and (ii) the derivative $f'(x)$ is piecewise continuous on $[-\pi, \pi)$, the Fourier series of $f(x)$ given by (2.75) is pointwise convergent and*

$$\lim_{n \to \infty} s_n(x) = \frac{f(x^-) + f(x^+)}{2} \quad \text{for all } x \in [-\pi, \pi) \tag{2.78}$$

where $s_n(x)$ is the corresponding nth partial sum given by (2.77), and $f(x^-)$ and $f(x^+)$ are the left-hand limit and the right-hand limit of $f(x)$, respectively.

See Appendix 2B for a review of terminologies of functions and see Appendix 2D for the proof of this theorem. From this theorem and (2.219), it follows that the Fourier series converges to $f(x)$ at the points of continuity and converges to $[f(x^-) + f(x^+)]/2$ at the points of discontinuity. Note that Theorem 2.36 is only a special case of the *Dirichlet Theorem*, for which the required conditions are known as the *Dirichlet conditions* [21, 22].[7]

Uniform Convergence

By using the Weierstrass M-test, we have the following theorem for uniform and absolute convergence of the Fourier series (Problem 2.17).

Theorem 2.37. *Suppose $\{c_k\}_{k=-\infty}^{\infty}$ is any absolutely summable sequence, i.e. $\sum_{k=-\infty}^{\infty} |c_k| < \infty$. Then the Fourier series $\sum_{k=-\infty}^{\infty} c_k e^{jkx}$ converges uniformly and absolutely to a continuous function of x on $[-\pi, \pi)$.*

Moreover, by using the Weierstrass M-test and the pointwise convergence theorem with more restrictive conditions on $f(x)$, we have another theorem regarding the uniform and absolute convergence [18, pp. 216–218].

Theorem 2.38. *Suppose $f(x)$ is a real or complex periodic function of period 2π. Then, under the conditions that (i) $f(x)$ is continuous on $[-\pi, \pi)$ and (ii) the derivative $f'(x)$ is piecewise continuous on $[-\pi, \pi)$, the Fourier series of $f(x)$ given by (2.75) converges uniformly and absolutely to $f(x)$ on $[-\pi, \pi)$.*

See Appendix 2E for the proof.

---

[7] The Dirichlet theorem due to P. L. Dirichlet (1829) was the first substantial progress on the convergence problem of Fourier series [13].

Global Behavior of Convergence

The Fourier series given by (2.75) is said to *converge in the mean-square (MS) sense* to $f(x)$ if

$$\lim_{n\to\infty} \int_{-\pi}^{\pi} |f(x) - s_n(x)|^2 \, dx = 0 \tag{2.79}$$

where $s_n(x)$ is the nth partial sum given by (2.77). Accordingly, with MS convergence, we can only get an overall picture about the convergence behavior over the entire interval. It reveals nothing about the detailed behavior of convergence at any point.

Recall that if $f(x)$ is in the $\mathcal{L}^2[-\pi, \pi)$ space, then MS convergence is equivalent to convergence in the norm. Correspondingly, Parseval's relation

$$\sum_{n=-\infty}^{\infty} |c_k|^2 = \frac{1}{2\pi} \int_{-\pi}^{\pi} |f(x)|^2 dx < \infty \tag{2.80}$$

holds and thus the sequence $\{c_k\}_{k=-\infty}^{\infty}$ is square summable. The converse is stated in the following theorem (Problem 2.18).

Theorem 2.39. *Suppose $\{c_k\}_{k=-\infty}^{\infty}$ is any square summable sequence, i.e. $\sum_{k=-\infty}^{\infty} |c_k|^2 < \infty$. Then the Fourier series $\sum_{k=-\infty}^{\infty} c_k e^{jkx}$ converges in the MS sense to a function in the $\mathcal{L}^2[-\pi, \pi)$ space.*

Furthermore, a more generalized theorem regarding the MS convergence is provided as follows. The proof is beyond the scope of this book; the reader can find it in [13, pp. 411–414] for the real case and [14, pp. 76–80] for the complex case.

Theorem 2.40. *Suppose $f(x)$ is a real or complex periodic function of period 2π. If the function $f(x)$ is bounded and integrable on $[-\pi, \pi)$, then the Fourier series of $f(x)$ given by (2.75) converges in the MS sense to $f(x)$ on $[-\pi, \pi)$.*

Compared with local convergence (pointwise convergence and uniform convergence), global convergence (MS convergence) requires even weaker conditions on the function $f(x)$ or the sequence $\{c_k\}_{k=-\infty}^{\infty}$ and so the existence of Fourier series is almost not an issue in practice.

Fourier Series of Generalized Functions

In some cases, we may need to deal with functions which are outside the ordinary scope of function theory. An important class of such functions is the one of *generalized functions* introduced by G. Temple (1953) [23]. Among this class, a representative is the so-called *impulse* or *Dirac delta function,* commonly denoted by $\delta(x)$.[8] It is mathematically defined by the following relations

$$\begin{cases} \delta(x) = 0 \quad \text{for } x \neq 0, \\ \int_{-\infty}^{\infty} \delta(x) dx = 1, \end{cases} \tag{2.81}$$

and possesses the following sifting property:

$$\int_{-\infty}^{\infty} \delta(x - \tau) f(x) dx = f(\tau). \tag{2.82}$$

Strictly speaking, a periodic function like

$$f(x) = \sum_{m=-\infty}^{\infty} 2\pi\delta(x + 2\pi m) \tag{2.83}$$

does not have a Fourier series. But, using (2.76) and (2.82), we can still mathematically define the Fourier series of $f(x)$ as

$$f(x) = \sum_{k=-\infty}^{\infty} e^{jkx} \quad (\text{i.e. } c_k = 1 \text{ for all } k) \tag{2.84}$$

and make use of this in many applications. In other words, the theory of Fourier series should be broadened for more extensive applications. The extended theory of Fourier series is, however, beyond the scope of this book; refer to [23, 24] for the details.

2.3 Optimization Theory

Consider that $J(\boldsymbol{\theta})$ is a real function of the $L \times 1$ vector

$$\boldsymbol{\theta} = (\theta_1, \theta_2, ..., \theta_L)^T \tag{2.85}$$

where θ_1, θ_2, ..., θ_L are real or complex unknown parameters to be determined. An optimization problem is to find (search for) a solution for $\boldsymbol{\theta}$ which minimizes or maximizes the function $J(\boldsymbol{\theta})$, referred to as the *objective function*. There are basically two types of optimization problems, *constrained optimization problems* and *unconstrained optimization problems* [12, 25, 26]. As the names indicate, the former type is subject to some constraints (e.g. equality constraints and inequality constraints), whereas the latter type does not involve any constraint. In the scope of the book, we are interested in unconstrained optimization problems, which along with the related theory are introduced in this section.

---

[8] The notation '$\delta(x)$' for Dirac delta function was first used by G. Kirchhoff, and then introduced into quantum mechanics by Dirac (1927) [23].

2.3.1 Vector Derivatives

As we will see, finding the solutions to the minima or maxima of the objective function $J(\boldsymbol{\theta})$ often involves manipulations of the following *first derivative* (with respect to $\boldsymbol{\theta}^*$)

$$\frac{\partial f(\boldsymbol{\theta})}{\partial \boldsymbol{\theta}^*} = \left(\frac{\partial f(\boldsymbol{\theta})}{\partial \theta_1^*}, \frac{\partial f(\boldsymbol{\theta})}{\partial \theta_2^*}, \ldots, \frac{\partial f(\boldsymbol{\theta})}{\partial \theta_L^*} \right)^T \tag{2.86}$$

where $f(\boldsymbol{\theta})$ is an arbitrary real or complex function of $\boldsymbol{\theta}$ and $\partial f(\boldsymbol{\theta})/\partial \theta_k^*$ is the *first partial derivative* of $f(\boldsymbol{\theta})$ with respect to the conjugate parameter θ_k^*.[9] However, the first derivative $\partial f(\boldsymbol{\theta})/\partial \boldsymbol{\theta}^*$, or equivalently the operator

$$\frac{\partial}{\partial \boldsymbol{\theta}^*} = \left(\frac{\partial}{\partial \theta_1^*}, \frac{\partial}{\partial \theta_2^*}, \ldots, \frac{\partial}{\partial \theta_L^*} \right)^T, \tag{2.87}$$

depends on whether $\boldsymbol{\theta}$ is real or complex, as discussed below.

Derivatives with Respect to a Real Vector

When $\boldsymbol{\theta}$ is real, applying the operator $\partial/\partial \boldsymbol{\theta}^*$ to $\boldsymbol{\theta}^T$ yields

$$\frac{\partial \boldsymbol{\theta}^T}{\partial \boldsymbol{\theta}^*} = \frac{\partial \boldsymbol{\theta}^T}{\partial \boldsymbol{\theta}} = \begin{pmatrix} \frac{\partial \theta_1}{\partial \theta_1} & \frac{\partial \theta_2}{\partial \theta_1} & \cdots & \frac{\partial \theta_L}{\partial \theta_1} \\ \frac{\partial \theta_1}{\partial \theta_2} & \frac{\partial \theta_2}{\partial \theta_2} & \cdots & \frac{\partial \theta_L}{\partial \theta_2} \\ \vdots & \vdots & \ddots & \vdots \\ \frac{\partial \theta_1}{\partial \theta_L} & \frac{\partial \theta_2}{\partial \theta_L} & \cdots & \frac{\partial \theta_L}{\partial \theta_L} \end{pmatrix} = \mathbf{I}, \tag{2.88}$$

which is useful to the derivation of $\partial f(\boldsymbol{\theta})/\partial \boldsymbol{\theta}$. In particular, if $f(\boldsymbol{\theta}) = \mathbf{b}^T \boldsymbol{\theta} = \boldsymbol{\theta}^T \mathbf{b}$ where the vector $\mathbf{b}$ is independent of $\boldsymbol{\theta}$, then

$$\frac{\partial f(\boldsymbol{\theta})}{\partial \boldsymbol{\theta}} = \left(\frac{\partial \boldsymbol{\theta}^T}{\partial \boldsymbol{\theta}} \right) \mathbf{b} = \mathbf{I}\mathbf{b} = \mathbf{b}. \tag{2.89}$$

Moreover, if $f(\boldsymbol{\theta}) = \boldsymbol{\theta}^T \mathbf{b}(\boldsymbol{\theta})$ where the vector $\mathbf{b}(\boldsymbol{\theta}) = \mathbf{A}\boldsymbol{\theta}$, then

$$\begin{aligned} \frac{\partial f(\boldsymbol{\theta})}{\partial \boldsymbol{\theta}} &= \left(\frac{\partial \boldsymbol{\theta}^T}{\partial \boldsymbol{\theta}} \right) \mathbf{b}(\boldsymbol{\theta}) + \frac{\partial \mathbf{b}^T(\boldsymbol{\theta})}{\partial \boldsymbol{\theta}} \boldsymbol{\theta} = \left(\frac{\partial \boldsymbol{\theta}^T}{\partial \boldsymbol{\theta}} \right) \mathbf{A}\boldsymbol{\theta} + \left(\frac{\partial \boldsymbol{\theta}^T}{\partial \boldsymbol{\theta}} \right) \mathbf{A}^T \boldsymbol{\theta} \\ &= \mathbf{I}\mathbf{A}\boldsymbol{\theta} + \mathbf{I}\mathbf{A}^T \boldsymbol{\theta} = (\mathbf{A} + \mathbf{A}^T)\boldsymbol{\theta}, \end{aligned} \tag{2.90}$$

which reduces to $\partial f(\boldsymbol{\theta})/\partial \boldsymbol{\theta} = 2\mathbf{A}\boldsymbol{\theta}$ when $\mathbf{A}$ is symmetric.

---

[9] Although utilization of $\partial f(\boldsymbol{\theta})/\boldsymbol{\theta}^*$ and that of $\partial f(\boldsymbol{\theta})/\boldsymbol{\theta}$ both lead to the same solutions for the optimization problems, Brandwood [27] has pointed out that the former gives rise to a slightly neater expression and thus is more convenient.

Derivatives with Respect to a Complex Vector

Now consider the case that $\boldsymbol{\theta} = (\theta_1, \theta_2, ..., \theta_L)^T$ is complex, i.e.

$$\theta_k = x_k + jy_k, \quad k = 1, 2, ..., L, \tag{2.91}$$

where $x_k = \text{Re}\{\theta_k\}$ is the real part of θ_k and $y_k = \text{Im}\{\theta_k\}$ is the imaginary part of θ_k. Naturally, one can derive $\partial f(\boldsymbol{\theta})/\partial \theta_k^*$ in terms of x_k and y_k. Alternatively, direct derivation of $\partial f(\boldsymbol{\theta})/\partial \theta_k^*$ (without involving x_k and y_k) is more appealing, but special care should be taken for the following reason. In conventional complex-variable theory, if $f(\boldsymbol{\theta})$ cannot be expressed in terms of only θ_k (i.e. it also consists of θ_k^*), then it is nowhere *differentiable* by θ_k and we say that $f(\boldsymbol{\theta})$ is not *analytic* [28]. The analytic problem, however, can be resolved by simply treating $f(\boldsymbol{\theta}) \equiv f(\boldsymbol{\theta}, \boldsymbol{\theta}^*)$ as a function of $2L$ independent variables $\theta_1, \theta_2, ..., \theta_L, \theta_1^*, \theta_2^*, ..., \theta_L^*$ [27]; see the following illustration.

Example 2.41
Consider the function $f(\theta) = \theta^*$ where $\theta = x + jy$, and x and y are real. According to the conventional complex-variable theory, the first derivative of $f(\theta)$ with respect to θ is given by [28]

$$\frac{df(\theta)}{d\theta} = \lim_{\Delta\theta \to 0} \frac{f(\theta + \Delta\theta) - f(\theta)}{\Delta\theta} = \lim_{\Delta\theta \to 0} \frac{\Delta\theta^*}{\Delta\theta}.$$

As illustrated in Fig. 2.5, if $\Delta\theta$ approaches zero along the real axis, i.e. $\Delta\theta = \Delta x \to 0$, then $df(\theta)/d\theta = 1$. If $\Delta\theta$ approaches zero along the imaginary axis, i.e. $\Delta\theta = j\Delta y \to 0$, then $df(\theta)/d\theta = -1$. As a result, there is no way to assign a unique value to $df(\theta)/d\theta$, and thus $f(\theta)$ is not differentiable. On the other hand, by treating $f(\theta) \equiv f(\theta, \theta^*)$ as a function of independent variables θ and θ^*, we obtain $\partial f(\theta, \theta^*)/\partial\theta = 0$ and $\partial f(\theta, \theta^*)/\partial\theta^* = 1$. That is, $f(\theta, \theta^*)$ is differentiable with respect to θ and θ^* independently.

□

With the treatment of independent variables θ_k and θ_k^*, we now proceed to derive the partial derivative $\partial f(\boldsymbol{\theta})/\partial \theta_k^*$. From (2.91), it follows that

$$x_k = \frac{1}{2}(\theta_k + \theta_k^*) \quad \text{and} \quad y_k = \frac{1}{2j}(\theta_k - \theta_k^*). \tag{2.92}$$

Differentiating x_k and y_k given by (2.92) with respect to θ_k and θ_k^* yields

$$\frac{\partial x_k}{\partial \theta_k} = \frac{1}{2}, \quad \frac{\partial x_k}{\partial \theta_k^*} = \frac{1}{2}, \quad \frac{\partial y_k}{\partial \theta_k} = \frac{1}{2j}, \quad \text{and} \quad \frac{\partial y_k}{\partial \theta_k^*} = -\frac{1}{2j}. \tag{2.93}$$

This, together with the *chain rule* [29], leads to

$$\frac{\partial f(\boldsymbol{\theta})}{\partial \theta_k} = \frac{\partial f(\boldsymbol{\theta})}{\partial x_k}\frac{\partial x_k}{\partial \theta_k} + \frac{\partial f(\boldsymbol{\theta})}{\partial y_k}\frac{\partial y_k}{\partial \theta_k} = \frac{1}{2}\left\{\frac{\partial f(\boldsymbol{\theta})}{\partial x_k} - j\frac{\partial f(\boldsymbol{\theta})}{\partial y_k}\right\} \tag{2.94}$$

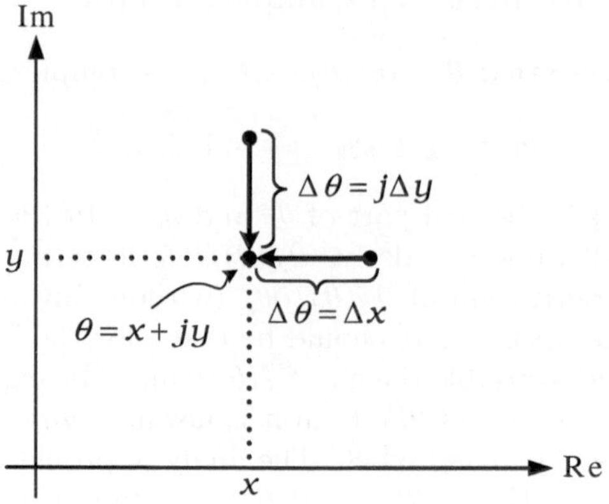

Fig. 2.5 Illustration of $\Delta\theta$ approaching zero along the real and imaginary axes

and

$$\frac{\partial f(\boldsymbol{\theta})}{\partial \theta_k^*} = \frac{\partial f(\boldsymbol{\theta})}{\partial x_k}\frac{\partial x_k}{\partial \theta_k^*} + \frac{\partial f(\boldsymbol{\theta})}{\partial y_k}\frac{\partial y_k}{\partial \theta_k^*} = \frac{1}{2}\left\{\frac{\partial f(\boldsymbol{\theta})}{\partial x_k} + j\frac{\partial f(\boldsymbol{\theta})}{\partial y_k}\right\}. \tag{2.95}$$

From (2.91), (2.94) and (2.95), it is clear that

$$\frac{\partial \theta_k^*}{\partial \theta_k} = \frac{\partial \theta_k}{\partial \theta_k^*} = 0 \quad \text{and} \quad \frac{\partial \theta_k}{\partial \theta_k} = \frac{\partial \theta_k^*}{\partial \theta_k^*} = 1. \tag{2.96}$$

By (2.96), we have

$$\frac{\partial \boldsymbol{\theta}^H}{\partial \boldsymbol{\theta}^*} = \mathbf{I} \quad \text{and} \quad \frac{\partial \boldsymbol{\theta}^T}{\partial \boldsymbol{\theta}^*} = \mathbf{0}, \tag{2.97}$$

which, again, are useful to the derivation of $\partial f(\boldsymbol{\theta})/\partial \boldsymbol{\theta}^*$. In particular, if $f(\boldsymbol{\theta}) = \mathbf{b}^H\boldsymbol{\theta}$ where $\mathbf{b}$ is independent of $\boldsymbol{\theta}$, then $\partial f(\boldsymbol{\theta})/\boldsymbol{\theta}^* = \mathbf{0}$, and if $f(\boldsymbol{\theta}) = \boldsymbol{\theta}^H\mathbf{b}$, then

$$\frac{\partial f(\boldsymbol{\theta})}{\partial \boldsymbol{\theta}^*} = \left(\frac{\partial \boldsymbol{\theta}^H}{\partial \boldsymbol{\theta}^*}\right)\mathbf{b} = \mathbf{b}. \tag{2.98}$$

If $f(\boldsymbol{\theta}) = \boldsymbol{\theta}^H\mathbf{A}\boldsymbol{\theta}$, then

$$\frac{\partial f(\boldsymbol{\theta})}{\partial \boldsymbol{\theta}^*} = \left(\frac{\partial \boldsymbol{\theta}^H}{\partial \boldsymbol{\theta}^*}\right)\mathbf{A}\boldsymbol{\theta} + \left(\frac{\partial \boldsymbol{\theta}^T}{\partial \boldsymbol{\theta}^*}\right)\left(\boldsymbol{\theta}^H\mathbf{A}\right)^T = \mathbf{A}\boldsymbol{\theta} \quad \text{(by (2.97))}. \tag{2.99}$$

Table 2.1 summarizes the vector derivatives for both real and complex cases.

Table 2.1 Summary of vector derivatives

Real Case				Complex Case			
$\frac{\partial \boldsymbol{\theta}^T}{\partial \boldsymbol{\theta}} = \mathbf{I}$				$\frac{\partial \boldsymbol{\theta}^H}{\partial \boldsymbol{\theta}^*} = \mathbf{I}$ and $\frac{\partial \boldsymbol{\theta}^T}{\partial \boldsymbol{\theta}^*} = \mathbf{0}$			
$f(\boldsymbol{\theta})$	$\boldsymbol{\theta}^T\mathbf{b}$	$\mathbf{b}^T\boldsymbol{\theta}$	$\boldsymbol{\theta}^T\mathbf{A}\boldsymbol{\theta}$	$f(\boldsymbol{\theta})$	$\boldsymbol{\theta}^H\mathbf{b}$	$\mathbf{b}^H\boldsymbol{\theta}$	$\boldsymbol{\theta}^H\mathbf{A}\boldsymbol{\theta}$
$\frac{\partial f(\boldsymbol{\theta})}{\partial \boldsymbol{\theta}}$	$\mathbf{b}$	$\mathbf{b}$	$(\mathbf{A}+\mathbf{A}^T)\boldsymbol{\theta}$	$\frac{\partial f(\boldsymbol{\theta})}{\partial \boldsymbol{\theta}^*}$	$\mathbf{b}$	$\mathbf{0}$	$\mathbf{A}\boldsymbol{\theta}$

2.3.2 Necessary and Sufficient Conditions for Solutions

From the foregoing discussions, we note that when the unknown parameter vector $\boldsymbol{\theta} = (\theta_1, \theta_2, ..., \theta_L)^T$ is complex, it is also more convenient to treat the real objective function $J(\boldsymbol{\theta}) \equiv J(\boldsymbol{\theta}, \boldsymbol{\theta}^*)$ as a function of independent variables θ_k and θ_k^*. As such, for notational convenience, let us reformulate the above-mentioned optimization problem into the equivalent problem of minimizing or maximizing the real objective function $J(\boldsymbol{\vartheta})$ where $\boldsymbol{\vartheta}$ is the real or complex unknown parameter vector defined as

$$\begin{cases} \boldsymbol{\vartheta} = (\vartheta_1, \vartheta_2, ..., \vartheta_L)^T = \boldsymbol{\theta} & \text{for real } \boldsymbol{\theta}, \\ \boldsymbol{\vartheta} = (\vartheta_1, \vartheta_2, ..., \vartheta_{2L})^T = (\boldsymbol{\theta}^T, \boldsymbol{\theta}^H)^T & \text{for complex } \boldsymbol{\theta}. \end{cases} \tag{2.100}$$

Several terminologies regarding $J(\boldsymbol{\vartheta})$ are introduced as follows.

The objective function $J(\boldsymbol{\vartheta})$ is said to have a *local minimum* or a *relative minimum* at the solution point $\widehat{\boldsymbol{\vartheta}}$ if there exists a real number $\varepsilon > 0$ such that

$$J(\widehat{\boldsymbol{\vartheta}}) \leq J(\boldsymbol{\vartheta}) \quad \text{for all } \boldsymbol{\vartheta} \text{ satisfying } \|\boldsymbol{\vartheta} - \widehat{\boldsymbol{\vartheta}}\| < \varepsilon. \tag{2.101}$$

The objective function $J(\boldsymbol{\vartheta})$ is said to have a *global minimum* or an *absolute minimum* at the solution point $\widehat{\boldsymbol{\vartheta}}$ if

$$J(\widehat{\boldsymbol{\vartheta}}) \leq J(\boldsymbol{\vartheta}) \quad \text{for all } \boldsymbol{\vartheta}. \tag{2.102}$$

Similarly, the objective function $J(\boldsymbol{\vartheta})$ is said to have a *local maximum* or a *relative maximum* at the solution point $\widehat{\boldsymbol{\vartheta}}$ if there exists a real number $\varepsilon > 0$ such that

$$J(\widehat{\boldsymbol{\vartheta}}) \geq J(\boldsymbol{\vartheta}) \quad \text{for all } \boldsymbol{\vartheta} \text{ satisfying } \|\boldsymbol{\vartheta} - \widehat{\boldsymbol{\vartheta}}\| < \varepsilon, \tag{2.103}$$

and have a *global maximum* or an *absolute maximum* at the solution point $\widehat{\boldsymbol{\vartheta}}$ if

$$J(\widehat{\boldsymbol{\vartheta}}) \geq J(\boldsymbol{\vartheta}) \quad \text{for all } \boldsymbol{\vartheta}. \tag{2.104}$$

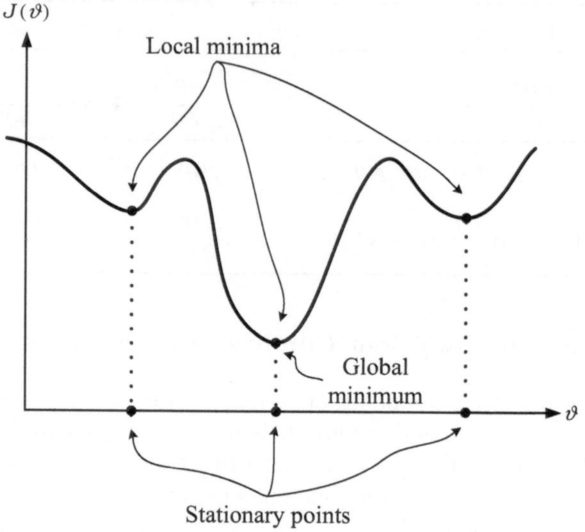

Fig. 2.6 Illustration of the solution points for the problem of minimizing $J(\vartheta)$ where ϑ is real

In other words, a global minimum (maximum) of $J(\boldsymbol{\vartheta})$ is also a local minimum (maximum) of $J(\boldsymbol{\vartheta})$. Figure 2.6 gives an illustration of these definitions.

Define the *gradient vector*, or simply the *gradient*, as[10]

$$\nabla J(\boldsymbol{\vartheta}) = \frac{\partial J(\boldsymbol{\vartheta})}{\partial \boldsymbol{\vartheta}^*} \tag{2.105}$$

(the physical meaning will be discussed later), where

$$\frac{\partial J(\boldsymbol{\vartheta})}{\partial \boldsymbol{\vartheta}^*} = \begin{cases} \dfrac{\partial J(\boldsymbol{\theta})}{\partial \boldsymbol{\theta}} & \text{for real } \boldsymbol{\theta}, \\ \left(\left[\dfrac{\partial J(\boldsymbol{\theta})}{\partial \boldsymbol{\theta}^*} \right]^T, \left[\dfrac{\partial J(\boldsymbol{\theta})}{\partial \boldsymbol{\theta}} \right]^T \right)^T & \text{for complex } \boldsymbol{\theta}. \end{cases} \tag{2.106}$$

A necessary condition for the local extrema (local minima or maxima) of $J(\boldsymbol{\vartheta})$ is as follows [26, p. 73].

Theorem 2.42 (Necessary Condition). *If the objective function $J(\boldsymbol{\vartheta})$ has an extremum at $\boldsymbol{\vartheta} = \widehat{\boldsymbol{\vartheta}}$ and if its first derivative $\partial J(\boldsymbol{\vartheta})/\partial \boldsymbol{\vartheta}^*$ exists at $\boldsymbol{\vartheta} = \widehat{\boldsymbol{\vartheta}}$, then its gradient*

[10] The gradient $\nabla J(\boldsymbol{\vartheta})$ defined as (2.105) is the same as that defined in [8, p. 894] except for a scale factor.

$$\nabla J(\widehat{\boldsymbol{\vartheta}}) \triangleq \nabla J(\boldsymbol{\vartheta})\Big|_{\boldsymbol{\vartheta} = \widehat{\boldsymbol{\vartheta}}} = \mathbf{0}. \tag{2.107}$$

The proof is left as an exercise (Problem 2.19). When $\widehat{\boldsymbol{\vartheta}}$ satisfies (2.107), it is said to be a *stationary point* of $J(\boldsymbol{\vartheta})$. Furthermore, a stationary point $\widehat{\boldsymbol{\vartheta}}$ is said to be a *saddle point* of $J(\boldsymbol{\vartheta})$ if it corresponds to a local minimum of $J(\boldsymbol{\vartheta})$ with respect to one direction on the hypersurface of $J(\boldsymbol{\vartheta})$ and a local maximum of $J(\boldsymbol{\vartheta})$ with respect to another direction [12, 26, 30]. In other words, a saddle point of $J(\boldsymbol{\vartheta})$ corresponds to an unstable equilibrium of $J(\boldsymbol{\vartheta})$, and thus it will typically not be obtained by optimization methods.

Example 2.43 (Saddle Point)
Consider the objective function $J(\boldsymbol{\vartheta}) = J(\vartheta_1, \vartheta_2) = -\vartheta_1^2 + \vartheta_2^2$ where $\boldsymbol{\vartheta} = (\vartheta_1, \vartheta_2)^T$, and ϑ_1 and ϑ_2 are real. Taking the first derivative of $J(\boldsymbol{\vartheta})$ with respect to $\boldsymbol{\vartheta}^*$ $(= \boldsymbol{\vartheta})$

$$\frac{\partial J(\boldsymbol{\vartheta})}{\partial \boldsymbol{\vartheta}} = \begin{pmatrix} \partial J(\boldsymbol{\vartheta})/\partial \vartheta_1 \\ \partial J(\boldsymbol{\vartheta})/\partial \vartheta_2 \end{pmatrix} = \begin{pmatrix} -2\vartheta_1 \\ 2\vartheta_2 \end{pmatrix}$$

and setting the result to zero, we obtain the stationary point $\widehat{\boldsymbol{\vartheta}} = (\widehat{\vartheta}_1, \widehat{\vartheta}_2)^T = (0,0)^T$. Figure 2.7 depicts the objective function $J(\vartheta_1, \vartheta_2)$ and the stationary point $(\widehat{\vartheta}_1, \widehat{\vartheta}_2) = (0,0)$. One can see from this figure that the function $J(\vartheta_1, \widehat{\vartheta}_2) = J(\vartheta_1, 0) = -\vartheta_1^2$ has a local maximum at $\vartheta_1 = \widehat{\vartheta}_1 = 0$, and the function $J(\widehat{\vartheta}_1, \vartheta_2) = J(0, \vartheta_2) = \vartheta_2^2$ has a local minimum at $\vartheta_2 = \widehat{\vartheta}_2 = 0$. This reveals that the stationary point $(\widehat{\vartheta}_1, \widehat{\vartheta}_2) = (0,0)$ is a saddle point. □

Let us emphasize that a stationary point may correspond to a local minimum point, a local maximum point, a saddle point, or a point of some other exotic category [12, pp. 217, 218]. Some categories of stationary points may be recognized by inspecting the Hermitian matrix

$$\mathbf{J}_2(\boldsymbol{\vartheta}) \triangleq \frac{\partial}{\partial \boldsymbol{\vartheta}}\left[\frac{\partial J(\boldsymbol{\vartheta})}{\partial \boldsymbol{\vartheta}}\right]^T = \frac{\partial}{\partial \boldsymbol{\theta}}\left[\frac{\partial J(\boldsymbol{\theta})}{\partial \boldsymbol{\theta}}\right]^T \tag{2.108}$$

for real $\boldsymbol{\theta}$, or the Hermitian matrix

$$\mathbf{J}_2(\boldsymbol{\vartheta}) \triangleq \frac{\partial}{\partial \boldsymbol{\vartheta}^*}\left[\frac{\partial J(\boldsymbol{\vartheta})}{\partial \boldsymbol{\vartheta}^*}\right]^H = \begin{pmatrix} \dfrac{\partial}{\partial \boldsymbol{\theta}^*}\left[\dfrac{\partial J(\boldsymbol{\theta})}{\partial \boldsymbol{\theta}^*}\right]^H & \dfrac{\partial}{\partial \boldsymbol{\theta}^*}\left[\dfrac{\partial J(\boldsymbol{\theta})}{\partial \boldsymbol{\theta}}\right]^H \\ \dfrac{\partial}{\partial \boldsymbol{\theta}}\left[\dfrac{\partial J(\boldsymbol{\theta})}{\partial \boldsymbol{\theta}^*}\right]^H & \dfrac{\partial}{\partial \boldsymbol{\theta}}\left[\dfrac{\partial J(\boldsymbol{\theta})}{\partial \boldsymbol{\theta}}\right]^H \end{pmatrix} \tag{2.109}$$

for complex $\boldsymbol{\theta}$, where the matrix $\mathbf{J}_2(\boldsymbol{\vartheta})$ is referred to as the *Hessian matrix* of $J(\boldsymbol{\vartheta})$. In particular, the local minimum points and local maximum points can be recognized by virtue of the Hessian matrix, as stated in the following theorem.

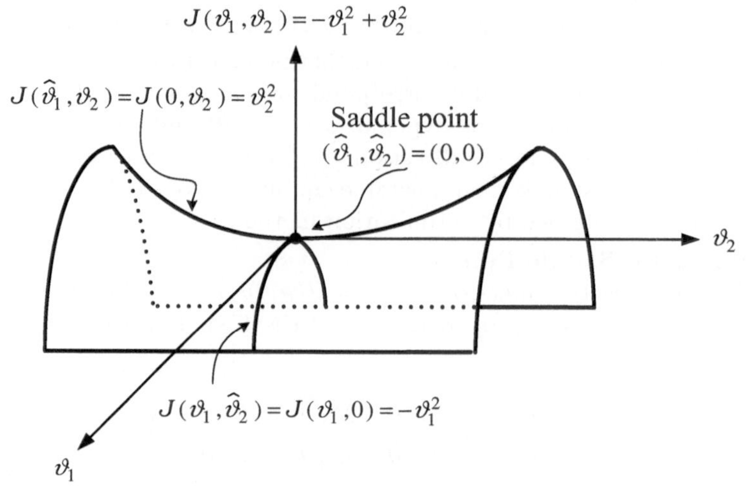

Fig. 2.7 Illustration of saddle point

Theorem 2.44 (Sufficient Conditions). *Suppose* $\widehat{\boldsymbol{\vartheta}}$ *is a stationary point of the objective function* $J(\boldsymbol{\vartheta})$*. If the Hessian matrix*

$$\mathbf{J}_2(\widehat{\boldsymbol{\vartheta}}) \triangleq \mathbf{J}_2(\boldsymbol{\vartheta})\Big|_{\boldsymbol{\vartheta} = \widehat{\boldsymbol{\vartheta}}} \tag{2.110}$$

is positive definite (negative definite), then $\widehat{\boldsymbol{\vartheta}}$ *corresponds to a local minimum (a local maximum) of* $J(\boldsymbol{\vartheta})$*.*

This theorem can be proved by virtue of the following *Taylor series* for $J(\boldsymbol{\vartheta})$ at $\boldsymbol{\vartheta} = \widehat{\boldsymbol{\vartheta}}$: (refer to [26, p. 71] for the real case)

$$J(\boldsymbol{\vartheta}) = J(\widehat{\boldsymbol{\vartheta}}) + (\boldsymbol{\vartheta} - \widehat{\boldsymbol{\vartheta}})^H \boldsymbol{\nabla} J(\widehat{\boldsymbol{\vartheta}}) + \frac{1}{2}(\boldsymbol{\vartheta} - \widehat{\boldsymbol{\vartheta}})^H \mathbf{J}_2(\widehat{\boldsymbol{\vartheta}})(\boldsymbol{\vartheta} - \widehat{\boldsymbol{\vartheta}}) + \cdots \tag{2.111}$$

We leave the proof of this theorem as an exercise (Problem 2.20).

2.3.3 Gradient-Type Optimization Methods

There are numerous types of optimization techniques available for solving the unconstrained optimization problem, among which we are interested in *gradient-type methods* for their efficiency as well as their wide scope of applications. Without loss of generality, we will introduce gradient-type methods in terms of the minimization problem of $J(\boldsymbol{\vartheta})$ because maximization of $J(\boldsymbol{\vartheta})$ is equivalent to minimization of $-J(\boldsymbol{\vartheta})$.

Iterative Procedure of Gradient-Type Methods

Let $\widehat{\boldsymbol{\vartheta}}$ denote a (local) minimum point of $J(\boldsymbol{\vartheta})$. Gradient-type methods are, in general, based on the following iterative procedure for searching for $\widehat{\boldsymbol{\vartheta}}$.

(S1) Set the iteration number $i = 0$.
(S2) Choose an appropriate initial condition $\boldsymbol{\vartheta}^{[0]}$ for $\widehat{\boldsymbol{\vartheta}}$ and an appropriate initial search direction $\mathbf{d}^{[0]}$.
(S3) Generate a new approximation to $\widehat{\boldsymbol{\vartheta}}$ via

$$\boldsymbol{\vartheta}^{[i+1]} = \boldsymbol{\vartheta}^{[i]} - \mu^{[i]}\mathbf{d}^{[i]} \tag{2.112}$$

where $\mu^{[i]} > 0$ is the step size which should be determined appropriately to make sure of the movement along the direction of a (local) minimum of $J(\boldsymbol{\vartheta})$.
(S4) Check the convergence of the procedure. If the procedure has not yet converged, then go to Step (S5); otherwise, obtain a (local) minimum point as $\widehat{\boldsymbol{\vartheta}} = \boldsymbol{\vartheta}^{[i+1]}$ and stop the procedure.
(S5) Find a new search direction $\mathbf{d}^{[i+1]}$ which points towards a (local) minimum of $J(\boldsymbol{\vartheta})$ in general.
(S6) Update the iteration number i by $(i + 1)$ and go to Step (S3).

This procedure is also depicted in Fig. 2.8 for clarity.

In Step (S3) of the iterative procedure, determination of the step size $\mu^{[i]}$ can be formulated into the problem of finding the parameter μ which minimizes the objective function $f(\mu) \triangleq J(\boldsymbol{\vartheta}^{[i]} - \mu\mathbf{d}^{[i]})$ (by (2.112)). Accordingly, this problem can be solved by using the class of *one-dimensional (1-D) minimization methods* such as the *1-D Newton method* (also known as the *Newton–Raphson method*), the *1-D quasi-Newton method*, and so on [26]. Alternatively, the step size $\mu^{[i]}$ can be simply chosen as the value of $\mu_0/2^k$ for a preassigned positive real number μ_0 and a certain (positive or negative) integer k such that $J(\boldsymbol{\vartheta}^{[i]} - (\mu_0/2^k)\mathbf{d}^{[i]}) < J(\boldsymbol{\vartheta}^{[i]})$. In Step (S4), the convergence criterion

$$\left| \frac{J(\boldsymbol{\vartheta}^{[i]}) - J(\boldsymbol{\vartheta}^{[i+1]})}{J(\boldsymbol{\vartheta}^{[i]})} \right| \le \zeta \tag{2.113}$$

can be used for testing the convergence of the iterative procedure where ζ is a small positive constant. Of course, other types of convergence criteria can also be applied. In Step (S5), the way of finding a new search direction $\mathbf{d}^{[i+1]}$ determines substantially the efficiency of gradient-type methods and thus leads to the main differences between the existing gradient-type methods. As indicated by the name "gradient-type method," the update of $\mathbf{d}^{[i+1]}$ involves the gradient $\boldsymbol{\nabla} J(\boldsymbol{\vartheta}^{[i+1]})$ and in some cases the Hessian matrix $\mathbf{J}_2(\boldsymbol{\vartheta}^{[i+1]})$. Note that the gradient-type methods that require only the gradient are referred to as *first-order methods*, while those requiring both the gradient and the

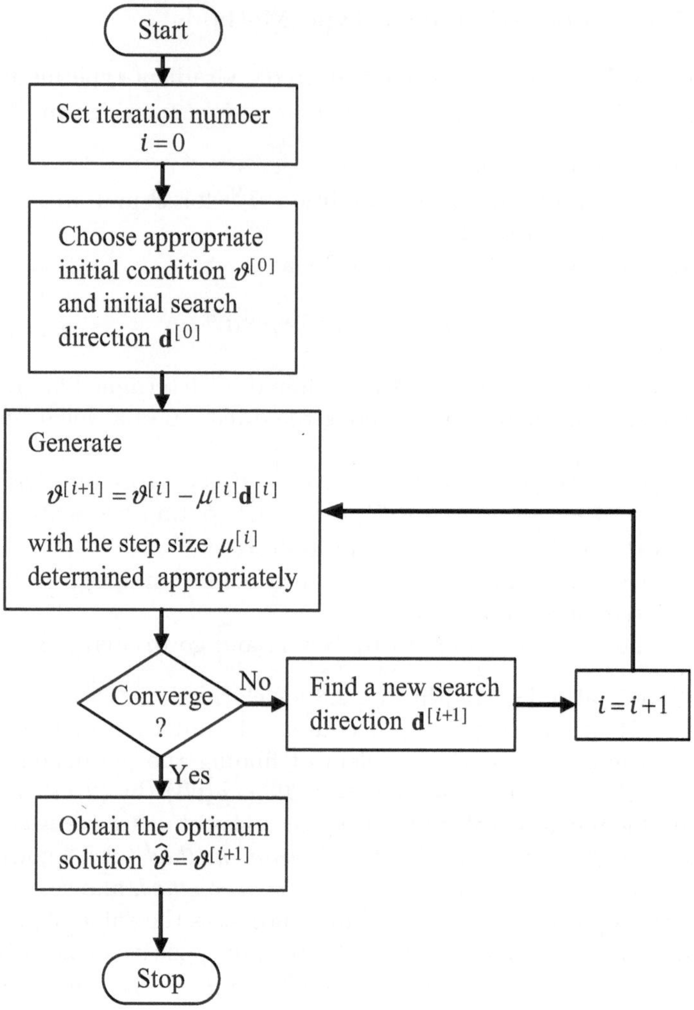

Fig. 2.8 Flow chart for the iterative procedure of gradient-type methods

Hessian matrix are referred to as *second-order methods.* As a final remark, all gradient-type methods are only guaranteed to find local minimum solutions due to the local property of the gradient nature.

Overview of Existing Gradient-Type Methods

Among the existing gradient-type methods for minimization of $J(\boldsymbol{\vartheta})$, the simplest is the so-called *steepest descent method*, which belongs to the category of first-order methods and is extremely important from a theoretical viewpoint. Convergence of the steepest descent method is more or less insensitive to the initial condition $\boldsymbol{\vartheta}^{[0]}$, but the convergence rate is excessively slow in the vicinity of minimum solution points [31, p. 91], thereby limiting its application scope. On the other hand, a well-known second-order method, the *Newton method*, exhibits a rather fast convergence rate in the vicinity of minimum solution points. The Newton method, however, requires the initial condition $\boldsymbol{\vartheta}^{[0]}$ to be sufficiently close to any one of the minimum solution points for convergence, and also requires the inverse Hessian matrix $\mathbf{J}_2^{-1}(\boldsymbol{\vartheta})$, whose computational complexity is in general quite high. To overcome the initial-condition problem of the Newton method, the *Marquardt method*, a combination of the steepest descent method and the Newton method, tries to share the merits of both methods. It performs as the steepest descent method at first and then performs as the Newton method when a minimum solution point is approached. Obviously, like the Newton method, the Marquardt method is a second-order method and also suffers from the problem of high computational complexity.

The motivation for reducing the computational complexity of Newton method further leads to the family of *quasi-Newton methods*. The idea behind quasi-Newton methods is to approximate either the Hessian matrix $\mathbf{J}_2(\boldsymbol{\vartheta})$ or its inverse $\mathbf{J}_2^{-1}(\boldsymbol{\vartheta})$ in terms of the gradient $\boldsymbol{\nabla}J(\boldsymbol{\vartheta})$. Clearly, quasi-Newton methods also belong to the category of first-order methods. A representative which approximates $\mathbf{J}_2(\boldsymbol{\vartheta})$ iteratively is the *Broyden–Fletcher–Goldfarb–Shanno (BFGS) method*, while a representative which approximates $\mathbf{J}_2^{-1}(\boldsymbol{\vartheta})$ iteratively is the *Davidon–Fletcher–Powell (DFP) method*. Known as the best quasi-Newton method, the BFGS method performs initially as the steepest descent method and then (after a number of iterations) performs as the Newton method. Our experience of computer simulation shows that the BFGS method is very efficient and numerically stable, and thus has been used for the simulation examples in this book. Next, let us give the detailed descriptions of some selected gradient-type methods, namely, the steepest descent method, the Newton method and the BFGS method.

Steepest Descent Method

At iteration i, the steepest descent method[11] updates the parameter vector $\boldsymbol{\vartheta}$ via

$$\boldsymbol{\vartheta}^{[i+1]} = \boldsymbol{\vartheta}^{[i]} - \mu^{[i]}\boldsymbol{\nabla}J(\boldsymbol{\vartheta}^{[i]}), \tag{2.114}$$

i.e. the search direction $\mathbf{d}^{[i]} = \boldsymbol{\nabla}J(\boldsymbol{\vartheta}^{[i]})$ (see (2.112)). The operation of (2.114) and the physical meaning of the gradient $\boldsymbol{\nabla}J(\boldsymbol{\vartheta})$ are interpreted as follows.

Let $\boldsymbol{\vartheta} + \Delta\boldsymbol{\vartheta}$ be a neighboring point of $\boldsymbol{\vartheta}$ and $\Delta J(\boldsymbol{\vartheta}) = J(\boldsymbol{\vartheta} + \Delta\boldsymbol{\vartheta}) - J(\boldsymbol{\vartheta})$ be the change in $J(\boldsymbol{\vartheta})$ due to $\Delta\boldsymbol{\vartheta}$. Then, by (2.111), we have

[11] The steepest descent method is also called the *Cauchy method* in recognition of the originator A. L. Cauchy (1847) [26].

$$\Delta J(\boldsymbol{\vartheta}) = (\Delta \boldsymbol{\vartheta})^H \, \nabla J(\boldsymbol{\vartheta}) \quad \text{as } \Delta \boldsymbol{\vartheta} \to \mathbf{0} \tag{2.115}$$

where we have ignored the second-order and other higher-order (≥ 3) terms. From (2.115) and the Cauchy–Schwartz inequality (Theorem 2.1), it follows that

$$|\Delta J(\boldsymbol{\vartheta})| \leq \|\Delta \boldsymbol{\vartheta}\| \cdot \|\nabla J(\boldsymbol{\vartheta})\| \quad \text{as } \Delta \boldsymbol{\vartheta} \to \mathbf{0} \tag{2.116}$$

and the equality holds only when $\Delta \boldsymbol{\vartheta} = \alpha \nabla J(\boldsymbol{\vartheta})$ where α is a real or complex scalar. This reveals that the change rate of $J(\boldsymbol{\vartheta})$ defined as

$$\lim_{\Delta \boldsymbol{\vartheta} \to \mathbf{0}} \frac{|\Delta J(\boldsymbol{\vartheta})|}{\|\Delta \boldsymbol{\vartheta}\|} \tag{2.117}$$

is upper bounded by $\|\nabla J(\boldsymbol{\vartheta})\|$, and that the gradient $\nabla J(\boldsymbol{\vartheta})$ represents the direction giving the maximum change rate of $J(\boldsymbol{\vartheta})$. Moreover, when $\Delta \boldsymbol{\vartheta} = -\mu \nabla J(\boldsymbol{\vartheta})$ for any real positive scalar μ, (2.115) reduces to

$$\Delta J(\boldsymbol{\vartheta}) = -\mu \|\nabla J(\boldsymbol{\vartheta})\|^2 \leq 0 \tag{2.118}$$

and thus

$$J(\boldsymbol{\vartheta} - \mu \nabla J(\boldsymbol{\vartheta})) = J(\boldsymbol{\vartheta} + \Delta \boldsymbol{\vartheta}) = J(\boldsymbol{\vartheta}) + \Delta J(\boldsymbol{\vartheta}) \leq J(\boldsymbol{\vartheta}), \tag{2.119}$$

which accounts for the operation of the update equation (2.114).

As a consequence of the preceding discussions, we come up with the following theorem to explain the physical meaning of the gradient $\nabla J(\boldsymbol{\vartheta})$.

Theorem 2.45. *The negative of the gradient, $-\nabla J(\boldsymbol{\vartheta})$, represents the direction giving the maximum change rate in reducing $J(\boldsymbol{\vartheta})$, i.e. the direction of steepest descent.*

Although the steepest descent method takes advantage of the gradient, the direction of steepest descent is only a local property (since $\Delta \boldsymbol{\vartheta} \to \mathbf{0}$) and thereby may vary from point to point. In fact, the steepest descent method quite often "zigzags" toward a local minimum, thereby requiring more and more steps of a smaller and smaller size when the minimum is approached [31, p. 91]. As such, it usually takes an enormous number of iterations to obtain an accurate solution.

Regarding the implementation of the update equation (2.114), it follows, from (2.105) and (2.106), that the update equation can be written as

$$\boldsymbol{\theta}^{[i+1]} = \boldsymbol{\theta}^{[i]} - \mu^{[i]} \cdot \left. \frac{\partial J(\boldsymbol{\theta})}{\partial \boldsymbol{\theta}} \right|_{\boldsymbol{\theta} = \boldsymbol{\theta}^{[i]}} \tag{2.120}$$

for real $\boldsymbol{\theta}$, and

$$\begin{pmatrix} \boldsymbol{\theta}^{[i+1]} \\ \boldsymbol{\theta}^{*[i+1]} \end{pmatrix} = \begin{pmatrix} \boldsymbol{\theta}^{[i]} \\ \boldsymbol{\theta}^{*[i]} \end{pmatrix} - \mu^{[i]} \cdot \left. \begin{pmatrix} \dfrac{\partial J(\boldsymbol{\theta})}{\partial \boldsymbol{\theta}^*} \\ \dfrac{\partial J(\boldsymbol{\theta})}{\partial \boldsymbol{\theta}} \end{pmatrix} \right|_{\boldsymbol{\theta} = \boldsymbol{\theta}^{[i]}} \tag{2.121}$$

for complex $\boldsymbol{\theta}$. One can easily see, from (2.121), that the update equation for $\boldsymbol{\theta}^{[i+1]}$ is equivalent to that for $\boldsymbol{\theta}^{*[i+1]}$ since $\mu^{[i]}$ is real, and thus only the former is actually needed. Table 2.2 summarizes the steepest descent method.

Table 2.2 Steepest descent method

	Update Equation
Generic form	At iteration i, update the parameter vector $\boldsymbol{\vartheta}$ via $$\boldsymbol{\vartheta}^{[i+1]} = \boldsymbol{\vartheta}^{[i]} - \mu^{[i]} \boldsymbol{\nabla} J(\boldsymbol{\vartheta}^{[i]})$$ where $\mu^{[i]} > 0$ is the step size and $\boldsymbol{\nabla} J(\boldsymbol{\vartheta}^{[i]})$ is the gradient at $\boldsymbol{\vartheta} = \boldsymbol{\vartheta}^{[i]}$.
Real case	At iteration i, update the real parameter vector $\boldsymbol{\theta}$ via $$\boldsymbol{\theta}^{[i+1]} = \boldsymbol{\theta}^{[i]} - \mu^{[i]} \cdot \left. \frac{\partial J(\boldsymbol{\theta})}{\partial \boldsymbol{\theta}} \right|_{\boldsymbol{\theta} = \boldsymbol{\theta}^{[i]}} .$$
Complex case	At iteration i, update the complex parameter vector $\boldsymbol{\theta}$ via $$\boldsymbol{\theta}^{[i+1]} = \boldsymbol{\theta}^{[i]} - \mu^{[i]} \cdot \left. \frac{\partial J(\boldsymbol{\theta})}{\partial \boldsymbol{\theta}^*} \right|_{\boldsymbol{\theta} = \boldsymbol{\theta}^{[i]}} .$$

Newton Method

Suppose that $\boldsymbol{\vartheta}_0$ is a guess for the parameter vector $\boldsymbol{\vartheta}$ and the Hessian matrix $\mathbf{J}_2(\boldsymbol{\vartheta}_0)$ is nonsingular. Replacing $\widehat{\boldsymbol{\vartheta}}$ in (2.111) by $\boldsymbol{\vartheta}_0$ and taking the first derivative of (2.111) with respect to $\boldsymbol{\vartheta}^*$ yields

$$\frac{\partial J(\boldsymbol{\vartheta})}{\partial \boldsymbol{\vartheta}^*} = \boldsymbol{\nabla} J(\boldsymbol{\vartheta}_0) + \alpha \mathbf{J}_2(\boldsymbol{\vartheta}_0)(\boldsymbol{\vartheta} - \boldsymbol{\vartheta}_0) \tag{2.122}$$

where all the higher-order terms (order ≥ 3) have been ignored and

$$\alpha = \begin{cases} 1 & \text{for real } \boldsymbol{\theta}, \\ 1/2 & \text{for complex } \boldsymbol{\theta}. \end{cases} \tag{2.123}$$

Setting (2.122) to zero, we obtain

$$\boldsymbol{\vartheta} = \boldsymbol{\vartheta}_0 - \frac{1}{\alpha}\mathbf{J}_2^{-1}(\boldsymbol{\vartheta}_0)\boldsymbol{\nabla}J(\boldsymbol{\vartheta}_0), \tag{2.124}$$

which reveals that $\boldsymbol{\vartheta}$ can be obtained from $\boldsymbol{\vartheta}_0$. However, since the higher-order terms that we have ignored may induce some errors in (2.124), it is suggested that (2.124) be used iteratively as follows: [26, pp. 389–391]

$$\boldsymbol{\vartheta}^{[i+1]} = \boldsymbol{\vartheta}^{[i]} - \mu^{[i]}\mathbf{J}_2^{-1}(\boldsymbol{\vartheta}^{[i]})\boldsymbol{\nabla}J(\boldsymbol{\vartheta}^{[i]}) \tag{2.125}$$

where $\boldsymbol{\vartheta}^{[i]}$ denotes the parameter vector $\boldsymbol{\vartheta}$ obtained at iteration i and $\mu^{[i]} > 0$ is the step size included to avoid divergence. As a result, the search direction for the Newton method is $\mathbf{d}^{[i]} = \mathbf{J}_2^{-1}(\boldsymbol{\vartheta}^{[i]})\boldsymbol{\nabla}J(\boldsymbol{\vartheta}^{[i]})$.

To further analyze the Newton method, let $\boldsymbol{\vartheta}^{[i]} = \boldsymbol{\vartheta}$, $\boldsymbol{\vartheta}^{[i+1]} = \boldsymbol{\vartheta} + \Delta\boldsymbol{\vartheta}$ and $\mu^{[i]} = \mu$ in (2.125). Then we have

$$\Delta\boldsymbol{\vartheta} = -\mu\mathbf{J}_2^{-1}(\boldsymbol{\vartheta})\boldsymbol{\nabla}J(\boldsymbol{\vartheta}), \quad \mu > 0. \tag{2.126}$$

Once again, by using (2.111), we have

$$J(\boldsymbol{\vartheta} + \Delta\boldsymbol{\vartheta}) = J(\boldsymbol{\vartheta}) + (\Delta\boldsymbol{\vartheta})^H \boldsymbol{\nabla}J(\boldsymbol{\vartheta}) + \frac{1}{2}(\Delta\boldsymbol{\vartheta})^H \mathbf{J}_2(\boldsymbol{\vartheta})\Delta\boldsymbol{\vartheta} + \cdots \tag{2.127}$$

where the higher-order (≥ 3) terms can be neglected as $\Delta\boldsymbol{\vartheta} \to \mathbf{0}$; this, in turn, requires that the step size μ be sufficiently small according to (2.126). From (2.126) and (2.127), it follows that the change $\Delta J(\boldsymbol{\vartheta}) \triangleq J(\boldsymbol{\vartheta} + \Delta\boldsymbol{\vartheta}) - J(\boldsymbol{\vartheta})$ can be written as

$$\Delta J(\boldsymbol{\vartheta}) = -\mu(1 - \frac{\mu}{2})\left[\boldsymbol{\nabla}J(\boldsymbol{\vartheta})\right]^H \mathbf{J}_2^{-1}(\boldsymbol{\vartheta})\boldsymbol{\nabla}J(\boldsymbol{\vartheta}) \quad \text{as } \Delta\boldsymbol{\vartheta} \to \mathbf{0}. \tag{2.128}$$

Accordingly, if $\mathbf{J}_2(\boldsymbol{\vartheta})$ is positive definite and $\mu < 2$, then the change $\Delta J(\boldsymbol{\vartheta}) \leq 0$ and

$$J(\boldsymbol{\vartheta}^{[i+1]}) = J(\boldsymbol{\vartheta} + \Delta\boldsymbol{\vartheta}) = J(\boldsymbol{\vartheta}) + \Delta J(\boldsymbol{\vartheta}) \leq J(\boldsymbol{\vartheta}) = J(\boldsymbol{\vartheta}^{[i]}). \tag{2.129}$$

That is, the search direction always points towards a (local) minimum of $J(\boldsymbol{\vartheta})$ when the Hessian matrix $\mathbf{J}_2(\boldsymbol{\vartheta})$ is positive definite, or equivalently $\mathbf{J}_2^{-1}(\boldsymbol{\vartheta})$ is positive definite (by Property 2.12), and the step size μ is chosen small enough. However, due to utilization of only the lower-order terms of the Taylor series in the derivation, the Newton method requires the initial condition $\boldsymbol{\vartheta}^{[0]}$ to be sufficiently close to the solution point. Moreover, it is generally difficult and sometimes almost impossible to compute $\mathbf{J}_2(\boldsymbol{\vartheta})$ as well as $\mathbf{J}_2^{-1}(\boldsymbol{\vartheta})$.

Regarding the implementation of the update equation (2.125), we note, from (2.105) and (2.106), that for real $\boldsymbol{\theta}$ the update equation is given by

$$\boldsymbol{\theta}^{[i+1]} = \boldsymbol{\theta}^{[i]} - \mu^{[i]} \mathbf{J}_2^{-1}(\boldsymbol{\theta}^{[i]}) \cdot \left. \frac{\partial J(\boldsymbol{\theta})}{\partial \boldsymbol{\theta}} \right|_{\boldsymbol{\theta} = \boldsymbol{\theta}^{[i]}} \tag{2.130}$$

where $\mathbf{J}_2(\boldsymbol{\theta}^{[i]}) \equiv \mathbf{J}_2(\boldsymbol{\vartheta}^{[i]})$, and for complex $\boldsymbol{\theta}$ it is given by

$$\begin{pmatrix} \boldsymbol{\theta}^{[i+1]} \\ \boldsymbol{\theta}^{*[i+1]} \end{pmatrix} = \begin{pmatrix} \boldsymbol{\theta}^{[i]} \\ \boldsymbol{\theta}^{*[i]} \end{pmatrix} - \mu^{[i]} \begin{pmatrix} \mathbf{A}^{[i]} & \mathbf{B}^{[i]} \\ \left(\mathbf{B}^{[i]}\right)^* & \left(\mathbf{A}^{[i]}\right)^* \end{pmatrix}^{-1} \cdot \left. \begin{pmatrix} \dfrac{\partial J(\boldsymbol{\theta})}{\partial \boldsymbol{\theta}^*} \\ \dfrac{\partial J(\boldsymbol{\theta})}{\partial \boldsymbol{\theta}} \end{pmatrix} \right|_{\boldsymbol{\theta} = \boldsymbol{\theta}^{[i]}} \tag{2.131}$$

where

$$\mathbf{A}^{[i]} = \left(\mathbf{A}^{[i]}\right)^H = \left. \frac{\partial}{\partial \boldsymbol{\theta}^*} \left[\frac{\partial J(\boldsymbol{\theta})}{\partial \boldsymbol{\theta}^*} \right]^H \right|_{\boldsymbol{\theta} = \boldsymbol{\theta}^{[i]}}, \tag{2.132}$$

$$\mathbf{B}^{[i]} = \left(\mathbf{B}^{[i]}\right)^T = \left. \frac{\partial}{\partial \boldsymbol{\theta}^*} \left[\frac{\partial J(\boldsymbol{\theta})}{\partial \boldsymbol{\theta}} \right]^H \right|_{\boldsymbol{\theta} = \boldsymbol{\theta}^{[i]}}. \tag{2.133}$$

Similar to the complex case of the steepest descent method, by (2.131), (2.132), (2.133) and Theorem 2.8, one can show that only the following update equation is needed for complex $\boldsymbol{\theta}$:

$$\boldsymbol{\theta}^{[i+1]} = \boldsymbol{\theta}^{[i]} - \mu^{[i]} \mathbf{C}^{[i]} \left\{ \left. \frac{\partial J(\boldsymbol{\theta})}{\partial \boldsymbol{\theta}^*} \right|_{\boldsymbol{\theta} = \boldsymbol{\theta}^{[i]}} - \mathbf{D}^{[i]} \cdot \left. \frac{\partial J(\boldsymbol{\theta})}{\partial \boldsymbol{\theta}} \right|_{\boldsymbol{\theta} = \boldsymbol{\theta}^{[i]}} \right\} \tag{2.134}$$

where

$$\mathbf{C}^{[i]} = \left\{ \mathbf{A}^{[i]} - \mathbf{B}^{[i]} \left[\left(\mathbf{A}^{[i]}\right)^* \right]^{-1} \left(\mathbf{B}^{[i]}\right)^* \right\}^{-1}, \tag{2.135}$$

$$\mathbf{D}^{[i]} = \mathbf{B}^{[i]} \left[\left(\mathbf{A}^{[i]}\right)^* \right]^{-1}. \tag{2.136}$$

Furthermore, one can simplify the update equation (2.134) by forcing $\mathbf{B}^{[i]} = \mathbf{0}$ for all iterations, and obtain the following "approximate" update equation for complex $\boldsymbol{\theta}$:

$$\boldsymbol{\theta}^{[i+1]} = \boldsymbol{\theta}^{[i]} - \mu^{[i]} \left(\mathbf{A}^{[i]}\right)^{-1} \cdot \left. \frac{\partial J(\boldsymbol{\theta})}{\partial \boldsymbol{\theta}^*} \right|_{\boldsymbol{\theta} = \boldsymbol{\theta}^{[i]}}. \tag{2.137}$$

We refer to the Newton method based on (2.137) as the *approximate Newton method.* Note that for the approximate Newton method, if the matrix $\mathbf{A}^{[i]}$ is positive definite, then the corresponding Hessian matrix approximated as

$$\mathbf{J}_2(\boldsymbol{\vartheta}^{[i]}) \approx \begin{pmatrix} \mathbf{A}^{[i]} & \mathbf{0} \\ \mathbf{0} & \left(\mathbf{A}^{[i]}\right)^* \end{pmatrix} \tag{2.138}$$

is positive definite, too. Accordingly, the above-mentioned interpretation for the operation of Newton method (see explanation of (2.129)) also applies to the approximate Newton method. Table 2.3 summarizes the Newton method and the approximate Newton method. Note that the approximate Newton method exists only for the complex case.

Table 2.3 Newton and approximate Newton methods

	Update Equation for the Newton Method
Generic form	At iteration i, update the parameter vector $\boldsymbol{\vartheta}$ via $$\boldsymbol{\vartheta}^{[i+1]} = \boldsymbol{\vartheta}^{[i]} - \mu^{[i]}\mathbf{J}_2^{-1}(\boldsymbol{\vartheta}^{[i]})\boldsymbol{\nabla} J(\boldsymbol{\vartheta}^{[i]})$$ where $\mu^{[i]} > 0$ is the step size, $\boldsymbol{\nabla} J(\boldsymbol{\vartheta}^{[i]})$ is the gradient at $\boldsymbol{\vartheta} = \boldsymbol{\vartheta}^{[i]}$, and $\mathbf{J}_2(\boldsymbol{\vartheta}^{[i]})$ is the Hessian matrix at $\boldsymbol{\vartheta} = \boldsymbol{\vartheta}^{[i]}$.
Real case	At iteration i, update the real parameter vector $\boldsymbol{\theta}$ via $$\boldsymbol{\theta}^{[i+1]} = \boldsymbol{\theta}^{[i]} - \mu^{[i]} \cdot \mathbf{J}_2^{-1}(\boldsymbol{\theta}^{[i]}) \cdot \left.\frac{\partial J(\boldsymbol{\theta})}{\partial \boldsymbol{\theta}}\right\|_{\boldsymbol{\theta} = \boldsymbol{\theta}^{[i]}}$$ where $\mathbf{J}_2(\boldsymbol{\theta}^{[i]}) = \mathbf{J}_2(\boldsymbol{\vartheta}^{[i]})$.
Complex case	At iteration i, update the complex parameter vector $\boldsymbol{\theta}$ via $$\boldsymbol{\theta}^{[i+1]} = \boldsymbol{\theta}^{[i]} - \mu^{[i]}\mathbf{C}^{[i]} \left\{ \left.\frac{\partial J(\boldsymbol{\theta})}{\partial \boldsymbol{\theta}^*}\right\|_{\boldsymbol{\theta} = \boldsymbol{\theta}^{[i]}} - \mathbf{D}^{[i]} \cdot \left.\frac{\partial J(\boldsymbol{\theta})}{\partial \boldsymbol{\theta}}\right\|_{\boldsymbol{\theta} = \boldsymbol{\theta}^{[i]}} \right\}$$ where $\mathbf{C}^{[i]}$ and $\mathbf{D}^{[i]}$ are given by (2.135) and (2.136), respectively.
	Update Equation for the Approximate Newton Method
Complex case	At iteration i, update the complex parameter vector $\boldsymbol{\theta}$ via $$\boldsymbol{\theta}^{[i+1]} = \boldsymbol{\theta}^{[i]} - \mu^{[i]} \left(\mathbf{A}^{[i]}\right)^{-1} \cdot \left.\frac{\partial J(\boldsymbol{\theta})}{\partial \boldsymbol{\theta}^*}\right\|_{\boldsymbol{\theta} = \boldsymbol{\theta}^{[i]}}$$ where $\mathbf{A}^{[i]}$ is given by (2.132).

Broyden–Fletcher–Goldfarb–Shanno Method

Recall that the idea behind the BFGS method is to approximate the inverse Hessian matrix $\mathbf{J}_2^{-1}(\boldsymbol{\vartheta}^{[i]})$ in (2.125) by virtue of the gradient $\boldsymbol{\nabla} J(\boldsymbol{\vartheta}^{[i]})$. Let $\mathbf{Q}^{[i]}$ be a Hermitian matrix, which will be obtained as an approximation to $\mathbf{J}_2^{-1}(\boldsymbol{\vartheta}^{[i]})$. Then, from (2.125), it follows that the update equation for the BFGS method is given by

$$\boldsymbol{\vartheta}^{[i+1]} = \boldsymbol{\vartheta}^{[i]} - \mu^{[i]}\mathbf{Q}^{[i]}\boldsymbol{\nabla} J(\boldsymbol{\vartheta}^{[i]}), \tag{2.139}$$

i.e. the search direction $\mathbf{d}^{[i]} = \mathbf{Q}^{[i]}\boldsymbol{\nabla} J(\boldsymbol{\vartheta}^{[i]})$. Next, let us present how to update $\mathbf{Q}^{[i+1]}$ from $\mathbf{Q}^{[i]}$, as well as how to choose an appropriate initial condition for $\mathbf{Q}^{[0]}$.

Update Equation for $\mathbf{Q}^{[i+1]}$

Let $\mathbf{P}^{[i]} = \left(\mathbf{Q}^{[i]}\right)^{-1}$, that is, $\mathbf{P}^{[i]}$ (a Hermitian matrix) is an approximation to $\mathbf{J}_2(\boldsymbol{\vartheta}^{[i]})$. We will first derive the update equation for $\mathbf{P}^{[i+1]}$ and then convert it to the one for $\mathbf{Q}^{[i+1]}$. By substituting $\boldsymbol{\vartheta} = \boldsymbol{\vartheta}^{[i]}$ and $\boldsymbol{\vartheta}_0 = \boldsymbol{\vartheta}^{[i+1]}$ into (2.122), we obtain

$$\mathbf{s}_{i+1} = \alpha\mathbf{J}_2(\boldsymbol{\vartheta}^{[i+1]})\mathbf{r}_{i+1} \tag{2.140}$$

where α is given by (2.123) and

$$\mathbf{r}_{i+1} = \boldsymbol{\vartheta}^{[i+1]} - \boldsymbol{\vartheta}^{[i]}, \tag{2.141}$$

$$\mathbf{s}_{i+1} = \boldsymbol{\nabla} J(\boldsymbol{\vartheta}^{[i+1]}) - \boldsymbol{\nabla} J(\boldsymbol{\vartheta}^{[i]}). \tag{2.142}$$

It follows that $\mathbf{P}^{[i+1]}$ should also satisfy (2.140) as follows:

$$\mathbf{s}_{i+1} = \alpha\mathbf{P}^{[i+1]}\mathbf{r}_{i+1}. \tag{2.143}$$

We note, from (2.100), (2.106), (2.141) and (2.142), that $\mathbf{r}_{i+1}$ and $\mathbf{s}_{i+1}$ are both $L \times 1$ vectors for real $\boldsymbol{\theta}$ and $(2L) \times 1$ vectors for complex $\boldsymbol{\theta}$. Also note, from (2.108) and (2.109), that $\mathbf{P}^{[i+1]}$ is an $L \times L$ symmetric matrix for real $\boldsymbol{\theta}$ and a $(2L) \times (2L)$ Hermitian matrix for complex $\boldsymbol{\theta}$. Therefore, the number of unknowns (to be determined) in $\mathbf{P}^{[i+1]}$ is more than the number of linear equations in (2.143), meaning that the solution satisfying (2.143) is not unique.

The general formula for updating $\mathbf{P}^{[i+1]}$ iteratively can be written as

$$\mathbf{P}^{[i+1]} = \mathbf{P}^{[i]} + \Delta\mathbf{P}^{[i]} \tag{2.144}$$

where, in theory, the matrix $\Delta\mathbf{P}^{[i]}$ can have rank as high as L for real $\boldsymbol{\theta}$ and $2L$ for complex $\boldsymbol{\theta}$, but rank 1 or rank 2 are more suitable in practice. By adopting the rank 2 update $\Delta\mathbf{P}^{[i]} = c_1\mathbf{z}_1\mathbf{z}_1^H + c_2\mathbf{z}_2\mathbf{z}_2^H$ (see [26, p. 398] for the real case), we have

$$\mathbf{P}^{[i+1]} = \mathbf{P}^{[i]} + c_1\mathbf{z}_1\mathbf{z}_1^H + c_2\mathbf{z}_2\mathbf{z}_2^H \tag{2.145}$$

where c_1 and c_2 are real or complex constants, and $\mathbf{z}_1$ and $\mathbf{z}_2$ are real or complex vectors to be determined. Substituting (2.145) into (2.143) yields

$$\mathbf{s}_{i+1} = \alpha\mathbf{P}^{[i]}\mathbf{r}_{i+1} + \alpha c_1(\mathbf{z}_1^H\mathbf{r}_{i+1})\mathbf{z}_1 + \alpha c_2(\mathbf{z}_2^H\mathbf{r}_{i+1})\mathbf{z}_2. \tag{2.146}$$

Equation (2.146) can be satisfied by choosing

$$\alpha c_1(\mathbf{z}_1^H\mathbf{r}_{i+1})\mathbf{z}_1 = \mathbf{s}_{i+1} \quad \text{and} \quad c_2(\mathbf{z}_2^H\mathbf{r}_{i+1})\mathbf{z}_2 = -\mathbf{P}^{[i]}\mathbf{r}_{i+1}, \tag{2.147}$$

which further leads to the following choice:

$$\mathbf{z}_1 = \mathbf{s}_{i+1} \tag{2.148}$$

$$\mathbf{z}_2 = \mathbf{P}^{[i]}\mathbf{r}_{i+1} \tag{2.149}$$

$$c_1 = \frac{1}{\alpha\mathbf{z}_1^H\mathbf{r}_{i+1}} = \frac{1}{\alpha\mathbf{s}_{i+1}^H\mathbf{r}_{i+1}} \tag{2.150}$$

$$c_2 = -\frac{1}{\mathbf{z}_2^H\mathbf{r}_{i+1}} = -\frac{1}{(\mathbf{P}^{[i]}\mathbf{r}_{i+1})^H\mathbf{r}_{i+1}}. \tag{2.151}$$

Substituting (2.148) through (2.151) into (2.145) gives rise to the following update equation for $\mathbf{P}^{[i+1]}$:

$$\mathbf{P}^{[i+1]} = \mathbf{P}^{[i]} + \frac{\mathbf{s}_{i+1}\mathbf{s}_{i+1}^H}{\alpha\mathbf{s}_{i+1}^H\mathbf{r}_{i+1}} - \frac{(\mathbf{P}^{[i]}\mathbf{r}_{i+1})(\mathbf{P}^{[i]}\mathbf{r}_{i+1})^H}{(\mathbf{P}^{[i]}\mathbf{r}_{i+1})^H\mathbf{r}_{i+1}}, \tag{2.152}$$

which is called the *Broyden–Fletcher–Goldfarb–Shanno (BFGS) formula* (refer to [26] for the real case).

To convert the update equation (2.152) into the one for $\mathbf{Q}^{[i+1]}$, let us re-express (2.152) as

$$\mathbf{P}^{[i+1]} = \mathbf{R} - \frac{(\mathbf{P}^{[i]}\mathbf{r}_{i+1})(\mathbf{P}^{[i]}\mathbf{r}_{i+1})^H}{(\mathbf{P}^{[i]}\mathbf{r}_{i+1})^H\mathbf{r}_{i+1}} \tag{2.153}$$

where

$$\mathbf{R} = \mathbf{P}^{[i]} + \frac{\mathbf{s}_{i+1}\mathbf{s}_{i+1}^H}{\alpha\mathbf{s}_{i+1}^H\mathbf{r}_{i+1}}. \tag{2.154}$$

Applying Woodbury's identity (Corollary 2.6) to (2.153) and (2.154) yields

$$\mathbf{Q}^{[i+1]} = \mathbf{R}^{-1} + \frac{\mathbf{R}^{-1}(\mathbf{P}^{[i]}\mathbf{r}_{i+1})(\mathbf{P}^{[i]}\mathbf{r}_{i+1})^H\mathbf{R}^{-1}}{(\mathbf{P}^{[i]}\mathbf{r}_{i+1})^H\mathbf{r}_{i+1} - (\mathbf{P}^{[i]}\mathbf{r}_{i+1})^H\mathbf{R}^{-1}(\mathbf{P}^{[i]}\mathbf{r}_{i+1})} \tag{2.155}$$

and

$$\mathbf{R}^{-1} = \mathbf{Q}^{[i]} - \frac{\mathbf{Q}^{[i]}\mathbf{s}_{i+1}\mathbf{s}_{i+1}^H\mathbf{Q}^{[i]}}{\alpha\mathbf{s}_{i+1}^H\mathbf{r}_{i+1} + \mathbf{s}_{i+1}^H\mathbf{Q}^{[i]}\mathbf{s}_{i+1}}, \tag{2.156}$$

respectively. By substituting (2.156) into (2.155) and after some algebraic manipulations, we obtain

$$\mathbf{Q}^{[i+1]} = \mathbf{Q}^{[i]} + \frac{1}{\mathbf{r}_{i+1}^H\mathbf{s}_{i+1}}\left\{(\alpha+\beta_i)\,\mathbf{r}_{i+1}\mathbf{r}_{i+1}^H - \mathbf{r}_{i+1}\mathbf{s}_{i+1}^H\mathbf{Q}^{[i]} - \mathbf{Q}^{[i]}\mathbf{s}_{i+1}\mathbf{r}_{i+1}^H\right\} \tag{2.157}$$

where

$$\beta_i = \frac{\mathbf{s}_{i+1}^H\mathbf{Q}^{[i]}\mathbf{s}_{i+1}}{\mathbf{s}_{i+1}^H\mathbf{r}_{i+1}} \tag{2.158}$$

is a real number. In the derivation of (2.157), we have used the facts that $\mathbf{P}^{[i]} = (\mathbf{P}^{[i]})^H$ and that $\mathbf{r}_{i+1}^H\mathbf{s}_{i+1} = \mathbf{s}_{i+1}^H\mathbf{r}_{i+1}$ is real (by (2.141), (2.142), (2.100), and (2.106)).

As a consequence, the BFGS method employs the update equation (2.139) for $\boldsymbol{\vartheta}^{[i+1]}$ along with the update equation (2.157) for $\mathbf{Q}^{[i+1]}$ to obtain the (local) minimum solution $\boldsymbol{\vartheta}$ without involving any second partial derivatives of $J(\boldsymbol{\vartheta})$.

Suggestion for the Initial Condition $\mathbf{Q}^{[0]}$

Since $\mathbf{J}_2^{-1}(\boldsymbol{\vartheta}^{[i+1]})$ is required to be positive definite in the Newton method, the Hermitian matrix $\mathbf{Q}^{[i+1]}$, as an approximation to $\mathbf{J}_2^{-1}(\boldsymbol{\vartheta}^{[i+1]})$, should also maintain the positive definite property. The following theorem reveals the conditions for maintaining the positive definite property of $\mathbf{Q}^{[i+1]}$ (refer to [32] for the real case).

Theorem 2.46. *If the matrix $\mathbf{Q}^{[i]}$ is positive definite and the step size $\mu^{[i]} > 0$ used in (2.139) is optimum, then the matrix $\mathbf{Q}^{[i+1]}$ generated from (2.157) is also positive definite where $\mathbf{r}_{i+1}$ and $\mathbf{s}_{i+1}$ defined as (2.141) and (2.142) are both nonzero vectors before convergence.*

See Appendix 2F for the proof. Theorem 2.46 suggests that $\mathbf{Q}^{[0]}$ be chosen as a positive definite matrix, in addition to utilization of an appropriate step size $\mu^{[i]}$. Usually, $\mathbf{Q}^{[0]} = \mathbf{I}$ is used. As such, the BFGS method performs initially as the steepest descent method because (2.139) reduces to (2.114) when $\mathbf{Q}^{[i]} = \mathbf{I}$. After a number of iterations, it performs as the Newton method because $\mathbf{Q}^{[i+1]}$ then appears as a good approximation to $\mathbf{J}_2^{-1}(\boldsymbol{\vartheta}^{[i+1]})$. On the other hand, numerical experience indicates that the BFGS method is less influenced by the error in determining $\mu^{[i]}$ [26, p. 406]. Nevertheless, in case the positive definite property of $\mathbf{Q}^{[i+1]}$ is violated due to this error, one may periodically reset $\mathbf{Q}^{[i+1]}$ via $\mathbf{Q}^{[i+1]} = \mathbf{I}$. The corresponding BFGS method then reverts to the steepest descent method at iteration $(i+1)$, but this time it has a much better initial condition $\boldsymbol{\vartheta}^{[i+1]}$.

Implementation of the BFGS Method

For the case of real $\boldsymbol{\theta}$, the update equation (2.139) can be written as

$$\boldsymbol{\theta}^{[i+1]} = \boldsymbol{\theta}^{[i]} - \mu^{[i]}\mathbf{Q}^{[i]} \cdot \left.\frac{\partial J(\boldsymbol{\theta})}{\partial \boldsymbol{\theta}}\right|_{\boldsymbol{\theta} = \boldsymbol{\theta}^{[i]}} \tag{2.159}$$

and the update equation (2.157) for $\mathbf{Q}^{[i+1]}$ reduces to

$$\mathbf{Q}^{[i+1]} = \mathbf{Q}^{[i]} + \frac{(1+\beta_i)\,\mathbf{r}_{i+1}\mathbf{r}_{i+1}^T - \mathbf{r}_{i+1}\mathbf{s}_{i+1}^T\mathbf{Q}^{[i]} - \mathbf{Q}^{[i]}\mathbf{s}_{i+1}\mathbf{r}_{i+1}^T}{\mathbf{r}_{i+1}^T\mathbf{s}_{i+1}} \tag{2.160}$$

(since $\alpha = 1$) where the initial condition $\mathbf{Q}^{[0]} = \mathbf{I}$ is suggested and

$$\beta_i = \frac{\mathbf{s}_{i+1}^T\mathbf{Q}^{[i]}\mathbf{s}_{i+1}}{\mathbf{s}_{i+1}^T\mathbf{r}_{i+1}}, \tag{2.161}$$

$$\mathbf{r}_{i+1} = \boldsymbol{\theta}^{[i+1]} - \boldsymbol{\theta}^{[i]}, \tag{2.162}$$

$$\mathbf{s}_{i+1} = \left\{ \left.\frac{\partial J(\boldsymbol{\theta})}{\partial \boldsymbol{\theta}}\right|_{\boldsymbol{\theta} = \boldsymbol{\theta}^{[i+1]}} \right\} - \left\{ \left.\frac{\partial J(\boldsymbol{\theta})}{\partial \boldsymbol{\theta}}\right|_{\boldsymbol{\theta} = \boldsymbol{\theta}^{[i]}} \right\}. \tag{2.163}$$

On the other hand, for the case of complex $\boldsymbol{\theta}$, the vectors $\mathbf{r}_{i+1}$ and $\mathbf{s}_{i+1}$ defined as (2.141) and (2.142) can be written as

$$\mathbf{r}_{i+1} = \left(\widetilde{\mathbf{r}}_{i+1}^T, \widetilde{\mathbf{r}}_{i+1}^H\right)^T \quad \text{and} \quad \mathbf{s}_{i+1} = \left(\widetilde{\mathbf{s}}_{i+1}^T, \widetilde{\mathbf{s}}_{i+1}^H\right)^T \tag{2.164}$$

where

$$\widetilde{\mathbf{r}}_{i+1} = \boldsymbol{\theta}^{[i+1]} - \boldsymbol{\theta}^{[i]}, \tag{2.165}$$

$$\widetilde{\mathbf{s}}_{i+1} = \left\{ \left.\frac{\partial J(\boldsymbol{\theta})}{\partial \boldsymbol{\theta}^*}\right|_{\boldsymbol{\theta} = \boldsymbol{\theta}^{[i+1]}} \right\} - \left\{ \left.\frac{\partial J(\boldsymbol{\theta})}{\partial \boldsymbol{\theta}^*}\right|_{\boldsymbol{\theta} = \boldsymbol{\theta}^{[i]}} \right\}. \tag{2.166}$$

By (2.152), (2.164) and Theorem 2.8, one can show that if the initial condition $\mathbf{Q}^{[0]} = \mathbf{I}$ is used, then the matrix $\mathbf{Q}^{[i]}$ obtained from the update equation (2.157) will have the following form:

$$\mathbf{Q}^{[i]} = \begin{pmatrix} \mathbf{Q}_A^{[i]} & \mathbf{Q}_B^{[i]} \\ \left(\mathbf{Q}_B^{[i]}\right)^* & \left(\mathbf{Q}_A^{[i]}\right)^* \end{pmatrix} \tag{2.167}$$

where $\mathbf{Q}_A^{[i]} = \left(\mathbf{Q}_A^{[i]}\right)^H$ and $\mathbf{Q}_B^{[i]} = \left(\mathbf{Q}_B^{[i]}\right)^T$ since $\mathbf{Q}^{[i]}$ is Hermitian. From (2.139), (2.157), (2.164) and (2.167), it follows that we need only the following update equation for complex $\boldsymbol{\theta}^{[i+1]}$:

$$\boldsymbol{\theta}^{[i+1]} = \boldsymbol{\theta}^{[i]} - \mu^{[i]} \left\{ \mathbf{Q}_A^{[i]} \cdot \left. \frac{\partial J(\boldsymbol{\theta})}{\partial \boldsymbol{\theta}^*} \right|_{\boldsymbol{\theta} = \boldsymbol{\theta}^{[i]}} + \mathbf{Q}_B^{[i]} \cdot \left. \frac{\partial J(\boldsymbol{\theta})}{\partial \boldsymbol{\theta}} \right|_{\boldsymbol{\theta} = \boldsymbol{\theta}^{[i]}} \right\} \tag{2.168}$$

and the following update equations for $\mathbf{Q}_A^{[i+1]}$ and $\mathbf{Q}_B^{[i+1]}$:

$$\begin{aligned} \mathbf{Q}_A^{[i+1]} = \mathbf{Q}_A^{[i]} + \frac{1}{2\mathrm{Re}\left\{\widetilde{\mathbf{r}}_{i+1}^H \widetilde{\mathbf{s}}_{i+1}\right\}} \Big\{ & (\alpha + \beta_i)\, \widetilde{\mathbf{r}}_{i+1} \widetilde{\mathbf{r}}_{i+1}^H - \widetilde{\mathbf{r}}_{i+1} \widetilde{\mathbf{s}}_{i+1}^H \mathbf{Q}_A^{[i]} \\ & - \mathbf{Q}_A^{[i]} \widetilde{\mathbf{s}}_{i+1} \widetilde{\mathbf{r}}_{i+1}^H - \widetilde{\mathbf{r}}_{i+1} \widetilde{\mathbf{s}}_{i+1}^T \left(\mathbf{Q}_B^{[i]}\right)^* - \mathbf{Q}_B^{[i]} \widetilde{\mathbf{s}}_{i+1}^* \widetilde{\mathbf{r}}_{i+1}^H \Big\}, \end{aligned} \tag{2.169}$$

$$\begin{aligned} \mathbf{Q}_B^{[i+1]} = \mathbf{Q}_B^{[i]} + \frac{1}{2\mathrm{Re}\left\{\widetilde{\mathbf{r}}_{i+1}^H \widetilde{\mathbf{s}}_{i+1}\right\}} \Big\{ & (\alpha + \beta_i)\, \widetilde{\mathbf{r}}_{i+1} \widetilde{\mathbf{r}}_{i+1}^T - \widetilde{\mathbf{r}}_{i+1} \widetilde{\mathbf{s}}_{i+1}^T \left(\mathbf{Q}_A^{[i]}\right)^* \\ & - \mathbf{Q}_A^{[i]} \widetilde{\mathbf{s}}_{i+1} \widetilde{\mathbf{r}}_{i+1}^T - \widetilde{\mathbf{r}}_{i+1} \widetilde{\mathbf{s}}_{i+1}^H \mathbf{Q}_B^{[i]} - \mathbf{Q}_B^{[i]} \widetilde{\mathbf{s}}_{i+1}^* \widetilde{\mathbf{r}}_{i+1}^T \Big\} \end{aligned} \tag{2.170}$$

where $\mathbf{Q}_A^{[0]} = \mathbf{I}$, $\mathbf{Q}_B^{[0]} = \mathbf{0}$, and

$$\beta_i = \frac{\mathrm{Re}\left\{\widetilde{\mathbf{s}}_{i+1}^H \mathbf{Q}_A^{[i]} \widetilde{\mathbf{s}}_{i+1} + \widetilde{\mathbf{s}}_{i+1}^H \mathbf{Q}_B^{[i]} \widetilde{\mathbf{s}}_{i+1}^*\right\}}{\mathrm{Re}\left\{\widetilde{\mathbf{s}}_{i+1}^H \widetilde{\mathbf{r}}_{i+1}\right\}}. \tag{2.171}$$

Furthermore, one can simplify the above update equations by forcing $\mathbf{Q}_B^{[i]} = \mathbf{0}$ for all iterations. The corresponding update equation for complex $\boldsymbol{\theta}^{[i+1]}$ is given by

$$\boldsymbol{\theta}^{[i+1]} = \boldsymbol{\theta}^{[i]} - \mu^{[i]} \mathbf{Q}_A^{[i]} \cdot \left. \frac{\partial J(\boldsymbol{\theta})}{\partial \boldsymbol{\theta}^*} \right|_{\boldsymbol{\theta} = \boldsymbol{\theta}^{[i]}} \tag{2.172}$$

and the corresponding update equation for $\mathbf{Q}_A^{[i+1]}$ is given by

$$\begin{aligned} \mathbf{Q}_A^{[i+1]} = \mathbf{Q}_A^{[i]} + \frac{1}{2\mathrm{Re}\left\{\widetilde{\mathbf{r}}_{i+1}^H \widetilde{\mathbf{s}}_{i+1}\right\}} \Big\{ & (\alpha + \beta_i)\, \widetilde{\mathbf{r}}_{i+1} \widetilde{\mathbf{r}}_{i+1}^H \\ & - \widetilde{\mathbf{r}}_{i+1} \widetilde{\mathbf{s}}_{i+1}^H \mathbf{Q}_A^{[i]} - \mathbf{Q}_A^{[i]} \widetilde{\mathbf{s}}_{i+1} \widetilde{\mathbf{r}}_{i+1}^H \Big\} \end{aligned} \tag{2.173}$$

where $\mathbf{Q}_A^{[0]} = \mathbf{I}$ and

$$\beta_i = \frac{\mathrm{Re}\left\{\widetilde{\mathbf{s}}_{i+1}^H \mathbf{Q}_A^{[i]} \widetilde{\mathbf{s}}_{i+1}\right\}}{\mathrm{Re}\left\{\widetilde{\mathbf{s}}_{i+1}^H \widetilde{\mathbf{r}}_{i+1}\right\}}. \tag{2.174}$$

Similarly, we refer to the BFGS method that is based on (2.172), (2.173) and (2.174) as the *approximate BFGS method.* As a result of the aforementioned discussions, the approximate BFGS method also maintains the positive definite property of the corresponding $\mathbf{Q}^{[i]}$, provided that the step size $\mu^{[i]}$ is chosen appropriately. Table 2.4 summarizes the BFGS method and the approximate BFGS method where the latter is only for the complex case.

2.4 Least-Squares Method

Many science and engineering problems require solving the following set of M linear equations in K unknowns:

$$\mathbf{A}\boldsymbol{\theta} = \mathbf{b} \tag{2.175}$$

where $\mathbf{A}$ is an $M \times K$ matrix, $\mathbf{b} = (b_1, b_2, ..., b_M)^T$ is an $M \times 1$ vector, and $\boldsymbol{\theta} = (\theta_1, \theta_2, ..., \theta_K)^T$ is a $K \times 1$ vector of unknown parameters to be solved. Let $\mathbf{A} = (\mathbf{a}_1, \mathbf{a}_2, ..., \mathbf{a}_K)$ where $\mathbf{a}_k$, $k = 1, 2, ..., K$, are the column vectors of $\mathbf{A}$. Then the set of linear equations (2.175) can be written as

$$\mathbf{b} = \sum_{k=1}^{K} \theta_k \mathbf{a}_k. \tag{2.176}$$

Usually, (2.175) has no exact solution because $\mathbf{b}$ is not ordinarily located in the column space of $\mathbf{A}$ [6, p. 221], i.e. $\mathbf{b}$ cannot be expressed as a linear combination of $\mathbf{a}_k$, $k = 1, 2, ..., K$, for any $\boldsymbol{\theta}$ (see (2.176)). The column space of $\mathbf{A}$ is often referred to as the *range space* of $\mathbf{A}$, whose dimension is equal to $\mathrm{rank}\{\mathbf{A}\}$. On the other hand, when $\mathbf{b} = \mathbf{0}$ (i.e. $\mathbf{A}\boldsymbol{\theta} = \mathbf{0}$), the corresponding set of solutions spans another subspace, referred to as the *null space* of $\mathbf{A}$. The dimension of the nullspace of $\mathbf{A}$, called the *nullity* of $\mathbf{A}$, is equal to $K - \mathrm{rank}\{\mathbf{A}\}$.

In practical applications, however, an approximate solution to (2.175) is still desired. Hence, let us change the original problem into the following approximation problem:

$$\mathbf{A}\boldsymbol{\theta} = \mathbf{b} - \boldsymbol{\varepsilon} \tag{2.177}$$

where

$$\boldsymbol{\varepsilon} = \mathbf{b} - \mathbf{A}\boldsymbol{\theta} = \mathbf{b} - \sum_{k=1}^{K} \theta_k \mathbf{a}_k \tag{2.178}$$

is the $M \times 1$ vector of approximation errors (equation errors). For the approximation problem, a widely used approach is to find $\boldsymbol{\theta}$ such that

$$\widehat{\mathbf{b}} = \mathbf{A}\boldsymbol{\theta} = \sum_{k=1}^{K} \theta_k \mathbf{a}_k \tag{2.179}$$

Table 2.4 BFGS and approximate BFGS methods

	Update Equation for the BFGS Method
Generic form	At iteration i, update the parameter vector $\boldsymbol{\vartheta}$ via $$\boldsymbol{\vartheta}^{[i+1]} = \boldsymbol{\vartheta}^{[i]} - \mu^{[i]}\mathbf{Q}^{[i]}\boldsymbol{\nabla} J(\boldsymbol{\vartheta}^{[i]})$$ and update the Hermitian matrix $\mathbf{Q}^{[i]}$ via (2.157) where $\mu^{[i]} > 0$ is the step size, $\boldsymbol{\nabla} J(\boldsymbol{\vartheta}^{[i]})$ is the gradient at $\boldsymbol{\vartheta} = \boldsymbol{\vartheta}^{[i]}$, and $\mathbf{Q}^{[0]} = \mathbf{I}$ is suggested. In the update equation (2.157), the parameters α and β_i are given by (2.123) and (2.158), respectively, and the vectors $\mathbf{r}_{i+1}$ and $\mathbf{s}_{i+1}$ are given by (2.141) and (2.142), respectively.
Real case	At iteration i, update the real parameter vector $\boldsymbol{\theta}$ via $$\boldsymbol{\theta}^{[i+1]} = \boldsymbol{\theta}^{[i]} - \mu^{[i]}\mathbf{Q}^{[i]} \cdot \left.\frac{\partial J(\boldsymbol{\theta})}{\partial \boldsymbol{\theta}}\right\|_{\boldsymbol{\theta} = \boldsymbol{\theta}^{[i]}}$$ and update the symmetric matrix $\mathbf{Q}^{[i]}$ via (2.160), in which β_i, $\mathbf{r}_{i+1}$ and $\mathbf{s}_{i+1}$ are given by (2.161), (2.162) and (2.163), respectively.
Complex case	At iteration i, update the complex parameter vector $\boldsymbol{\theta}$ via $$\boldsymbol{\theta}^{[i+1]} = \boldsymbol{\theta}^{[i]} - \mu^{[i]}\left\{\mathbf{Q}_A^{[i]} \cdot \left.\frac{\partial J(\boldsymbol{\theta})}{\partial \boldsymbol{\theta}^*}\right\|_{\boldsymbol{\theta} = \boldsymbol{\theta}^{[i]}} + \mathbf{Q}_B^{[i]} \cdot \left.\frac{\partial J(\boldsymbol{\theta})}{\partial \boldsymbol{\theta}}\right\|_{\boldsymbol{\theta} = \boldsymbol{\theta}^{[i]}}\right\}$$ and update the Hermitian matrix $\mathbf{Q}_A^{[i]}$ and the matrix $\mathbf{Q}_B^{[i]}$ via (2.169) and (2.170), in which $\mathbf{Q}_A^{[0]} = \mathbf{I}$, $\mathbf{Q}_B^{[0]} = \mathbf{0}$, and β_i, $\widetilde{\mathbf{r}}_{i+1}$ and $\widetilde{\mathbf{s}}_{i+1}$ are given by (2.171), (2.165) and (2.166), respectively.

	Update Equation for the Approximate BFGS Method
Complex case	At iteration i, update the complex parameter vector $\boldsymbol{\theta}$ via $$\boldsymbol{\theta}^{[i+1]} = \boldsymbol{\theta}^{[i]} - \mu^{[i]}\mathbf{Q}_A^{[i]} \cdot \left.\frac{\partial J(\boldsymbol{\theta})}{\partial \boldsymbol{\theta}^*}\right\|_{\boldsymbol{\theta} = \boldsymbol{\theta}^{[i]}}$$ and update the matrix $\mathbf{Q}_A^{[i]}$ via (2.173), in which $\mathbf{Q}_A^{[0]} = \mathbf{I}$ and β_i, $\widetilde{\mathbf{r}}_{i+1}$ and $\widetilde{\mathbf{s}}_{i+1}$ are given by (2.174), (2.165) and (2.166), respectively.

approximates $\mathbf{b}$ in the sense of minimizing the objective function

$$J_{\mathrm{LS}}(\boldsymbol{\theta}) = \|\boldsymbol{\varepsilon}\|^2 = \sum_{m=1}^{M} |\varepsilon_m|^2 \tag{2.180}$$

where ε_m is the mth entry of $\boldsymbol{\varepsilon}$. The problem of minimizing the sum of squared errors given by (2.180) is called the *least-squares (LS) problem* and the corresponding solution is called the *least-squares (LS) solution.*

2.4.1 Full-Rank Overdetermined Least-Squares Problem

Consider that $M \geq K$ and $\mathbf{A}$ is of full rank, i.e. $\mathrm{rank}\{\mathbf{A}\} = K$. The LS solution is derived as follows. Taking the first derivative of $J_{\mathrm{LS}}(\boldsymbol{\theta})$ given by (2.180) with respect to $\boldsymbol{\theta}^*$ and setting the result to zero yields

$$\frac{\partial J_{\mathrm{LS}}(\boldsymbol{\theta})}{\partial \boldsymbol{\theta}^*} = -\mathbf{A}^H (\mathbf{b} - \mathbf{A}\boldsymbol{\theta}) = -\mathbf{A}^H \boldsymbol{\varepsilon} = \mathbf{0}, \tag{2.181}$$

which gives rise to

$$\mathbf{A}^H \mathbf{A} \boldsymbol{\theta} = \mathbf{A}^H \mathbf{b}. \tag{2.182}$$

From (2.181), it follows that

$$\mathbf{a}_k^H \boldsymbol{\varepsilon} = 0 \qquad \text{for } k = 1, 2, ..., K. \tag{2.183}$$

That is, the error vector $\boldsymbol{\varepsilon}$ is orthogonal to the column vectors $\mathbf{a}_k$, thereby leading to the name "*normal equations*" for the set of equations (2.182) [2]. As illustrated in Fig. 2.9 (by (2.178) and (2.179)), $\boldsymbol{\varepsilon}$ has the minimum norm only when it is orthogonal (perpendicular) to the range space of $\mathbf{A}$ (the plane). This observation indicates that the solution obtained from (2.182) corresponds to the global minimum of $J_{\mathrm{LS}}(\boldsymbol{\theta})$.

Since $\mathbf{A}$ is of full rank, $\mathbf{A}^H \mathbf{A}$ is a nonsingular $K \times K$ matrix and thus there is only a unique solution to (2.182) as follows:

$$\widehat{\boldsymbol{\theta}}_{\mathrm{LS}} = (\mathbf{A}^H \mathbf{A})^{-1} \mathbf{A}^H \mathbf{b} \tag{2.184}$$

where $\widehat{\boldsymbol{\theta}}_{\mathrm{LS}}$ represents the LS solution for $\boldsymbol{\theta}$. Substituting (2.184) into (2.179) gives

$$\widehat{\mathbf{b}} = \mathbf{P}_A \mathbf{b} \tag{2.185}$$

where

$$\mathbf{P}_A = \mathbf{A}(\mathbf{A}^H \mathbf{A})^{-1} \mathbf{A}^H \tag{2.186}$$

is an $M \times M$ matrix. From Fig. 2.9, one can observe that $\widehat{\mathbf{b}}$ corresponds to the projection of $\mathbf{b}$ onto the range space of $\mathbf{A}$. For this reason, $\mathbf{P}_A$ is called the *projection matrix* of $\mathbf{A}$. It has the following properties.

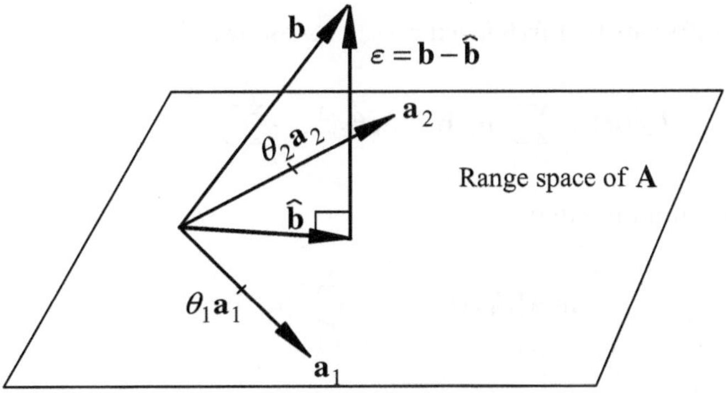

Fig. 2.9 Geometrical explanation of the LS method for $K = 2$

- Idempotent property: $\mathbf{P}_A \mathbf{P}_A = \mathbf{P}_A$.
- Hermitian property: $\mathbf{P}_A^H = \mathbf{P}_A$.

The idempotent property implies that $\mathbf{P}_A \widehat{\mathbf{b}} = \widehat{\mathbf{b}}$, i.e. the projection of $\widehat{\mathbf{b}}$ onto the range space is still $\widehat{\mathbf{b}}$. This can also be observed from Fig. 2.9 where $\widehat{\mathbf{b}}$ is already in the range space. When $M = K$, (2.184) reduces to

$$\widehat{\boldsymbol{\theta}}_{\text{LS}} = \mathbf{A}^{-1}\mathbf{b} \tag{2.187}$$

and the corresponding $\mathbf{P}_A = \mathbf{I}$. That is, there is no need for any projection because $\mathbf{b}$ is already in the range space for this case.

2.4.2 Generic Least-Squares Problem

Now consider the general case that M can be less than K and $\text{rank}\{\mathbf{A}\} = r \le \min\{M, K\}$, i.e. $\mathbf{A}$ can be rank deficient. The SVD of $\mathbf{A}$ is given by

$$\mathbf{A} = \mathbf{U}\boldsymbol{\Sigma}\mathbf{V}^H = \mathbf{U}\begin{pmatrix} \boldsymbol{\Lambda} & \mathbf{0} \\ \mathbf{0} & \mathbf{0} \end{pmatrix}\mathbf{V}^H \tag{2.188}$$

where $\mathbf{U} = (\mathbf{u}_1, \mathbf{u}_2, ..., \mathbf{u}_M)$ and $\mathbf{V} = (\mathbf{v}_1, \mathbf{v}_2, ..., \mathbf{v}_K)$ are $M \times M$ and $K \times K$ unitary matrices, respectively, and $\boldsymbol{\Lambda} = \text{diag}\{\lambda_1, \lambda_2, ..., \lambda_r\}$. The vectors $\mathbf{u}_k$ and $\mathbf{v}_k$ are the left and right singular vectors of $\mathbf{A}$, respectively, and λ_1, λ_2, ..., λ_r are the real positive singular values. By (2.178), (2.180) and (2.188),

$$\begin{aligned} J_{\text{LS}}(\boldsymbol{\theta}) &= \|\mathbf{b} - \mathbf{A}\boldsymbol{\theta}\|^2 = \|\mathbf{U}^H(\mathbf{b} - \mathbf{A}\boldsymbol{\theta})\|^2 \\ &= \|\mathbf{U}^H\mathbf{b} - \mathbf{U}^H\mathbf{A}\mathbf{V}\mathbf{V}^H\boldsymbol{\theta}\|^2 = \|\mathbf{U}^H\mathbf{b} - \boldsymbol{\Sigma}\widetilde{\boldsymbol{\theta}}\|^2 \end{aligned} \tag{2.189}$$

where

$$\widetilde{\boldsymbol{\theta}} = (\widetilde{\theta}_1, \widetilde{\theta}_2, ..., \widetilde{\theta}_K)^T \triangleq \mathbf{V}^H\boldsymbol{\theta}. \tag{2.190}$$

Equation (2.189) can be further expressed as follows:

$$J_{\mathrm{LS}}(\boldsymbol{\theta}) = \sum_{k=1}^{r} |\mathbf{u}_k^H \mathbf{b} - \lambda_k \widetilde{\theta}_k|^2 + \sum_{k=r+1}^{M} |\mathbf{u}_k^H \mathbf{b}|^2. \tag{2.191}$$

Clearly, the minimum value

$$\min\{J_{\mathrm{LS}}(\boldsymbol{\theta})\} = \sum_{k=r+1}^{M} |\mathbf{u}_k^H \mathbf{b}|^2$$

is attained when the first r entries of $\widetilde{\boldsymbol{\theta}}$ satisfy

$$\widetilde{\theta}_k = \frac{\mathbf{u}_k^H \mathbf{b}}{\lambda_k} \quad \text{for } k = 1, 2, ..., r, \tag{2.192}$$

regardless of what the remaining entries $\widetilde{\theta}_k$, $k = r+1, r+2, ..., K$, are. This indicates that there are infinitely many solutions to the generic LS problem.

Among these solutions, the LS solution $\widehat{\boldsymbol{\theta}}_{\mathrm{LS}}$ is always chosen as the one with the minimum norm. It is therefore also referred to as the *minimum-norm solution*. Because $\|\widetilde{\boldsymbol{\theta}}\|^2 = \boldsymbol{\theta}^H \mathbf{V}\mathbf{V}^H \boldsymbol{\theta} = \|\boldsymbol{\theta}\|^2$, the minimum-norm solution $\widehat{\boldsymbol{\theta}}_{\mathrm{LS}}$ corresponds to letting $\widetilde{\theta}_k = 0$ for $k = r+1, r+2, ..., K$. This, together with (2.190) and (2.192), therefore gives

$$\widehat{\boldsymbol{\theta}}_{\mathrm{LS}} = \mathbf{V}\widetilde{\boldsymbol{\theta}} = \sum_{k=1}^{K} \mathbf{v}_k \widetilde{\theta}_k = \sum_{k=1}^{r} \mathbf{v}_k \left(\frac{\mathbf{u}_k^H \mathbf{b}}{\lambda_k} \right). \tag{2.193}$$

The solution given by (2.193) is also equivalent to the form

$$\widehat{\boldsymbol{\theta}}_{\mathrm{LS}} = \mathbf{A}^+ \mathbf{b} \tag{2.194}$$

where

$$\mathbf{A}^+ = \sum_{k=1}^{r} \frac{1}{\lambda_k} \mathbf{v}_k \mathbf{u}_k^H = \mathbf{V}\boldsymbol{\Sigma}^+ \mathbf{U}^H \tag{2.195}$$

is a $K \times M$ matrix in which

$$\boldsymbol{\Sigma}^+ = \begin{pmatrix} \boldsymbol{\Lambda}^{-1} & \mathbf{0} \\ \mathbf{0} & \mathbf{0} \end{pmatrix} \tag{2.196}$$

is also a $K \times M$ matrix. The matrix $\mathbf{A}^+$ is called the *Moore–Penrose generalized inverse* or the *pseudoinverse* of $\mathbf{A}$. By substituting (2.194) into (2.179), we also obtain $\widehat{\mathbf{b}}$ as given by (2.185) with the generic form of projection matrix

$$\mathbf{P}_A = \mathbf{A}\mathbf{A}^+. \tag{2.197}$$

Table 2.5 gives a summary of the LS method. When $\mathbf{A}$ is of full rank and $M \geq K$ (i.e. the full-rank overdetermined LS problem), the pseudoinverse $\mathbf{A}^{+} = (\mathbf{A}^H\mathbf{A})^{-1}\mathbf{A}^H$ (Problem 2.21), and thus the LS solution given by (2.194) reduces to the one given by (2.184). Nevertheless, if computational complexity is not of major concern, it is preferred to use (2.194) to obtain the LS solution due to the better numerical properties of SVD. On the other hand, for the case of $\mathbf{A}$ having a special structure such as the Toeplitz structure, it may be better to use other algorithms that take advantage of the special structure to solve the LS problem.

Table 2.5 Least-squares (LS) method

Problem	Find a $K \times 1$ vector $\boldsymbol{\theta}$ to solve the set of linear equations $$\mathbf{A}\boldsymbol{\theta} = \mathbf{b}$$ by minimizing the sum of squared errors $$J_{\text{LS}}(\boldsymbol{\theta}) = \|\boldsymbol{\varepsilon}\|^2$$ where $\mathbf{A}$ is an $M \times K$ matrix with the SVD $\mathbf{A} = \mathbf{U}\boldsymbol{\Sigma}\mathbf{V}^H$ and $\boldsymbol{\varepsilon} = \mathbf{b} - \mathbf{A}\boldsymbol{\theta}$ is the error vector.
Generic solution	The (minimum-norm) LS solution $$\widehat{\boldsymbol{\theta}}_{\text{LS}} = \mathbf{A}^{+}\mathbf{b}$$ where $\mathbf{A}^{+}$ is the pseudoinverse of $\mathbf{A}$ given by $$\mathbf{A}^{+} = \mathbf{V}\boldsymbol{\Sigma}^{+}\mathbf{U}^H = \sum_{k=1}^{r} \frac{1}{\lambda_k}\mathbf{v}_k\mathbf{u}_k^H.$$
Special cases	(i) $M \geq K$ and $\text{rank}\{\mathbf{A}\} = K$: $$\widehat{\boldsymbol{\theta}}_{\text{LS}} = (\mathbf{A}^H\mathbf{A})^{-1}\mathbf{A}^H\mathbf{b}$$ (ii) $M = K$ and $\text{rank}\{\mathbf{A}\} = K$: $$\widehat{\boldsymbol{\theta}}_{\text{LS}} = \mathbf{A}^{-1}\mathbf{b}$$

2.5 Summary

We have reviewed the definitions of vectors, vector spaces, matrices, and some special forms of matrices. Several useful formulas and properties of matrices as well as matrix decomposition including eigendecomposition and the SVD were described. The SVD was then applied to the derivation of a minimum-norm solution to the generic LS problem. Regarding the mathematical analysis, we have dealt with the convergence of sequences and series including the Fourier series, as well as sequence and function spaces. As for the optimization theory, we have introduced the necessary and sufficient conditions for solutions, carefully dealt with the first derivative of the objective function with respect to a complex vector, and provided an overview of gradient-type optimization methods. Three popular gradient-type methods were introduced in terms of a complex-valued framework since they are often applied to blind equalization problems. Vector differentiation was then applied to find the solution to the full-rank overdetermined LS problem.

Appendix 2A
Proof of Theorem 2.15

The theorem can be proved by either of the following two approaches.

Approach I: Derivation from the Matrix $\mathbf{A}^H\mathbf{A}$

According to Properties 2.13 and 2.11, the eigenvalues of the $K \times K$ matrix $\mathbf{A}^H\mathbf{A}$ are all real nonnegative. Therefore, let λ_1, λ_2, ..., λ_K be nonnegative real numbers, and λ_1^2, λ_2^2, ..., λ_K^2 be the eigenvalues of $\mathbf{A}^H\mathbf{A}$. Furthermore, let these eigenvalues be arranged in the following order:

$$\lambda_1^2 \geq \lambda_2^2 \geq \cdots \geq \lambda_r^2 > 0$$

and

$$\lambda_{r+1}^2 = \lambda_{r+2}^2 = \cdots = \lambda_K^2 = 0 \tag{2.198}$$

where the second equation follows from the fact that $\text{rank}\{\mathbf{A}^H\mathbf{A}\} = \text{rank}\{\mathbf{A}\} = r$. Let $\mathbf{v}_1, \mathbf{v}_2, ..., \mathbf{v}_K$ be the orthonormal eigenvectors of $\mathbf{A}^H\mathbf{A}$ corresponding to the eigenvalues λ_1^2, λ_2^2, ..., λ_K^2, respectively. That is,

$$\mathbf{A}^H\mathbf{A}\mathbf{v}_i = \lambda_i^2\mathbf{v}_i, \quad i = 1, 2, ..., K \tag{2.199}$$

and

$$\mathbf{v}_i^H\mathbf{v}_j = \begin{cases} 1, & \text{for } i = j, \\ 0, & \text{for } i \neq j. \end{cases} \tag{2.200}$$

Let $\mathbf{V} = (\mathbf{V}_1 \ \mathbf{V}_2)$ be a $K \times K$ matrix where $\mathbf{V}_1 = (\mathbf{v}_1, \mathbf{v}_2, ..., \mathbf{v}_r)$ is a $K \times r$ matrix and $\mathbf{V}_2 = (\mathbf{v}_{r+1}, \mathbf{v}_{r+2}, ..., \mathbf{v}_K)$ is a $K \times (K - r)$ matrix. Then, from (2.200), it follows that

$$\mathbf{V}^H\mathbf{V} = \begin{pmatrix} \mathbf{V}_1^H \\ \mathbf{V}_2^H \end{pmatrix} (\mathbf{V}_1 \ \mathbf{V}_2) = \begin{pmatrix} \mathbf{V}_1^H\mathbf{V}_1 & \mathbf{V}_1^H\mathbf{V}_2 \\ \mathbf{V}_2^H\mathbf{V}_1 & \mathbf{V}_2^H\mathbf{V}_2 \end{pmatrix} = \mathbf{I},$$

i.e. $\mathbf{V}$ is a unitary matrix. By (2.198) and (2.199), we have

$$\begin{aligned}\mathbf{A}^H\mathbf{A}\mathbf{V}_2 &= (\mathbf{A}^H\mathbf{A}\mathbf{v}_{r+1}, \mathbf{A}^H\mathbf{A}\mathbf{v}_{r+2}, ..., \mathbf{A}^H\mathbf{A}\mathbf{v}_K) \\ &= (\lambda_{r+1}^2\mathbf{v}_{r+1}, \lambda_{r+2}^2\mathbf{v}_{r+2}, ..., \lambda_K^2\mathbf{v}_K) = \mathbf{0},\end{aligned}$$

implying that

$$(\mathbf{A}\mathbf{V}_2)^H(\mathbf{A}\mathbf{V}_2) = \mathbf{V}_2^H(\mathbf{A}^H\mathbf{A}\mathbf{V}_2) = \mathbf{0}$$

or

$$\mathbf{A}\mathbf{V}_2 = \mathbf{0}. \tag{2.201}$$

In the same way, by (2.199), we have

$$\mathbf{A}^H\mathbf{A}\mathbf{V}_1 = (\lambda_1^2\mathbf{v}_1, \lambda_2^2\mathbf{v}_2, ..., \lambda_r^2\mathbf{v}_r) = \mathbf{V}_1\mathbf{\Lambda}^2 \tag{2.202}$$

where $\mathbf{\Lambda}^2 = \text{diag}\{\lambda_1^2, \lambda_2^2, ..., \lambda_r^2\}$. Equation (2.202) gives rise to

$$\mathbf{V}_1^H\mathbf{A}^H\mathbf{A}\mathbf{V}_1 = \mathbf{V}_1^H\mathbf{V}_1\mathbf{\Lambda}^2 = \mathbf{\Lambda}^2,$$

implying that

$$(\mathbf{A}\mathbf{V}_1\mathbf{\Lambda}^{-1})^H(\mathbf{A}\mathbf{V}_1\mathbf{\Lambda}^{-1}) = \mathbf{\Lambda}^{-1}(\mathbf{V}_1^H\mathbf{A}^H\mathbf{A}\mathbf{V}_1)\mathbf{\Lambda}^{-1} = \mathbf{I} \tag{2.203}$$

where we have used the fact that $\mathbf{\Lambda}^{-H} = \mathbf{\Lambda}^{-1}$ since the λ_i are real. Let the $M \times r$ matrix $\mathbf{A}\mathbf{V}_1\mathbf{\Lambda}^{-1} = \mathbf{U}_1$, i.e.

$$\mathbf{U}_1 = (\mathbf{u}_1, \mathbf{u}_2, ..., \mathbf{u}_r) = \left(\frac{\mathbf{A}\mathbf{v}_1}{\lambda_1}, \frac{\mathbf{A}\mathbf{v}_2}{\lambda_2}, ..., \frac{\mathbf{A}\mathbf{v}_r}{\lambda_r} \right). \tag{2.204}$$

From (2.203), we obtain $\mathbf{U}_1^H\mathbf{U}_1 = \mathbf{I}$ which gives

$$\mathbf{U}_1^H(\mathbf{A}\mathbf{V}_1\mathbf{\Lambda}^{-1}) = \mathbf{I}$$

or

$$\mathbf{A}\mathbf{V}_1 = \mathbf{U}_1\mathbf{\Lambda}. \tag{2.205}$$

Choose an $M \times (M - r)$ matrix $\mathbf{U}_2 = (\mathbf{u}_{r+1}, \mathbf{u}_{r+2}, ..., \mathbf{u}_M)$ such that $\mathbf{U} = (\mathbf{U}_1 \ \mathbf{U}_2)$ is an $M \times M$ unitary matrix, i.e. $\mathbf{U}_2^H\mathbf{U}_2 = \mathbf{I}$, $\mathbf{U}_2^H\mathbf{U}_1 = \mathbf{0}$, and $\mathbf{U}_1^H\mathbf{U}_2 = \mathbf{0}$. Then

$$\mathbf{U}^H\mathbf{A}\mathbf{V} = \begin{pmatrix} \mathbf{U}_1^H \\ \mathbf{U}_2^H \end{pmatrix} \mathbf{A}(\mathbf{V}_1\ \mathbf{V}_2) = \begin{pmatrix} \mathbf{U}_1^H\mathbf{A}\mathbf{V}_1 & \mathbf{U}_1^H\mathbf{A}\mathbf{V}_2 \\ \mathbf{U}_2^H\mathbf{A}\mathbf{V}_1 & \mathbf{U}_2^H\mathbf{A}\mathbf{V}_2 \end{pmatrix}$$
$$= \begin{pmatrix} \mathbf{U}_1^H\mathbf{U}_1\mathbf{\Lambda} & \mathbf{0} \\ \mathbf{U}_2^H\mathbf{U}_1\mathbf{\Lambda} & \mathbf{0} \end{pmatrix} \quad \text{(by (2.201) and (2.205))}$$
$$= \begin{pmatrix} \mathbf{\Lambda} & \mathbf{0} \\ \mathbf{0} & \mathbf{0} \end{pmatrix} = \mathbf{\Sigma}. \tag{2.206}$$

This, together with the fact that both $\mathbf{U}$ and $\mathbf{V}$ are unitary, therefore proves the theorem.

Approach II: Derivation from the Matrix $\mathbf{A}\mathbf{A}^H$

According to Properties 2.13 and 2.11, the eigenvalues of the $M \times M$ matrix $\mathbf{A}\mathbf{A}^H$ are all real nonnegative. Therefore, let λ_1, λ_2, ..., λ_M be nonnegative real numbers, and λ_1^2, λ_2^2, ..., λ_M^2 be the eigenvalues of $\mathbf{A}\mathbf{A}^H$. Furthermore, let these eigenvalues be arranged in the following order:

$$\lambda_1^2 \geq \lambda_2^2 \geq \cdots \geq \lambda_r^2 > 0$$

and

$$\lambda_{r+1}^2 = \lambda_{r+2}^2 = \cdots = \lambda_M^2 = 0. \tag{2.207}$$

Let $\mathbf{u}_1$, $\mathbf{u}_2$, ..., $\mathbf{u}_M$ be the orthonormal eigenvectors of $\mathbf{A}\mathbf{A}^H$ corresponding to the eigenvalues λ_1^2, λ_2^2, ..., λ_M^2, respectively. That is,

$$\mathbf{A}\mathbf{A}^H\mathbf{u}_i = \lambda_i^2\mathbf{u}_i, \quad i = 1, 2, ..., M \tag{2.208}$$

and

$$\mathbf{u}_i^H\mathbf{u}_j = \begin{cases} 1, & \text{for } i = j, \\ 0, & \text{for } i \neq j. \end{cases} \tag{2.209}$$

Let $\mathbf{U} = (\mathbf{U}_1\ \mathbf{U}_2)$ be an $M \times M$ matrix where $\mathbf{U}_1 = (\mathbf{u}_1, \mathbf{u}_2, ..., \mathbf{u}_r)$ is an $M \times r$ matrix and $\mathbf{U}_2 = (\mathbf{u}_{r+1}, \mathbf{u}_{r+2}, ..., \mathbf{u}_M)$ is an $M \times (M - r)$ matrix. Then, from (2.209), it follows that

$$\mathbf{U}^H\mathbf{U} = \begin{pmatrix} \mathbf{U}_1^H \\ \mathbf{U}_2^H \end{pmatrix} (\mathbf{U}_1\ \mathbf{U}_2) = \begin{pmatrix} \mathbf{U}_1^H\mathbf{U}_1 & \mathbf{U}_1^H\mathbf{U}_2 \\ \mathbf{U}_2^H\mathbf{U}_1 & \mathbf{U}_2^H\mathbf{U}_2 \end{pmatrix} = \mathbf{I},$$

i.e. $\mathbf{U}$ is a unitary matrix. By (2.207) and (2.208), we have

$$\mathbf{A}\mathbf{A}^H\mathbf{U}_2 = (\lambda_{r+1}^2\mathbf{u}_{r+1}, \lambda_{r+2}^2\mathbf{u}_{r+2}, ..., \lambda_M^2\mathbf{u}_M) = \mathbf{0},$$

implying that

$$(\mathbf{U}_2^H\mathbf{A})(\mathbf{U}_2^H\mathbf{A})^H = \mathbf{U}_2^H(\mathbf{A}\mathbf{A}^H\mathbf{U}_2) = \mathbf{0}$$

or

$$\mathbf{U}_2^H\mathbf{A} = \mathbf{0}. \tag{2.210}$$

In the same way, by (2.208), we have

$$\mathbf{A}\mathbf{A}^H\mathbf{U}_1 = (\lambda_1^2\mathbf{u}_1, \lambda_2^2\mathbf{u}_2, ..., \lambda_r^2\mathbf{u}_r) = \mathbf{U}_1\mathbf{\Lambda}^2 \tag{2.211}$$

or

$$\mathbf{U}_1^H\mathbf{A}\mathbf{A}^H\mathbf{U}_1 = \mathbf{U}_1^H\mathbf{U}_1\mathbf{\Lambda}^2 = \mathbf{\Lambda}^2,$$

implying that

$$(\mathbf{A}^H\mathbf{U}_1\mathbf{\Lambda}^{-1})^H(\mathbf{A}^H\mathbf{U}_1\mathbf{\Lambda}^{-1}) = \mathbf{\Lambda}^{-1}(\mathbf{U}_1^H\mathbf{A}\mathbf{A}^H\mathbf{U}_1)\mathbf{\Lambda}^{-1} = \mathbf{I}. \tag{2.212}$$

Let the $K \times r$ matrix $\mathbf{A}^H\mathbf{U}_1\mathbf{\Lambda}^{-1} = \mathbf{V}_1$, i.e.

$$\mathbf{V}_1 = (\mathbf{v}_1, \mathbf{v}_2, ..., \mathbf{v}_r) = \left(\frac{\mathbf{A}^H\mathbf{u}_1}{\lambda_1}, \frac{\mathbf{A}^H\mathbf{u}_2}{\lambda_2}, ..., \frac{\mathbf{A}^H\mathbf{u}_r}{\lambda_r}\right). \tag{2.213}$$

From (2.212), we obtain $\mathbf{V}_1^H\mathbf{V}_1 = \mathbf{I}$, which gives

$$(\mathbf{A}^H\mathbf{U}_1\mathbf{\Lambda}^{-1})^H\mathbf{V}_1 = \mathbf{I}$$

or

$$\mathbf{U}_1^H\mathbf{A} = \mathbf{\Lambda}\mathbf{V}_1^H. \tag{2.214}$$

Choose a $K\times(K-r)$ matrix $\mathbf{V}_2 = (\mathbf{v}_{r+1}, \mathbf{v}_{r+2}, ..., \mathbf{v}_K)$ such that $\mathbf{V} = (\mathbf{V}_1\ \mathbf{V}_2)$ is a $K \times K$ unitary matrix, i.e. $\mathbf{V}_2^H\mathbf{V}_2 = \mathbf{I}$, $\mathbf{V}_2^H\mathbf{V}_1 = \mathbf{0}$, and $\mathbf{V}_1^H\mathbf{V}_2 = \mathbf{0}$. Then

$$\begin{aligned}\mathbf{U}^H\mathbf{A}\mathbf{V} &= \begin{pmatrix}\mathbf{U}_1^H\\ \mathbf{U}_2^H\end{pmatrix}\mathbf{A}(\mathbf{V}_1\ \mathbf{V}_2) = \begin{pmatrix}\mathbf{U}_1^H\mathbf{A}\mathbf{V}_1 & \mathbf{U}_1^H\mathbf{A}\mathbf{V}_2\\ \mathbf{U}_2^H\mathbf{A}\mathbf{V}_1 & \mathbf{U}_2^H\mathbf{A}\mathbf{V}_2\end{pmatrix}\\ &= \begin{pmatrix}\mathbf{\Lambda}\mathbf{V}_1^H\mathbf{V}_1 & \mathbf{\Lambda}\mathbf{V}_1^H\mathbf{V}_2\\ \mathbf{0} & \mathbf{0}\end{pmatrix} \quad \text{(by (2.210) and (2.214))}\\ &= \begin{pmatrix}\mathbf{\Lambda} & \mathbf{0}\\ \mathbf{0} & \mathbf{0}\end{pmatrix} = \mathbf{\Sigma}.\end{aligned} \tag{2.215}$$

This, together with the fact that both $\mathbf{U}$ and $\mathbf{V}$ are unitary, proves this theorem, too.

Q.E.D.

Appendix 2B
Some Terminologies of Functions

A *function* written as $y = f(x)$ is a rule that assigns to each element x in a set A one and only one element y in a set B. The set A is called the *domain* of $f(x)$, while the set B is called the *range* of $f(x)$. The symbol x representing an arbitrary element in A is called an *independent variable.* Some terminologies for $f(x)$ defined on an interval $[x_{\rm L}, x_{\rm U}]$ (the domain of $f(x)$) are as follows.

- A function $f(x)$ is said to be *even* (*odd*) if $f(-x) = f(x)$ $(f(-x) = -f(x))$ for all $x \in [x_{\rm L}, x_{\rm U}]$.
- A function $f(x)$ is said to be *periodic with period* T if $f(x + kT) = f(x)$ for all $x \in [x_{\rm L}, x_{\rm U}]$ and any nonzero integer k.
- A function $f(x)$ is said to be *bounded* if $|f(x)| \leq M$ for all $x \in [x_{\rm L}, x_{\rm U}]$ where M is a positive constant.
- A function $f(x)$ is said to be *increasing* (*decreasing*) or, briefly, *monotonic* if $f(x_0) \leq f(x_1)$ $(f(x_0) \geq f(x_1))$ for all x_0, $x_1 \in [x_{\rm L}, x_{\rm U}]$ and $x_0 < x_1$.
- A function $f(x)$ is said to be *strictly increasing* (*strictly decreasing*) if $f(x_0) < f(x_1)$ $(f(x_0) > f(x_1))$ for all x_0, $x_1 \in [x_{\rm L}, x_{\rm U}]$ and $x_0 < x_1$.

Continuity of Functions

A function $f(x)$ is said to be *continuous at a point* x_0 if $\lim_{x \to x_0} f(x) = f(x_0)$, i.e. for every real number $\varepsilon > 0$ there exists a real number $\Delta x > 0$ such that

$$|f(x) - f(x_0)| < \varepsilon \quad \text{whenever } 0 < |x - x_0| < \Delta x \tag{2.216}$$

where Δx is, in general, dependent on ε and x_0. Furthermore, define the *left-hand limit* of $f(x)$ at a point x_0 as

$$f(x_0^-) = \lim_{x \to x_0^-} f(x) = \lim_{\substack{x \to x_0 \\ x < x_0}} f(x) \tag{2.217}$$

and the *right-hand limit* of $f(x)$ at x_0 as

$$f(x_0^+) = \lim_{x \to x_0^+} f(x) = \lim_{\substack{x \to x_0 \\ x > x_0}} f(x). \tag{2.218}$$

Then $f(x)$ is continuous at x_0 if and only if [29, 33]

$$f(x_0^-) = f(x_0^+) = f(x_0). \tag{2.219}$$

On the other hand, a function $f(x)$ is said to be *discontinuous at a point* x_0 if it fails to be continuous at x_0.

A function $f(x)$ is said to be *continuous on an open interval* $(x_{\rm L}, x_{\rm U})$ if it is continuous at every point $x \in (x_{\rm L}, x_{\rm U})$. A function $f(x)$ is said to be *continuous on a closed interval* $[x_{\rm L}, x_{\rm U}]$ if it is continuous on $(x_{\rm L}, x_{\rm U})$ and, meanwhile, $f(x_L^+) = f(x_L)$ and $f(x_U^-) = f(x_U)$. Furthermore, as illustrated in Fig. 2.10, a function $f(x)$ is said to be *piecewise continuous on an interval* $[x_L, x_U]$ if there are at most a finite number of points $x_L = x_1 < x_2 < \cdots < x_n = x_U$ such that (i) $f(x)$ is continuous on each subinterval (x_k, x_{k+1}) for $k = 1, 2, ..., n-1$ and (ii) both $f(x_k^-)$ and $f(x_k^+)$ exist for $k = 1, 2, ..., n$ [13, 18, 34]. In a word, a piecewise continuous function has a finite number of finite discontinuities. Moreover, a continuous function is merely a special case of a piecewise continuous function.

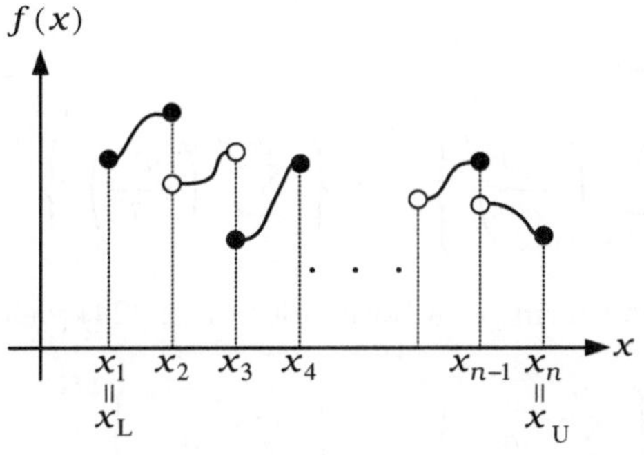

Fig. 2.10 A piecewise continuous function $f(x)$ on an interval $[x_L, x_U]$

Continuity of Derivatives

The *derivative* of a function $f(x)$ at a point x_0 is defined as

$$f'(x_0) = \left.\frac{df(x)}{dx}\right|_{x=x_0} = \lim_{\Delta x \to 0} \frac{f(x_0 + \Delta x) - f(x_0)}{\Delta x} \tag{2.220}$$

provided that the limit exists. Define the *left-hand derivative* of $f(x)$ at x_0 as

$$f'(x_0^-) = \lim_{\Delta x \to 0^-} \frac{f(x_0 + \Delta x) - f(x_0^-)}{\Delta x} \tag{2.221}$$

and the *right-hand derivative* of $f(x)$ at x_0 as

$$f'(x_0^+) = \lim_{\Delta x \to 0+} \frac{f(x_0 + \Delta x) - f(x_0^+)}{\Delta x}. \tag{2.222}$$

Then the derivative $f'(x)$ is said to be *piecewise continuous on an interval* $[x_L, x_U]$ if $f(x)$ is piecewise continuous on $[x_L, x_U]$ and, meanwhile, there are at most a finite number of points $x_L = x_1 < x_2 < \cdots < x_n = x_U$ such that (i) $f'(x)$ exists and is continuous on each subinterval (x_k, x_{k+1}) for $k = 1, 2, ..., n-1$ and (ii) both $f'(x_k^-)$ and $f'(x_k^+)$ exist for $k = 1, 2, ..., n$ [13, 18]. Note that if $f'(x)$ exists at a point x_0, then $f(x)$ is continuous at x_0.

Appendix 2C
Proof of Theorem 2.33

From the assumption that $\sum_{n=-\infty}^{\infty} |a_n|^s < \infty$, it follows that $|a_n|$ is bounded above. Let $\beta = \max\{|a_n|, n = -N \sim N\}$. Since $\max\{|a_n|/\beta, n = -N \sim N\} = 1$ and $l > s \geq 1$, one can easily infer that

$$1 \leq \sum_{n=-N}^{N} \left(\frac{|a_n|}{\beta}\right)^l \leq \sum_{n=-N}^{N} \left(\frac{|a_n|}{\beta}\right)^s, \tag{2.223}$$

which further leads to

$$\left\{\sum_{n=-N}^{N} \left(\frac{|a_n|}{\beta}\right)^l\right\}^{1/l} \leq \left\{\sum_{n=-N}^{N} \left(\frac{|a_n|}{\beta}\right)^s\right\}^{1/s}. \tag{2.224}$$

Canceling the common term β on both sides of (2.224) yields

$$\left\{\sum_{n=-N}^{N} |a_n|^l\right\}^{1/l} \leq \left\{\sum_{n=-N}^{N} |a_n|^s\right\}^{1/s}. \tag{2.225}$$

Since (2.225) holds for any N and $\left\{\sum_{n=-\infty}^{\infty} |a_n|^s\right\}^{1/s} < \infty$, letting $N \to \infty$ therefore gives (2.59). Thus, what remains to prove is the equality condition of (2.59).

Suppose that there are M terms of the sequence $\{a_n\}$ corresponding to $|a_n| = \beta$, and that $|a_n| < \beta$ for $n \in \Omega$ where Ω is a set of indices. It can be seen that the equality of (2.59) requires the equality of (2.223) and the equality of (2.224) for $N \to \infty$. From the equality of (2.223) for $N \to \infty$, we have

$$M + \sum_{n\in\Omega} \left(\frac{|a_n|}{\beta}\right)^l = M + \sum_{n\in\Omega} \left(\frac{|a_n|}{\beta}\right)^s,$$

implying that $|a_n| = 0$ for $n \in \Omega$. From this result and the equality of (2.224) for $N \to \infty$, we have

$$M^{1/l} = M^{1/s},$$

implying that $M = 1$. This therefore completes the proof.

Q.E.D.

Appendix 2D
Proof of Theorem 2.36

Since $s_n(x)$ is periodic with period 2π, substituting (2.76) into (2.77) yields

$$\begin{aligned} s_n(x) &= \frac{1}{2\pi}\int_{-\pi}^{\pi} f(x-t)D_n(t)dt = \frac{1}{2\pi}\int_{-\pi}^{\pi} f(x+t)D_n(t)dt \\ &= g_n(x) + \widetilde{g}_n(x) \end{aligned} \tag{2.226}$$

where

$$D_n(t) = \sum_{k=-n}^{n} e^{jkt} = \frac{\sin\left[(2n+1)t/2\right]}{\sin(t/2)} \tag{2.227}$$

is the so-called *nth Dirichlet kernel* [14] and

$$g_n(x) = \frac{1}{2\pi}\int_{0}^{\pi} f(x+t)D_n(t)dt, \tag{2.228}$$

$$\widetilde{g}_n(x) = \frac{1}{2\pi}\int_{-\pi}^{0} f(x+t)D_n(t)dt. \tag{2.229}$$

By expressing $D_n(t) = 1 + 2\sum_{k=1}^{n}\cos kt$, we obtain the integrations

$$\frac{1}{2\pi}\int_{0}^{\pi} D_n(t)dt = \frac{1}{2} \quad \text{and} \quad \frac{1}{2\pi}\int_{-\pi}^{0} D_n(t)dt = \frac{1}{2}. \tag{2.230}$$

Further express $g_n(x)$ given by (2.228) as

$$g_n(x) = \frac{1}{2\pi}\int_{0}^{\pi} f(x^+)D_n(t)dt + \frac{1}{2\pi}\int_{0}^{\pi} [f(x+t) - f(x^+)]D_n(t)dt$$

which, together with (2.227) and (2.230), gives

$$\begin{aligned} g_n(x) - \frac{f(x^+)}{2} &= \frac{1}{2\pi}\int_{-\pi}^{\pi} h(t)\sin\frac{(2n+1)t}{2}dt \\ &= \frac{1}{2\pi}\int_{-\pi}^{\pi} h_1(t)\sin(nt)dt + \frac{1}{2\pi}\int_{-\pi}^{\pi} h_2(t)\cos(nt)dt \end{aligned} \tag{2.231}$$

where

$$h(t) = \begin{cases} \dfrac{f(x+t) - f(x^+)}{\sin(t/2)}, & 0 \le t < \pi, \\ 0, & -\pi \le t < 0, \end{cases}$$

and $h_1(t) = h(t)\cos(t/2)$, $h_2(t) = h(t)\sin(t/2)$. By definition, the left-hand limit $h(0^-) = \lim_{t\to 0^-} h(t) = 0$, and the right-hand limit

$$h(0^+) = \lim_{t \to 0^+} h(t) = \lim_{t \to 0^+} \left[\frac{f(x+t) - f(x^+)}{t} \right] \cdot \left[\frac{t}{\sin(t/2)} \right] = 2f'(x^+)$$

exists since $f'(x^+)$ exists. From this and the condition that $f(x)$ is piecewise continuous on $[-\pi, \pi)$, it follows that $h(t)$ is piecewise continuous on $[-\pi, \pi)$ and, thus, square integrable on $[-\pi, \pi)$. In other words, $h(t)$ is in $\mathcal{L}^2[-\pi, \pi)$ and so are $h_1(t)$ and $h_2(t)$. Accordingly, the two terms in the second line of (2.231) are identical to zero as $n \to \infty$ (by Problem 2.16) and therefore

$$\lim_{n \to \infty} g_n(x) = \frac{f(x^+)}{2}. \tag{2.232}$$

In a similar way, by expressing $\widetilde{g}_n(x)$ given by (2.229) as

$$\widetilde{g}_n(x) = \frac{1}{2\pi} \int_{-\pi}^{0} f(x^-) D_n(t) dt + \frac{1}{2\pi} \int_{-\pi}^{0} [f(x+t) - f(x^-)] D_n(t) dt$$

and with the condition that $f'(x^-)$ exists and $f(x)$ is piecewise continuous on $[-\pi, \pi)$, we also have

$$\lim_{n \to \infty} \widetilde{g}_n(x) = \frac{f(x^-)}{2}. \tag{2.233}$$

Equation (2.78) then follows from (2.226), (2.232) and (2.233).

Q.E.D.

Appendix 2E
Proof of Theorem 2.38

Since $f'(x)$ is piecewise continuous on $[-\pi, \pi)$, it is integrable on $[-\pi, \pi)$ and has the Fourier series

$$f'(x) \sim \sum_{k=-\infty}^{\infty} \widetilde{c}_k e^{jkx}$$

where

$$\widetilde{c}_0 = \frac{1}{2\pi} \int_{-\pi}^{\pi} f'(x) dx = \frac{f(\pi) - f(-\pi)}{2\pi} = 0, \tag{2.234}$$

$$\begin{aligned} \widetilde{c}_k &= \frac{1}{2\pi} \int_{-\pi}^{\pi} f'(x) e^{-jkx} dx = \left. \frac{f(x) e^{-jkx}}{2\pi} \right|_{-\pi}^{\pi} + \frac{jk}{2\pi} \int_{-\pi}^{\pi} f(x) e^{-jkx} dx \\ &= jkc_k, \quad \text{for } |k| = 1 \sim \infty. \end{aligned} \tag{2.235}$$

By (2.235) and the Cauchy–Schwartz inequality (Theorem 2.32), we have

$$\sum_{k=-n}^{n} |c_k| = |c_0| + \sum_{k=-n,k\neq 0}^{n} |k|^{-1} \cdot |\widetilde{c}_k|$$
$$\leq |c_0| + \left\{ \sum_{k=-n,k\neq 0}^{n} |k|^{-2} \right\}^{1/2} \left\{ \sum_{k=-n,k\neq 0}^{n} |\widetilde{c}_k|^2 \right\}^{1/2}$$
$$= |c_0| + \sqrt{2} \left\{ \sum_{k=1}^{n} k^{-2} \right\}^{1/2} \left\{ \sum_{k=1}^{n} \left(|\widetilde{c}_k|^2 + |\widetilde{c}_{-k}|^2 \right) \right\}^{1/2}. \quad (2.236)$$

As shown in Example 2.26, the series $\sum_{k=1}^{\infty} k^{-2}$ converges, indicating that

$$\sum_{k=1}^{\infty} k^{-2} < \infty. \quad (2.237)$$

Moreover, since $f'(x)$ is piecewise continuous on $[-\pi, \pi)$, it is square integrable on $[-\pi, \pi)$ and therefore is in $\mathcal{L}^2[-\pi, \pi)$. Accordingly, by Bessel's inequality (2.69) and (2.234),

$$\sum_{k=-\infty,k\neq 0}^{\infty} |\widetilde{c}_k|^2 \leq \frac{1}{2\pi} \int_{-\pi}^{\pi} |f'(x)|^2 \, dx < \infty. \quad (2.238)$$

As a consequence of (2.236), (2.237) and (2.238), $\sum_{k=-\infty}^{\infty} |c_k| < \infty$ and, by Theorem 2.37, the Fourier series given by (2.75) is uniformly and absolutely convergent on $[-\pi, \pi)$. From this and by the pointwise convergence theorem, it then follows that the Fourier series given by (2.75) converges uniformly and absolutely to $f(x)$ since $f(x)$ is continuous on $[-\pi, \pi)$.

Q.E.D.

Appendix 2F
Proof of Theorem 2.46

According to Property 2.12, the proof is equivalent to showing that $\mathbf{P}^{[i+1]}$ is positive definite under the conditions that (i) both $\mathbf{P}^{[i]}$ and $\mathbf{Q}^{[i]}$ are positive definite, (ii) $\mathbf{r}_{i+1} \neq \mathbf{0}$, (iii) $\mathbf{s}_{i+1} \neq \mathbf{0}$, and (iv) $\mu^{[i]}$ is optimum for iteration i.

By (2.152), we can express the Hermitian form of $\mathbf{P}^{[i+1]}$ as follows:

$$\mathbf{x}^H \mathbf{P}^{[i+1]} \mathbf{x} = \mathbf{x}^H \mathbf{P}^{[i]} \mathbf{x} + \frac{\left|\mathbf{s}_{i+1}^H \mathbf{x}\right|^2}{\alpha \mathbf{s}_{i+1}^H \mathbf{r}_{i+1}} - \frac{\left|\mathbf{r}_{i+1}^H \mathbf{P}^{[i]} \mathbf{x}\right|^2}{\mathbf{r}_{i+1}^H \mathbf{P}^{[i]} \mathbf{r}_{i+1}}$$
$$= \frac{\left(\mathbf{x}^H \mathbf{P}^{[i]} \mathbf{x}\right)\left(\mathbf{r}_{i+1}^H \mathbf{P}^{[i]} \mathbf{r}_{i+1}\right) - \left|\mathbf{r}_{i+1}^H \mathbf{P}^{[i]} \mathbf{x}\right|^2}{\mathbf{r}_{i+1}^H \mathbf{P}^{[i]} \mathbf{r}_{i+1}} + \frac{\left|\mathbf{s}_{i+1}^H \mathbf{x}\right|^2}{\alpha \mathbf{s}_{i+1}^H \mathbf{r}_{i+1}} \quad (2.239)$$

for any $\mathbf{x} \neq \mathbf{0}$. Let the SVD of the Hermitian matrix $\mathbf{P}^{[i]}$ be written as

$$\mathbf{P}^{[i]} = \sum_{k=1}^{\widetilde{L}} \lambda_k \mathbf{u}_k \mathbf{u}_k^H \quad \text{(see (2.42))} \tag{2.240}$$

where

$$\widetilde{L} = \begin{cases} L & \text{for real } \boldsymbol{\theta}, \\ 2L & \text{for complex } \boldsymbol{\theta} \end{cases} \tag{2.241}$$

and $\mathbf{u}_k$ is the orthonormal eigenvector of $\mathbf{P}^{[i]}$ associated with the eigenvalue λ_k. Since $\mathbf{P}^{[i]}$ is positive definite, all the eigenvalues λ_k are (real) positive. Then, by the Cauchy–Schwartz inequality (Theorem 2.1), we have

$$\begin{aligned} \left|\mathbf{r}_{i+1}^H \mathbf{P}^{[i]} \mathbf{x}\right|^2 &= \left| \sum_{k=1}^{\widetilde{L}} \left(\sqrt{\lambda_k} \mathbf{r}_{i+1}^H \mathbf{u}_k\right) \left(\sqrt{\lambda_k} \mathbf{u}_k^H \mathbf{x}\right) \right|^2 \\ &\le \left\{ \sum_{k=1}^{\widetilde{L}} \lambda_k \left|\mathbf{r}_{i+1}^H \mathbf{u}_k\right|^2 \right\} \left\{ \sum_{k=1}^{\widetilde{L}} \lambda_k \left|\mathbf{u}_k^H \mathbf{x}\right|^2 \right\} \\ &= \left\{ \sum_{k=1}^{\widetilde{L}} \lambda_k \mathbf{r}_{i+1}^H \mathbf{u}_k \mathbf{u}_k^H \mathbf{r}_{i+1} \right\} \left\{ \sum_{k=1}^{\widetilde{L}} \lambda_k \mathbf{x}^H \mathbf{u}_k \mathbf{u}_k^H \mathbf{x} \right\} \\ &= \left(\mathbf{r}_{i+1}^H \mathbf{P}^{[i]} \mathbf{r}_{i+1}\right) \left(\mathbf{x}^H \mathbf{P}^{[i]} \mathbf{x}\right). \end{aligned} \tag{2.242}$$

From (2.239), (2.242), and the fact that $\mathbf{r}_{i+1}^H \mathbf{P}^{[i]} \mathbf{r}_{i+1} > 0$ (since $\mathbf{P}^{[i]}$ is positive definite and $\mathbf{r}_{i+1} \neq \mathbf{0}$), it follows that

$$\mathbf{x}^H \mathbf{P}^{[i+1]} \mathbf{x} \ge \frac{\left|\mathbf{s}_{i+1}^H \mathbf{x}\right|^2}{\alpha \mathbf{s}_{i+1}^H \mathbf{r}_{i+1}} \quad \text{for any } \mathbf{x} \neq \mathbf{0} \tag{2.243}$$

and the equality holds only when $\mathbf{x} = c \cdot \mathbf{r}_{i+1}$ for any nonzero scalar c.

On the other hand, since

$$\boldsymbol{\vartheta}^{[i+1]} = \boldsymbol{\vartheta}^{[i]} - \mu^{[i]} \mathbf{d}^{[i]} \tag{2.244}$$

where $\mathbf{d}^{[i]} = \mathbf{Q}^{[i]} \boldsymbol{\nabla} J(\boldsymbol{\vartheta}^{[i]})$, the necessary condition for the optimum $\mu^{[i]}$ can be derived by minimizing the objective function $f(\mu^{[i]}) \triangleq J(\boldsymbol{\vartheta}^{[i]} - \mu^{[i]} \mathbf{d}^{[i]})$. More specifically, by using the chain rule, we obtain

$$\frac{df(\mu^{[i]})}{d\mu^{[i]}} = \left[\frac{\partial J(\boldsymbol{\vartheta})}{\partial \boldsymbol{\vartheta}}\right]^H \Bigg|_{\boldsymbol{\vartheta} = \boldsymbol{\vartheta}^{[i+1]}} \cdot \frac{d\boldsymbol{\vartheta}^{[i+1]}}{d\mu^{[i]}} = -\left[\boldsymbol{\nabla} J(\boldsymbol{\vartheta}^{[i+1]})\right]^H \mathbf{d}^{[i]} = 0. \tag{2.245}$$

This result, together with (2.141), (2.142) and (2.244), therefore leads to

$$\mathbf{s}_{i+1}^H \mathbf{r}_{i+1} = \mu^{[i]} \left[\nabla J(\boldsymbol{\vartheta}^{[i]}) \right]^H \mathbf{d}^{[i]}$$
$$= \mu^{[i]} \left[\nabla J(\boldsymbol{\vartheta}^{[i]}) \right]^H \mathbf{Q}^{[i]} \nabla J(\boldsymbol{\vartheta}^{[i]}) > 0 \qquad (2.246)$$

since $\mu^{[i]} > 0$, $\mathbf{Q}^{[i]}$ is positive definite and $\nabla J(\boldsymbol{\vartheta}^{[i]}) \neq \mathbf{0}$ before convergence.

As a consequence of (2.243) and (2.246),

$$\mathbf{x}^H \mathbf{P}^{[i+1]} \mathbf{x} \geq \frac{\left| \mathbf{s}_{i+1}^H \mathbf{x} \right|^2}{\alpha \mathbf{s}_{i+1}^H \mathbf{r}_{i+1}} \geq 0 \quad \text{for any } \mathbf{x} \neq \mathbf{0}. \qquad (2.247)$$

The first equality of (2.247) holds only when $\mathbf{x} = c \cdot \mathbf{r}_{i+1}$ for any $c \neq 0$, while the second equality of (2.247) holds only when $\mathbf{s}_{i+1}^H \mathbf{x} = 0$. In other words, for any $\mathbf{x} \neq \mathbf{0}$, $\mathbf{x}^H \mathbf{P}^{[i+1]} \mathbf{x} = 0$ happens only when $\mathbf{s}_{i+1}^H (c \cdot \mathbf{r}_{i+1}) = 0$, that contradicts (2.246). As a result, the Hermitian form $\mathbf{x}^H \mathbf{P}^{[i+1]} \mathbf{x} > 0$ for any $\mathbf{x} \neq \mathbf{0}$ and accordingly $\mathbf{P}^{[i+1]}$ is positive definite.

Q.E.D.

Problems

2.1. Prove Theorem 2.1.

2.2. Prove Theorem 2.2. (*Hint*: Use the Cauchy–Schwartz inequality.)

2.3. Prove Theorem 2.5.

2.4. Prove Theorem 2.7. (*Hint*: Express **A** as a multiplication of a lower triangular matrix and an upper triangular matrix.)

2.5. Prove Theorem 2.8.

2.6. Prove Properties 2.9 to 2.12.

2.7. Prove Property 2.13.

2.8. Prove Property 2.14. (*Hint*: Use Property 2.10.)

2.9. (a) Find the eigenvalues and the normalized eigenvectors of the matrix

$$\mathbf{A} = \begin{pmatrix} 3 & 1 \\ 1 & 3 \end{pmatrix}.$$

(b) Use part (a) to find the eigendecomposition of the matrix **A**.

2.10. Prove Corollary 2.16.

2.11. Prove Theorem 2.17.

2.12. Prove Theorem 2.18.

2.13. Prove Theorem 2.21.

2.14. Prove Theorem 2.25.

2.15. Prove Theorem 2.30.

2.16. Suppose $\{\phi_n(x), n = -\infty \sim \infty\}$ is a set of real or complex orthogonal functions in $\mathcal{L}^2[x_\mathrm{L}, x_\mathrm{U}]$. Show that if $f(x)$ is a real or complex function in $\mathcal{L}^2[x_\mathrm{L}, x_\mathrm{U}]$, then

$$\lim_{|n|\to\infty} \int_{x_\mathrm{L}}^{x_\mathrm{U}} f(x)\phi_n^*(x)dx = 0.$$

(*Hint*: Use Bessel's inequality.)

2.17. Prove Theorem 2.37. (*Hint*: Use the Weierstrass M-test and Theorem 2.27.)

2.18. Prove Theorem 2.39. (*Hint*: Use Theorem 2.30 to show that the sequence $\{\sum_{k=-n}^{n} c_k e^{jkx}\}_{n=1}^{\infty}$ is a Cauchy sequence in $\mathcal{L}^2[-\pi, \pi)$.)

2.19. Prove Theorem 2.42.

2.20. Prove Theorem 2.44.

2.21. Show that if $\mathbf{A}$ is a full-rank $M \times K$ matrix and $M \geq K$, then its pseudoinverse $\mathbf{A}^+ = (\mathbf{A}^H\mathbf{A})^{-1}\mathbf{A}^H$.

2.22. Find the LS solution to the set of linear equations $\mathbf{A}\boldsymbol{\theta} = \mathbf{b}$ where

$$\mathbf{A} = \begin{pmatrix} 1 & 2 \\ 2 & -1 \\ 5 & 2 \\ 3 & -4 \end{pmatrix} \quad \text{and} \quad \mathbf{b} = \begin{pmatrix} 2 \\ -1 \\ 1 \\ 3 \end{pmatrix}.$$

2.23. Consider the set of linear equations $\mathbf{A}\boldsymbol{\theta} = \mathbf{b}$ where

$$\mathbf{A} = \begin{pmatrix} 0.6 & -1.6 & 0 \\ 0.8 & 1.2 & 0 \\ 0 & 0 & 0 \\ 0 & 0 & 0 \end{pmatrix} \quad \text{and} \quad \mathbf{b} = \begin{pmatrix} -0.5 \\ 2 \\ 0 \\ 0 \end{pmatrix}.$$

(a) Find the SVD of $\mathbf{A}$.
(b) Find the LS solution for $\boldsymbol{\theta}$.

Computer Assignments

2.1. Suppose $f(x)$ is a periodic function of period 2π and

$$f(x) = \begin{cases} 1, & |x| \leq \pi/2, \\ 0, & \pi/2 < |x| \leq \pi. \end{cases}$$

(a) Let $s_n(x)$ denote the nth partial sum of the Fourier series of $f(x)$. Find the Fourier series of $f(x)$ and the partial sum $\lim_{n\to\infty} s_n(x)$.

(b) Plot the partial sums $s_1(x)$, $s_3(x)$ and $s_{23}(x)$, and specify what phenomenon you observe.

References

1. R. A. Horn and C. R. Johnson, *Matrix Analysis.* New York: Cambridge University Press, 1990.
2. G. Strang, *Linear Algebra and Its Applications.* New York: Harcourt Brace Jovanovich, 1988.
3. S. H. Friedberg, A. J. Insel, and L. E. Spence, *Linear Algebra.* New Jersey: Prentice-Hall, 1989.
4. E. Kreyszig, *Advanced Engineering Mathematics.* New York: John Wiley & Sons, 1999.
5. H. Stark and Y. Yang, *Vector Space Projections: A Numerical Approach to Signal and Image Processing, Neural Nets, and Optics.* New York: John Wiley & Sons, 1998.
6. G. H. Golub and C. F. van Loan, *Matrix Computations.* London: The Johns Hopkins University Press, 1989.
7. S. M. Kay, *Fundamentals of Statistical Signal Processing: Estimation Theory.* New Jersey: Prentice-Hall, 1993.
8. S. Haykin, *Adaptive Filter Theory.* New Jersey: Prentice-Hall, 1996.
9. J. M. Mendel, *Lessons in Estimation Theory for Signal Processing, Communications, and Control.* New Jersey: Prentice-Hall, 1995.
10. V. C. Klema and A. J. Laub, "The singular value decomposition: Its computation and some applications," *IEEE Trans. Automatic Control*, vol. AC-25, no. 2, pp. 164–176, Apr. 1980.
11. D. Kincaid and W. Cheney, *Numerical Analysis: Mathematics of Scientific Computing.* California: Brooks/Cole, 1996.
12. K. G. Binmore, *Mathematical Analysis: A Straightforward Approach.* New York: Cambridge University Press, 1982.
13. M. Stoll, *Introduction to Real Analysis.* Boston: Addison Wesley Longman, 2001.
14. E. M. Stein and R. Shakarchi, *Fourier Analysis: An Introduction.* New Jersey: Princeton University Press, 2003.
15. R. G. Bartle, *The Elements of Real Analysis.* New York: John Wiley & Sons, 1970.

16. W. R. Wade, *An Introduction to Analysis.* New Jersey: Prentice-Hall, 1995.
17. H. L. Royden, *Real Analysis.* New Jersey: Prentice-Hall, 1988.
18. E. Haug and K. K. Choi, *Methods of Engineering Mathematics.* New Jersey: Prentice-Hall, 1993.
19. C.-Y. Chi and M.-C. Wu, "Inverse filter criteria for blind deconvolution and equalization using two cumulants," *Signal Processing*, vol. 43, no. 1, pp. 55–63, Apr. 1995.
20. G. H. Hardy, J. E. Littlewood, and G. Polya, *Inequalities.* London: Cambridge University Press, 1934.
21. A. V. Oppenheim and A. S. Willsky with I. T. Young, *Signals and Systems.* New Jersey: Prentice-Hall, 1983.
22. O. K. Ersoy, *Fourier-related Transforms, Fast Algorithms and Applications.* New Jersey: Prentice-Hall, 1997.
23. R. N. Bracewell, *The Fourier Transform and Its Applications.* Boston: McGraw-Hill, 2000.
24. M. J. Lighthill, *Introduction to Fourier Analysis and Generalised Functions.* Cambridge: Cambridge University Press, 1958.
25. D. G. Luenberger, *Linear and Nonlinear Programming.* California: Addison-Wesley Publishing, 1984.
26. S. S. Rao, *Engineering Optimization: Theory and Practice.* New York: John Wiley & Sons, 1996.
27. D. H. Brandwood, "A complex gradient operator and its application in adaptive array theory," *IEE Proc.*, vol. 130, pts. F and H, no. 1, pp. 11–16, Feb. 1983.
28. E. B. Saff and A. D. Snider, *Fundamentals of Complex Analysis for Mathematics, Science, and Engineering.* New Jersey: Prentice-Hall, 1993.
29. E. W. Swokowski, *Calculus with Analytic Geometry.* Boston: Prindle, Weber & Schmidt, 1983.
30. J. W. Bandler, "Optimization methods for computer-aided design," *IEEE Trans. Microwave Theory and Techniques*, vol. MTT-17, no. 8, pp. 533–552, Aug. 1969.
31. B. S. Gottfried and J. Weisman, *Introduction to Optimization Theory.* New Jersey: Prentice-Hall, 1973.
32. H. Y. Huang, "Unified approach to quadratically convergent algorithms for function minimization," *Journal of Optimization Theory and Applications*, vol. 5, pp. 405–423, 1970.
33. J. Stewart, *Calculus: Early Transcendentals.* New York: International Thomson, 1999.
34. M. R. Spiegel, *Theory and Problems of Fourier Analysis with Applications to Boundary Value Problems.* New York: McGraw-Hill, 1974.

3

Fundamentals of Statistical Signal Processing

This chapter deals with several fundamental topics of statistical signal processing, including discrete-time signals and systems, random variables, random processes, and estimation theory.

3.1 Discrete-Time Signals and Systems

We start by reviewing the fundamentals of 1-D SISO discrete-time signals and systems including the definitions, notation, time-domain and transform-domain characterization along with some useful transformation tools. The treatment of 1-D MIMO signals and systems and that of 2-D SISO signals and systems are left to Parts III and IV, respectively.

3.1.1 Time-Domain Characterization

Discrete-Time Signals

A *discrete-time signal*, denoted by $x[n]$, is a real or complex function defined only on discrete-time points $n = 0, \pm 1, \pm 2, ...$, and thus is commonly represented as a sequence of real or complex numbers. In practice, a discrete-time signal $x[n]$ is often derived from periodically sampling a corresponding *continuous-time signal*, denoted by $x(t)$, which is a function defined along a continuum of time. Specifically, the discrete-time signal $x[n]$ is obtained via

$$x[n] = x(t = nT), \quad n = 0, \pm 1, \pm 2, ... \tag{3.1}$$

where T is called the *sampling period*. Accordingly, the discrete-time index n is also referred to as the *sample number* of $x[n]$. A representative of discrete-time signals is the *unit sample sequence* defined as

$$\delta[n] = \begin{cases} 1, & n = 0, \\ 0, & n \neq 0. \end{cases} \tag{3.2}$$

It is also referred to as the *(discrete-time) impulse*, the *Kronecker delta function*, or the *delta function*. One can see that the Kronecker delta function $\delta[n]$ is well defined. It does not suffer from the mathematical complications as the Dirac delta function $\delta(t)$ (see (2.81) and (2.82)).

Discrete-Time Systems

A *discrete-time system*, also referred to as a *discrete-time filter*, is a system whose input and output are both discrete-time signals. The response of a discrete-time system to the impulse $\delta[n]$ is called the *impulse response* of the system. A discrete-time system is said to be *linear time-invariant (LTI)* or *linear shift-invariant (LSI)* if it is completely characterized by its impulse response. Specifically, let $x[n]$ and $y[n]$ be, respectively, the input and output of a discrete-time LTI system whose impulse response is denoted by $h[n]$. Then, for any input $x[n]$, the output $y[n]$ is completely determined by the discrete-time convolutional model

$$y[n] = h[n] \star x[n] = \sum_{k=-\infty}^{\infty} h[k]x[n-k] \tag{3.3}$$

where the notation '$\star$' represents *linear convolution* or briefly *convolution*. Similarly, a discrete-time LTI system $h[n]$ defined by (3.3) may be derived from a corresponding continuous-time LTI system, $h(t)$, defined by the continuous-time convolutional model

$$y(t) = h(t) \star x(t) = \int_{-\infty}^{\infty} h(\tau)x(t-\tau)d\tau \tag{3.4}$$

where $x(t)$ and $y(t)$ are the input and output of $h(t)$, respectively.

Moreover, let $x[n]$ and $y[n]$ be the input and output of an LTI system $h[n]$, respectively, and $h_1[n]$ and $h_2[n]$ are two other LTI systems. As depicted in Fig. 3.1, if $h[n] = h_1[n] \star h_2[n]$, then

$$y[n] = (h_1[n] \star h_2[n]) \star x[n] = h_1[n] \star (h_2[n] \star x[n]) = h_2[n] \star (h_1[n] \star x[n])$$

since the convolution operation is associative and commutative. On the other hand, as depicted in Fig. 3.2, if $h[n] = h_1[n] + h_2[n]$, then

$$y[n] = (h_1[n] + h_2[n]) \star x[n] = h_1[n] \star x[n] + h_2[n] \star x[n]$$

since the convolution operation is distributive.

Several terminologies for LTI systems are described as follows.

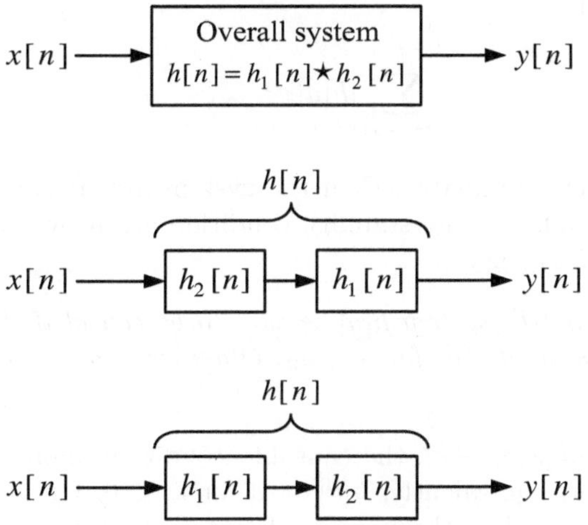

Fig. 3.1 Equivalence between an overall system $h[n]$ and cascaded connections of two systems $h_1[n]$ and $h_2[n]$

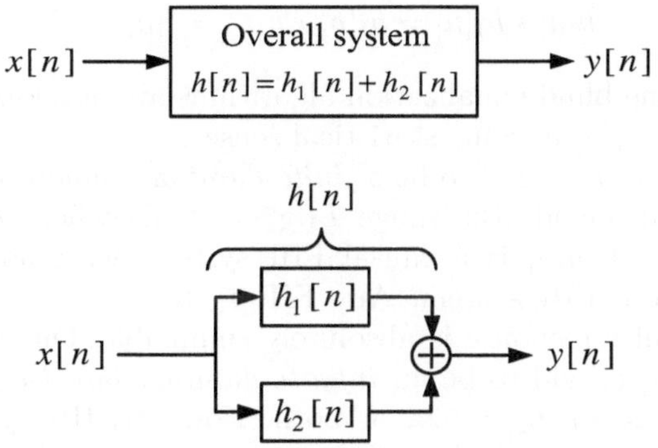

Fig. 3.2 Equivalence between an overall system $h[n]$ and a parallel connection of two systems $h_1[n]$ and $h_2[n]$

Definition 3.1. *An LTI system $h[n]$ is said to be stable in the bounded-input bounded-output (BIBO) sense or, briefly, stable if for every bounded input its output is also bounded. Otherwise, the system $h[n]$ is said to be unstable.*

According to this definition, for any bounded input $|x[n]| \le B_x < \infty$, the output of $h[n]$ is bounded by

$$|y[n]| = \left| \sum_{k=-\infty}^{\infty} h[k]x[n-k] \right| \le \sum_{k=-\infty}^{\infty} |h[k]| \cdot |x[n-k]| \le B_x \sum_{k=-\infty}^{\infty} |h[k]|.$$

This implies that if

$$\sum_{n=-\infty}^{\infty} |h[n]| < \infty, \tag{3.5}$$

i.e. $h[n]$ is absolutely summable, then the system $h[n]$ is stable. In fact, as reported in [1, pp. 30, 31], the stability condition given by (3.5) is not only sufficient but also necessary.

Definition 3.2. *An LTI system $h[n]$ is said to be causal if its output $y[n_0]$ depends only on its input $x[n]$ for $n \leq n_0$. Otherwise, the system $h[n]$ is said to be noncausal.*

Since $y[n_0] = \sum_k h[k]x[n_0 - k]$, the causality condition implies that $h[k] = 0$ for $n_0 - k > n_0$ or, equivalently, $h[k] = 0$ for $k < 0$. On the other hand, if $h[k] = 0$ for $k \geq 0$, then the noncausal system $h[n]$ is further said to be *anticausal* [2].

Definition 3.3. *An LTI system $h_{\mathrm{I}}[n]$ is called the inverse system of an LTI system $h[n]$ if*

$$h[n] \star h_{\mathrm{I}}[n] = h_{\mathrm{I}}[n] \star h[n] = \delta[n]. \tag{3.6}$$

As we will see, some blind equalization algorithms are developed to estimate the inverse system $h_{\mathrm{I}}[n]$ in some statistical sense.

An LTI system $h[n]$ is said to be a *finite-duration impulse response (FIR)* system if $h[n] = 0$ outside the range $L_1 \leq n \leq L_2$ where L_1 and L_2 are integers. As $L_1 = 0$, $h[n]$ is a causal FIR system; whereas as $L_2 = -1$, $h[n]$ is an anticausal FIR system. Any FIR system is always stable since, obviously, its impulse response is absolutely summable. On the other hand, an LTI system $h[n]$ is said to be an *infinite-duration impulse response (IIR)* system if $L_1 = -\infty$ or $L_2 = \infty$, or both. However, IIR systems may be unstable.

3.1.2 Transformation Tools

z-Transform

The *z-transform* of a discrete-time signal $x[n]$ is defined as

$$X(z) = \mathcal{Z}\{x[n]\} = \sum_{n=-\infty}^{\infty} x[n]z^{-n} \tag{3.7}$$

where $\mathcal{Z}\{\cdot\}$ stands for z-transform and z is a complex variable. From (3.7), it follows that

$$|X(z)| \leq \sum_{n=-\infty}^{\infty} |x[n]| \cdot |z|^{-n}, \tag{3.8}$$

which implies that the convergence of the series $\sum_{n=-\infty}^{\infty} x[n]z^{-n}$ depends on $|z|$. In other words, as illustrated in Fig. 3.3, the *region of convergence (ROC)* for the existence of $X(z)$ is typically a ring (the shaded region) on the complex z-plane where $z = re^{j\omega}$. If the ROC of $X(z)$ includes the unit circle, then $X(z)$ exists at $|z| = |e^{j\omega}| = 1$.

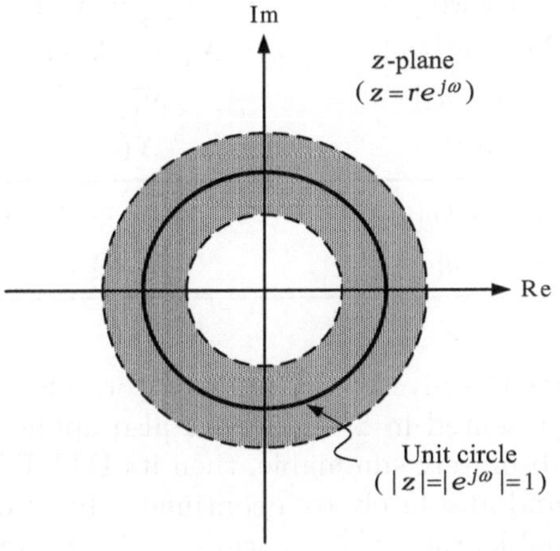

Fig. 3.3 A typical region of convergence

Table 3.1 lists some useful properties of the z-transform where the ROC should be determined accordingly. The proofs of these properties can be found, for example, in [1, pp. 119–126].

Discrete-Time Fourier Transform

The *discrete-time Fourier transform (DTFT)* of a discrete-time signal $x[n]$ is defined as

$$X(\omega) = \mathcal{F}\{x[n]\} = \sum_{n=-\infty}^{\infty} x[n]e^{-j\omega n} \tag{3.9}$$

where $\mathcal{F}\{\cdot\}$ denotes the DTFT operator. One can see that $X(\omega)$ is a continuous-time periodic function with period 2π. Furthermore, from (3.9) and (2.75), it follows that one can think of the DTFT as a Fourier series. As such, according to (2.76), the *inverse DTFT* of $X(\omega)$ is given by

$$x[n] = \mathcal{F}^{-1}\{X(\omega)\} = \frac{1}{2\pi}\int_{-\pi}^{\pi} X(\omega)e^{j\omega n}d\omega \tag{3.10}$$

Table 3.1 Properties of z-transform

Discrete-time Signal	z-Transform
$x[n]$	$X(z) = \mathcal{Z}\{x[n]\}$
$y[n]$	$Y(z) = \mathcal{Z}\{y[n]\}$
$x[-n]$	$X(1/z)$
$x^*[n]$	$X^*(z^*)$
Real $x[n]$	$X(z) = X^*(z^*)$
$x[n] = x^*[-n]$	$X(z) = X^*(1/z^*)$
$x[n-n_0]$	$X(z)z^{-n_0}$
$x[n]z_0^n$	$X(z/z_0)$
$ax[n] + by[n]$	$aX(z) + bY(z)$
$x[n] \star y[n]$	$X(z)Y(z)$

where $\mathcal{F}^{-1}\{\cdot\}$ denotes the inverse DTFT operator. The convergence analysis of Fourier series presented in Section 2.2.4 also applies to the DTFT. In particular, if $x[n]$ is absolutely summable, then its DTFT $\sum_{n=-\infty}^{\infty} x[n]e^{-j\omega n}$ converges uniformly and absolutely to a continuous function of ω on $[-\pi, \pi)$. If $x[n]$ is square summable, then $\sum_{n=-\infty}^{\infty} x[n]e^{-j\omega n}$ converges in the MS sense to a function in the $\mathcal{L}^2[-\pi, \pi)$ space. If $x[n] = 1$ for all n, then its DTFT

$$X(\omega) = \sum_{m=-\infty}^{\infty} 2\pi\delta(\omega + 2\pi m) \tag{3.11}$$

where $\delta(\omega)$ is the Dirac delta function of ω. Moreover, comparing (3.9) with (3.7), one can see that $X(\omega)$ corresponds to $X(z)$ evaluated at $z = e^{j\omega}$. This implies that if the ROC of $X(z)$ includes the unit circle, then $X(\omega) = X(z = e^{j\omega})$ exists.

Table 3.2 lists some useful properties of the DTFT, for which the proofs can be found in [1, pp. 55–62]. Note, from this table, that

$$X(\omega) \otimes Y(\omega) \triangleq \frac{1}{2\pi} \int_{-\pi}^{\pi} X(\theta)Y(\omega - \theta)d\theta \tag{3.12}$$

is called the *periodic convolution* of $X(\omega)$ and $Y(\omega)$ since both $X(\omega)$ and $Y(\omega)$ are periodic with period 2π. The notation '$\otimes$' represents periodic convolution. Parseval's relation given by (2.80) is restated for the DTFT as follows:

$$\sum_{n=-\infty}^{\infty} |x[n]|^2 = \frac{1}{2\pi} \int_{-\pi}^{\pi} |X(\omega)|^2 d\omega. \tag{3.13}$$

Table 3.2 Properties of DTFT

Discrete-time Signal	Discrete-time Fourier Transform
$x[n]$	$X(\omega) = \mathscr{F}\{x[n]\}$
$y[n]$	$Y(\omega) = \mathscr{F}\{y[n]\}$
$x[-n]$	$X(-\omega)$
$x^*[n]$	$X^*(-\omega)$
Real $x[n]$	$X(\omega) = X^*(-\omega)$
$x[n] = x^*[-n]$	Real $X(\omega)$
$x[n - n_0]$	$X(\omega)e^{-j\omega n_0}$
$x[n]e^{j\omega_0 n}$	$X(\omega - \omega_0)$
$ax[n] + by[n]$	$aX(\omega) + bY(\omega)$
$x[n] \star y[n]$	$X(\omega)Y(\omega)$
$x[n]y[n]$	$X(\omega) \otimes Y(\omega)$

Discrete Fourier Transform

While the DTFT and inverse DTFT are powerful in theoretical analyses and algorithm developments, they are not easily computed by means of digital computations due to the nature of continuous radian frequency ω. For this problem, let us assume that $x[n]$, $n = 0, 1, ..., N-1$, is a finite-duration sequence of length N. The DTFT of $x[n]$, $X(\omega)$, can be computed for $\omega = 2\pi k/N$, $k = 0, 1, ..., N-1$, via the following *discrete Fourier transform (DFT)*:

$$\begin{aligned} X[k] = \text{DFT}\{x[n]\} &= \sum_{n=0}^{N-1} x[n]e^{-j2\pi kn/N} \\ &= \sum_{n=0}^{N-1} x[n]W_N^{kn}, \quad 0 \le k \le N-1 \end{aligned} \tag{3.14}$$

where DFT$\{\cdot\}$ denotes the DFT operator and $W_N = e^{-j2\pi/N}$ is called the *twiddle factor*. Clearly, $X[k]$ always exists. The *inverse DFT* of the finite-duration sequence $X[k]$, $k = 0, 1, ..., N-1$, is given by

$$\begin{aligned} x[n] = \text{IDFT}\{X[k]\} &= \frac{1}{N}\sum_{k=0}^{N-1} X[k]e^{j2\pi kn/N} \\ &= \frac{1}{N}\sum_{k=0}^{N-1} X[k]W_N^{-kn}, \quad 0 \le n \le N-1 \end{aligned} \tag{3.15}$$

where IDFT$\{\cdot\}$ denotes the inverse DFT operator.

Table 3.3 lists some useful properties of the DFT, for which the proofs can be found in [1, pp. 546–551]. Note, from this table, that the notation '$((n))_N$' represents '(n modulo N)' and the notation '$\circledast$' represents the operation of *circular convolution* defined as

$$x[n] \circledast y[n] = \sum_{m=0}^{N-1} x[m]y[((n-m))_N], \quad 0 \le n \le N-1. \tag{3.16}$$

One can see, from (3.14) and (3.15), that computation of the DFT as well as that of the IDFT requires numbers of complex multiplications and complex additions in the order of N^2. This indicates that the computational complexity of the DFT and IDFT is extraordinarily large when N is large. In practice, many fast algorithms are available for efficiently computing the DFT and IDFT [1, 3]. They are collectively called *fast Fourier transform (FFT)* and *inverse fast Fourier transform (IFFT)* algorithms, respectively. The basic idea behind the FFT and IFFT is to take advantage of the following two properties of the twiddle factor W_N.

- Complex conjugate symmetry: $W_N^{-kn} = \left(W_N^{kn}\right)^*$.
- Periodicity: $W_N^{k(n+N)} = W_N^{(k+N)n} = W_N^{kn}$.

As a consequence, the FFT (IFFT) requires numbers of complex multiplications and complex additions in the order of $N \log_2 N$.

Table 3.3 Properties of DFT

Discrete-time Signal	Discrete Fourier Transform
$x[n]$	$X[k] = \text{DFT}\{x[n]\}$
$y[n]$	$Y[k] = \text{DFT}\{y[n]\}$
$x[((-n))_N]$	$X[((-k))_N]$
$x^*[n]$	$X^*[((-k))_N]$
Real $x[n]$	$X[k] = X^*[((-k))_N]$
$x[n] = x^*[((-n))_N]$	Real $X[k]$
$x[((n-n_0))_N]$	$x[k]e^{-j2\pi k n_0/N}$
$x[n]e^{j2\pi k_0 n/N}$	$X[((k-k_0))_N]$
$ax[n] + by[n]$	$aX[k] + bY[k]$
$x[n] \circledast y[n]$	$X[k]Y[k]$
$x[n]y[n]$	$\frac{1}{N}X[k] \circledast Y[k]$

By means of the FFT and IFFT, the linear convolution of two finite-duration sequences can be computed more efficiently via their circular convolution. Specifically, let the sequence $x_1[n] = 0$ outside the range $0 \leq n \leq L-1$, and the sequence $x_2[n] = 0$ outside the range $0 \leq n \leq P-1$. Then the linear convolution

$$y_{\mathrm{L}}[n] = x_1[n] \star x_2[n] \tag{3.17}$$

can be computed via the following procedure.

(S1) Use an N-point FFT algorithm to compute the N-point DFTs of the sequences $x_1[n]$ and $x_2[n]$, denoted by $X_1[k]$ and $X_2[k]$.
(S2) Compute $Y_{\mathrm{C}}[k] = X_1[k] \cdot X_2[k]$ for $0 \leq k \leq N-1$.
(S3) Use an N-point IFFT algorithm to compute the N-point IDFT of $Y_{\mathrm{C}}[k]$, denoted by $y_{\mathrm{C}}[n]$.
(S4) Obtain the sequence $\widehat{y}_{\mathrm{L}}[n] = y_{\mathrm{C}}[n]$ for $0 \leq n \leq N-1$.

Next, let us derive the condition for this procedure such that $\widehat{y}_{\mathrm{L}}[n] = y_{\mathrm{L}}[n]$. Since $Y_{\mathrm{L}}(\omega) = X_1(\omega) \cdot X_2(\omega)$ (by taking the DTFT of (3.17)), we can treat $Y_{\mathrm{C}}[k]$ in (S2) as a sequence obtained from sampling $Y_{\mathrm{L}}(\omega)$ at $\omega = 2\pi k/N$, $k = 0, 1, ..., N-1$. Correspondingly, the time-domain sequence is given by

$$y_{\mathrm{C}}[n] = x_1[n] \circledast x_2[n] = \begin{cases} \sum_{m=-\infty}^{\infty} y_{\mathrm{L}}[n+mN], & 0 \leq n \leq N-1, \\ 0, & \text{otherwise.} \end{cases}$$

As illustrated in Fig. 3.4, if the FFT (IFFT) size N is greater than or equal to the length of $y_{\mathrm{L}}[n]$, i.e. $N \geq L+P-1$, then $y_{\mathrm{C}}[n] = y_{\mathrm{L}}[n]$ for $0 \leq n \leq N-1$; otherwise, $y_{\mathrm{C}}[n]$ is an aliased version of $y_{\mathrm{L}}[n]$. As a result, the condition $N \geq L+P-1$ ensures that $\widehat{y}_{\mathrm{L}}[n] = y_{\mathrm{L}}[n]$.

3.1.3 Transform-Domain Characterization

Frequency Responses

From the foregoing discussions, we note that if an LTI system $h[n]$ is stable, i.e. $h[n]$ is absolutely summable, then its DTFT $\mathscr{F}\{h[n]\}$ is guaranteed to exist and converge uniformly and absolutely to a continuous function of ω on $[-\pi, \pi)$. The DTFT $\mathscr{F}\{h[n]\}$, denoted by $H(\omega)$, is called the *frequency response* of $h[n]$. Since $H(\omega)$ is complex, it can be expressed in polar form as

$$H(\omega) = |H(\omega)| \cdot e^{j\measuredangle H(\omega)} \tag{3.18}$$

where $|H(\omega)|$, the magnitude of $H(\omega)$, is called the *magnitude response* of $h[n]$ and $\measuredangle H(\omega)$, the phase of $H(\omega)$, is called the *phase response* of $h[n]$. The phase $\measuredangle H(\omega)$ is also usually denoted by $\arg[H(\omega)]$.

An LTI system $h[n]$ is said to be *linear phase* if its phase response

$$\arg[H(\omega)] = \alpha\omega + \beta \tag{3.19}$$

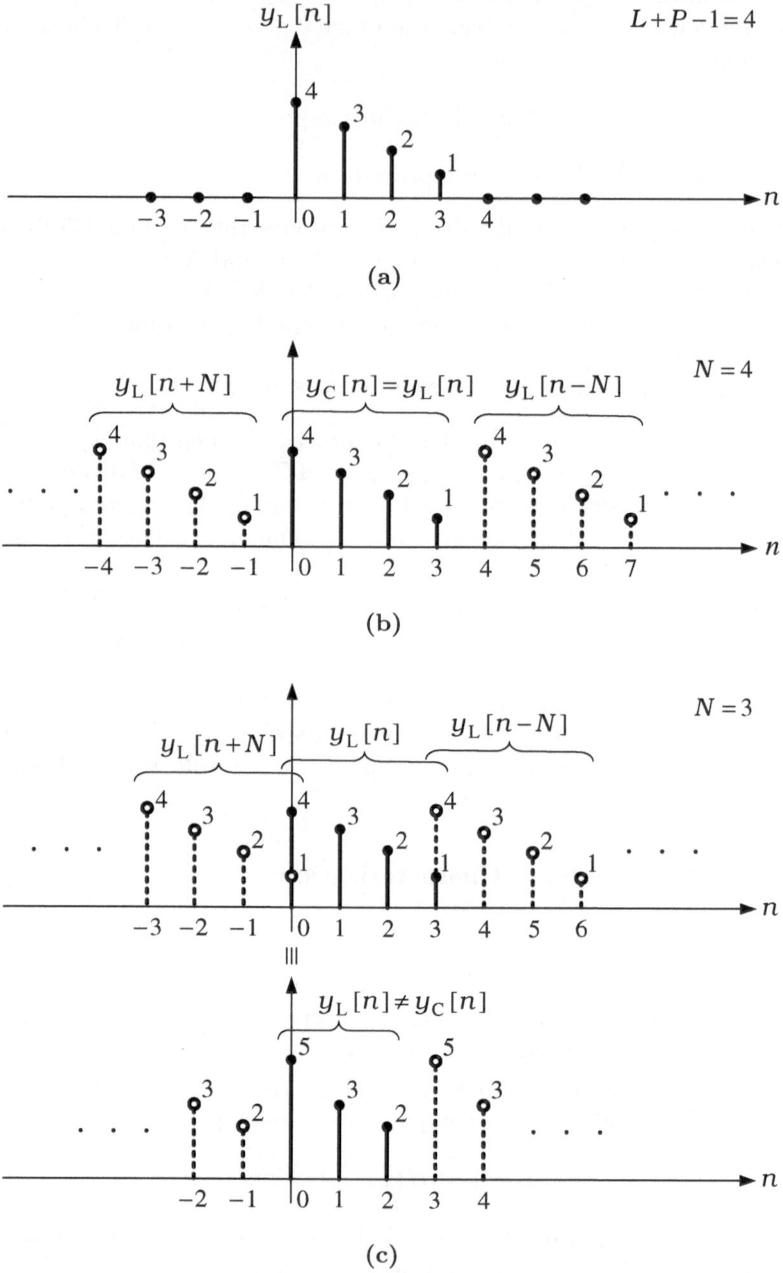

Fig. 3.4 Illustration of the equivalence between linear and circular convolution. (**a**) The sequence $y_L[n]$ with length $L + P - 1 = 4$, (**b**) the sequence $y_C[n] = y_L[n]$ for $N = 4 = L + P - 1$, and (**c**) $y_C[n] \neq y_L[n]$ for $N = 3 < L + P - 1$

where α and β are real constants.[1] Furthermore, if $\alpha = 0$ and $\beta = k\pi$ for any integer k (i.e. $H(\omega) = |H(\omega)|$ or $-|H(\omega)|$), the system $h[n]$ is called a *zero-phase system*, and correspondingly $h[n] = h^*[-n]$ since $H(\omega)$ is real.

Autoregressive Moving-Average Models

For an LTI system $h[n]$, its z-transform, denoted by $H(z)$, is called the *system function* or the *transfer function* of $h[n]$, that is often modeled as a parametric rational function:

$$H(z) = \frac{B(z)}{A(z)} = \frac{\sum_{k=0}^{q} b_k z^{-k}}{\sum_{k=0}^{p} a_k z^{-k}} = \frac{b_0}{a_0} \cdot \frac{\prod_{k=1}^{q}(1 - d_k z^{-1})}{\prod_{k=1}^{p}(1 - c_k z^{-1})} \tag{3.20}$$

where a_k, b_k, c_k, and d_k are real or complex constants. The model given by (3.20) for $H(z)$ is called an *autoregressive moving-average (ARMA) model* or an *ARMA*(p, q) *model*. It reduces to an *autoregressive (AR) model* or an *AR*(p) *model* when $B(z) = b_0$ (i.e. $q = 0$), and reduces to a *moving-average (MA) model* or an *MA*(q) *model* when $A(z) = a_0$ (i.e. $p = 0$). Moreover, one can see, from (3.20), that $H(z) = 0$ when $z = d_k$ (a root of $B(z)$), which is therefore referred to as a *zero* of the system. On the other hand, $z = c_k$ (a root of $A(z)$) is referred to as a *pole* of the system since $H(z = c_k) = \infty$. This implies that the ROC of any stable system $H(z)$ should not contain any pole. Accordingly, if $h[n]$ is a right-sided sequence, the ROC should be outside the outermost pole of the system; whereas if $h[n]$ is a left-sided sequence, the ROC should be inside the innermost pole of the system [1, p. 105]. Problem 3.1 gives an example illuminating this fact.

From the above discussion, a causal stable LTI system $h[n]$ requires that the ROC of $H(z)$ be outside the outermost pole of the system. Furthermore, we have learnt that for stable $h[n]$, its frequency response $H(\omega) = H(z = e^{j\omega})$ exists (since $h[n]$ is absolutely summable) and thereby the ROC of $H(z)$ includes the unit circle. As a consequence, a causal stable LTI system $h[n]$ requires that all the poles of the system be inside the unit circle. Moreover, let $h_{\mathrm{I}}[n]$ denote the inverse system of $h[n]$. Then it has the transfer function

$$H_{\mathrm{I}}(z) = \frac{1}{H(z)} \tag{3.21}$$

(by (3.6)), revealing that the zeros of $H(z)$ are the poles of $H_{\mathrm{I}}(z)$. Accordingly, the requirement for both $h[n]$ and $h_{\mathrm{I}}[n]$ being causal stable is that all the poles and zeros of the system $h[n]$ be inside the unit circle. Such a system is called a *minimum-phase system*; the reason for the name will be clear later on. On the other hand, an LTI system which is not a minimum-phase system is called

[1] More precisely, an LTI system $h[n]$ with phase response given by (3.19) is called a generalized linear phase system and reduces to a linear phase system as $\beta = 0$ [1]. The distinction, however, is immaterial to the purpose of this book.

a *nonminimum-phase system*, while the one having all its poles and zeros outside the unit circle is called a *maximum-phase system.* As a remark, stable inverse systems may not exist and many LTI systems in practical applications are nonminimum-phase systems.

An LTI system $h[n]$ is said to be an *allpass system* if its magnitude response $|H(\omega)| = c$ where c is a constant. A real allpass system may be modeled as the following ARMA(p,p) filter:

$$H_{\text{AP}}(z) = \frac{z^{-p}A(z^{-1})}{A(z)} = \frac{a_0 z^{-p} + a_1 z^{-p+1} + \cdots + a_p}{a_0 + a_1 z^{-1} + \cdots + a_p z^{-p}} \tag{3.22}$$

where a_k is a real constant. It can easily be seen, from (3.22), that if a is a pole of $H_{\text{AP}}(z)$, then $1/a$ must be a zero of $H_{\text{AP}}(z)$. When $A(z)$ is a minimum-phase system, i.e. all the roots of $A(z)$ are inside the unit circle, $H_{\text{AP}}(z)$ is a causal stable allpass filter; when $A(z)$ is a maximum-phase system, i.e. all the roots of $A(z)$ are outside the unit circle, $H_{\text{AP}}(z)$ is an anticausal stable allpass filter. Note that $H_{\text{AP}}(z)$ cannot have zeros on the unit circle; otherwise, it becomes unstable. Furthermore, when $H_{\text{AP}}(z)$ is causal (anticausal) stable, its inverse system $1/H_{\text{AP}}(z)$, also an ARMA allpass filter, is anticausal (causal) stable. Moreover, any real causal allpass system $H_{\text{AP}}(z)$ possesses the following phase response property: [1, p. 278]

$$\arg[H_{\text{AP}}(\omega)] \leq 0 \quad \text{for } 0 \leq \omega < \pi. \tag{3.23}$$

From (3.20), it follows that any nonminimum-phase rational system function $H(z)$ can be expressed as a minimum-phase-allpass (MP-AP) decomposition given by

$$H(z) = H_{\text{MP}}(z) \cdot H_{\text{AP}}(z), \tag{3.24}$$

and thus $-\arg[H(\omega)] = -\arg[H_{\text{MP}}(\omega)] - \arg[H_{\text{AP}}(\omega)]$. Since $-\arg[H_{\text{AP}}(\omega)] \geq 0$ for any real causal allpass system $H_{\text{AP}}(z)$, the real minimum-phase system $H_{\text{MP}}(z)$ exhibits the minimum-phase lag among all the systems having the magnitude response $|H(\omega)|$. This thereby leads to the name *minimum-phase system* or, precisely, *minimum-phase-lag system* for $H_{\text{MP}}(z)$.

Fourier Series Based Models

Suppose $h[n]$ is an LTI system with frequency response $H(\omega)$. Since both $\ln|H(\omega)|$ and $\angle H(\omega)$ are periodic with period 2π, they can be approximated by the Fourier series. In particular, by assuming that $h[n]$ is real and $h[0] = 1$ for simplicity, $H(\omega)$ can be approximated by a parametric *Fourier series based model (FSBM)* written as the *magnitude–phase (MG-PS) decomposition*

$$H(\omega) = H_{\text{MG}}(\omega) \cdot H_{\text{PS}}(\omega) \tag{3.25}$$

where

$$H_{\mathrm{MG}}(\omega) = |H(\omega)| = \exp\left\{ \sum_{i=1}^{p} \alpha_i \cos(i\omega) \right\} \tag{3.26}$$

$$H_{\mathrm{PS}}(\omega) = e^{j\measuredangle H(\omega)} = \exp\left\{ j \sum_{i=1}^{q} \beta_i \sin(i\omega) \right\} \tag{3.27}$$

in which α_i and β_i are real.[2] The FSBM ensures that $H(\omega) = H^*(-\omega)$ since $h[n]$ is real, and is specifically referred to as an *FSBM*(p,q). The system magnitude and phase are characterized by the magnitude parameters α_i and phase parameters β_i, respectively, for the FSBM, whereas they are simultaneously characterized by the poles and zeros for the ARMA model. When $\alpha_i = 0$ for $i = 1, 2, ..., p$, the FSBM is an allpass system; when $\beta_i = 0$ for $i = 1, 2, ..., q$, the FSBM is a zero-phase system.

By (3.25), (3.26) and (3.27), we have the following inverse DTFT:

$$\widetilde{h}[n] \triangleq \mathscr{F}^{-1}\{\ln H(\omega)\} = \begin{cases} \frac{1}{2}(\alpha_n - \beta_n), & 1 \le n \le \max\{p,q\}, \\ \frac{1}{2}(\alpha_{-n} + \beta_{-n}), & -\max\{p,q\} \le n \le -1, \\ 0, & \text{otherwise}, \end{cases} \tag{3.28}$$

which is known as the *complex cepstrum* of $h[n]$. Note that $\widetilde{h}[n] = 0$ for $n < 0$ ($n > 0$) if and only if $h[n]$ is minimum phase (maximum phase) [9, pp. 781–782]. Accordingly, we obtain the following result.

Fact 3.4. *The FSBM given by (3.25), (3.26) and (3.27) is a minimum-phase (maximum-phase) system if* $p = q$ *and* $\beta_i = -\alpha_i$ ($\beta_i = \alpha_i$) *for* $i = 1, 2, ..., p$.

In light of this fact, the FSBM(p,q) can also be written as the *minimum-phase allpass (MP-AP) decomposition*

$$H(\omega) = H_{\mathrm{MP}}(\omega) \cdot H_{\mathrm{AP}}(\omega) \tag{3.29}$$

where

$$H_{\mathrm{MP}}(\omega) = \exp\left\{ \sum_{i=1}^{p} \alpha_i e^{-j\omega i} \right\} \tag{3.30}$$

$$H_{\mathrm{AP}}(\omega) = \exp\left\{ j \sum_{i=1}^{\max\{p,q\}} (\alpha_i + \beta_i) \sin(i\omega) \right\} \tag{3.31}$$

in which $\alpha_i = 0$ for $i > p$ and $\beta_i = 0$ for $i > q$.

From (3.25), (3.26) and (3.27) or from (3.29), (3.30) and (3.31), it follows that both $H(\omega)$ and $dH(\omega)/d\omega$ are continuous functions on $[-\pi, \pi)$ for finite

[2] Dianat and Raghuveer [4] proposed an MG-PS FSBM for reconstruction of the magnitude and phase responses of 1-D and 2-D signals, that was then generalized by Chi *et al.* [5–8] with more applications to statistical signal processing.

p and q. Hence, by Theorem 2.38, the DTFT $\sum_{n=-\infty}^{\infty} h[n]e^{-j\omega n}$ converges uniformly to $H(\omega)$ on $[-\pi, \pi)$ and it also converges absolutely, i.e.

$$\sum_{n=-\infty}^{\infty} \left|h[n]e^{-j\omega n}\right| = \sum_{n=-\infty}^{\infty} |h[n]| < \infty.$$

Note that this may not hold true for infinite p or infinite q or both, as exemplified in Example 2.20. As a consequence, the FSBM(p, q) of finite p and q always corresponds to a stable LTI system no matter whether it is causal or noncausal. This brings some benefits to utilization of the FSBM. For example, in system identification, phase-estimation algorithms based on the FSBM with all $\alpha_i = 0$ (i.e. an allpass model) are generally simpler than those based on the ARMA allpass model because the stability issue never exists for the former with finite q [5].

As a remark, the FSBM(p, q) for real $h[n]$ can be extended to the case of complex $h[n]$; see [7] for the details.

3.2 Random Variables

This section briefly reviews real and complex random variables including their statistical characterization, moments and cumulants, and summarizes some probability distributions that are useful to statistical signal processing.

3.2.1 Statistical Characterization

Real Random Variables

In an experiment involving randomness, a single indecomposable outcome is called a *sample point* and the set of all possible sample points is called a *sample space*, whose subset is called an *event*. A *real random variable* is a rule (a function) to assign every outcome of an experiment to a real number.[3] Specifically, as illustrated in Fig. 3.5, a real random variable x defined on a sample space S is a function which assigns each sample point $s_i \in S$ a real number $\alpha_i = x(s_i)$. The function $x(\cdot)$ is deterministic, but the sample point s_i and, accordingly, the real number α_i are random in nature, i.e. they cannot be specified exactly before conducting the experiment.

Example 3.5

In a single trial of tossing a die, the sample space $S = \{1, 2, 3, 4, 5, 6\}$. The event of occurrence of even integers is the set $\{2, 4, 6\}$, and the event of occurrence of odd integers is the set $\{1, 3, 5\}$. A random variable $x \triangleq 0.1s_i$ is a function which maps each sample point $s_i \in S$ into a real number $\alpha_i \in \{0.1, 0.2, 0.3, 0.4, 0.5, 0.6\}$.

□

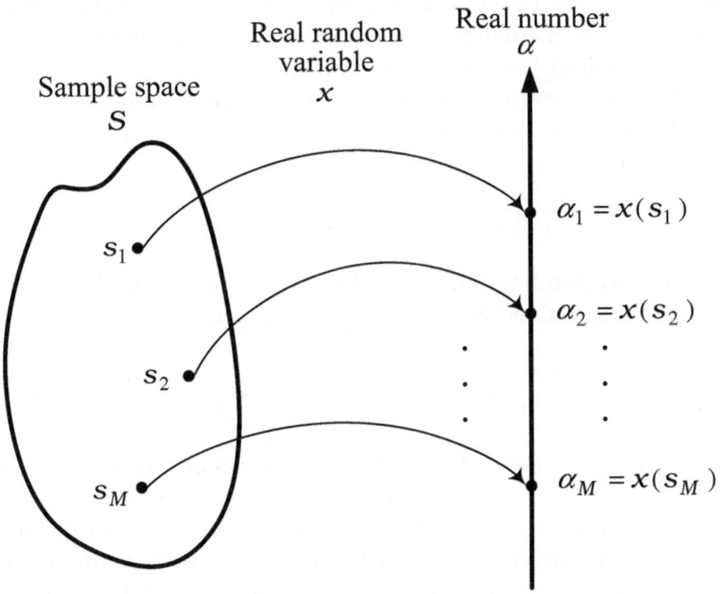

Fig. 3.5 Illustration of a real random variable x

Let $\Pr\{A\}$ denote the probability of an event A where $0 \leq \Pr\{A\} \leq 1$. A real random variable x can be characterized statistically by the *cumulative distribution function* defined as

$$F_x(\alpha) = \Pr\{x \leq \alpha\} \tag{3.32}$$

where α is a real number, $F_x(-\infty) = 0$, and $F_x(\infty) = 1$. The function $F_x(\alpha)$ is also commonly referred to as the *distribution function* or, briefly, the *distribution* of x. Alternatively, the real random variable x can also be characterized statistically by the *probability density function (pdf)* defined as

$$f_x(\alpha) = \frac{d}{d\alpha} F_x(\alpha) \tag{3.33}$$

or, equivalently,

$$F_x(u) = \int_{-\infty}^{u} f_x(\alpha) d\alpha \tag{3.34}$$

where $\int_{-\infty}^{\infty} f_x(\alpha) d\alpha = F_x(\infty) = 1$, and $f_x(\alpha) \geq 0$ since $F_x(\alpha)$ is a nondecreasing function of α.

[3] The terminology "random variable" is a misnomer since it actually represents a function of sample points, not a variable [10].

For combined experiments or repeated trials of a single experiment, we may need to deal with multiple random variables. Consider that x_1, x_2, ..., x_k are k real random variables. The vector $\mathbf{x} = (x_1, x_2, ..., x_k)^T$ is referred to as a *real random vector* since its entries are real random variables. The *(multivariate) distribution function* of $\mathbf{x}$ is defined as the *joint distribution function* of x_1, x_2, ..., x_k given by

$$F_{\mathbf{x}}(\boldsymbol{\alpha}) \equiv F_{x_1,x_2,...,x_k}(\alpha_1, \alpha_2, ..., \alpha_k) = \Pr\{x_1 \le \alpha_1, x_2 \le \alpha_2, ..., x_k \le \alpha_k\}$$

where α_i are real numbers and $\boldsymbol{\alpha} = (\alpha_1, \alpha_2, ..., \alpha_k)^T$. Similarly, the *(multivariate) pdf* of $\mathbf{x}$ is defined as the *joint pdf* of x_1, x_2, ..., x_k given by

$$f_{\mathbf{x}}(\boldsymbol{\alpha}) \equiv f_{x_1,x_2,...,x_k}(\alpha_1, \alpha_2, ..., \alpha_k) = \frac{\partial^k F_{x_1,x_2,...,x_k}(\alpha_1, \alpha_2, ..., \alpha_k)}{\partial \alpha_1 \partial \alpha_2 \cdots \partial \alpha_k}.$$

Note that $f_{\mathbf{x}}(\boldsymbol{\alpha}) \ge 0$ since $F_{\mathbf{x}}(\boldsymbol{\alpha})$ is a nondecreasing function, $F_{\mathbf{x}}(\boldsymbol{\alpha}) = 1$ as $\alpha_1 = \cdots = \alpha_k = \infty$, and $F_{\mathbf{x}}(\boldsymbol{\alpha}) = 0$ as $\alpha_l = -\infty$ for $l \in \{1, 2, ..., k\}$. Because (joint) pdfs represent the same information as the corresponding (joint) distribution functions, we will hereafter deal with random variables in terms of only (joint) pdfs for brevity. Moreover, for notational convenience, we will use the notation $f(\mathbf{x})$ for $f_{\mathbf{x}}(\boldsymbol{\alpha})$ whenever there is no ambiguity.

From the pdf $f(\mathbf{x})$, one can obtain the so-called *marginal pdf* of x_l:

$$f(x_l) = \int_{-\infty}^{\infty} \cdots \int_{-\infty}^{\infty} f(\mathbf{x}) dx_1 \cdots dx_{l-1} dx_{l+1} \cdots dx_k. \tag{3.35}$$

The pdf of x_l conditioned on $x_m = \alpha_m$ is defined as the *conditional pdf*

$$f(x_l | x_m = \alpha_m) \equiv f(x_l | x_m) = \frac{f(x_l, x_m)}{f(x_m)}, \tag{3.36}$$

provided that $f(x_m) > 0$. The conditional pdf, together with the marginal pdf, leads to *Bayes' rule* [11]

$$f(x_l | x_m) = \frac{f(x_m | x_l) f(x_l)}{f(x_m)} = \frac{f(x_m | x_l) f(x_l)}{\int_{-\infty}^{\infty} f(x_m | x_l) f(x_l) dx_l}, \tag{3.37}$$

provided that $f(x_l) > 0$ and $f(x_m) > 0$.

Definition 3.6. *Two real random variables x_l and x_m are said to be statistically independent if their joint pdf satisfies*

$$f(x_l, x_m) = f(x_l) f(x_m). \tag{3.38}$$

Equation (3.38) is equivalent to $f(x_l | x_m) = f(x_l)$, revealing that the statistical characteristics of x_l are not affected by any value of x_m.

Complex Random Variables

A *complex random variable* $x = x_\mathrm{R} + jx_\mathrm{I}$ is defined in terms of two real random variables x_R and x_I. Specifically, its pdf $f(x)$ is specified in terms of the joint pdf of x_R and x_I as

$$f(x) = f(x_\mathrm{R}, x_\mathrm{I}). \tag{3.39}$$

More generally, let x_1, x_2, ..., x_k be k complex random variables where $x_{l,\mathrm{R}} = \mathrm{Re}\{x_l\}$ and $x_{l,\mathrm{I}} = \mathrm{Im}\{x_l\}$ are real random variables. The pdf of the vector $\mathbf{x} = (x_1, x_2, ..., x_k)^T$, known as a *complex random vector*, is specified in terms of the joint pdf of the $2k$ real random variables $x_{l,\mathrm{R}}$ and $x_{l,\mathrm{I}}$ as

$$f(\mathbf{x}) = f(x_{1,\mathrm{R}}, x_{2,\mathrm{R}}, ..., x_{k,\mathrm{R}}, x_{1,\mathrm{I}}, x_{2,\mathrm{I}}, ..., x_{k,\mathrm{I}}). \tag{3.40}$$

The pdf of x_l conditioned on x_m is defined as

$$f(x_l|x_m) = \frac{f(x_{l,\mathrm{R}}, x_{l,\mathrm{I}}, x_{m,\mathrm{R}}, x_{m,\mathrm{I}})}{f(x_{m,\mathrm{R}}, x_{m,\mathrm{I}})} = \frac{f(x_l, x_m)}{f(x_m)}, \tag{3.41}$$

provided that $f(x_m) > 0$. This implies that Bayes' rule and the condition of statistical independence are applicable to the case of complex random variables without any change.

3.2.2 Moments

Statistical Averages

Let x be a real or complex random variable with pdf $f(x)$, and $g(x)$ be an arbitrary deterministic function of x. Then the *statistical average* or *expected value* of $g(x)$ is defined as

$$E\{g(x)\} = \int_{-\infty}^{\infty} g(x)f(x)dx \tag{3.42}$$

where $E\{\cdot\}$ denotes the *expectation* operator.[4] Equation (3.42) is known as the *fundamental theorem of expectation* [10]. As a special case of (3.42), if x is of discrete type, namely, it takes a value α_m with probability $\Pr\{x = \alpha_m\} = p_m$, then its pdf

$$f(x) = \sum_m p_m \delta(x - \alpha_m), \tag{3.43}$$

[4] More precisely, for a complex random variable $x = x_\mathrm{R} + jx_\mathrm{I}$ the expectation

$$E\{g(x)\} = \int_{-\infty}^{\infty} \int_{-\infty}^{\infty} g(x)f(x)dx_\mathrm{R}dx_\mathrm{I}.$$

and substituting (3.43) into (3.42) produces

$$E\{g(x)\} = \sum_m p_m g(\alpha_m). \tag{3.44}$$

As an extension of (3.42), let $\mathbf{x} = (x_1, x_2, ..., x_k)^T$ be a real or complex random vector with pdf $f(\mathbf{x})$, and let $g(\mathbf{x})$ be any deterministic function of $\mathbf{x}$. Then the expected value of $g(\mathbf{x})$ is defined as

$$\begin{aligned} E\{g(\mathbf{x})\} &= \int_{-\infty}^{\infty} g(\mathbf{x}) f(\mathbf{x}) d\mathbf{x} \\ &= \int_{-\infty}^{\infty} \cdots \int_{-\infty}^{\infty} g(x_1, ..., x_k) f(x_1, ..., x_k) dx_1 dx_2 \cdots dx_k \end{aligned} \tag{3.45}$$

where we have used the shorthand notations '$\int_{-\infty}^{\infty}$' and '$d\mathbf{x}$' to stand for '$\int_{-\infty}^{\infty} \cdots \int_{-\infty}^{\infty}$' and '$dx_1 dx_2 \cdots dx_k$,' respectively. The following two theorems immediately follow from (3.45).

Theorem 3.7 (Complex Conjugation). *Let $g(\mathbf{x})$ be a deterministic function of a real or complex random vector $\mathbf{x}$. Then*

$$E\{g^*(\mathbf{x})\} = [E\{g(\mathbf{x})\}]^*. \tag{3.46}$$

Theorem 3.8 (Linearity). *Let $g_m(\mathbf{x})$ be a deterministic function of a real or complex random vector $\mathbf{x}$. Then*

$$E\left\{\sum_m a_m g_m(\mathbf{x})\right\} = \sum_m a_m E\{g_m(\mathbf{x})\} \tag{3.47}$$

where a_m is an arbitrary real or complex constant.

Consider that x and y are two real or complex random variables, and $g(x, y)$ is a deterministic function of x and y. Given that $x = \alpha$ where α is a constant, the function $g(x, y) = g(\alpha, y)$ depends only on the random variable y, and the corresponding expectation, called the *conditional expectation*, is defined as

$$E\{g(x, y)|x\} = \int_{-\infty}^{\infty} g(x, y) f(y|x) dy. \tag{3.48}$$

With the help of the conditional expectation, one may compute $E\{g(x, y)\}$ more easily by virtue of the following equation.

$$\begin{aligned} E\{g(x, y)\} &= \int_{-\infty}^{\infty} \left[\int_{-\infty}^{\infty} g(x, y) f(y|x) dy\right] f(x) dx \qquad \text{(by (3.36))} \\ &= E_x\left\{E_{y|x}\{g(x, y)|x\}\right\} \qquad \text{(by (3.48) and (3.42))} \end{aligned} \tag{3.49}$$

where in the second line the subscripts are added for clarity.

Moments of Random Variables

Let $x_1, x_2, ..., x_k$ be real or complex random variables, and $\mathbf{x} = (x_1, x_2, ..., x_k)^T$. Then the *kth-order (joint) moment* of x_1, x_2, ..., x_k is defined as

$$E\{x_1 x_2 \cdots x_k\} = \int_{-\infty}^{\infty} x_1 x_2 \cdots x_k \cdot f(\mathbf{x}) d\mathbf{x}. \tag{3.50}$$

For a random variable x, letting $k = 1$ and $x_1 = x$ in (3.50) gives the first-order moment $E\{x\}$, which is called the *mean* of x. Letting $k = 2$, $x_1 = x$ and $x_2 = x^*$ in (3.50) gives the second-order moment $E\{|x|^2\}$, which is referred to as the *mean-square value* of x. For random variables x and y, letting $k = 2$, $x_1 = x$ and $x_2 = y^*$ in (3.50) gives the second-order joint moment $E\{xy^*\}$, which is called the *correlation* of x and y.

Similar to (3.50), the *kth-order (joint) central moment* of x_1, x_2, ..., x_k is defined as

$$E\left\{(x_1 - m_1)(x_2 - m_2) \cdots (x_k - m_k)\right\} = \int_{-\infty}^{\infty} \prod_{i=1}^{k} (x_i - m_i) \cdot f(\mathbf{x}) d\mathbf{x} \tag{3.51}$$

where m_i is the mean of x_i. Equation (3.51) reduces to (3.50) as x_1, x_2, ..., x_k are zero-mean. For a random variable x with mean m_x, letting $k = 1$ and $x_1 = x$ in (3.51) gives the first-order central moment $E\{x - m_x\} = 0$. Letting $k = 2$, $x_1 = x$ and $x_2 = x^*$ in (3.51) gives the second-order central moment

$$\sigma_x^2 \triangleq E\{|x - m_x|^2\} = E\{|x|^2\} - |m_x|^2, \tag{3.52}$$

which is called the *variance* of x. The positive square root of σ_x^2 is called the *standard deviation* of x, that is commonly used as a measure of the dispersion of x. For random variables x and y with mean values m_x and m_y, respectively, letting $k = 2$, $x_1 = x$ and $x_2 = y^*$ in (3.51) gives the second-order joint central moment

$$c_{xy} \triangleq E\{(x - m_x)(y - m_y)^*\} = E\{xy^*\} - m_x m_y^*, \tag{3.53}$$

which is called the *covariance* of x and y.

Related to the covariance c_{xy}, a statistical version of the *Cauchy–Schwartz inequality* is provided as follows (Problem 3.2).

Theorem 3.9 (Cauchy–Schwartz Inequality). *Let x and y be real or complex random variables with $E\{|x|^2\} > 0$ and $E\{|y|^2\} > 0$. Then*

$$|E\{xy^*\}| \leq \left(E\{|x|^2\}\right)^{1/2} \cdot \left(E\{|y|^2\}\right)^{1/2} \tag{3.54}$$

and the equality holds if and only if $x = \alpha y$ where $\alpha \neq 0$ is an arbitrary real or complex constant.

Viewing $(x - m_x)$ and $(y - m_y)$ as two zero-mean random variables and using (3.54), we obtain

$$|c_{xy}| \leq \sigma_x \cdot \sigma_y. \tag{3.55}$$

Definition 3.10. *As a measure of the interdependence between x and y, the normalized covariance defined as*

$$\rho_{xy} \triangleq \frac{c_{xy}}{\sigma_x \cdot \sigma_y} \tag{3.56}$$

is called the correlation coefficient of x and y where $0 \leq |\rho_{xy}| \leq 1$.

Note that $|\rho_{xy}| = 1$ when $x - m_x = \alpha(y - m_y)$ for any nonzero constant α.

Definition 3.11. *The random variables x and y are said to be uncorrelated if $\rho_{xy} = 0$ or, by (3.53),*

$$E\{xy^*\} = E\{x\}E\{y^*\}. \tag{3.57}$$

From (3.38), it follows that if x and y are statistically independent, then they are surely uncorrelated. The converse, however, is not necessarily true. Moreover, the random variables x and y are said to be *orthogonal* if

$$E\{xy^*\} = 0, \tag{3.58}$$

which can be interpreted by viewing x and y as two vectors with inner product defined as $E\{xy^*\}$; see [11, p. 154] for further details.

Moments of Random Vectors

Let $\mathbf{x} = (x_1, x_2, ..., x_k)^T$ where x_1, x_2, ..., x_k are real or complex random variables. The moments of the random vector $\mathbf{x}$ can be defined in terms of the joint moments of x_1, x_2, ..., x_k. In particular, the *mean* of $\mathbf{x}$ is defined as

$$\mathbf{m}_x \triangleq E\{\mathbf{x}\} = (E\{x_1\}, E\{x_2\}, ..., E\{x_k\})^T. \tag{3.59}$$

The *correlation matrix* of $\mathbf{x}$ is defined as

$$\mathbf{R}_x \triangleq E\{\mathbf{x}\mathbf{x}^H\} = \begin{pmatrix} E\{|x_1|^2\} & E\{x_1 x_2^*\} & \cdots & E\{x_1 x_k^*\} \\ E\{x_2 x_1^*\} & E\{|x_2|^2\} & \cdots & E\{x_2 x_k^*\} \\ \vdots & \vdots & \ddots & \vdots \\ E\{x_k x_1^*\} & E\{x_k x_2^*\} & \cdots & E\{|x_k|^2\} \end{pmatrix} \tag{3.60}$$

and the *covariance matrix* of $\mathbf{x}$ is defined as

$$\mathbf{C}_x \triangleq E\{(\mathbf{x} - \mathbf{m}_x)(\mathbf{x} - \mathbf{m}_x)^H\} = \mathbf{R}_x - \mathbf{m}_x \mathbf{m}_x^H. \tag{3.61}$$

One can see that both $\mathbf{R}_x$ and $\mathbf{C}_x$ are Hermitian.

Characteristic Functions

Let x_1, x_2, ..., x_k be k real random variables and $\mathbf{x} = (x_1, x_2, ..., x_k)^T$. Then the *(joint) characteristic function* of x_1, x_2, ..., x_k is defined as

$$\Phi(\omega_1, \omega_2, ..., \omega_k) \equiv \Phi(\boldsymbol{\omega}) = E\left\{\exp\left(j\boldsymbol{\omega}^T\mathbf{x}\right)\right\} \tag{3.62}$$

where ω_1, ω_2, ..., ω_k are real variables and $\boldsymbol{\omega} = (\omega_1, \omega_2, ..., \omega_k)^T$. From (3.62), it follows that

$$\frac{\partial^k\Phi(\omega_1, \omega_2, ..., \omega_k)}{\partial\omega_1\partial\omega_2\cdots\partial\omega_k} = (j)^k E\left\{x_1 x_2 \cdots x_k \cdot \exp\left(j\boldsymbol{\omega}^T\mathbf{x}\right)\right\},$$

which further leads to

$$E\{x_1 x_2 \cdots x_k\} = (-j)^k \left.\frac{\partial^k\Phi(\omega_1, \omega_2, ..., \omega_k)}{\partial\omega_1\partial\omega_2\cdots\partial\omega_k}\right|_{\boldsymbol{\omega}=\mathbf{0}}. \tag{3.63}$$

This indicates that the kth-order (joint) moment of real random variables x_1, x_2, ..., x_k can also be defined in terms of their (joint) characteristic function $\Phi(\boldsymbol{\omega})$. Accordingly, $\Phi(\boldsymbol{\omega})$ is also referred to as the *moment generating function* of x_1, x_2, ..., x_k [12].

Similar to (3.62), if $x_l = x_{l,\mathrm{R}} + jx_{l,\mathrm{I}}$, $l = 1, 2, ..., k$, are complex random variables, then their (joint) characteristic function is defined as

$$\Phi(\omega_{1,\mathrm{R}}, ..., \omega_{k,\mathrm{R}}, \omega_{1,\mathrm{I}}, ..., \omega_{k,\mathrm{I}}) = E\left\{\exp\left(j\sum_{l=1}^{k}\omega_{l,\mathrm{R}}x_{l,\mathrm{R}} + \omega_{l,\mathrm{I}}x_{l,\mathrm{I}}\right)\right\} \tag{3.64}$$

where $\omega_{l,\mathrm{R}}$ and $\omega_{l,\mathrm{I}}$ are real variables. Let $\mathbf{x} = (x_1, x_2, ..., x_k)^T$ and $\boldsymbol{\omega} = (\omega_1, \omega_2, ..., \omega_k)^T$ where $\omega_l = \omega_{l,\mathrm{R}} + j\omega_{l,\mathrm{I}}$. Then (3.64) is equivalent to

$$\Phi(\boldsymbol{\omega}) = E\left\{\exp\left[j\mathrm{Re}(\boldsymbol{\omega}^H\mathbf{x})\right]\right\} = E\left\{\exp\left[\frac{j}{2}\left(\boldsymbol{\omega}^H\mathbf{x} + \boldsymbol{\omega}^T\mathbf{x}^*\right)\right]\right\}, \tag{3.65}$$

which can be viewed as the (joint) characteristic function of two independent sets of random variables $\{x_1, x_2, ..., x_k\}$ and $\{x_1^*, x_2^*, ..., x_k^*\}$ [13]. Accordingly, by treating $\{\omega_1, \omega_2, ..., \omega_k\}$ and $\{\omega_1^*, \omega_2^*, ..., \omega_k^*\}$ as two independent sets of variables and taking the kth partial derivative of (3.65), one can obtain the kth-order (joint) moment

$$E\{x_1 \cdots x_l x_{l+1}^* \cdots x_k^*\} = (-2j)^k \left.\frac{\partial^k\Phi(\omega_1, \omega_2, ..., \omega_k)}{\partial\omega_1^*\cdots\partial\omega_l^*\partial\omega_{l+1}\cdots\partial\omega_k}\right|_{\boldsymbol{\omega}=\mathbf{0}}. \tag{3.66}$$

3.2.3 Cumulants

Definitions of Cumulants

Consider two statistically independent real random variables x and y whose characteristic functions are denoted by $\Phi_x(\omega)$ and $\Phi_y(\omega)$, respectively, where the subscripts are added for clarity. From (3.38) and (3.62), it follows that the characteristic function of the random variable $(x+y)$ is given by

$$\Phi_{x+y}(\omega) = \Phi_x(\omega) \cdot \Phi_y(\omega).$$

By defining

$$\Psi_x(\omega) = \ln\Phi_x(\omega) = \ln E\{\exp(j\omega x)\}$$

which is known as the *second characteristic function* of x, we obtain the second characteristic function of $(x+y)$ as

$$\Psi_{x+y}(\omega) = \Psi_x(\omega) + \Psi_y(\omega).$$

That is, the second characteristic function of the sum $(x+y)$ is simply the sum of the respective second characteristic functions of x and y. For this reason, the second characteristic function is also called the *cumulative function* [14].

Similar to (3.63) and (3.66) for defining a set of statistical quantities — moments, taking the partial derivatives of the cumulative functions also leads to another set of statistical quantities, called *cumulants.*[5] To be more specific and more general, let x_1, x_2, ..., x_k be k real or complex random variables. Then their *kth-order (joint) cumulant* is defined as

$$\text{cum}\{x_1, x_2, ..., x_k\} = (-j)^k \left. \frac{\partial^k \Psi(\omega_1, \omega_2, ..., \omega_k)}{\partial\omega_1 \partial\omega_2 \cdots \partial\omega_k} \right|_{\boldsymbol{\omega}=\mathbf{0}} \tag{3.67}$$

for the real case and

$$\text{cum}\{x_1, ..., x_l, x_{l+1}^*, ..., x_k^*\} = (-2j)^k \left. \frac{\partial^k \Psi(\omega_1, \omega_2, ..., \omega_k)}{\partial\omega_1^* \cdots \partial\omega_l^* \partial\omega_{l+1} \cdots \partial\omega_k} \right|_{\boldsymbol{\omega}=\mathbf{0}} \tag{3.68}$$

for the complex case where $\boldsymbol{\omega} = (\omega_1, \omega_2, ..., \omega_k)^T$,

$$\Psi(\omega_1, \omega_2, ..., \omega_k) \equiv \Psi(\boldsymbol{\omega}) = \ln\Phi(\boldsymbol{\omega}) = \ln E\left\{\exp\left[j\text{Re}(\boldsymbol{\omega}^H \mathbf{x})\right]\right\} \tag{3.69}$$

is the *(joint) second characteristic function* of x_1, x_2, ..., x_k, and $\mathbf{x} = (x_1, x_2, ..., x_k)^T$. Accordingly, $\Psi(\boldsymbol{\omega})$ is also referred to as the *cumulant generating function* of x_1, x_2, ..., x_k [12].

[5] Originated from the cumulative function, the word "cumulant" is due to E. A. Cornish and R. A. Fisher (1937) [14], while its earlier names "semi-variant" and "half-invariant" were introduced by T. N. Thiele (1889) [12].

From (3.63) and (3.66) to (3.69), one may notice that cumulants seem to relate to moments. Indeed, the kth-order cumulant of real or complex random variables x_1, x_2, ..., x_k can be expressed in terms of the moments of x_1, x_2, ..., x_k of orders up to k. In particular, for $k = 1$, the first-order cumulant

$$\text{cum}\{x_1\} = E\{x_1\}. \tag{3.70}$$

Furthermore, if x_1, x_2, x_3, and x_4 are zero-mean, real or complex random variables, then

$$\text{cum}\{x_1, x_2\} = E\{x_1 x_2\}, \tag{3.71}$$

$$\text{cum}\{x_1, x_2, x_3\} = E\{x_1 x_2 x_3\}, \tag{3.72}$$

$$\begin{aligned}\text{cum}\{x_1, x_2, x_3, x_4\} &= E\{x_1 x_2 x_3 x_4\} - E\{x_1 x_2\}E\{x_3 x_4\} \\ &\quad - E\{x_1 x_3\}E\{x_2 x_4\} - E\{x_1 x_4\}E\{x_2 x_3\}.\end{aligned} \tag{3.73}$$

And if x_1, x_2, x_3, and x_4 are nonzero-mean random variables, then one can replace x_i in the right-hand sides of the formulas (3.71) to (3.73) by $(x_i - E\{x_i\})$ [15]. As a remark, the fundamental difference between cumulants and (central) moments appears as cumulant order ≥ 4.

The generic form of the relationship between cumulants and moments is shown in Appendix 3A. From this relationship and Theorem 3.7, one can derive the following theorem (Problem 3.3).

Theorem 3.12 (Complex Conjugation). *For real or complex random variables x_1, x_2, ..., x_k, the kth-order cumulant*

$$\text{cum}\{x_1^*, ..., x_l^*, x_{l+1}, ..., x_k\} = \left(\text{cum}\{x_1, ..., x_l, x_{l+1}^*, ..., x_k^*\}\right)^*. \tag{3.74}$$

Let the $(p+q)$th-order cumulant of a real or complex random variable x be denoted by

$$C_{p,q}\{x\} = \text{cum}\{x : p, x^* : q\} \tag{3.75}$$

where

$$\text{cum}\{.., x : l, ...\} \triangleq \text{cum}\{..., \underbrace{x, x, ..., x}_{l \text{ terms}}, ...\}. \tag{3.76}$$

Then, by Theorem 3.12, $C_{p,q}\{x\} = (C_{q,p}\{x\})^*$. Moreover, when x is complex, its $(p+q)$th-order cumulant has more than one definition, depending on the choice of (p, q). For example, there are three possible definitions for the second-order cumulant of x, namely, $C_{2,0}\{x\}$, $C_{0,2}\{x\}$ and $C_{1,1}\{x\}$ where $C_{0,2}\{x\} = (C_{2,0}\{x\})^*$. The third-order cumulant of x has four possible definitions, namely, $C_{3,0}\{x\} = (C_{0,3}\{x\})^*$ and $C_{2,1}\{x\} = (C_{1,2}\{x\})^*$, while the fourth-order cumulant of x has five possible definitions, namely, $C_{4,0}\{x\} = (C_{0,4}\{x\})^*$, $C_{3,1}\{x\} = (C_{1,3}\{x\})^*$, and $C_{2,2}\{x\}$.

Although the notation in (3.75) is unified for both real and complex cases, it is sometimes more convenient and specific to represent the $(p+q)$th-order cumulant of a real random variable x as

$$C_{p+q}\{x\} = \text{cum}\{x : p+q\}. \tag{3.77}$$

Unlike the complex case, the $(p+q)$th-order cumulant of real x has only one definition.

Properties of Cumulants

In the following, we list some fundamental properties of cumulants with the proofs left as an exercise (Problem 3.4) [15–19] .

Property 3.13. *Cumulants are symmetric in their arguments. Namely, for real or complex random variables x_1, x_2, ..., x_k, the kth-order cumulant*

$$\text{cum}\{x_{l_1}, x_{l_2}, ..., x_{l_k}\} = \text{cum}\{x_1, x_2, ..., x_k\} \tag{3.78}$$

where the sequence $\{l_1, l_2, ..., l_k\}$ is a permutation of the sequence $\{1, 2, ..., k\}$.

Property 3.14. *For real or complex random variables x_1, x_2, ..., x_k, the kth-order cumulant*

$$\text{cum}\{\alpha_1 x_1, \alpha_2 x_2, ..., \alpha_k x_k\} = \alpha_1 \alpha_2 \cdots \alpha_k \cdot \text{cum}\{x_1, x_2, ..., x_k\} \tag{3.79}$$

where α_1, α_2, ..., α_k are real or complex constants.

Property 3.15. *If x_1, x_2, ..., x_k are real or complex random variables and α is a real or complex constant, then the kth-order cumulant*

$$\text{cum}\{x_1 + \alpha, x_2, ..., x_k\} = \text{cum}\{x_1, x_2, ..., x_k\} \quad \textit{for } k \geq 2. \tag{3.80}$$

Property 3.16. *Cumulants are additive in their arguments. Namely, if x_1, x_2, ..., x_k and y are real or complex random variables, then the kth-order cumulant*

$$\text{cum}\{x_1 + y, x_2, ..., x_k\} = \text{cum}\{x_1, x_2, ..., x_k\} + \text{cum}\{y, x_2, ..., x_k\}, \tag{3.81}$$

no matter whether x_1 and y are statistically independent or not.

Property 3.17. *If the set of real or complex random variables $\{x_1$, x_2, ..., $x_k\}$ are statistically independent of another set of real or complex random variables $\{y_1$, y_2, ..., $y_k\}$, then the kth-order cumulant*

$$\begin{aligned} &\text{cum}\{x_1 + y_1, x_2 + y_2, ..., x_k + y_k\} \\ &\quad = \text{cum}\{x_1, x_2, ..., x_k\} + \text{cum}\{y_1, y_2, ..., y_k\}. \end{aligned} \tag{3.82}$$

Property 3.18. *If a subset of real or complex random variables* $\{x_1, x_2, ..., x_k\}$ *are statistically independent of the rest, then the kth-order cumulant*

$$\text{cum}\{x_1, x_2, ..., x_k\} = 0. \tag{3.83}$$

Let us emphasize that higher-order (≥ 3) moments, in general, do not possess the counterparts of Properties 3.15 through 3.18 [18, pp. 12–14]. On the other hand, according to the above-mentioned properties, one can treat "cumulant" as an operator, just like treating "expectation" as an operator [19]. Thus, it is preferred to work with higher-order cumulants, whenever needed, instead of higher-order moments.

Physical Meanings of Cumulants

To explain the physical meanings of cumulants, consider that x is a real random variable with mean m_x and variance σ_x^2, and assume that its pdf $f(x)$ is unimodal, i.e. $f(x)$ has only one peak. Note that many distributions encountered in practice are unimodal [20, p. 29]. Obviously, the first-order cumulant $C_1\{x\} = E\{x\} = m_x$ (see (3.70)) indicates the central location of $f(x)$, while the second-order cumulant $C_2\{x\} = E\{(x - m_x)^2\} = \sigma_x^2$ (see (3.71)) indicates the spread of $f(x)$ about the central location m_x. As detailed below, the third-order cumulant $C_3\{x\}$ and the fourth-order cumulant $C_4\{x\}$ give a sketch about the shape of $f(x)$.

By (3.72), the third-order cumulant $C_3\{x\} = E\{(x-m_x)^3\}$, which is called the *skewness* of x [18, 19]. Invariant to any real scale factor, the *normalized skewness* or the *normalized third-order cumulant*

$$\gamma_3\{x\} \triangleq \frac{C_3\{x\}}{(\sigma_x^2)^{3/2}} = \frac{E\{(x - m_x)^3\}}{[E\{(x - m_x)^2\}]^{3/2}} \tag{3.84}$$

can be used as a dimensionless measure of the skewness or symmetry of $f(x)$ about the central location m_x. From (3.42), it follows that $\gamma_3\{x\} = 0$ if $f(x)$ is symmetric about m_x (see Fig. 3.6a) since $(x - m_x)^3$ is antisymmetric about m_x. On the other hand, a positive value of $\gamma_3\{x\}$ is usually found when $f(x)$ is *skewed to the right* or *positively skewed* [20–22]. As illustrated in Fig. 3.6b, one can find a region Ω_2 such that

$$\int_{x\in\Omega_2} (x - m_x)^3 f(x)dx = -\int_{x\in\Omega_1} (x - m_x)^3 f(x)dx,$$

which leads to

$$C_3\{x\} = \int_{x\in\Omega_1\cup\Omega_2\cup\Omega_3} (x - m_x)^3 f(x)dx = \int_{x\in\Omega_3} (x - m_x)^3 f(x)dx > 0$$

and thus $\gamma_3\{x\} > 0$. Based on a similar reasoning, a negative value of $\gamma_3\{x\}$ usually happens when $f(x)$ is *skewed to the left* or *negatively skewed* (see Fig. 3.6c).

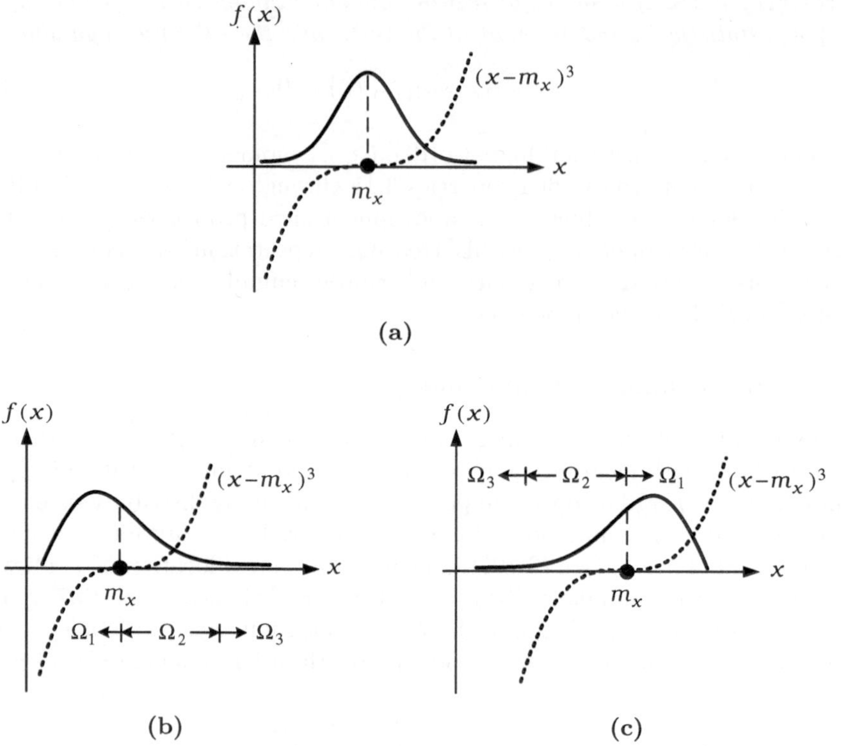

Fig. 3.6 The pdfs for normalized skewness (**a**) $\gamma_3\{x\} = 0$ ($f(x)$ symmetric), (**b**) $\gamma_3\{x\} > 0$ ($f(x)$ skewed to the right), and (**c**) $\gamma_3\{x\} < 0$ ($f(x)$ skewed to the left)

By (3.73), the fourth-order cumulant

$$C_4\{x\} = E\{(x - m_x)^4\} - 3\left[E\{(x - m_x)^2\}\right]^2, \tag{3.85}$$

which is called the *kurtosis* of x [18, 19]. Invariant to any real scale factor, the *normalized kurtosis* or the *normalized fourth-order cumulant*[6]

$$\gamma_4\{x\} \triangleq \frac{C_4\{x\}}{(\sigma_x^2)^2} = \frac{E\{(x - m_x)^4\}}{\left[E\{(x - m_x)^2\}\right]^2} - 3 \tag{3.86}$$

can be used as a dimensionless measure of the relative peakedness or flatness of $f(x)$, compared with a reference pdf. The reference pdf typically adopted is the well-known Gaussian pdf (to be introduced later) whose normalized

---

[6] K. Pearson (1905) introduced the term "kurtosis," and proposed the moment ratio $E\{(x - m_x)^4\}/\sigma_x^4$ as a measure of the relative kurtosis or peakedness of $f(x)$ [12], that has been widely adopted in statistics [20–23]. This book, however, follows the convention of higher-order statistical signal processing reported in [18, 19].

kurtosis is known to be zero. As depicted in Fig. 3.7, if $\gamma_4\{x\} > 0$, the pdf $f(x)$ is said to be more peaked than the Gaussian pdf; whereas if $\gamma_4\{x\} < 0$, $f(x)$ is said to be less peaked or flatter than the Gaussian pdf [20–22].

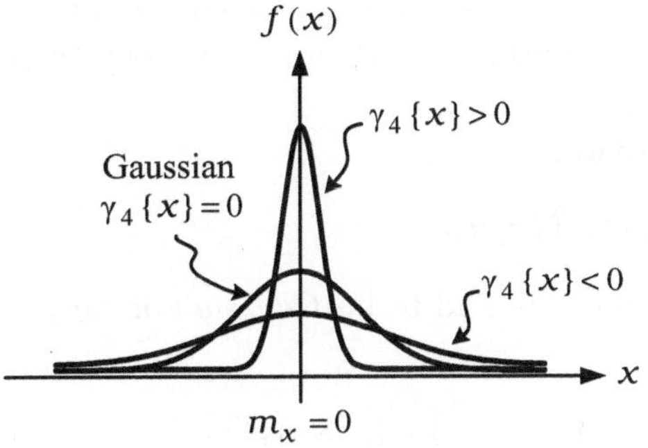

Fig. 3.7 The pdfs for different values of normalized kurtosis $\gamma_4\{x\}$

As for the case of complex random variable x with mean m_x and variance σ_x^2, the kurtosis of x is defined as the fourth-order cumulant

$$C_{2,2}\{x\} = E\left\{|x-m_x|^4\right\} - 2\left(E\{|x-m_x|^2\}\right)^2 - \left|E\left\{(x-m_x)^2\right\}\right|^2 \quad (3.87)$$

and the *normalized kurtosis* or the *normalized fourth-order cumulant*

$$\gamma_{2,2}\{x\} \triangleq \frac{C_{2,2}\{x\}}{(\sigma_x^2)^2} = \frac{E\left\{|x-m_x|^4\right\} - \left|E\left\{(x-m_x)^2\right\}\right|^2}{\left(E\left\{|x-m_x|^2\right\}\right)^2} - 2. \quad (3.88)$$

Note that there is no corresponding definition for skewness. More generally, the *normalized* $(p+q)$*th-order cumulant*[7]

$$\gamma_{p,q}\{x\} \triangleq \frac{C_{p,q}\{x\}}{(\sigma_x^2)^{(p+q)/2}}, \quad (3.89)$$

whose physical meaning for cumulant order $(p+q) \geq 3$ is not yet clear.

3.2.4 Some Useful Distributions

Distribution of random variables can be divided into two categories, *Gaussian distribution* and *non-Gaussian distribution.* Gaussian distribution is suitable

[7] For the real case, the normalized cumulants of orders greater than two are called the *gamma coefficients* or the *g-statistics* in statistics [12].

for modeling such physical phenomena as noise sources and can be handled mathematically, thereby leading to its widespread applications in almost all fields of science and engineering. On the other hand, non-Gaussian distribution includes uniform distribution, Laplace distribution, exponential distribution, Bernoulli distribution, etc. It represents the probabilistic characteristics of most signals in such applications as digital communications, speech processing, seismic exploration, image processing, biomedical signal processing, radar, sonar, astronomy, oceanography, plasma physics, and so on.

Gaussian Distribution

Real Gaussian Random Variables

A real random variable x is said to be *Gaussian* or *normal* if its pdf is given by (see Fig. 3.8)

$$f(x) = \frac{1}{\sqrt{2\pi\sigma_x^2}} \exp\left\{ -\frac{(x-m_x)^2}{2\sigma_x^2} \right\}, \quad -\infty < x < \infty \tag{3.90}$$

where m_x and σ_x^2 are the mean and variance of x, respectively. It is said to be *standard Gaussian* or *standard normal* if $m_x = 0$ and $\sigma_x^2 = 1$.

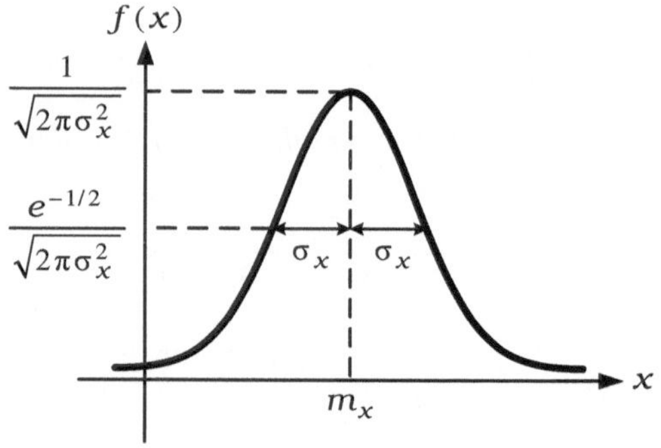

Fig. 3.8 The pdf of a real Gaussian random variable x

More generally, a real random vector $\mathbf{x} = (x_1, x_2, ..., x_N)^T$ is said to be *Gaussian* or, equivalently, the set of real random variables x_1, x_2, ..., x_N is said to be *jointly Gaussian* if for any nonzero vector $\mathbf{a} = (a_1, a_2, ..., a_N)^T$ the random variable

$$\mathbf{a}^T\mathbf{x} = a_1x_1 + a_2x_2 + \cdots + a_Nx_N$$

is Gaussian [11, 24]. According to this definition, it can be shown (Problem 3.6) that if $\mathbf{x}$ is real Gaussian, then its pdf is given by

$$f(\mathbf{x}) = \frac{1}{(2\pi)^{\frac{N}{2}} \cdot |\mathbf{C}_x|^{\frac{1}{2}}} \exp\left\{-\frac{1}{2}(\mathbf{x} - \mathbf{m}_x)^T \mathbf{C}_x^{-1} (\mathbf{x} - \mathbf{m}_x)\right\} \tag{3.91}$$

where $\mathbf{m}_x$ is the mean of $\mathbf{x}$ and $\mathbf{C}_x$ is the covariance matrix of $\mathbf{x}$ that has to be positive definite. One can see that the Gaussian pdf $f(\mathbf{x})$ is completely determined by $\mathbf{m}_x$ and $\mathbf{C}_x$, and thereby is commonly denoted by $\mathcal{N}(\mathbf{m}_x, \mathbf{C}_x)$.

Complex Gaussian Random Variables

Consider a complex random vector

$$\mathbf{x} = (x_1, x_2, ..., x_N)^T = \mathbf{x}_\mathrm{R} + j\mathbf{x}_\mathrm{I}$$

where $\mathbf{x}_\mathrm{R} = \mathrm{Re}\{\mathbf{x}\} = (x_{1,\mathrm{R}}, x_{2,\mathrm{R}}, ..., x_{N,\mathrm{R}})^T$ and $\mathbf{x}_\mathrm{I} = \mathrm{Im}\{\mathbf{x}\} = (x_{1,\mathrm{I}}, x_{2,\mathrm{I}}, ..., x_{N,\mathrm{I}})^T$. The complex random vector $\mathbf{x}$ is said to be *Gaussian* or *normal* if the $2N$ real random variables $x_{1,\mathrm{R}}, x_{2,\mathrm{R}}, ..., x_{N,\mathrm{R}}, x_{1,\mathrm{I}}, x_{2,\mathrm{I}}, ..., x_{N,\mathrm{I}}$ are jointly Gaussian. A further result is given as follows [11, 25] (Problem 3.7).

Theorem 3.19 (Goodman's Theorem). *If $\mathbf{x}$ is a complex Gaussian random vector which satisfies the condition*[8]

$$E\{(\mathbf{x} - \mathbf{m}_x)(\mathbf{x} - \mathbf{m}_x)^T\} = \mathbf{0}, \tag{3.92}$$

then its pdf is given by

$$f(\mathbf{x}) = \frac{1}{\pi^N \cdot |\mathbf{C}_x|} \exp\left\{-(\mathbf{x} - \mathbf{m}_x)^H \mathbf{C}_x^{-1} (\mathbf{x} - \mathbf{m}_x)\right\} \tag{3.93}$$

where $\mathbf{m}_x$ is the mean of $\mathbf{x}$ and $\mathbf{C}_x$ is the covariance matrix of $\mathbf{x}$.

Let us emphasize that the condition (3.92) also frequently holds for complex random vectors in such applications as digital communications and radar. Accordingly, we assume the condition (3.92) is always satisfied for simplicity.

As a special case of Theorem 3.19, the pdf of a complex Gaussian random variable x is given by

$$f(x) = \frac{1}{\pi\sigma_x^2} \exp\left\{-\frac{|x - m_x|^2}{\sigma_x^2}\right\} \tag{3.94}$$

where m_x and σ_x^2 are the mean and variance of x.

---

[8] Complex Gaussian random vectors which satisfy the condition (3.92) are also called *circularly Gaussian random vectors* [13].

Properties of Gaussian Random Variables

Some useful properties of Gaussian random variables are given as follows (Problem 3.8).

Property 3.20. *Linear transformations of real or complex jointly Gaussian random variables are also jointly Gaussian.*

Property 3.21. *If real or complex jointly Gaussian random variables are uncorrelated, then they are statistically independent.*

Property 3.22. *If real or complex random variables x_1, x_2, ..., x_k are jointly Gaussian, then their kth-order cumulant*

$$\text{cum}\{x_1, x_2, ..., x_k\} = 0 \quad \text{for } k \geq 3. \tag{3.95}$$

Property 3.23. *If real or complex random variables x_1, x_2, x_3, and x_4 are zero-mean and jointly Gaussian, then the fourth-order moment*

$$\begin{aligned} E\{x_1x_2x_3x_4\} = &\ E\{x_1x_2\}E\{x_3x_4\} + E\{x_1x_3\}E\{x_2x_4\} \\ &+ E\{x_1x_4\}E\{x_2x_3\}. \end{aligned} \tag{3.96}$$

As a summary, Table 3.4 lists some statistical parameters of Gaussian random variable.

Uniform Distribution

A real random variable x is said to be *uniformly distributed* or, briefly, *uniform* in the interval $[\alpha, \beta]$ if its pdf is given by (see Fig. 3.9)

$$f(x) = \begin{cases} \dfrac{1}{\beta - \alpha}, & \alpha \leq x \leq \beta, \\ 0, & \text{otherwise}. \end{cases} \tag{3.97}$$

Table 3.5 lists some statistical parameters of the uniform random variable x. From this table, one can see that $\gamma_3\{x\} = 0$ since $f(x)$ is symmetric about the central location m_x, while $\gamma_4\{x\} < 0$ since uniform pdf is obviously much flatter than Gaussian pdf.

Laplace Distribution

A real random variable x is said to be *Laplace* if its pdf is given by

$$f(x) = \frac{1}{2\beta} \exp\left\{ -\frac{|x - m_x|}{\beta} \right\}, \quad -\infty < x < \infty \tag{3.98}$$

Table 3.4 Summary of Gaussian distribution

Statistical Parameters of Real Gaussian Random Variable x	
pdf	$f(x) = \frac{1}{\sqrt{2\pi\sigma_x^2}} \exp\left\{-\frac{(x-m_x)^2}{2\sigma_x^2}\right\}, \ -\infty < x < \infty$
Mean	m_x
Variance	σ_x^2
kth-order cumulant	$C_k\{x\} = 0, \ k \geq 3$
Normalized skewness	$\gamma_3\{x\} = 0$
Normalized kurtosis	$\gamma_4\{x\} = 0$

Statistical Parameters of Complex Gaussian Random Variable x	
pdf	$f(x) = \frac{1}{\pi\sigma_x^2} \exp\left\{-\frac{\lvert x-m_x\rvert^2}{\sigma_x^2}\right\}, \ -\infty < x < \infty$
Mean	m_x
Variance	σ_x^2
$(p+q)$th-order cumulant	$C_{p,q}\{x\} = 0, \ p+q \geq 3$
Normalized kurtosis	$\gamma_{2,2}\{x\} = 0$

† For complex random variable $x = x_R + jx_I$, the notation '$-\infty < x < \infty$' represents '$-\infty < x_R < \infty$ and $-\infty < x_I < \infty$.'

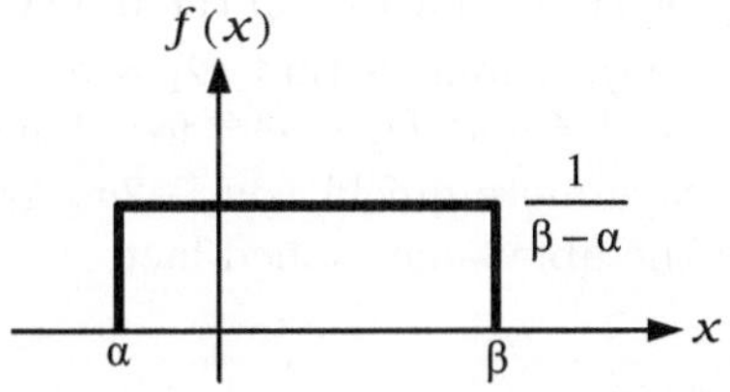

Fig. 3.9 The pdf of a uniform random variable x

Table 3.5 Summary of uniform distribution

Statistical Parameters of Uniform Random Variable x	
pdf	$f(x) = \begin{cases} \dfrac{1}{\beta-\alpha}, & \alpha \le \beta \\ 0, & \text{otherwise} \end{cases}$
Mean	$m_x = (\beta - \alpha)/2$
Variance	$\sigma_x^2 = (\beta - \alpha)^2/12$
Skewness	$C_3\{x\} = 0$
Kurtosis	$C_4\{x\} = -(\beta - \alpha)^4/120$
Normalized Skewness	$\gamma_3\{x\} = 0$
Normalized Kurtosis	$\gamma_4\{x\} = -6/5$

where m_x is the mean of x and the parameter $\beta > 0$. Figure 3.10 depicts the pdf $f(x)$ and Table 3.6 lists some statistical parameters of the Laplace random variable x. From this table, one can see that $\gamma_3\{x\} = 0$ since $f(x)$ is symmetric about m_x, while $\gamma_4\{x\} > 0$, indicating that the Laplace pdf is more peaked than the Gaussian pdf. To realize this latter fact, we observe that both Laplace and Gaussian pdfs exhibit double-sided exponential decay, and their decay rate can be compared by simply examining the two values

$$D_1 \triangleq \frac{f(x = m_x + \sigma_x)}{f(x = m_x)} \quad \text{and} \quad D_2 \triangleq \frac{f(x = m_x + 2\sigma_x)}{f(x = m_x)}$$

since most of the energy of $f(x)$ is located in the interval $[-2\sigma_x, 2\sigma_x]$ for both pdfs. From (3.98) and (3.90), it follows that $D_1 = e^{-\sqrt{2}}$ and $D_2 = e^{-2\sqrt{2}}$ for Laplace pdf, while $D_1 = e^{-1/2}$ and $D_2 = e^{-2}$ for Gaussian pdf. This reveals that the Laplace pdf decays more quickly on $[-2\sigma_x, 2\sigma_x]$ than the Gaussian pdf, thereby supporting the above-mentioned fact.

Exponential Distribution

A real random variable x is said to be *exponential* if its pdf is given by

$$f(x) = \begin{cases} \dfrac{1}{\sigma_x} \exp\left\{ -\dfrac{(x-\alpha)}{\sigma_x} \right\}, & \alpha \le x < \infty, \\ 0, & \text{otherwise} \end{cases} \tag{3.99}$$

where $\sigma_x > 0$ is the standard deviation of x. Figure 3.11 depicts the pdf $f(x)$ and Table 3.7 lists some statistical parameters of the exponential random

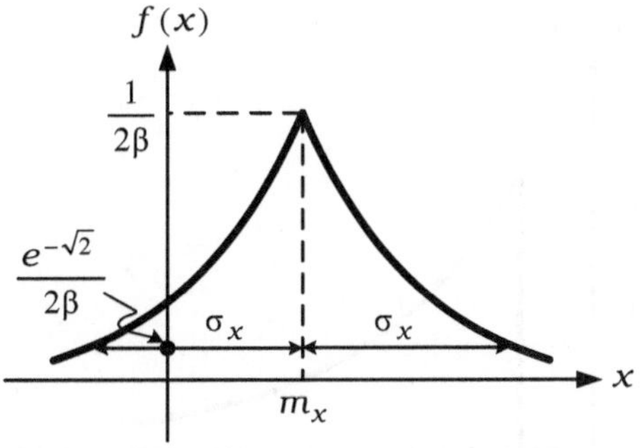

Fig. 3.10 The pdf of a Laplace random variable x

Table 3.6 Summary of Laplace distribution

Statistical Parameters of Laplace Random Variable x	
pdf	$f(x) = \frac{1}{2\beta}\exp\left\{-\frac{\lvert x-m_x\rvert}{\beta}\right\}, \quad -\infty < x < \infty,\ \beta > 0$
Mean	m_x
Variance	$\sigma_x^2 = 2\beta^2$
Skewness	$C_3\{x\} = 0$
Kurtosis	$C_4\{x\} = 12\beta^4$
Normalized skewness	$\gamma_3\{x\} = 0$
Normalized kurtosis	$\gamma_4\{x\} = 3$

variable x. From this table, one can see that $\gamma_3\{x\} > 0$ since $f(x)$ is skewed to the right (see Fig. 3.11), while $\gamma_4\{x\} > 0$ for the same reason as explained for the Laplace pdf. Furthermore, because the exponential pdf (one-sided curve) is more peaked than the Laplace pdf (double-sided curve), the normalized kurtosis of the former ($\gamma_4\{x\} = 6$) is, as expected, greater than that of the latter ($\gamma_4\{x\} = 3$).

Bernoulli Distribution

A real random variable x is said to be *Bernoulli* if it is of discrete type with the following probability assignment [24]

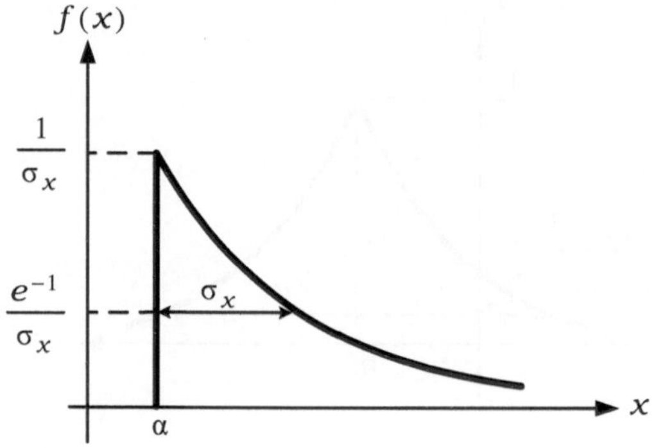

Fig. 3.11 The pdf of an exponential random variable x

Table 3.7 Summary of exponential distribution

Statistical Parameters of Exponential Random Variable x	
pdf	$f(x) = \begin{cases} \frac{1}{\sigma_x} \exp\left\{-\frac{(x-\alpha)}{\sigma_x}\right\}, & \alpha \le x < \infty,\ \sigma_x > 0 \\ 0, & \text{otherwise} \end{cases}$
Mean	$m_x = \alpha + \sigma_x$
Variance	σ_x^2
Skewness	$C_3\{x\} = 2\sigma_x^3$
Kurtosis	$C_4\{x\} = 6\sigma_x^4$
Normalized skewness	$\gamma_3\{x\} = 2$
Normalized kurtosis	$\gamma_4\{x\} = 6$

$$\Pr\{x\} = \begin{cases} p, & x = \alpha, \\ 1-p, & x = \beta, \\ 0, & \text{otherwise} \end{cases} \tag{3.100}$$

or, equivalently, the pdf

$$f(x) = p\delta(x-\alpha) + (1-p)\delta(x-\beta), \quad -\infty < x < \infty \tag{3.101}$$

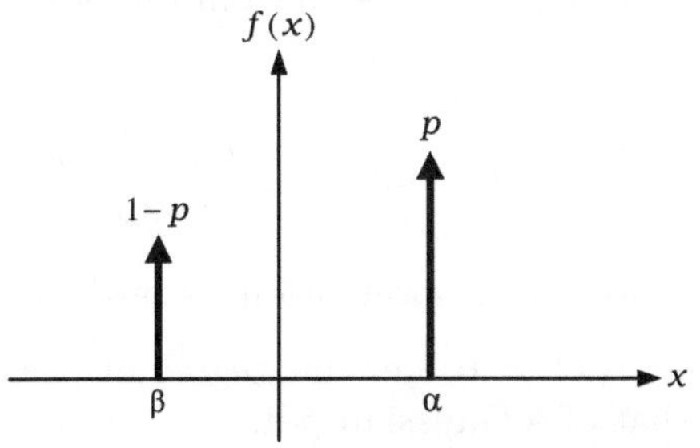

Fig. 3.12 The pdf of a Bernoulli random variable x

where $0 \le p \le 1$.[9] Figure 3.12 depicts the pdf $f(x)$ and Table 3.8 lists some statistical parameters of the Bernoulli random variable x.

Table 3.8 Summary of Bernoulli distribution

Statistical Parameters of Bernoulli Random Variable x	
pdf	$f(x) = p\delta(x-\alpha) + (1-p)\delta(x-\beta)$, $-\infty < x < \infty,\ 0 \le p \le 1$
Mean	$m_x = p(\alpha - \beta) + \beta$
Variance	$\sigma_x^2 = p(1-p)(\alpha-\beta)^2$
Skewness	$C_3\{x\} = p(1-p)(1-2p)(\alpha-\beta)^3$
Kurtosis	$C_4\{x\} = p(1-p)(1-6p+6p^2)(\alpha-\beta)^4$
Normalized skewness	$\gamma_3\{x\} = \dfrac{1-2p}{\sqrt{p(1-p)}}$
Normalized kurtosis	$\gamma_4\{x\} = \dfrac{1-6p+6p^2}{p(1-p)}$

From Table 3.8, we have the following observations regarding $\gamma_3\{x\}$.

- If $p = 1/2$, then $\gamma_3\{x\} = 0$, i.e. $f(x)$ is symmetric about m_x.
- If $p > 1/2$, then $\gamma_3\{x\} < 0$, i.e. $f(x)$ is skewed to the left.
- If $p < 1/2$, then $\gamma_3\{x\} > 0$, i.e. $f(x)$ is skewed to the right.

[9] The name *Bernoulli distribution* is due to J. Bernoulli (1713) [12]. In statistics, the Bernoulli distribution is defined as a "*binomial distribution*" with pdf $f(x) = \sum_{k=0}^{n} \binom{n}{k} p^k (1-p)^{n-k} \delta(x-k)$.

The three observations can easily be verified from Fig. 3.12. Moreover, letting $\gamma_4\{x\} = 0$ gives rise to

$$p = \frac{1}{2} - \frac{1}{2\sqrt{3}} \triangleq p_1 \quad \text{and} \quad p = \frac{1}{2} + \frac{1}{2\sqrt{3}} \triangleq p_2.$$

Accordingly, we have the following observations regarding $\gamma_4\{x\}$.

- If $p = p_1$ or p_2, then $\gamma_4\{x\} = 0$, i.e. the degree of peakedness of $f(x)$ is roughly the same as that of a Gaussian pdf.
- If $p_1 < p < p_2$, then $\gamma_4\{x\} < 0$, i.e. $f(x)$ is flatter than a Gaussian pdf.
- If $0 \le p < p_1$ or $p_2 < p \le 1$, then $\gamma_4\{x\} > 0$, i.e. $f(x)$ is more peaked than a Gaussian pdf.

To help understanding of the latter two observations, let us consider two extreme cases: (i) $p = 1/2$ (i.e. $p_1 < p < p_2$) and (ii) $p = 0$ (i.e. $p < p_1$) or $p = 1$ ($p_2 < p$). For case (i), $f(x)$ can be viewed as a uniform pdf of discrete type and thus is flatter than a Gaussian pdf. For case (ii), $f(x)$ reduces to a Dirac delta function and, obviously, is more peaked than a Gaussian pdf.

Symbol Distributions in Digital Communications

In digital communications, source information to be transmitted is typically represented by a symbol sequence along with the assumption that each symbol is drawn from a finite set of symbols (an alphabet) with equal probability. The finite set of symbols is referred to as a *constellation.* Three popular constellations as well as their associated statistical parameters are described below where all the third-order cumulants are equal to zero and all the kurtosis are negative. Note that distributions with negative kurtosis are called *sub-Gaussian distributions*, whereas distributions with positive kurtosis are called *super-Gaussian distributions* [26, pp. 16–17].

Pulse Amplitude Modulation (PAM)

For an M-ary PAM (M-PAM) symbol x, its probability assignment

$$\Pr\{x\} = \begin{cases} 1/M, & x = \pm(2m+1)d \quad \text{for } m = 0, 1, ..., (M/2) - 1, \\ 0, & \text{otherwise.} \end{cases} \tag{3.102}$$

Figure 3.13a depicts the *constellation diagram* for $M = 8$ and Table 3.9 lists some statistical parameters of the M-PAM symbol x.

Phase Shift Keying (PSK)

For an M-ary PSK (M-PSK) symbol x, its probability assignment

$$\Pr\{x\} = \begin{cases} 1/M, & x = d \cdot \exp(j2\pi m/M) \quad \text{for } m = 0, 1, ..., M-1, \\ 0, & \text{otherwise.} \end{cases} \tag{3.103}$$

Figure 3.13b depicts the constellation diagram for $M = 8$ and Table 3.9 lists some statistical parameters of the M-PSK symbol x. Note that 2-PSK is the same as 2-PAM, and 4-PSK is often referred to as *quadrature PSK (QPSK).*

Quadrature Amplitude Modulation (QAM)

For simplicity, we consider only the case of M^2-ary QAM (M^2-QAM) symbol $x = x_\mathrm{R} + jx_\mathrm{I}$, which can be treated as a combination of two statistically independent M-PAM symbols x_R and x_I. Accordingly, the probability assignment is given by $\Pr\{x\} = \Pr\{x_\mathrm{R}\} \cdot \Pr\{x_\mathrm{I}\}$ where $\Pr\{x_\mathrm{R}\}$ and $\Pr\{x_\mathrm{I}\}$ are defined as (3.102). Figure 3.13c depicts the constellation diagram of 16-QAM ($M = 4$) and Table 3.9 lists some statistical parameters of the M^2-QAM symbol x.

3.3 Random Processes

This section reviews the fundamentals of discrete-time random processes with emphasis on stationary processes and an important class of nonstationary processes, cyclostationary processes.

3.3.1 Statistical Characterization

A *random process* or a *stochastic process*[10] is a rule to assign every outcome of an experiment a function of time [10, 11]. In particular, for a discrete-time random process $x[n]$ as illustrated in Fig. 3.14, each sample point s in the sample space S is assigned a *sample function* of time n. A sample function of $x[n]$ is also called a *realization* of $x[n]$, while a complete collection of all the sample functions of $x[n]$ is called an *ensemble* of $x[n]$. On the other hand, the value of the random process $x[n]$ at a specific n is a random variable, implying that $x[n]$ can also be viewed as a sequence of random variables. As such, a real random process $x[n]$ is completely characterized by the joint pdf of the random variables $x[n_1]$, $x[n_2]$, ..., $x[n_k]$ for all n_1, n_2, ..., n_k and any k. As an example, $x[n]$ is said to be a *Gaussian process* if the random variables $x[n_1]$, $x[n_2]$, ..., $x[n_k]$ are jointly Gaussian for all n_1, n_2, ..., n_k and any k. Moreover, two real random processes $x[n]$ and $y[n]$ are completely characterized by the joint pdf of the random variables $x[n_1]$, $x[n_2]$, ..., $x[n_k]$,

[10] The word "stochastic" comes from a Greek word meaning "to guess at" [10].

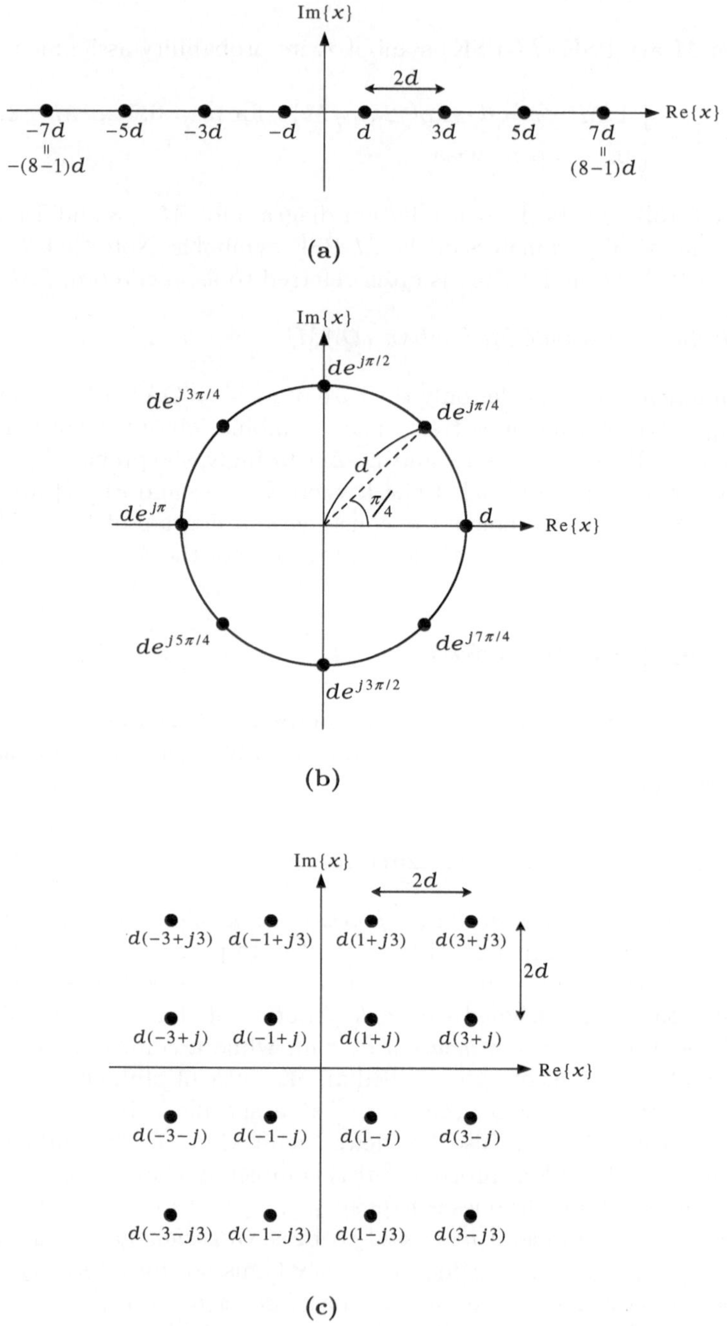

Fig. 3.13 Constellation diagram of (**a**) 8-PAM, (**b**) 8-PSK and (**c**) 16-QAM

Table 3.9 Statistical parameters of PAM/PSK/QAM symbols

M-PAM Symbol x, $M = 2, 4, 6, 8, 10, ...$	
Mean	$m_x = 0$
Variance	$\sigma_x^2 = (M^2 - 1)d^2/3$
Kurtosis	$C_4\{x\} = -2(M^4 - 1)d^4/15$
Normalized kurtosis	$\gamma_4\{x\} = -\dfrac{6(M^4 - 1)}{5(M^2 - 1)^2}$
M-PSK Symbol x, $M = 4, 8, 16, ...$	
Mean	$m_x = 0$
Variance	$\sigma_x^2 = d^2$
Kurtosis	$C_{2,2}\{x\} = -d^4$
Normalized kurtosis	$\gamma_{2,2}\{x\} = -1$
M^2-QAM Symbol x, $M = 2, 4, 8, 16, ...$	
Mean	$m_x = 0$
Variance	$\sigma_x^2 = 2(M^2 - 1)d^2/3$
Kurtosis	$C_{2,2}\{x\} = -4(M^4 - 1)d^4/15$
Normalized kurtosis	$\gamma_{2,2}\{x\} = -\dfrac{3(M^4 - 1)}{5(M^2 - 1)^2}$

† All the other second-, third- and fourth-order cumulants not listed above are equal to zero.

$y[i_1]$, $y[i_2]$, ..., $y[i_l]$, while a complex random process $x[n] = x_R[n] + jx_I[n]$ is specified in terms of the joint statistical properties of the real random processes $x_R[n]$ and $x_I[n]$. Accordingly, for random processes $x_1[n]$, $x_2[n]$, ..., $x_k[n]$, their *kth-order (joint) moment function* is defined as

$$E\{x_1[n_1]x_2[n_2]\cdots x_k[n_k]\}$$
$$= \int_{-\infty}^{\infty} \cdots \int_{-\infty}^{\infty} \alpha_1 \cdots \alpha_k f_{x_1[n_1],...,x_k[n_k]}(\alpha_1, ..., \alpha_k) d\alpha_1 \cdots d\alpha_k \tag{3.104}$$

where the expectation operator $E\{\cdot\}$ is also referred to as *ensemble average*. Their *kth-order (joint) cumulant function* $\text{cum}\{x_1[n_1], x_2[n_2], ..., x_k[n_k]\}$ is defined similarly.

Some statistics of a random process $x[n]$ are defined as follows.

- *Mean function:*

$$m_x[n] = E\{x[n]\}. \tag{3.105}$$

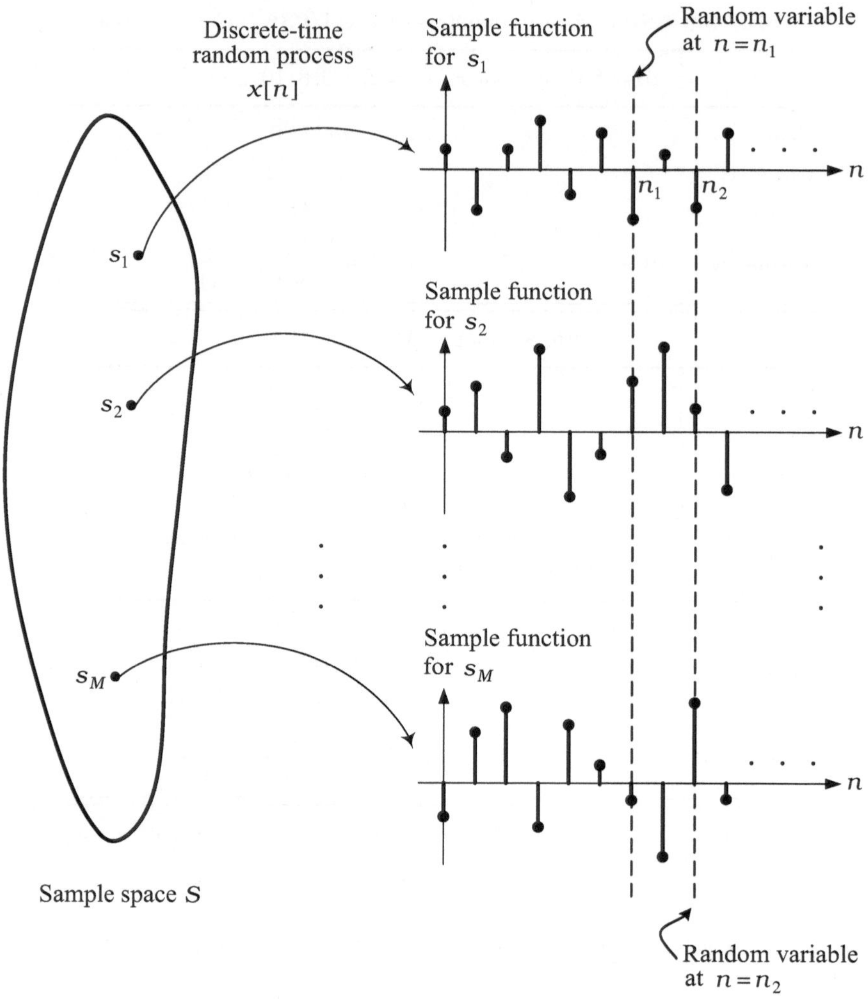

Fig. 3.14 Illustration of a discrete-time random process $x[n]$

- *Autocorrelation function*:

$$r_x[n_1, n_2] = E\{x[n_1]x^*[n_2]\}. \tag{3.106}$$

- *Autocovariance function*:

$$\begin{aligned} c_x[n_1, n_2] &= E\{(x[n_1] - m_x[n_1])(x[n_2] - m_x[n_2])^*\} \\ &= r_x[n_1, n_2] - m_x[n_1](m_x[n_2])^*. \end{aligned} \tag{3.107}$$

Definition 3.24. *A random process $x[n]$ is said to be a white process or a white noise if its mean function equals zero and the random variables $x[n_i]$ and $x[n_l]$ are uncorrelated for all $n_i \neq n_l$.*[11]

Definition 3.25. *A random process $x[n]$ is said to be independently and identically distributed (i.i.d.) if the random variables $x[n_1]$, $x[n_2]$, ..., $x[n_k]$ are statistically independent and have the same pdf (or distribution) for all n_1, n_2, ..., n_k and any k [11].*

Obviously, a white process is a special case of zero-mean, i.i.d. random process.

Moreover, two random processes $x[n]$ and $y[n]$ are said to be

- *orthogonal* if their *cross-correlation function*

$$r_{xy}[n_1, n_2] = E\{x[n_1]y^*[n_2]\} = 0 \quad \text{for all } n_1,\ n_2, \tag{3.108}$$

- *uncorrelated* if their *cross-covariance function*

$$\begin{aligned} c_{xy}[n_1, n_2] &= E\{(x[n_1] - m_x[n_1])(y[n_2] - m_y[n_2])^*\} \\ &= r_{xy}[n_1, n_2] - m_x[n_1](m_y[n_2])^* = 0 \quad \text{for all } n_1,\ n_2, \end{aligned} \tag{3.109}$$

  and
- *statistically independent* if the set of random variables $\{x[n_1], x[n_2], ..., x[n_k]\}$ is statistically independent of the set of random variables $\{y[i_1], y[i_2], ..., y[i_l]\}$ for all n_1, n_2, ..., n_k, i_1, i_2, ..., i_l and any k, l.

When $m_x[n] = m_y[n] = 0$ for all n, $c_x[n_1, n_2] = r_x[n_1, n_2]$ and $c_{xy}[n_1, n_2] = r_{xy}[n_1, n_2]$ for all n_1, n_2.

In statistical signal processing, the mean function of random processes is generally not utilized in developing algorithms and thus is often removed before any further processing. As such, in what follows, we always assume that the random processes to be dealt with are zero-mean. Hence, we are concerned with only autocorrelation and cross-correlation functions, collectively called *correlation functions*, as well as cumulant functions.

3.3.2 Stationary Processes

A real or complex random process $x[n]$ is said to be *strict-sense stationary (SSS)* or, briefly, *stationary*[12] if its statistical properties are invariant to an integer time shift of the origin, i.e.

$$f_{x[n_1],x[n_2],...,x[n_k]}(\alpha_1, ..., \alpha_k) = f_{x[n_1-\tau],x[n_2-\tau],...,x[n_k-\tau]}(\alpha_1, ..., \alpha_k) \tag{3.110}$$

---

[11] The white process defined here is precisely referred to as a "second-order" white process in [18, 19, 27], which also give the definitions of "higher-order" white processes.

[12] A discrete-time stationary process does not imply that its continuous-time counterpart is also stationary [28].

for any integer τ and any order k. Otherwise, it is said to be *nonstationary* [29]. Furthermore, two random processes $x[n]$ and $y[n]$ are said to be *jointly SSS* if all the joint statistical properties of $x[n]$ and $y[n]$ are the same as those of $x[n-\tau]$ and $y[n-\tau]$ for any integer τ.

A random process $x[n]$ is said to be *pth-order stationary* if (3.110) holds for $k = p$. Substituting (3.110) into (3.35) leads to the fact that if (3.110) holds for $k = p$, then it also holds for all $k < p$. This says that a pth-order stationary process $x[n]$ is also kth-order stationary for $k = 1, 2, ..., p-1$. Obviously, an SSS process is surely pth-order stationary for any p. From (3.110), it follows that for a first-order stationary process $x[n]$, its mean function satisfies

$$m_x[n-\tau] = \int_{-\infty}^{\infty} \alpha f_{x[n-\tau]}(\alpha) d\alpha = \int_{-\infty}^{\infty} \alpha f_{x[n]}(\alpha) d\alpha = m_x[n] \qquad (3.111)$$

for all n and any integer τ. That is, the mean function $m_x[n]$ is merely a constant and thus is also referred to as the *mean* of $x[n]$. Furthermore, for a second-order stationary process $x[n]$, its mean satisfies (3.111) and its autocorrelation function satisfies

$$\begin{aligned} r_x[n_1-\tau, n_2-\tau] &= \int_{-\infty}^{\infty}\int_{-\infty}^{\infty} \alpha_1\alpha_2^* f_{x[n_1-\tau],x^*[n_2-\tau]}(\alpha_1,\alpha_2)d\alpha_1 d\alpha_2 \\ &= \int_{-\infty}^{\infty}\int_{-\infty}^{\infty} \alpha_1\alpha_2^* f_{x[n_1],x^*[n_2]}(\alpha_1,\alpha_2)d\alpha_1 d\alpha_2 \\ &= r_x[n_1, n_2] \end{aligned} \qquad (3.112)$$

for all n_1, n_2 and any integer τ. That is, $r_x[n_1, n_2]$ depends only on the time difference $(n_1 - n_2)$ and thus can be represented as

$$r_x[l] = E\{x[n]x^*[n-l]\}. \qquad (3.113)$$

In a similar way, a necessary condition for a $(p+q)$th-order stationary process $x[n]$ is that its $(p+q)$th-order cumulant function

$$\text{cum}\{x[n_1], ..., x[n_p], x^*[n_{p+1}], ..., x^*[n_{p+q}]\}$$

depends only on the time differences $(n_1 - n_2)$, ..., $(n_1 - n_{p+q})$, and thus can be represented as

$$\begin{aligned} &C_{p,q}^x[l_1, l_2, ..., l_{p+q-1}] \\ &= \text{cum}\{x[n], x[n-l_1], ..., x[n-l_p], x^*[n-l_{p+1}], ..., x^*[n-l_{p+q-1}]\} \end{aligned} \qquad (3.114)$$

where $C_{p,q}^x[l_1, l_2, ..., l_{p+q-1}]$ is specifically denoted by $C_{p+q}^x[l_1, l_2, ..., l_{p+q-1}]$ when $x[n]$ is real. Note that $C_{1,1}^x[l] = E\{x[n]x^*[n-l]\} = r_x[l]$.

A random process $x[n]$ is said to be *wide-sense stationary (WSS)* if it satisfies (3.111) and (3.112). According to the definition, a pth-order stationary process for $p \geq 2$ is also WSS; the converse, however, may not be true.

Note that a WSS Gaussian process is also SSS since its pdf is completely determined by its mean and autocorrelation function. Moreover, two random processes $x[n]$ and $y[n]$ are said to be *jointly WSS* if both $x[n]$ and $y[n]$ are WSS and their cross-correlation function satisfies

$$r_{xy}[n_1 - \tau, n_2 - \tau] = r_{xy}[n_1, n_2] \tag{3.115}$$

for all n_1, n_2 and any integer τ. That is, $r_{xy}[n_1, n_2]$ depends only on the time difference $(n_1 - n_2)$ and thus can be represented as

$$r_{xy}[l] = E\{x[n]y^*[n-l]\}. \tag{3.116}$$

Next, let us further deal with the statistics of stationary processes.

Second-Order Statistics (SOS)

Second-order statistics include autocorrelation functions and cross-correlation functions as well as power spectra and cross-power spectra defined below.

The *power spectrum* or *spectrum* of a WSS process $x[n]$ is defined as

$$S_x(\omega) = \mathcal{F}\{r_x[l]\} = \sum_{l=-\infty}^{\infty} r_x[l]e^{-j\omega l} \tag{3.117}$$

where $r_x[l]$ is the autocorrelation function of $x[n]$. This equation is known as the *Wiener–Khintchine relation* or the *Einstein–Wiener–Khintchine relation* [24]. In other words, $r_x[l]$ can be obtained from $S_x(\omega)$ as follows:

$$r_x[l] = \mathcal{F}^{-1}\{S_x(\omega)\} = \frac{1}{2\pi}\int_{-\pi}^{\pi} S_x(\omega)e^{j\omega l}d\omega. \tag{3.118}$$

From this equation, it follows that the power of $x[n]$ can be written as

$$E\{|x[n]|^2\} = r_x[0] = \frac{1}{2\pi}\int_{-\pi}^{\pi} S_x(\omega)d\omega \tag{3.119}$$

and thereby $S_x(\omega)$ is also called the *power spectral density* of $x[n]$. On the other hand, an alternative definition of $S_x(\omega)$ is shown as [30, p. 59]

$$S_x(\omega) = \lim_{N\to\infty} E\left\{\frac{1}{2N+1}\left|\sum_{n=-N}^{N} x[n]e^{-j\omega n}\right|^2\right\} \tag{3.120}$$

provided that $\sum_{l=-\infty}^{\infty} |r_x[l]| < \infty$ (Problem 3.11).

Moreover, the *cross-power spectrum* or *cross-spectrum* of two jointly WSS processes $x[n]$ and $y[n]$ is defined as

$$S_{xy}(\omega) = \mathcal{F}\{r_{xy}[l]\} = \sum_{l=-\infty}^{\infty} r_{xy}[l]e^{-j\omega l} \tag{3.121}$$

where $r_{xy}[l]$ is the cross-correlation function of $x[n]$ and $y[n]$.

Properties of Second-Order Statistics

Suppose $x[n]$ is a WSS process. Its autocorrelation function $r_x[l]$ possesses the following two properties.

Property 3.26 (Hermitian Symmetry). *The autocorrelation function* $r_x[l] = r_x^*[-l]$ *for all* l.

Property 3.27. *The autocorrelation function* $r_x[l]$ *is bounded by* $|r_x[l]| \leq r_x[0]$ *for all* l.

Property 3.26 follows directly from the definition, while Property 3.27 can be obtained by using the Cauchy–Schwartz inequality (Theorem 3.9).

Let $\mathbf{x}[n] = (x[n], x[n-1], ..., x[n-L+1])^T$ be an $L \times 1$ random vector of $x[n]$. By Property 3.26, the correlation matrix of $\mathbf{x}[n]$ is given by

$$\mathbf{R}_x = E\{\mathbf{x}[n]\mathbf{x}^H[n]\} = \begin{pmatrix} r_x[0] & r_x[1] & \cdots & r_x[L-1] \\ r_x^*[1] & r_x[0] & \cdots & r_x[L-2] \\ \vdots & \vdots & \ddots & \vdots \\ r_x^*[L-1] & r_x^*[L-2] & \cdots & r_x[0] \end{pmatrix}, \tag{3.122}$$

which possesses the following properties.

Property 3.28. *The correlation matrix* $\mathbf{R}_x$ *is a Hermitian Toeplitz matrix.*

Property 3.29. *The correlation matrix* $\mathbf{R}_x$ *is a positive semidefinite matrix.*

Property 3.28 directly follows from (3.122), while the proof of Property 3.29 is left as an exercise (Problem 3.12). Let us emphasize that for most applications the correlation matrix is positive definite (nonsingular) [31, pp. 102, 103]. Moreover, an autocorrelation function $r_x[l]$ is said to be *positive semidefinite (definite)* if the corresponding correlation matrix $\mathbf{R}_x$ defined as (3.122) is positive semidefinite (definite) for all L.

On the other hand, by assuming that $\sum_{l=-\infty}^{\infty} |r_x[l]| < \infty$, the power spectrum $S_x(\omega)$ possesses the following two properties.

Property 3.30. *The power spectrum* $S_x(\omega)$ *is real.*

Property 3.31. *The power spectrum* $S_x(\omega)$ *is nonnegative.*

Property 3.30 directly follows from Property 3.26, while the proof of Property 3.31 is left as an exercise (Problem 3.13(a)). The fact that $\mathbf{R}_x$ is positive definite for most applications implies that $S_x(\omega) > 0$ (Problem 3.13(b)). Moreover, it is important to note that Properties 3.30 and 3.31 provide the necessary and sufficient condition for a sequence that can be treated as a legitimate autocorrelation function [24, p. 147].

Moreover, suppose $x[n]$ and $y[n]$ are joint WSS processes. Their cross-correlation function $r_{xy}[l]$ and cross-correlation matrix $\mathbf{R}_{xy} = E\{\mathbf{x}[n]\mathbf{y}^H[n]\}$ possess the following properties.

Property 3.32 (Hermitian Symmetry). *The cross-correlation function* $r_{xy}[l] = r^*_{yx}[-l]$ *for all* l.

Property 3.33. *The cross-correlation function* $r_{xy}[l]$ *is bounded by* $|r_{xy}[l]| \leq (r_x[0] \cdot r_y[0])^{1/2}$ *for all* l.

Property 3.34. *The cross-correlation matrix* $\mathbf{R}_{xy} = \mathbf{R}^H_{yx}$.

Properties 3.32 and 3.34 directly follow from the definitions, while Property 3.33 can be obtained by using the Cauchy–Schwartz inequality (Theorem 3.9). Furthermore, suppose the cross-power spectrum $S_{xy}(\omega)$ of $x[n]$ and $y[n]$ exists for all ω. Then Property 3.32 implies the following property.

Property 3.35. *The cross-power spectrum* $S_{xy}(\omega) = S^*_{yx}(\omega)$.

LTI Systems Driven by WSS Processes

As depicted in Fig. 3.15, consider that $x[n]$ is a random signal generated from the following convolutional model:

$$x[n] = h[n] \star u[n] = \sum_{k=-\infty}^{\infty} h[k]u[n-k] \tag{3.123}$$

where $u[n]$ is the source signal (the driving input) and $h[n]$ is an SISO LTI system. For the signal model, let us make the following assumptions.

(SOS-1) The SISO LTI system $h[n]$ is stable, i.e. $h[n]$ is absolutely summable.
(SOS-2) The source signal $u[n]$ is a WSS white process with variance σ_u^2.

Under Assumptions (SOS-1) and (SOS-2), the random signal $x[n]$ given by (3.123) is called an *ARMA process* if $h[n]$ is an ARMA system. It reduces to an *AR process* and an *MA process* when $h[n]$ is an AR system and an MA system, respectively. Moreover, by definition, Assumption (SOS-2) implies that the autocorrelation function of $u[n]$ is given by

$$r_u[l] = \sigma_u^2 \cdot \delta[l] \tag{3.124}$$

and correspondingly the power spectrum of $u[n]$ is given by

$$S_u(\omega) = \sigma_u^2 \quad \text{for all } \omega. \tag{3.125}$$

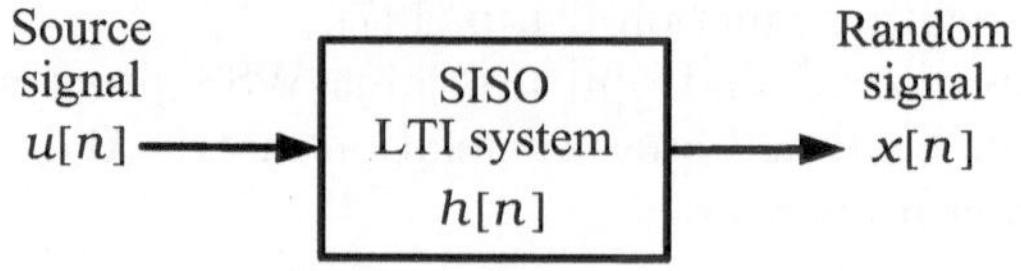

Fig. 3.15 Signal model for generating the random signal $x[n]$

That is, $u[n]$ has a flat power spectrum, thereby leading to the name *white process*; otherwise, it is said to be a *colored process.*

Using (3.123) and (3.124), we obtain the autocorrelation function of $x[n]$ as follows:

$$\begin{aligned} E\{x[n_1]x^*[n_2]\} &= \sum_{k=-\infty}^{\infty} \sum_{m=-\infty}^{\infty} h[k]h^*[m]r_u[-k+n_1-n_2+m] \\ &= \sigma_u^2 \sum_{k=-\infty}^{\infty} h[k]h^*[k-(n_1-n_2)]. \end{aligned}$$

This reveals that $E\{x[n_1]x^*[n_2]\}$ depends only on the time difference $(n_1 - n_2)$ and thus can be written as

$$r_x[l] = \sigma_u^2 \sum_{k=-\infty}^{\infty} h[k]h^*[k-l]. \tag{3.126}$$

In other words, under (SOS-2), $x[n]$ is also WSS provided that $|r_x[l]| < \infty$ for all l or, by Property 3.27, $r_x[0] < \infty$. It can be shown that under (SOS-1) and (SOS-2), $\sum_{l=-\infty}^{\infty} |r_x[l]| < \infty$ (a special case of Problem 3.14) and accordingly $r_x[0] < \infty$. As a consequence, (SOS-1) and (SOS-2) ensure that $x[n]$ is WSS.

Taking the DTFT of (3.126) gives the power spectrum of $x[n]$ as

$$S_x(\omega) = \sigma_u^2 \cdot |H(\omega)|^2. \tag{3.127}$$

By Theorem 2.37, the absolute summability of $r_x[l]$ implies that the DTFT of $r_x[l]$ converges uniformly and absolutely to a continuous function, namely, $S_x(\omega)$, on $[-\pi, \pi)$. In other words, the existence of $S_x(\omega)$ for all ω is guaranteed under (SOS-1) and (SOS-2). Moreover, an important observation from (3.127) is that the phase information about the system $h[n]$ has been completely lost in $S_x(\omega)$ or equivalently in $r_x[l]$, and thus the second-order statistics $r_x[l]$ and $S_x(\omega)$ are said to be *phase blind.* This implies that one cannot use only second-order statistics to extract or identify a system's phase.

Example 3.36

Consider the following two cases for the signal model (3.123).

- Case I: The system $H(z) = H_1(z) = (1 - az^{-1})(1 - bz^{-1})$ and the corresponding system output $x[n]$ is denoted by $x_1[n]$ where a and b are real numbers, $|a| < 1$ and $|b| < 1$.
- Case II: The system $H(z) = H_2(z) = (1 - az^{-1})(1 - bz)$ and the corresponding system output $x[n]$ is denoted by $x_2[n]$.

Both systems $H_1(z)$ and $H_2(z)$ are FIR, thereby satisfying (SOS-1). Moreover, $H_1(z)$ is minimum phase with two zeros at $z = a$ and $z = b$ (both inside the unit circle), whereas $H_2(z)$ is nonminimum phase with two zeros at $z = a$ (inside the unit circle) and $z = 1/b$ (outside the unit circle). Figures 3.16a, b plot the magnitude and phase responses of $H_1(z)$, respectively, and those of $H_2(z)$ for $a = 1/2$ and $b = 1/4$. One can see that $H_1(z)$ and $H_2(z)$ have identical magnitude responses but different phase responses.

Suppose $u[n]$ satisfies (SOS-2). Then, using (3.127), we obtain the same power spectra of $x_1[n]$ and $x_2[n]$ as follows:

$$S_{x_1}(\omega) = S_{x_2}(\omega) = \sigma_u^2(1 + a^2 - 2a\cos\omega)(1 + b^2 - 2b\cos\omega).$$

Figure 3.17 plots $S_{x_1}(\omega)$ (solid line) and $S_{x_2}(\omega)$ (dashed line) for $a = 1/2$, $b = 1/4$ and $\sigma_u^2 = 1$ where the solid and dashed lines overlap. These results reveal that $H_1(z)$ and $H_2(z)$ cannot be identified with only the power spectrum. □

Next, let us further consider the noisy signal model shown in Fig. 3.18 where

$$y[n] = x[n] + w[n] \tag{3.128}$$

is the noisy signal, $w[n]$ is the additive noise, and $x[n]$ is the noise-free signal generated from (3.123) with Assumptions (SOS-1) and (SOS-2). For the noisy signal model, two more assumptions are made as follows.

(SOS-3) The noise $w[n]$ is a zero-mean WSS (white or colored) process.
(SOS-4) The source signal $u[n]$ is statistically independent of the noise $w[n]$.

Assumption (SOS-4) indicates that $x[n]$ given by (3.123) is also statistically independent of $w[n]$. From (3.128), (SOS-3) and (SOS-4), it follows that the autocorrelation function of $y[n]$ is given by

$$r_y[l] = r_x[l] + r_w[l] \tag{3.129}$$

and the power spectrum of $y[n]$ is given by

$$S_y(\omega) = S_x(\omega) + S_w(\omega) \tag{3.130}$$

where $r_w[l]$ and $S_w(\omega)$ are the autocorrelation function and power spectrum of $w[n]$, respectively. Define the *signal-to-noise ratio (SNR)* associated with $y[n]$ as

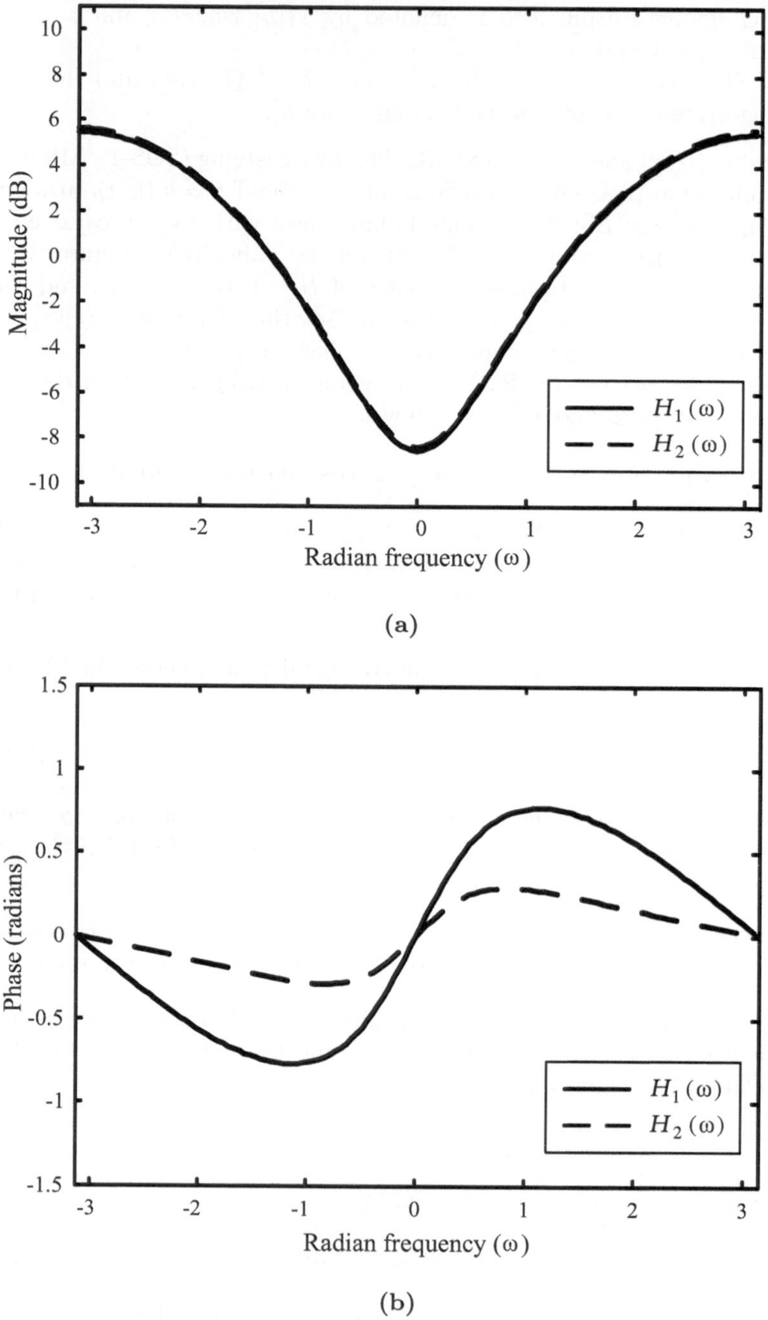

Fig. 3.16 (**a**) The magnitude responses and (**b**) the phase responses of $H_1(z)$ and $H_2(z)$ for $a = 1/2$ and $b = 1/4$

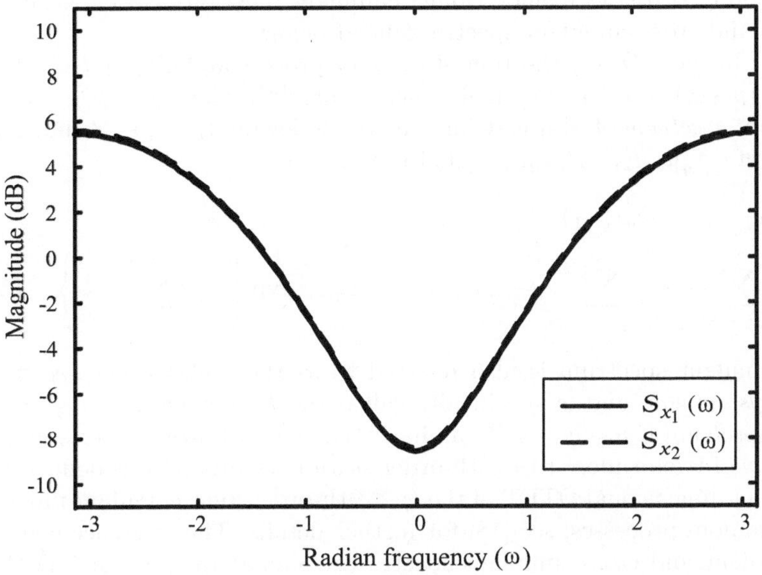

Fig. 3.17 The power spectra $S_{x_1}(\omega)$ and $S_{x_2}(\omega)$

$$\text{SNR} = \frac{E\{|x[n]|^2\}}{E\{|w[n]|^2\}}. \tag{3.131}$$

Then, from (3.129), (3.130) and Property 3.27, we note that in theory, the situation that $r_y[l] = r_x[l]$ for all l or, equivalently, $S_y(\omega) = S_x(\omega)$ for all ω happens only when the noise power (variance) $r_w[0] = E\{|w[n]|^2\} = 0$, i.e. SNR $= \infty$. In practice, however, the SNR is always finite and thus the performance of most SOS based algorithms is sensitive to additive WSS noise.

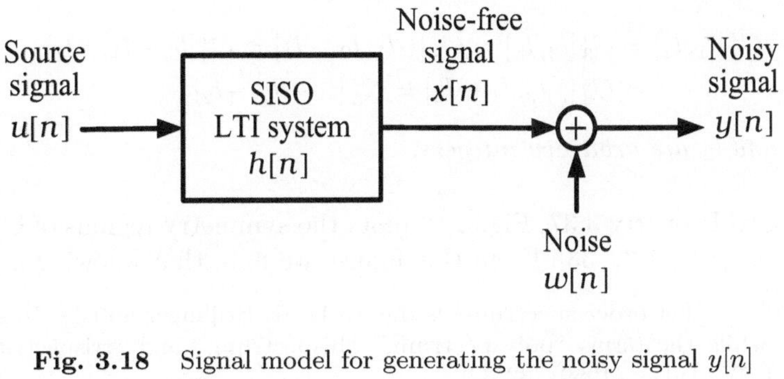

Fig. 3.18 Signal model for generating the noisy signal $y[n]$

Higher-Order Statistics (HOS)

Higher-order statistics include (joint) cumulant functions as well as cumulant spectra and cross-cumulant spectra defined below.

Let $x[n]$ be a $(p+q)$th-order stationary process and $C^x_{p,q}[l_1, l_2, ..., l_{p+q-1}]$ be the $(p+q)$th-order cumulant function of $x[n]$. Then the $(p+q)$*th-order cumulant spectrum* of $x[n]$ is defined as the following $(p+q-1)$-dimensional DTFT of $C^x_{p,q}[l_1, l_2, ..., l_{p+q-1}]$: [16, 18, 19, 32, 33]

$$\begin{aligned} & S^x_{p,q}(\omega_1, \omega_2, ..., \omega_{p+q-1}) \\ &= \sum_{l_1=-\infty}^{\infty} \cdots \sum_{l_{p+q-1}=-\infty}^{\infty} C^x_{p,q}[l_1, l_2, ..., l_{p+q-1}] \exp\left\{ -j \sum_{i=1}^{p+q-1} \omega_i l_i \right\}. \end{aligned} \tag{3.132}$$

The cumulant spectrum is also referred to as the *higher-order spectrum* or the *polyspectrum*, and is specifically called the *bispectrum* if $p+q=3$ and the *trispectrum* if $p+q=4$.[13] Similarly, the $(p+q)$*th-order cross-cumulant spectrum* of two or more $(p+q)$th-order stationary processes is defined as the $(p+q-1)$-dimensional DTFT of the $(p+q)$th-order joint cumulant function of these random processes; see [18] for further details. The existence conditions of cumulant and cross-cumulant spectra are similar to those of 1-D DTFT, which we have discussed in Section 3.1.2.

Properties of Higher-Order Statistics

From Property 3.13, it follows that the third-order cumulant function of a stationary process $x[n]$ possesses the following property.

Property 3.37 (Symmetry). *For complex $x[n]$ the third-order cumulant function*

$$\begin{aligned} C^x_{2,1}[l_1, l_2] &= C^x_{2,1}[-l_1, l_2 - l_1] \\ &= \left(C^x_{1,2}[-l_2, l_1 - l_2]\right)^* = \left(C^x_{1,2}[l_1 - l_2, -l_2]\right)^*, \end{aligned} \tag{3.133}$$

while for real $x[n]$ the third-order cumulant function

$$\begin{aligned} C^x_3[l_1, l_2] &= C^x_3[l_2, l_1] = C^x_3[-l_1, l_2 - l_1] = C^x_3[l_2 - l_1, -l_1] \\ &= C^x_3[-l_2, l_1 - l_2] = C^x_3[l_1 - l_2, -l_2] \end{aligned} \tag{3.134}$$

where l_1 and l_2 are arbitrary integers.

According to Property 3.37, Fig. 3.19 plots the symmetry regions of $C^x_3[l_1, l_2]$ for real $x[n]$ [18, 19, 32, 33]. From this figure, we note that knowing $C^x_3[l_1, l_2]$

[13] The term "higher-order spectrum" is due to D. R. Brillinger and M. Rosenblatt (1967), while the terms "polyspectrum," "bispectrum," and "trispectrum" are due to J. W. Tukey (1959) [32, 34].

in any one of the six regions is equivalent to knowing the entire $C_3^x[l_1, l_2]$. In light of this property, computational complexity for algorithms involving the entire $C_3^x[l_1, l_2]$ can be significantly reduced. Furthermore, also by Property 3.13, one can easily show that $C_{p,q}^x[l_1, l_2, ..., l_{p+q-1}]$ for $p + q > 3$ exhibits more symmetry regions than $C_{2,1}^x[l_1, l_2]$. It is, however, tedious to specify all the symmetry properties for $p + q > 3$, which are therefore omitted here.

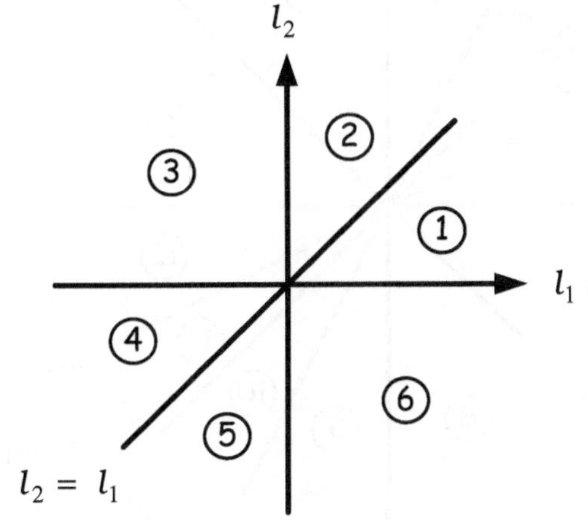

Fig. 3.19 Six symmetry regions of $C_3^x[l_1, l_2]$ for real $x[n]$

On the other hand, suppose the bispectrum $S_{2,1}^x(\omega_1, \omega_2)$ of $x[n]$ exists for all ω_1, ω_2. Then the following property directly follows from Property 3.37.

Property 3.38 (Symmetry). *For complex $x[n]$ the bispectrum*

$$\begin{aligned} S_{2,1}^x(\omega_1, \omega_2) &= S_{2,1}^x(-\omega_1 - \omega_2, \omega_2) \\ &= \left[S_{1,2}^x(\omega_1 + \omega_2, -\omega_1)\right]^* = \left[S_{1,2}^x(-\omega_1, \omega_1 + \omega_2)\right]^*, \end{aligned} \tag{3.135}$$

while for real $x[n]$ the bispectrum

$$\begin{aligned} S_3^x(\omega_1, \omega_2) &= [S_3^x(-\omega_1, -\omega_2)]^* = S_3^x(\omega_2, \omega_1) \\ &= S_3^x(-\omega_1 - \omega_2, \omega_2) = S_3^x(\omega_2, -\omega_1 - \omega_2) \\ &= S_3^x(-\omega_1 - \omega_2, \omega_1) = S_3^x(\omega_1, -\omega_1 - \omega_2) \end{aligned} \tag{3.136}$$

where $-\pi \leq \omega_1 < \pi$ and $-\pi \leq \omega_2 < \pi$.

According to Property 3.38, Fig. 3.20 plots the symmetry regions of $S_3^x(\omega_1, \omega_2)$ for real $x[n]$ [18, 19, 32, 33]. Similarly, knowing $S_3^x(\omega_1, \omega_2)$ in any one of the twelve regions is equivalent to knowing the entire $S_3^x(\omega_1, \omega_2)$, thereby leading to significant simplification of bispectrum computation. Furthermore, the trispectrum and other higher-order cumulant spectra of $x[n]$ also possess similar symmetry properties, which are omitted here for brevity.

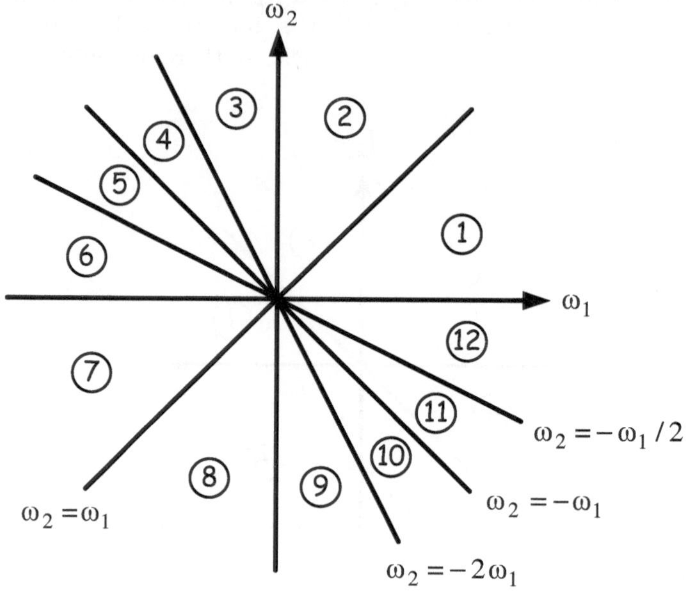

Fig. 3.20 Twelve symmetry regions of $S_3^x(\omega_1, \omega_2)$ for real $x[n]$

LTI Systems Driven by Stationary Processes

For the noisy signal model shown in Fig. 3.18, algorithms using HOS are generally based on the following assumptions.

(HOS-1) The SISO LTI system $h[n]$ is stable.
(HOS-2) The source signal $u[n]$ is a zero-mean, i.i.d., stationary non-Gaussian process with $(p+q)$th-order cumulant $C_{p,q}\{u[n]\} \neq 0$.
(HOS-3) The noise $w[n]$ is a zero-mean WSS (white or colored) Gaussian process.
(HOS-4) The source signal $u[n]$ is statistically independent of the noise $w[n]$.

Under Assumptions (HOS-1) and (HOS-2), the random signal $x[n]$ given by (3.123) is called a *linear process* [35]. Moreover, by Property 3.18, Assumption (HOS-2) implies that the $(p+q)$th-order cumulant function of $u[n]$ is given by [15, 33]

$$C_{p,q}^u[l_1, l_2, ..., l_{p+q-1}] = C_{p,q}\{u[n]\} \cdot \delta[l_1]\delta[l_2] \cdots \delta[l_{p+q-1}]. \tag{3.137}$$

Correspondingly, the $(p+q)$th-order cumulant spectrum of $u[n]$ is given by

$$S_{p,q}^{u}(\omega_1, \omega_2, ..., \omega_{p+q-1}) = C_{p,q}\{u[n]\} \quad \text{for all } \omega_1, \omega_2, ..., \omega_{p+q-1}, \tag{3.138}$$

that is, the polyspectrum is flat for all frequencies.

Using (3.123), (3.137) and Properties 3.14 and 3.16, one can express the $(p+q)$th-order cumulant function of $x[n]$ as follows:

$$\begin{aligned}
&\mathrm{cum}\{x[n_1], ..., x[n_p], x^*[n_{p+1}], ..., x^*[n_{p+q}]\} \\
&= \sum_{k_1=-\infty}^{\infty} \cdots \sum_{k_{p+q}=-\infty}^{\infty} h[k_1]\cdots h[k_p]h^*[k_{p+1}]\cdots h^*[k_{p+q}] \\
&\quad \cdot \mathrm{cum}\{u[n_1-k_1], ..., u[n_p-k_p], u^*[n_{p+1}-k_{p+1}], ..., u^*[n_{p+q}-k_{p+q}]\} \\
&= C_{p,q}\{u[n]\} \cdot \sum_{k_1=-\infty}^{\infty} h[k_1]h[k_1-(n_1-n_2)]\cdots h[k_1-(n_1-n_p)] \\
&\qquad\qquad \cdot h^*[k_1-(n_1-n_{p+1})]\cdots h^*[k_1-(n_1-n_{p+q})].
\end{aligned}$$

This reveals that $\mathrm{cum}\{x[n_1], ..., x[n_p], x^*[n_{p+1}], ..., x^*[n_{p+q}]\}$ depends only on the time differences $(n_1 - n_2)$, ..., $(n_1 - n_{p+q})$, and thus can be written as

$$\begin{aligned}
C_{p,q}^{x}[l_1, l_2, ..., l_{p+q-1}] = C_{p,q}\{u[n]\} \cdot \sum_{k=-\infty}^{\infty} h[k]h[k-l_1]\cdots h[k-l_{p-1}] \\
\cdot h^*[k-l_p]\cdots h^*[k-l_{p+q-1}].
\end{aligned} \tag{3.139}$$

In other words, under (HOS-2), $x[n]$ satisfies the necessary condition of $(p+q)$th-order stationarity provided that $|C_{p,q}^{x}[l_1, l_2, ..., l_{p+q-1}]| < \infty$ for all l_1, l_2, ..., l_{p+q-1}. It can be shown [35, p. 302] that under (HOS-1) and (HOS-2),

$$\sum_{l_1=-\infty}^{\infty} \cdots \sum_{l_{p+q-1}=-\infty}^{\infty} \left|C_{p,q}^{x}[l_1, l_2, ..., l_{p+q-1}]\right| < \infty \tag{3.140}$$

(Problem 3.14) and accordingly $|C_{p,q}^{x}[l_1, l_2, ..., l_{p+q-1}]| < \infty$ for all l_1, l_2, ..., l_{p+q-1}. In fact, under (HOS-1) and if $u[n]$ is stationary, then $x[n]$ is stationary, too [11, p. 309]. As a consequence, (HOS-1) and (HOS-2) ensure that $x[n]$ is stationary. Moreover, for ease of use in subsequent chapters, a useful formula for generalizing (3.139) is summarized in the following theorem (Problem 3.15).

Theorem 3.39. *Suppose $z_1[n]$, $z_2[n]$, ..., $z_{p+q}[n]$ are real or complex random processes modeled as $z_i[n] = h_i[n] \star u[n]$, $i = 1, 2, ..., p+q$, where $u[n]$ is a stationary process satisfying Assumption* (HOS-2) *and $h_i[n]$ are arbitrary stable LTI systems. Then*

$$\begin{aligned}
&\mathrm{cum}\{z_1[n-l_1], ..., z_p[n-l_p], z_{p+1}^*[n-l_{p+1}], ..., z_{p+q}^*[n-l_{p+q}]\} \\
&= C_{p,q}\{u[n]\} \cdot \sum_{k=-\infty}^{\infty} h_1[k-l_1]\cdots h_p[k-l_p]h_{p+1}^*[k-l_{p+1}]\cdots h_{p+q}^*[k-l_{p+q}]
\end{aligned}$$

for cumulant order $p+q \geq 2$.

Taking the $(p+q-1)$-dimensional DTFT of (3.139) yields[14]

$$\begin{aligned} S^x_{p,q}(\omega_1,\omega_2,...,\omega_{p+q-1}) &= C_{p,q}\{u[n]\}\cdot\sum_k h[k] \\ &\quad \cdot\prod_{i=1}^{p-1}\left(\sum_{l_i} h[k-l_i]e^{-j\omega_i l_i}\right)\cdot\prod_{i=p}^{p+q-1}\left(\sum_{l_i} h^*[k-l_i]e^{-j\omega_i l_i}\right) \\ &= C_{p,q}\{u[n]\}\cdot H\left(\sum_{i=1}^{p+q-1}\omega_i\right)\cdot\prod_{i=1}^{p-1}H(-\omega_i)\cdot\prod_{i=p}^{p+q-1}H^*(\omega_i). \end{aligned} \tag{3.141}$$

Under (HOS-1) and (HOS-2), the existence and continuity of the cumulant spectrum $S^x_{p,q}(\omega_1,\omega_2,...,\omega_{p+q-1})$ follow from Theorem 2.37 and (3.140). Two important special cases of (3.141) are as follows.

- Bispectrum: $(p,q)=(2,1)$

$$S^x_{2,1}(\omega) = C_{2,1}\{u[n]\}\cdot H(\omega_1+\omega_2)H(-\omega_1)H^*(\omega_2). \tag{3.142}$$

- Trispectrum: $(p,q)=(2,2)$

$$S^x_{2,2}(\omega) = C_{2,2}\{u[n]\}\cdot H(\omega_1+\omega_2+\omega_3)H(-\omega_1)H^*(\omega_2)H^*(\omega_3). \tag{3.143}$$

An example regarding the bispectrum is as follows.

Example 3.40
Consider, again, the two cases in Example 3.36 where $u[n]$ is assumed to satisfy (HOS-2). Using (3.142), we obtain the bispectrum of $x_1[n]$, $S^{x_1}_3(\omega_1,\omega_2)$, displayed in Fig. 3.21 and the bispectrum of $x_2[n]$, $S^{x_2}_3(\omega_1,\omega_2)$, displayed in Fig. 3.22 for $a=1/2$, $b=1/4$ and $C_3\{u[n]\}=1$. One can see that $S^{x_1}_3(\omega_1,\omega_2)$ and $S^{x_2}_3(\omega_1,\omega_2)$ have identical magnitudes but different phases, which therefore provides the distinguishability of $H_1(z)$ and $H_2(z)$.

□

Unlike second-order statistics, higher-order (≥ 3) statistics (cumulant functions or polyspectra) of the non-Gaussian signal $x[n]$ contain not only magnitude information but also phase information about the system $h[n]$. It is for this reason that most blind equalization/identification algorithms either explicitly or implicitly employ higher-order (≥ 3) statistics, provided that the signals involved are non-Gaussian.

As for the noisy signal $y[n]$ given by (3.128), by (HOS-3), (HOS-4) and Property 3.17, the $(p+q)$th-order cumulant function of $y[n]$ is given by

$$\begin{aligned} C^y_{p,q}[l_1,l_2,...,l_{p+q-1}] &= C^x_{p,q}[l_1,l_2,...,l_{p+q-1}] + C^w_{p,q}[l_1,l_2,...,l_{p+q-1}] \\ &= C^x_{p,q}[l_1,l_2,...,l_{p+q-1}] \end{aligned} \tag{3.144}$$

[14] As reported by Mendel [15], a major generation of (3.126) and (3.127) to (3.139) and (3.141) was established by M. S. Bartlett (1955) and by D. R. Brillinger and M. Rosenblatt (1967).

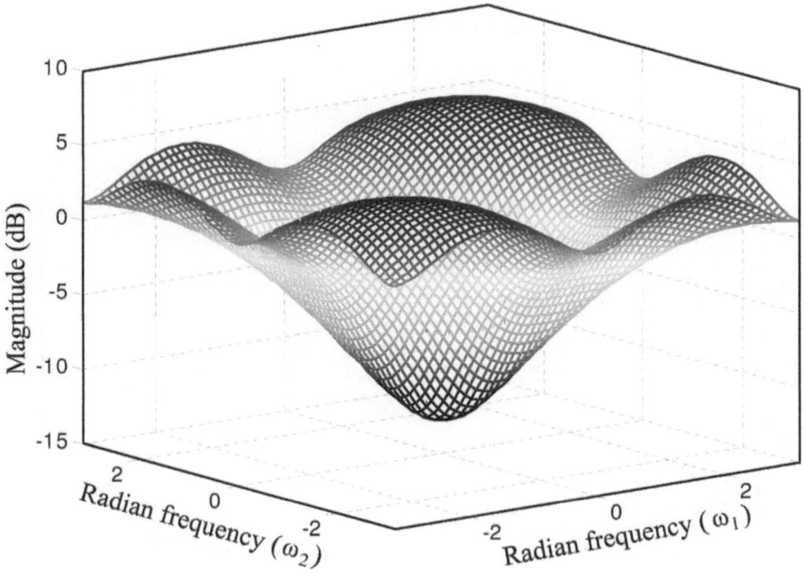

(a)

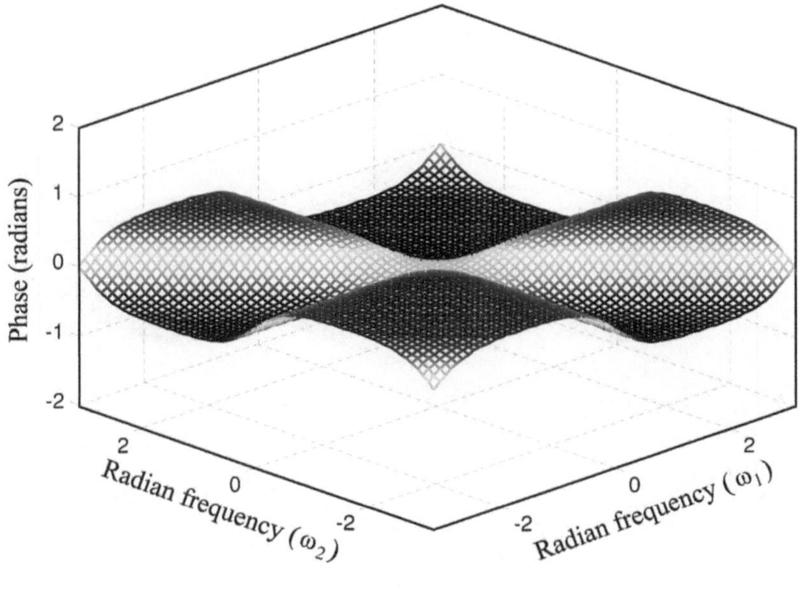

(b)

Fig. 3.21 (**a**) The magnitude and (**b**) the phase of the bispectrum $S_3^{x_1}(\omega_1, \omega_2)$

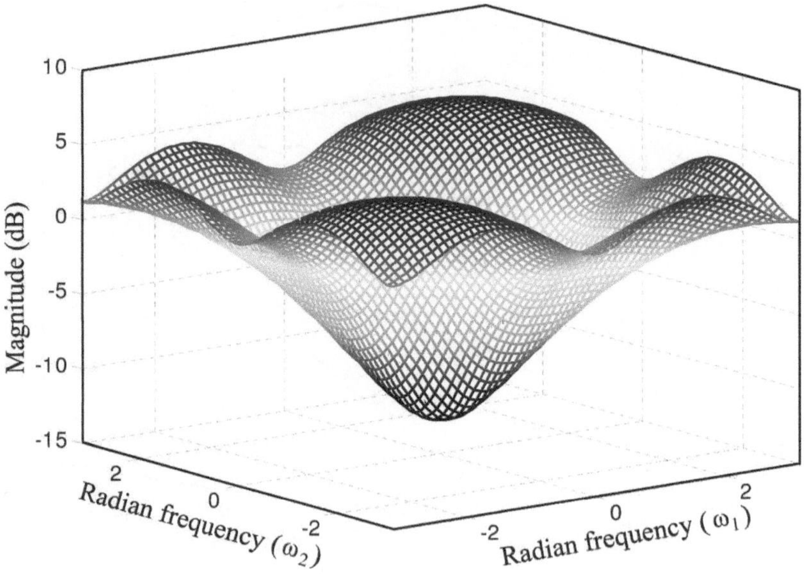

(a)

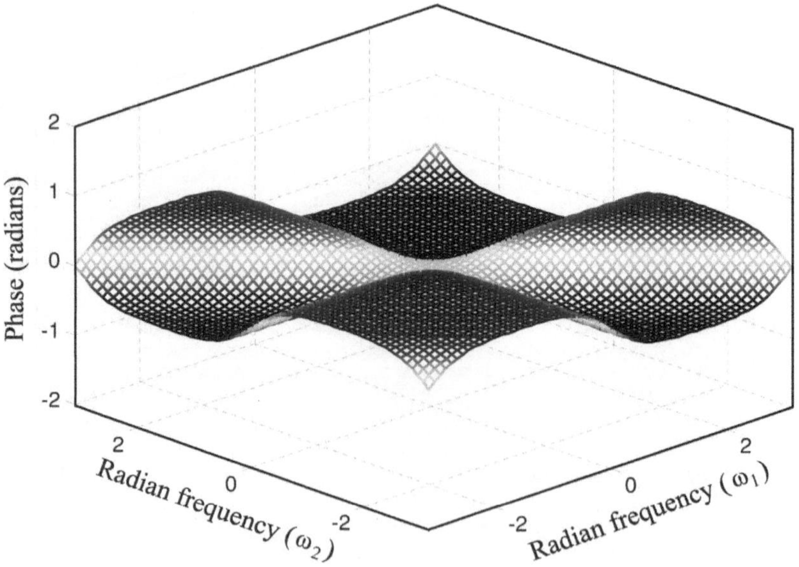

(b)

Fig. 3.22 (**a**) The magnitude and (**b**) the phase of the bispectrum $S_3^{x_2}(\omega_1, \omega_2)$

where $p+q \geq 3$ and

$$C_{p,q}^{w}[l_1, l_2, ..., l_{p+q-1}] = 0 \quad \text{for all } l_1, l_2, ..., l_{p+q-1} \tag{3.145}$$

regardless of what the SNR is. Correspondingly, the $(p+q)$th-order cumulant spectrum of $y[n]$ is given by

$$\begin{aligned} S_{p,q}^{y}(\omega_1, \omega_2, ..., \omega_{p+q-1}) &= S_{p,q}^{x}(\omega_1, \omega_2, ..., \omega_{p+q-1}) + S_{p,q}^{w}(\omega_1, \omega_2, ..., \omega_{p+q-1}) \\ &= S_{p,q}^{x}(\omega_1, \omega_2, ..., \omega_{p+q-1}) \end{aligned} \tag{3.146}$$

where $p+q \geq 3$ and

$$S_{p,q}^{w}(\omega_1, \omega_2, ..., \omega_{p+q-1}) = 0 \quad \text{for all } \omega_1, \omega_2, ..., \omega_{p+q-1}. \tag{3.147}$$

As a consequence of (3.144) and (3.146), the performance of algorithms using only HOS (order ≥ 3) are insensitive to Gaussian noise no matter whether the noise is white or colored. Incidently, (3.145) and (3.147) imply that for an arbitrary zero-mean stationary process $z[n]$, the function $\left|C_{p,q}^{z}[l_1, l_2, ..., l_{p+q-1}]\right|$ or $\left|S_{p,q}^{z}(\omega_1, \omega_2, ..., \omega_{p+q-1})\right|$ can be used as a measure of how far $z[n]$ deviates from a zero-mean WSS Gaussian process having the same SOS as $z[n]$ [15, 18, 19, 36].

Over the past several decades, HOS have been applied to a wide variety of science and engineering areas including communications, sonar, radar, speech, image, geophysics, astronomy, biomedicine, optics, mechanics, and so on. A comprehensive bibliography by Swami *et al.* [37] offers a collection of 1423 references regarding HOS in the period 1984–1994, while another bibliography by Delaney and Walsh [38] also offers a collection of about 280 references in a similar time period.

3.3.3 Cyclostationary Processes

A real or complex random process $x[n]$ is said to be *strict-sense cyclostationary (SSCS)* or, briefly, *cyclostationary with period M* if its kth-order pdf satisfies

$$f_{x[n_1],x[n_2],...,x[n_k]}(\alpha_1, ..., \alpha_k) = f_{x[n_1-\tau M],x[n_2-\tau M],...,x[n_k-\tau M]}(\alpha_1, ..., \alpha_k)$$

for any integer τ and any order k where M is an integer.[15] Furthermore, two random processes $x[n]$ and $y[n]$ are said to be *jointly SSCS with period M* if all the joint statistical properties of $x[n]$ and $y[n]$ are the same as those of $x[n-\tau M]$ and $y[n-\tau M]$ for any integer τ. Clearly, if $M=1$, then SSCS (jointly SSCS) processes reduce to SSS (jointly SSS) processes.

[15] As reported by Gardner and Franks [39], the term "cyclostationary" was introduced by W.R. Bennett (1958), while some other investigators have used the terms "periodically stationary," "periodically correlated," and "periodic nonstationary."

A random process $x[n]$ is said to be *wide-sense cyclostationary (WSCS) with period M* if (i) its mean function satisfies

$$m_x[n - \tau M] = m_x[n] \tag{3.148}$$

for all n and any integer τ and (ii) its autocorrelation function satisfies

$$r_x[n - \tau M, n - l - \tau M] = r_x[n, n - l] \tag{3.149}$$

for all n, l and any integer τ. That is, the mean and autocorrelation functions are both periodic in n with period M. Note that an SSCS process with period M is also WSCS with period M; the converse, however, may not be true. Furthermore, two random processes $x[n]$ and $y[n]$ are said to be *jointly WSCS with period M* if both $x[n]$ and $y[n]$ are WSCS with period M and their cross-correlation function satisfies

$$r_{xy}[n - \tau M, n - l - \tau M] = r_{xy}[n, n - l] \tag{3.150}$$

for all n, l and any integer τ. That is, $r_{xy}[n, n-l]$ is periodic in n with period M. Clearly, if $M = 1$, then WSCS (jointly WSCS) processes reduce to WSS (jointly WSS) processes.

Moreover, given a WSCS process, the following theorem shows that an associated WSS process can be derived by applying a random integer shift to the WSCS process [11,39–41] (Problem 3.16).

Theorem 3.41. *Suppose $x[n]$ is a WSCS process with period M and η is a random integer uniformly distributed in $[0, M-1]$, and η is statistically independent of $x[n]$. Then the randomly shifted process $y[n] = x[n-\eta]$ is WSS with mean*

$$m_y = \frac{1}{M} \sum_{n=0}^{M-1} m_x[n] \tag{3.151}$$

and autocorrelation function

$$r_y[l] = \frac{1}{M} \sum_{n=0}^{M-1} r_x[n, n-l] \tag{3.152}$$

where $m_x[n]$ and $r_x[n, n-l]$ are the mean and autocorrelation functions of $x[n]$, respectively.

Second-Order Cyclostationary Statistics (SOCS)

Suppose $x[n]$ is a WSCS process with period M. Since the autocorrelation function $r_x[n, n-l]$ of $x[n]$ is periodic in n with period M, it can be expanded as a Fourier series with the αth Fourier coefficient [42,43]

$$r_x^{[\alpha]}[l] = \frac{1}{M}\sum_{n=0}^{M-1} r_x[n, n-l]e^{-j2\pi\alpha n/M}, \quad \alpha = 0, 1, ..., M-1. \tag{3.153}$$

Note that (3.153) is a discrete-time version of (2.76) by taking the discrete-time nature of $r_x[n, n-l]$ into account. The Fourier coefficient $r_x^{[\alpha]}[l]$ is known as the *cyclic autocorrelation function* of $x[n]$ indexed by the *cycle frequency* parameter α [10,40,42]. Moreover, the *cyclic spectrum* or *cyclic spectral density* of $x[n]$ is defined as the DTFT of $r_x^{[\alpha]}[l]$ (with respect to l) given by

$$S_x^{[\alpha]}(\omega) = \mathscr{F}\{r_x^{[\alpha]}[l]\} = \sum_{l=-\infty}^{\infty} r_x^{[\alpha]}[l]e^{-j\omega l}, \quad \alpha = 0, 1, ..., M-1. \tag{3.154}$$

Note that (3.153) reduces to (3.152) for $\alpha = 0$, implying that $r_x^{[0]}[l]$ and $S_x^{[0]}(\omega)$ correspond, respectively, to the autocorrelation function and the power spectrum of a WSS process derived from $x[n]$ (see Theorem 3.41). In other words, like second-order statistics, $r_x^{[0]}[l]$ and $S_x^{[0]}(\omega)$ are also phase blind. However, this is generally not true for $r_x^{[\alpha]}[l]$ and $S_x^{[\alpha]}(\omega)$ for $\alpha \geq 1$ (to be revealed below).

Second-order cyclostationary statistics (SOCS) are of particular importance in digital communications where cyclostationarity arises from the periodic formation of digitally modulated signals. The treatment of SOCS for digitally modulated signals will be given next, but the topic of higher-order cyclostationary statistics (HOCS) will not be pursued in this book. The reader can refer to [44, 45] and the references therein for the treatment of HOCS.

Cyclostationarity of Digital Communication Signals

Consider the digital communication system depicted in Fig. 3.23. The symbol $u[k]$ drawn from a finite set of alphabets is transmitted at time kT where T is the *symbol period*, and the continuous-time received signal is given by

$$y(t) = \sum_{k=-\infty}^{\infty} u[k]h(t-kT) + w(t) \tag{3.155}$$

where $h(t)$ is the continuous-time SISO LTI channel accounting for actual channel response as well as transmitting and receiving filters, and $w(t)$ is the continuous-time channel noise. The continuous-time received signal $y(t)$ is then sampled at time $t = n\widetilde{T}$ where $\widetilde{T} = T/M$ is the *sampling period* and $M \geq 2$ is an integer referred to as the *oversampling factor*. Note that $1/T$ is called the *symbol rate* (symbols per second) or the *baud rate*,[16] while $1/\widetilde{T}$

[16] The term "baud," named after E. Baudot (1874) in recognition of his pioneering work in telegraphy, is the measure of modulation rate, i.e. the number of transitions per second. Its usage is clarified in [46].

is called the *sampling rate* (samples per second). After sampling $y(t)$, the discrete-time received signal is obtained as

$$y[n] = y(n\widetilde{T}) = \sum_{k=-\infty}^{\infty} u[k]h([n-kM]\widetilde{T}) + w(n\widetilde{T}). \tag{3.156}$$

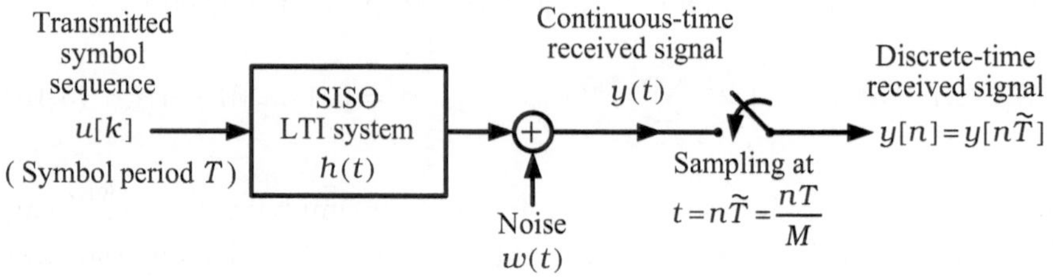

Fig. 3.23 A digital communication system

Let $h[n] = h(n\widetilde{T})$ and $w[n] = w(n\widetilde{T})$, that is, $h[n]$ and $w[n]$ are the discrete-time SISO LTI channel and the discrete-time channel noise corresponding to $h(t)$ and $w(t)$, respectively. Then $y[n]$ can be further expressed as

$$y[n] = x[n] + w[n] \tag{3.157}$$

where $x[n]$ is the discrete-time noise-free received signal given by

$$x[n] = \sum_{k=-\infty}^{\infty} u[k]h[n-kM]. \tag{3.158}$$

The input–output relation given by (3.158) is exactly a description of the multirate system [47] shown in Fig. 3.24 where the expanded signal

$$u_{\rm E}[n] = \begin{cases} u[n/M], & n = 0, \pm M, \pm 2M, ..., \\ 0, & \text{otherwise}. \end{cases} \tag{3.159}$$

This clearly indicates that $u[n]$ and $x[n]$ operate at different rates.

With regard to the SOCS of $y[n]$, let us make the following assumptions.

(CS-1) The SISO LTI system $h[n]$ is stable.
(CS-2) The symbol sequence $u[n]$ is a WSS white process with variance σ_u^2.
(CS-3) The noise $w[n]$ is a zero-mean WSS (white or colored) process.
(CS-4) The symbol sequence $u[n]$ is statistically independent of the noise $w[n]$.

Assumption (CS-2) implies that the autocorrelation function $r_u[l]$ of $u[n]$ is given as (3.124), while Assumption (CS-4) implies that $x[n]$ is statistically

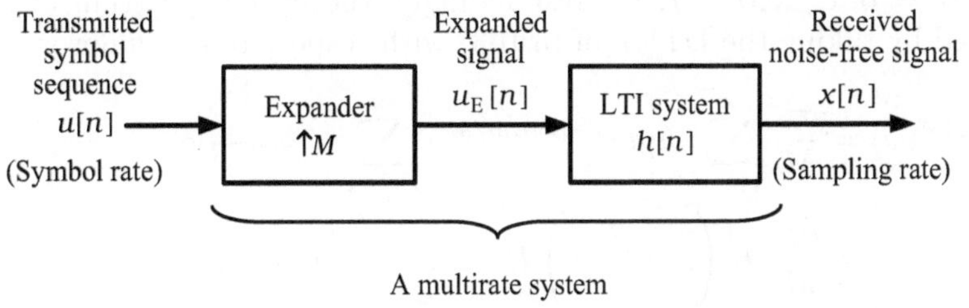

Fig. 3.24 Explanation of the relation between $x[n]$ and $u[n]$ via a multirate system

independent of $w[n]$. Using (3.158) and (3.124), one can express the autocorrelation function of $x[n]$ as follows:

$$r_x[n, n-l] = \sum_{k=-\infty}^{\infty} \sum_{m=-\infty}^{\infty} h[n-kM]h^*[n-l-mM]r_u[k-m]$$
$$= \sigma_u^2 \sum_{k=-\infty}^{\infty} h[n-kM]h^*[n-l-kM]. \tag{3.160}$$

It then follows that

$$r_x[n-\tau M, n-l-\tau M] = \sigma_u^2 \sum_{k=-\infty}^{\infty} h[n-(k+\tau)M]h^*[n-l-(k+\tau)M]$$
$$= r_x[n, n-l] \quad \text{for all } n, l \text{ and any integer } \tau. \tag{3.161}$$

In other words, under (CS-2), $r_x[n, n-l]$ is periodic in n with period M and thus $x[n]$ is a WSCS process with period M provided that $|r_x[n, n-l]| < \infty$ for all n, l. It can be shown (Problem 3.17) that under (CS-1) and (CS-2),

$$\sum_{l=-\infty}^{\infty} \sum_{n=0}^{M-1} |r_x[n, n-l]| < \infty \tag{3.162}$$

and accordingly $|r_x[n, n-l]| < \infty$ for all n, l. As a consequence, (CS-1) and (CS-2) ensure that $x[n]$ is WSCS.

Substituting (3.160) into (3.153) yields

$$r_x^{[\alpha]}[l] = \frac{\sigma_u^2}{M} \sum_{k=-\infty}^{\infty} \sum_{n=0}^{M-1} h[n-kM]h^*[n-kM-l]e^{-j2\pi\alpha n/M}$$
$$= \frac{\sigma_u^2}{M} \sum_{m=-\infty}^{\infty} h[m]h^*[m-l]e^{-j2\pi\alpha m/M} \tag{3.163}$$

where $\alpha = 0, 1, ..., M-1$. Correspondingly, the cyclic spectrum of $x[n]$ is obtained by taking the DTFT of (3.163) with respect to l as follows: [43, 48]

$$S_x^{[\alpha]}(\omega) = \frac{\sigma_u^2}{M} \sum_{m=-\infty}^{\infty} h[m]e^{-j2\pi\alpha m/M} \cdot \sum_{l=-\infty}^{\infty} h^*[m-l]e^{-j\omega l}$$
$$= \frac{\sigma_u^2}{M} \cdot H\left(\omega + \frac{2\pi\alpha}{M}\right) H^*(\omega), \quad \alpha = 0, 1, ..., M-1. \tag{3.164}$$

Under Assumptions (CS-1) and (CS-2), equations (3.153) and (3.162) lead to

$$\sum_{l=-\infty}^{\infty} \left| r_x^{[\alpha]}[l] \right| \le \frac{1}{M} \sum_{l=-\infty}^{\infty} \sum_{n=0}^{M-1} |r_x[n, n-l]| < \infty \tag{3.165}$$

and accordingly the existence and continuity of the cyclic spectrum $S_x^{\alpha}(\omega)$ follow from Theorem 2.37.

Incidentally, equalization of a digital communication system is said to be *fractionally spaced equalization* if the sampling period $\widetilde{T}$ is less than the symbol period T (i.e. oversampling), and is said to be *baud-spaced equalization* if $\widetilde{T} = T$ (i.e. the sampling rate equals the baud rate). From (3.164) and (3.127), one can easily see that $S_x^{[0]}(\omega) = S_x(\omega)/M$ is phase blind, whereas $S_x^{[\alpha]}(\omega)$ (and accordingly $r_x^{[\alpha]}[l]$) for $\alpha \ge 1$ contain not only the magnitude but also phase information about the channel $h[n]$. This important discovery, due to Tong, Xu and Kailath [48], therefore opens up the possibility of blind fractionally spaced equalization and identification using only SOCS under some certain conditions [40, 43, 48–55] (to name a few). The channel identifiability with only $S_x^{[\alpha]}(\omega)$ is illustrated in the following example.

Example 3.42

For the signal model (3.158), the transfer functions $H_1(z)$ and $H_2(z)$ given in Example 3.36 are considered for the system $h[n]$, and the symbol sequence $u[n]$ is assumed to satisfy (CS-2). The noise-free signal $x[n]$ corresponds to $H(z) = H_1(z)$ is denoted by $x_1[n]$, while that corresponds to $H(z) = H_2(z)$ is denoted by $x_2[n]$. By using (3.164) with $M = 2$ and $\alpha = 1$, we obtain the cyclic spectrum of $x_1[n]$ as

$$S_{x_1}^{[1]}(\omega) = \frac{\sigma_u^2}{2}(1 - a^2 - j2a\sin\omega)(1 - b^2 - j2b\sin\omega)$$

and the cyclic spectrum of $x_2[n]$ as

$$S_{x_2}^{[1]}(\omega) = \frac{\sigma_u^2}{2}(1 - a^2 - j2a\sin\omega)(1 - b^2 + j2b\sin\omega).$$

Figures 3.25a, b display the magnitude and phase of $S_{x_1}^{[1]}(\omega)$ (solid lines), respectively, and those of $S_{x_2}^{[1]}(\omega)$ (dashed lines) for $a = 1/2$, $b = 1/4$ and

$\sigma_u^2 = 1$ where the solid and dashed lines in Fig. 3.25a overlap. The phase difference in Fig. 3.25b indicates that it is possible to distinguish between $H_1(z)$ and $H_2(z)$ with only cyclic spectra.

□

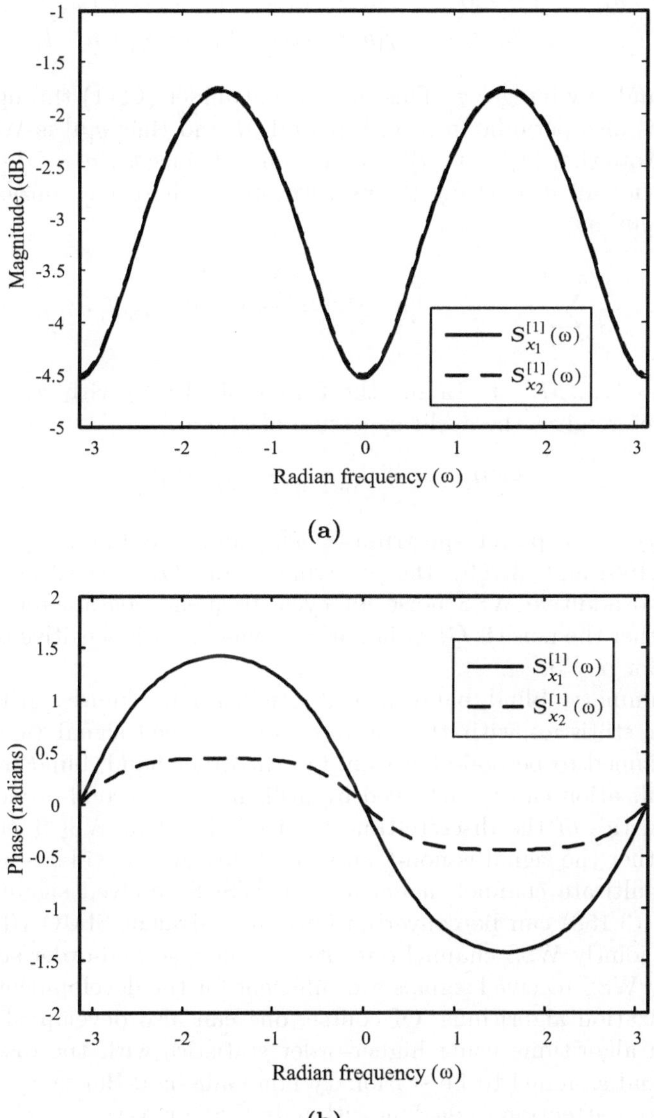

Fig. 3.25 (**a**) The magnitudes and (**b**) the phases of $S_{x_1}^{[1]}(\omega)$ and $S_{x_2}^{[1]}(\omega)$

Returning to the noisy received signal $y[n]$ given by (3.157), by (CS-3) and (CS-4), one can express the autocorrelation function of $y[n]$ as

$$r_y[n, n-l] = r_x[n, n-l] + r_w[l] \tag{3.166}$$

where $r_w[l]$ is the autocorrelation function of $w[n]$. By (3.166) and (3.161),

$$\begin{aligned} r_y[n-\tau M, n-l-\tau M] &= r_x[n-\tau M, n-l-\tau M] + r_w[l] \\ &= r_x[n, n-l] + r_w[l] = r_y[n, n-l] \end{aligned} \tag{3.167}$$

for all n, l and any integer τ. This means that under (CS-1) through (CS-4), $r_y[n, n-l]$ is also periodic in n with period M and thus $y[n]$ is WSCS with period M. Note that $|r_y[n, n-l]| < \infty$ for all n, l since $|r_x[n, n-l]| < \infty$ and $|r_w[l]| < \infty$ for all n, l. Using (3.166), we obtain the cyclic autocorrelation function of $y[n]$ as

$$r_y^{[\alpha]}[l] = \frac{1}{M} \sum_{n=0}^{M-1} r_y[n, n-l] e^{-j2\pi\alpha n/M} = r_x^{[\alpha]}[l] + r_w[l] \cdot \delta[\alpha] \tag{3.168}$$

where $\alpha = 0, 1, ..., M-1$. Taking the DTFT of (3.168) with respect to the index l therefore gives the cyclic spectrum of $y[n]$ as

$$S_y^{[\alpha]}(\omega) = S_x^{[\alpha]}(\omega) + S_w(\omega) \cdot \delta[\alpha] \tag{3.169}$$

where $S_w(\omega)$ is the power spectrum of $w[n]$ and $\alpha = 0, 1, ..., M-1$. As a result of (3.168) and (3.169), the performance of SOCS based algorithms is insensitive to additive WSS noise for cycle frequency parameter $\alpha \geq 1$ no matter whether the noise is Gaussian or not, whereas it is sensitive to additive WSS noise for $\alpha = 0$.

As a summary, blind baud spaced equalization requires utilization of higher-order statistics with the discrete-time received signal (a stationary process) assumed to be non-Gaussian. On the other hand, blind fractionally spaced equalization can be achieved by utilizing only second-order cyclostationary statistics of the discrete-time received signal (a WSCS process) no matter whether the signal is non-Gaussian or not. As we will see in Part III, the SISO multirate channel model for the WSCS received signal given by (3.157) and (3.158) can be converted into an equivalent SIMO LTI channel model with jointly WSS channel outputs. As such, second-order statistics of these jointly WSS received signals are sufficient for the development of SIMO blind equalization algorithms. Of course, one can also develop SIMO blind equalization algorithms using higher-order statistics with the discrete-time received signal assumed to be stationary non-Gaussian. Hence, we will hereafter direct our attention to dealing with only WSS or stationary signals using second-order and/or higher-order statistics.

3.4 Estimation Theory

This section deals with the problem of estimating a set of unknown parameters as well as the associated methods. Basically, these methods can be categorized into the following three classes.

- *Methods for deterministic parameters.* The estimation methods are developed under the assumption that the unknown parameters to be estimated are deterministic. These methods are sometimes called the *classical estimation methods.* Two representatives are the *maximum-likelihood (ML) estimation method* and the *method of moments.*
- *Methods for random parameters.* The estimation methods are developed under the assumption that the unknown parameters to be estimated are random. These methods are sometimes called the *Bayesian estimation methods.* Two representatives are the *minimum mean-square-error (MMSE) estimation method* and the *maximum a posteriori (MAP) method.*
- *Methods for both deterministic and random parameters.* The estimation methods are developed regardless of whether the unknown parameters to be estimated are deterministic or random. In other words, they are applicable to both deterministic and random parameters. A representative is the *least-squares (LS) estimation method.*

Among the numerous estimation methods, only the ML, MMSE and LS estimation methods as well as the method of moments are to be presented below. For a complete and excellent exposition of estimation methods, we encourage the reader to consult [17, 56].

3.4.1 Estimation Problem

Single Parameter

Consider that $x[n]$ is a random process which depends on an unknown parameter θ. The problem of *parameter estimation* is to estimate the unknown parameter θ from a finite set of measurements $\{x[0],\ x[1],\ ...,\ x[N-1]\}$ as

$$\widehat{\theta} = \phi(x[0], x[1], ..., x[N-1]) \tag{3.170}$$

where $\phi(\cdot)$ is a deterministic function (transformation) to be determined. Note that θ is either deterministic or random depending on the type of problem, while $\widehat{\theta}$ is always random since it is a transformation of random variables $x[0], x[1], ..., x[N-1]$. The random variable $\widehat{\theta}$ is called an *estimator* of θ, while a realization of the random variable $\widehat{\theta}$ is called an *estimate* for θ. For notational convenience, we also use the same notation $\widehat{\theta}$ to represent an estimate for θ without confusion.

Example 3.43
Suppose we are given a set of measurements $\{x[0], x[1], ..., x[N-1]\}$ where $x[n]$ is a stationary process whose mean $m_x = E\{x[n]\}$ is the unknown parameter to be estimated. We naturally have the following estimator of m_x:

$$\widehat{m}_x = \frac{1}{N}\sum_{n=0}^{N-1} x[n], \tag{3.171}$$

i.e. simply replacing the ensemble average $E\{x[n]\}$ by its time average. The estimator $\widehat{m}_x$ is called the *sample-mean estimator* of m_x, and the corresponding estimate is called the *sample mean* for m_x.

□

Let us provide a further insight into the problem of parameter estimation as follows. Let $\mathbf{x} = (x[0], x[1], ..., x[N-1])^T$. When θ is deterministic, $\mathbf{x}$ is related to θ in the way that its pdf denoted by $f(\mathbf{x})$ has a certain form governed by θ, but $\mathbf{x}$ is not a deterministic function of θ since $\mathbf{x}$ is a random vector. For instance, for a zero-mean Gaussian variable $x[0]$ with unknown variance θ, its pdf $f(x[0])$ is a Gaussian shape with spread determined by θ. As such, we also use the notation $f(\mathbf{x};\theta)$ for $f(\mathbf{x})$ to emphasize its dependence on θ. Since $\widehat{\theta}$ is a deterministic function of $\mathbf{x}$, it is also statistically dependent on θ but not a deterministic function of θ. In other words, $\partial\widehat{\theta}/\partial\theta^* = 0$. On the other hand, when θ is random, $\mathbf{x}$ and θ are related via their joint pdf denoted by $f(\mathbf{x},\theta)$ and, similarly, $\widehat{\theta}$ is statistically dependent on θ but not a deterministic function of θ. Figure 3.26 summarizes the relations, terminologies and notation for the problem of estimating the single parameter θ.

Multiple Parameters

Now consider that $x[n]$ is a random process which depends on L unknown parameters $\theta[0]$, $\theta[1]$, ..., $\theta[L-1]$. The problem of *parameter estimation* here is to estimate the unknown parameter vector $\boldsymbol{\theta} = (\theta[0], \theta[1], ..., \theta[L-1])^T$ from a finite set of measurements $\{x[0], x[1], ..., x[N-1]\}$ as

$$\widehat{\boldsymbol{\theta}} = \boldsymbol{\phi}(x[0], x[1], ..., x[N-1]) \tag{3.172}$$

where $\boldsymbol{\phi}(\cdot)$ is a vector of L deterministic functions to be determined. The parameter vector $\boldsymbol{\theta}$ is either deterministic or random, while the estimator $\widehat{\boldsymbol{\theta}}$ is always random. A realization of the estimator $\widehat{\boldsymbol{\theta}}$ is an estimate for $\boldsymbol{\theta}$, which is also denoted by $\widehat{\boldsymbol{\theta}}$ for notational convenience. Similarly, $\mathbf{x} = (x[0], x[1], ..., x[N-1])^T$ is related to $\boldsymbol{\theta}$ via its pdf denoted by $f(\mathbf{x}) \equiv f(\mathbf{x};\boldsymbol{\theta})$ for deterministic $\boldsymbol{\theta}$, while $\mathbf{x}$ and $\boldsymbol{\theta}$ are related via their joint pdf denoted by $f(\mathbf{x},\boldsymbol{\theta})$ for random $\boldsymbol{\theta}$. As such, $\widehat{\boldsymbol{\theta}}$ is statistically dependent on $\boldsymbol{\theta}$ but not a vector of deterministic functions of $\boldsymbol{\theta}$, i.e. $\partial\widehat{\boldsymbol{\theta}}^H/\partial\boldsymbol{\theta}^* = \partial\widehat{\boldsymbol{\theta}}^T/\partial\boldsymbol{\theta}^* = \mathbf{0}$. Next, let us define some desirable statistical properties of estimators.

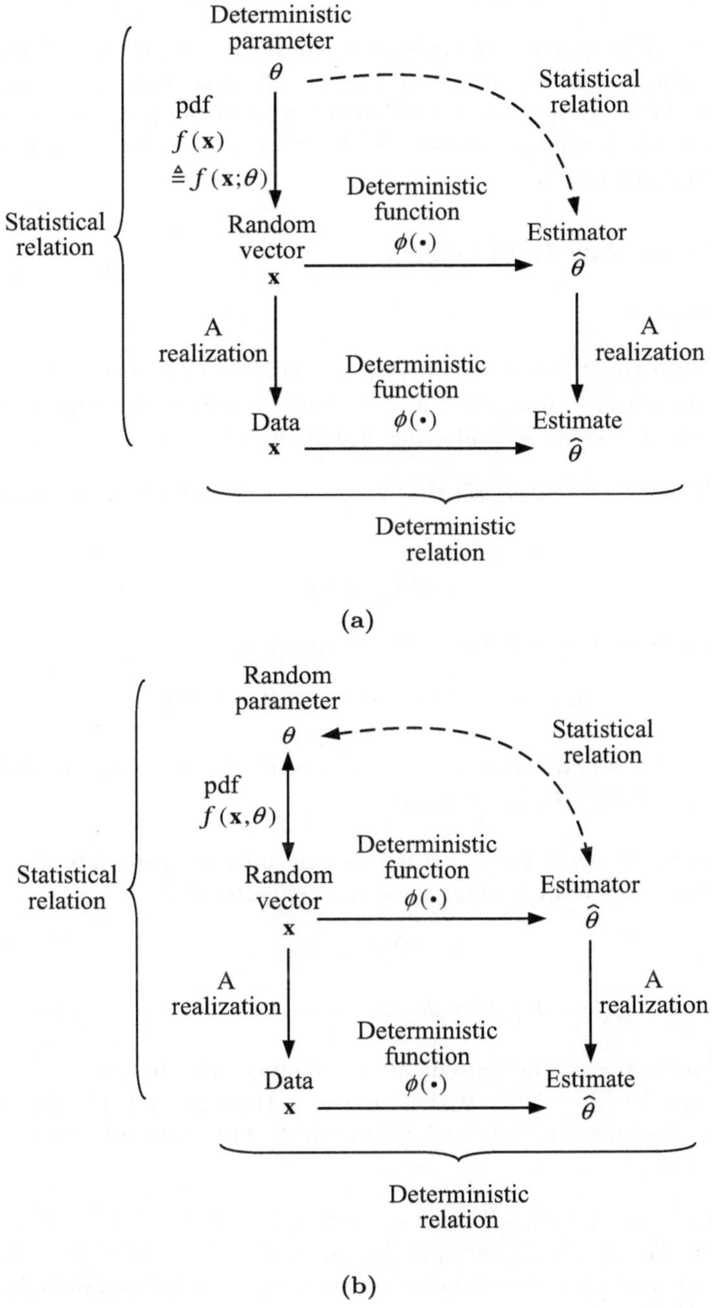

Fig. 3.26 Summary of the problem of estimating (**a**) deterministic parameter θ and (**b**) random parameter θ

3.4.2 Properties of Estimators

The statistical properties of estimators to be discussed are divided into two types: those valid for any data length and those valid for infinite data length. The latter properties are called the *asymptotic properties* of estimators. Clearly, the former properties also hold for infinite data length, but the converse may not be true.

Properties for Any Data Length

Single Parameter

Although an estimator $\widehat{\theta}$ of unknown parameter θ can be completely characterized by its pdf $f(\widehat{\theta})$, it is generally sufficient and relatively simple to analyze $\widehat{\theta}$ only via the first- and second-order statistics of $\widehat{\theta}$.

Definition 3.44. *An estimator $\widehat{\theta}$ of unknown parameter θ is said to be unbiased if*

$$E\{\widehat{\theta}\} = E\{\theta\}; \tag{3.173}$$

otherwise, it is said to be biased with bias given by

$$\text{Bias}(\widehat{\theta}) = E\{\theta - \widehat{\theta}\} = E\{\theta\} - E\{\widehat{\theta}\}. \tag{3.174}$$

Note that when θ is deterministic, (3.173) and (3.174) reduce to $E\{\widehat{\theta}\} = \theta$ and $\text{Bias}(\widehat{\theta}) = \theta - E\{\widehat{\theta}\}$, respectively.

Definition 3.45. *An unbiased estimator $\widehat{\theta}$ of unknown parameter θ is said to be more efficient*[17] *than another unbiased estimator $\widetilde{\theta}$ if*

$$\text{Var}(\widehat{\theta}) \leq \text{Var}(\widetilde{\theta}) \tag{3.175}$$

where $\text{Var}(z) = E\{|z - E\{z\}|^2\}$ *denotes the variance of random variable z.*

That is, $\widehat{\theta}$ makes use of measurements more efficiently than $\widetilde{\theta}$.

Moreover, let $\varepsilon = \theta - \widehat{\theta}$ denote the estimation error. Obviously, it is desirable to have an unbiased estimator $\widehat{\theta}$ with smallest error variance $\text{Var}(\varepsilon) = \text{Var}(\widehat{\theta})$.

Definition 3.46. *Among all unbiased estimators of θ, if the unbiased estimator $\widehat{\theta}$ has the minimum variance for all θ, then it is called the minimum-variance unbiased (MVU) estimator or the uniformly minimum-variance unbiased (UMVU) estimator.*

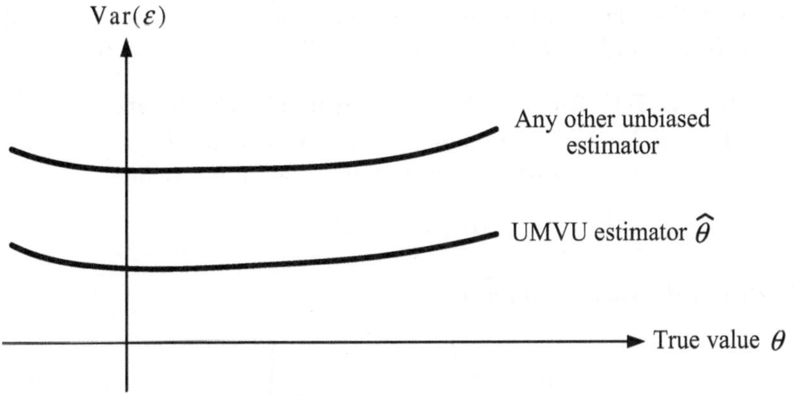

Fig. 3.27 Illustration of the error variance for the UMVU estimator $\widehat{\theta}$

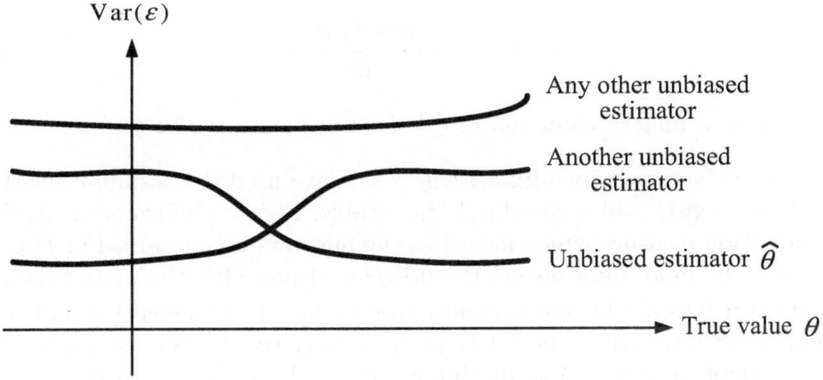

Fig. 3.28 Illustration of the error variance dependent on θ

Note that the terminology "UMVU" is used to emphasize that the condition of minimum variance holds for all θ, as illustrated in Fig. 3.27. In other words, the UMVU estimator is more efficient than any other unbiased estimator. However, as illustrated in Fig. 3.28, the UMVU estimator does not always exist [56, p. 20].

Given an unbiased estimator $\widehat{\theta}$ obtained with some kind of estimation method, we may wish to know about its efficiency of utilizing data and whether it is the UMVU estimator or not. One simple approach, if possible to apply,

[17] The concept of efficiency is due to R. A. Fisher (1922) [12], for whom a short biography can be found in [17, pp. 85–86].

is to test the following well-known lower bound on the variance of $\widehat{\theta}$, called the *Cramér–Rao bound (CRB)* or the *Cramér–Rao lower bound (CRLB)*.

Theorem 3.47 (CRB for Single Deterministic Parameter). *Suppose* $\mathbf{x} = (x[0], x[1], ..., x[N-1])^T$ *is a vector of* N *measurements. If* $\widehat{\theta}$ *is an unbiased estimator of unknown deterministic parameter* θ *based on* $\mathbf{x}$*, then*

$$\mathrm{Var}(\widehat{\theta}) \geq F^{-1}(\theta) \tag{3.176}$$

where $\mathrm{Var}(\widehat{\theta})$ *is the variance of* $\widehat{\theta}$ *and*

$$F(\theta) = E\left\{ \left| \frac{\partial \ln f(\mathbf{x};\theta)}{\partial \theta^*} \right|^2 \right\} \tag{3.177}$$

is the so-called Fisher's information in which $f(\mathbf{x};\theta)$ *is the pdf of* $\mathbf{x}$*. The equality of (3.176) holds if and only if*

$$\widehat{\theta} - \theta = \beta(\theta) \frac{\partial \ln f(\mathbf{x};\theta)}{\partial \theta^*} \tag{3.178}$$

where $\beta(\theta)$ *is a nonzero function of* θ *but not a function of* $\mathbf{x}$ *or* $\widehat{\theta}$*.*

The proof is given in Appendix 3B where we have used the assumptions that $\partial f(\mathbf{x};\theta)/\partial\theta^*$ exists and is absolutely integrable. Fisher's information $F(\theta)$ is an information measure which measures the information contained in $f(\mathbf{x};\theta)$. Intuitively, the more information the pdf $f(\mathbf{x};\theta)$ provides, the more accurate the estimator $\widehat{\theta}$ would be and therefore the smaller the variance $\mathrm{Var}(\widehat{\theta})$ would be. This is exactly what the CRB (3.176) says. By further assuming that $\partial^2 f(\mathbf{x};\theta)/\partial\theta\partial\theta^*$ exists and is absolutely integrable, we have another form of $F(\theta)$ as follows: (Problem 3.18)

$$F(\theta) = -E\left\{ \frac{\partial^2 \ln f(\mathbf{x};\theta)}{\partial \theta \partial \theta^*} \right\}. \tag{3.179}$$

Note that Theorem 3.47 can be applied to the estimators of random θ by replacing the pdf $f(\mathbf{x};\theta)$ with the joint pdf $f(\mathbf{x},\theta)$ together with some modifications of the required assumptions; see [57, p. 177] for further details. Also it can be generalized for biased estimators [24, p. 292].

For an unbiased estimator $\widehat{\theta}$, we can compare the variance of $\widehat{\theta}$ with the CRB (if available) so as to know about its efficiency of utilizing data.

Definition 3.48. *If* $\widehat{\theta}$ *is an unbiased estimator of* θ *and attains the CRB given by (3.176) for all* θ*, it is an efficient estimator.*

According to this definition, an efficient estimator is also the UMVU estimator. As illustrated in Fig. 3.29, the converse may not be true, however. Let us emphasize that there exist other lower bounds which are tighter than the CRB. This means that if an unbiased estimator does not attain the CRB, it still may be the UMVU estimator. Consequently, the CRB provides only a partial solution to the problem of determining whether an unbiased estimator is the UMVU estimator or not. Alternatively, this problem can be completely resolved by means of the so-called *completeness-sufficiency approach*; see [56] for the details.

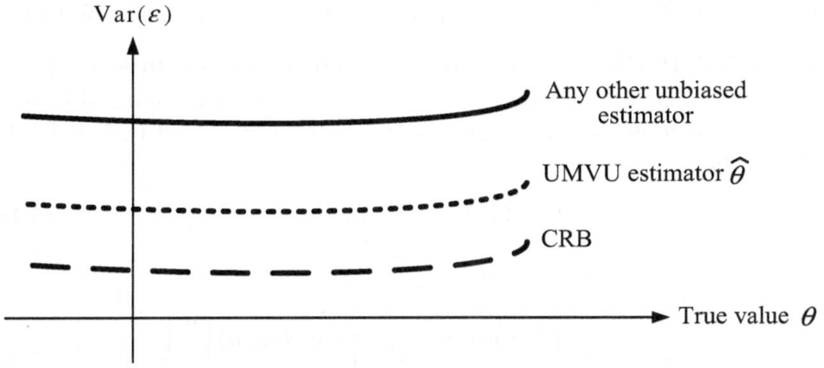

Fig. 3.29 Illustration of the difference between the CRB and the variance of the UMVU estimator $\widehat{\theta}$

Multiple Parameters

Next, let us discuss the statistical properties of estimators of multiple parameters for any data length.

Definition 3.49. *An estimator* $\widehat{\boldsymbol{\theta}}$ *of unknown parameter vector* $\boldsymbol{\theta} = (\theta[0], \theta[1], ..., \theta[L-1])^T$ *is said to be unbiased if*

$$E\{\widehat{\boldsymbol{\theta}}\} = E\{\boldsymbol{\theta}\}; \tag{3.180}$$

otherwise, it is said to be biased with bias given by

$$\text{Bias}(\widehat{\boldsymbol{\theta}}) = E\{\boldsymbol{\theta} - \widehat{\boldsymbol{\theta}}\} = E\{\boldsymbol{\theta}\} - E\{\widehat{\boldsymbol{\theta}}\}. \tag{3.181}$$

For ease of later use, let

$$\mathbf{C}(\mathbf{z}) = E\left\{(\mathbf{z} - E\{\mathbf{z}\})(\mathbf{z} - E\{\mathbf{z}\})^H\right\} \tag{3.182}$$

denote the covariance matrix of random vector $\mathbf{z}$, and the notation '$\mathbf{A} > \mathbf{B}$' ('$\mathbf{A} \geq \mathbf{B}$') be used to represent that $(\mathbf{A} - \mathbf{B})$ is positive definite (positive semidefinite).

Definition 3.50. *An unbiased estimator* $\widehat{\boldsymbol{\theta}}$ *is said to be more efficient than another unbiased estimator* $\widetilde{\boldsymbol{\theta}}$ *if*

$$\mathbf{C}(\widehat{\boldsymbol{\theta}}) \leq \mathbf{C}(\widetilde{\boldsymbol{\theta}}). \tag{3.183}$$

Definition 3.51. *Among all unbiased estimators of* $\boldsymbol{\theta}$*, the unbiased estimator* $\widehat{\boldsymbol{\theta}}$ *is said to be the UMVU estimator of* $\boldsymbol{\theta}$ *if* $\mathrm{Var}(\widehat{\theta}[k])$*,* $k = 0, 1, ..., L-1$*, are minimum for all* $\boldsymbol{\theta}$ *where* $\widehat{\theta}[k]$ *is the kth entry of* $\widehat{\boldsymbol{\theta}}$ *[56, p. 23].*

That is, each estimator $\widehat{\theta}[k]$ is more efficient than any other unbiased estimator of $\theta[k]$ if $\widehat{\boldsymbol{\theta}}$ is the UMVU estimator of $\boldsymbol{\theta}$.

Similarly, the CRB for multiple deterministic parameters is as follows.

Theorem 3.52 (CRB for Multiple Deterministic Parameters). *Suppose* $\mathbf{x} = (x[0], x[1], ..., x[N-1])^T$ *is a vector of N measurements. If* $\widehat{\boldsymbol{\theta}}$ *is an unbiased estimator of deterministic parameter vector* $\boldsymbol{\theta} = (\theta[0], \theta[1], ..., \theta[L-1])^T$ *based on* $\mathbf{x}$*, then*

$$\mathbf{C}(\widehat{\boldsymbol{\theta}}) \geq \mathbf{F}^{-1}(\boldsymbol{\theta}) \tag{3.184}$$

where $\mathbf{C}(\widehat{\boldsymbol{\theta}})$ *is the covariance matrix of* $\widehat{\boldsymbol{\theta}}$ *and*

$$\mathbf{F}(\boldsymbol{\theta}) = E\left\{\left[\frac{\partial \ln f(\mathbf{x};\boldsymbol{\theta})}{\partial \boldsymbol{\theta}^*}\right] \cdot \left[\frac{\partial \ln f(\mathbf{x};\boldsymbol{\theta})}{\partial \boldsymbol{\theta}^*}\right]^H\right\} \tag{3.185}$$

is Fisher's information matrix in which $f(\mathbf{x};\boldsymbol{\theta})$ *is the pdf of* $\mathbf{x}$*. The equality of (3.184) holds if and only if*

$$\widehat{\boldsymbol{\theta}} - \boldsymbol{\theta} = \beta(\boldsymbol{\theta}) \cdot \mathbf{F}^{-1}(\boldsymbol{\theta}) \frac{\partial \ln f(\mathbf{x};\boldsymbol{\theta})}{\partial \boldsymbol{\theta}^*} \tag{3.186}$$

where $\beta(\boldsymbol{\theta})$ *is a nonzero function of* $\boldsymbol{\theta}$ *but not a function of* $\mathbf{x}$ *or* $\widehat{\boldsymbol{\theta}}$*.*

The proof is given in Appendix 3C where we have used the assumptions that $\partial f(\mathbf{x};\boldsymbol{\theta})/\partial \boldsymbol{\theta}^*$ exists and is absolutely integrable, and $\mathbf{F}(\boldsymbol{\theta})$ is nonsingular. Theorem 3.52 can also be applied to the estimators of random $\boldsymbol{\theta}$ by replacing the pdf $f(\mathbf{x};\boldsymbol{\theta})$ with the joint pdf $f(\mathbf{x},\boldsymbol{\theta})$ together with some modifications of the required assumptions; see [58, pp. 99, 100] for further details. Moreover, Theorem 3.52 implies the following corollary (Problem 3.19).

Corollary 3.53. *Suppose* $\mathbf{x} = (x[0], x[1], ..., x[N-1])^T$ *is a vector of N measurements. If* $\widehat{\boldsymbol{\theta}}$ *is an unbiased estimator of deterministic parameter vector* $\boldsymbol{\theta} = (\theta[0], \theta[1], ..., \theta[L-1])^T$ *based on* $\mathbf{x}$*, then*

$$\mathrm{Var}(\widehat{\theta}[k]) \geq \left[\mathbf{F}^{-1}(\boldsymbol{\theta})\right]_{k,k} \quad \text{for } k = 0, 1, ..., L-1 \tag{3.187}$$

where $\widehat{\theta}[k]$ *is the kth entry of* $\widehat{\boldsymbol{\theta}}$ *and* $\mathbf{F}(\boldsymbol{\theta})$ *is the Fisher's information matrix defined as (3.185).*

By the CRB, we have the following definition of efficiency.

Definition 3.54. *If $\widehat{\boldsymbol{\theta}}$ is an unbiased estimator of $\boldsymbol{\theta}$ and attains the CRB given by (3.184) for all $\boldsymbol{\theta}$, it is an efficient estimator.*

According to this definition, if $\widehat{\boldsymbol{\theta}}$ is an efficient estimator, then $\widehat{\theta}[k]$ attains the CRB given by (3.187) for all $\theta[k]$, and thus $\widehat{\boldsymbol{\theta}}$ is the UMVU estimator of $\boldsymbol{\theta}$. As a result, the CRB for multiple parameters also provides a partial solution to the problem of determining whether an unbiased estimator $\widehat{\boldsymbol{\theta}}$ is the UMVU estimator or not.

Properties for Infinite Data Length

Not all the properties of estimators hold for any data length N. Some properties do depend on the data length N. To emphasize this dependence, let us further denote an estimator $\widehat{\boldsymbol{\theta}}$ based on N measurements by $\widehat{\boldsymbol{\theta}}_N$. In other words, we have a sequence of infinitely many estimators $\{\widehat{\boldsymbol{\theta}}_N\}_{N=1}^{\infty}$, for which only the asymptotic properties of $\widehat{\boldsymbol{\theta}}_N$ are considered here. Since $\{\widehat{\boldsymbol{\theta}}_N\}_{N=1}^{\infty}$ is a sequence of random vectors, its convergence needs to be defined in some statistical sense before the treatment of its asymptotic properties.

Stochastic Convergence

With regard to an infinite sequence of random vectors, four popular definitions of stochastic convergence are given as follows.

Convergence with Probability One. A random sequence $\{\mathbf{z}_n\}_{n=1}^{\infty}$ is said to converge with probability one (or converge almost everywhere or converge almost surely) to a random vector $\mathbf{z}$ if for each sample point s_m in sample space,

$$\Pr\left\{\lim_{n\to\infty} \mathbf{z}_n(s_m) = \mathbf{z}(s_m)\right\} = 1 \tag{3.188}$$

where $\mathbf{z}_n(s_m)$ and $\mathbf{z}(s_m)$ are, respectively, the values of the random vectors $\mathbf{z}_n$ and $\mathbf{z}$ corresponding to the sample point s_m. We use the shorthand notation $\mathbf{z}_n \xrightarrow{\text{w.p.1}} \mathbf{z}$ for this case.

Mean-Square (MS) Convergence. A random sequence $\{\mathbf{z}_n\}_{n=1}^{\infty}$ is said to converge in the MS sense to a random vector $\mathbf{z}$ if

$$\lim_{n\to\infty} E\left\{\|\mathbf{z}_n - \mathbf{z}\|^2\right\} = 0, \tag{3.189}$$

and we write $\mathbf{z}_n \xrightarrow{\text{MS}} \mathbf{z}$.

Convergence in Probability. A random sequence $\{\mathbf{z}_n\}_{n=1}^{\infty}$ is said to converge in probability to a random vector $\mathbf{z}$ if for any real number $\varepsilon > 0$

$$\lim_{n\to\infty} \Pr\left\{\|\mathbf{z}_n - \mathbf{z}\| > \varepsilon\right\} = 0, \tag{3.190}$$

and we write $\mathbf{z}_n \xrightarrow{\text{P}} \mathbf{z}$.

Convergence in Distribution. A random sequence $\{\mathbf{z}_n\}_{n=1}^{\infty}$ with distribution function $F_n(\mathbf{z})$ is said to converge in distribution (or converge in law) to a random vector $\mathbf{z}$ with distribution function $F(\mathbf{z})$ if

$$\lim_{n\to\infty} F_n(\mathbf{z}) = F(\mathbf{z}) \tag{3.191}$$

for all continuity points of $F(\mathbf{z})$. The shorthand notation $\mathbf{z}_n \xrightarrow{\text{d}} \mathbf{z}$ is used for this case.

Convergence in distribution is concerned with only the convergence of a sequence of distribution functions rather than the convergence of a random sequence. Obviously, it is the weakest among the four types of stochastic convergence since different random sequences may correspond to the same distribution function. Moreover, the relation between convergence in probability and MS convergence can be observed by using *Tchebycheff's inequality* [59, p. 205].

Theorem 3.55 (Tchebycheff's Inequality). *Suppose* $\mathbf{z}$ *is an arbitrary random vector. Then for any real number* $\varepsilon > 0$

$$\Pr\{\|\mathbf{z}\| > \varepsilon\} \leq \frac{E\{\|\mathbf{z}\|^2\}}{\varepsilon^2}. \tag{3.192}$$

The proof is left as an exercise (Problem 3.20). Replacing $\mathbf{z}$ in (3.192) by $(\mathbf{z}_n - \mathbf{z})$ and letting $n \to \infty$ yield

$$\lim_{n\to\infty} \Pr\{\|\mathbf{z}_n - \mathbf{z}\| > \varepsilon\} \leq \lim_{n\to\infty} \frac{E\{\|\mathbf{z}_n - \mathbf{z}\|^2\}}{\varepsilon^2}. \tag{3.193}$$

This clearly indicates that MS convergence implies convergence in probability. In fact, convergence with probability one also implies convergence in probability; see [59, pp. 380, 381] for the proof. As a summary, Fig. 3.30 clarifies the relations for the four types of stochastic convergence [11, p. 210].

Asymptotic Properties of Estimators

Definition 3.56. *An estimator* $\widehat{\boldsymbol{\theta}}_N$ *of unknown parameter vector* $\boldsymbol{\theta}$ *based on a set of* N *measurements is said to be asymptotically unbiased if*

$$\lim_{N\to\infty} E\{\widehat{\boldsymbol{\theta}}_N\} = E\{\boldsymbol{\theta}\}. \tag{3.194}$$

Clearly, an unbiased estimator $\widehat{\boldsymbol{\theta}}_N$ is always asymptotically unbiased, while a biased estimator $\widehat{\boldsymbol{\theta}}_N$ may also be asymptotically unbiased.

Definition 3.57. *An estimator* $\widehat{\boldsymbol{\theta}}_N$ *of unknown parameter vector* $\boldsymbol{\theta}$ *is said to be weakly consistent if* $\widehat{\boldsymbol{\theta}}_N \xrightarrow{\text{P}} \boldsymbol{\theta}$ *and is said to be strongly consistent if* $\widehat{\boldsymbol{\theta}}_N \xrightarrow{\text{w.p.1}} \boldsymbol{\theta}$.

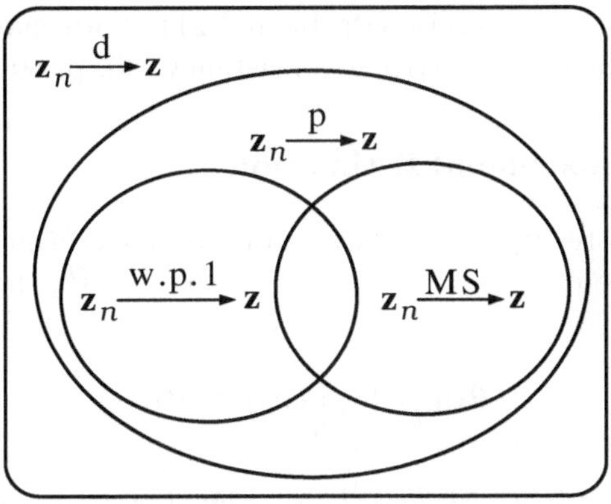

Fig. 3.30 Relations for the four types of stochastic convergence

Note that the consistency of $\widehat{\boldsymbol{\theta}}_N$ does not require that $\widehat{\boldsymbol{\theta}}_N$ be unbiased or asymptotically unbiased. In principle, strong consistency is more desirable than weak consistency, but, in practice, analysis for the former is generally much harder than for the latter. For this reason, unless stated otherwise, we will always refer to consistent estimators in the sense of weak consistency.

From the foregoing discussion of stochastic convergence, it follows that if $\widehat{\boldsymbol{\theta}}_N \xrightarrow{\text{MS}} \boldsymbol{\theta}$, then $\widehat{\boldsymbol{\theta}}_N \xrightarrow{\text{p}} \boldsymbol{\theta}$ and thus $\widehat{\boldsymbol{\theta}}_N$ is consistent. This therefore suggests the following method to determine whether $\widehat{\boldsymbol{\theta}}_N$ is a consistent estimator or not. When $\boldsymbol{\theta}$ is deterministic, the mean-square-error (MSE)

$$\begin{aligned} E\left\{\left\|\widehat{\boldsymbol{\theta}}_N - \boldsymbol{\theta}\right\|^2\right\} &= E\left\{\left\|\left(\widehat{\boldsymbol{\theta}}_N - E\{\widehat{\boldsymbol{\theta}}_N\}\right) + \left(E\{\widehat{\boldsymbol{\theta}}_N\} - \boldsymbol{\theta}\right)\right\|^2\right\} \\ &= E\left\{\left\|\widehat{\boldsymbol{\theta}}_N - E\{\widehat{\boldsymbol{\theta}}_N\}\right\|^2\right\} + \left\|E\{\widehat{\boldsymbol{\theta}}_N\} - \boldsymbol{\theta}\right\|^2 \\ &= \mathrm{tr}\left\{\mathbf{C}(\widehat{\boldsymbol{\theta}}_N)\right\} + \left\|\mathrm{Bias}(\widehat{\boldsymbol{\theta}}_N)\right\|^2 \end{aligned} \tag{3.195}$$

where $\mathbf{C}(\widehat{\boldsymbol{\theta}}_N)$ is the covariance matrix of $\widehat{\boldsymbol{\theta}}_N$ and $\mathrm{Bias}(\widehat{\boldsymbol{\theta}}_N)$ is the bias of $\widehat{\boldsymbol{\theta}}_N$. From (3.195), it follows that if both $\mathrm{Bias}(\widehat{\boldsymbol{\theta}}_N)$ and $\mathrm{tr}\{\mathbf{C}(\widehat{\boldsymbol{\theta}}_N)\}$ approach zero as $N \to \infty$, then $\widehat{\boldsymbol{\theta}}_N \xrightarrow{\text{MS}} \boldsymbol{\theta}$ and thus $\widehat{\boldsymbol{\theta}}_N$ is a consistent estimator of $\boldsymbol{\theta}$. As a special case of this method, for the single-parameter case, if the bias $\mathrm{Bias}(\widehat{\theta}_N) \to 0$ and the variance $\mathrm{Var}(\widehat{\theta}_N) \to 0$ as $N \to \infty$, then $\widehat{\theta}_N$ is a consistent estimator. Once $\widehat{\boldsymbol{\theta}}_N$ is known to be consistent, the following carry-over property of consistency can then be applied [60, p. 57] or [35, p. 80].

Theorem 3.58 (Slutsky's Theorem). *Suppose $\widehat{\boldsymbol{\theta}}_N$ is a consistent estimator of unknown parameter vector $\boldsymbol{\theta}$. Then $\phi(\widehat{\boldsymbol{\theta}}_N)$ is also a consistent estimator of $\phi(\boldsymbol{\theta})$ where $\phi(\widehat{\boldsymbol{\theta}}_N)$ is an arbitrary continuous function of $\widehat{\boldsymbol{\theta}}_N$.*

We leave the proof as an exercise (Problem 3.21). Note that Theorem 3.58 is also applicable in the sense of strong consistency [35, p. 80].

3.4.3 Maximum-Likelihood Estimation

Suppose the parameter vector $\boldsymbol{\theta}$ to be estimated is deterministic. Based on a vector of N measurements $\mathbf{x} = (x[0], x[1], ..., x[N-1])^T$, the ML estimator is defined as follows:

$$\widehat{\boldsymbol{\theta}}_{\mathrm{ML}} = \underset{\boldsymbol{\theta}}{\operatorname{argmax}}\, f(\mathbf{x}; \boldsymbol{\theta}) \tag{3.196}$$

where $f(\mathbf{x}; \boldsymbol{\theta})$ is called the *likelihood function.* Let us emphasize that when $f(\mathbf{x}; \boldsymbol{\theta})$ is regarded as the likelihood function of $\boldsymbol{\theta}$, we treat $\boldsymbol{\theta}$ as a vector of independent variables and treat $\mathbf{x}$ as a vector of fixed quantities; whereas when $f(\mathbf{x}; \boldsymbol{\theta})$ is regarded as the pdf of $\mathbf{x}$, we treat $\mathbf{x}$ as a vector of independent variables and treat $\boldsymbol{\theta}$ as a vector of fixed quantities. The basic idea behind the ML method is explained as follows [24, pp. 280, 281]. Consider the case of single measurement $x[0]$ with pdf $f(x[0]; m_x)$ where m_x is the mean of the random variable $x[0]$. Figure 3.31 depicts the pdf $f(x[0]; m_x)$ for $m_x = m_1$ and $m_x = m_2$. Given that $m_x = m_i$, the probability of observing that $x[0] = \chi$ can be expressed as

$$\begin{aligned} \Pr\{x[0] = \chi | m_i\} &= \lim_{\varepsilon \to 0} \int_{\chi-\varepsilon}^{\chi+\varepsilon} f(x[0]; m_i) dx[0] \\ &\approx \lim_{\varepsilon \to 0} 2\varepsilon \cdot f(x[0] = \chi; m_i), \quad i = 1 \text{ or } 2. \end{aligned}$$

This means that the larger the likelihood function $f(x[0] = \chi; m_i)$ with respect to m_i, the more likely the event of observing that $x[0] = \chi$ happens. This therefore leads to the generic form of ML estimator as given by (3.196).

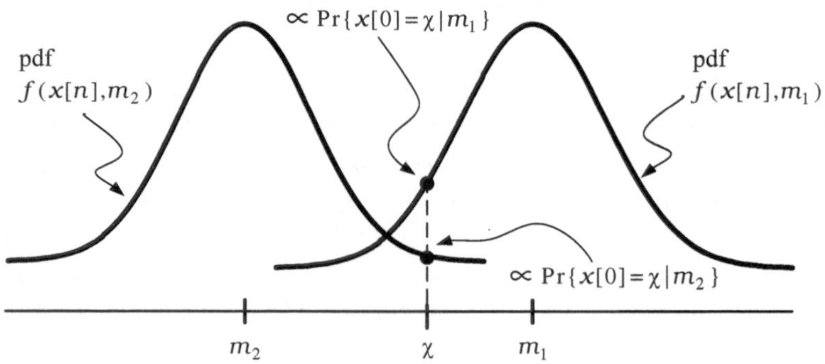

Fig. 3.31 Explanation of the ML estimation method

In many applications, it is more convenient to obtain an ML estimator via

$$\widehat{\boldsymbol{\theta}}_{\mathrm{ML}} = \underset{\boldsymbol{\theta}}{\operatorname{argmax}} \{\ln f(\mathbf{x}; \boldsymbol{\theta})\} \tag{3.197}$$

since the natural logarithm is a monotonic function. The function $\ln f(\mathbf{x}; \boldsymbol{\theta})$ is called the *log-likelihood function.*

Linear Gaussian Model

As a special case of the ML estimation problem, suppose the vector of measurements $\mathbf{x}$ is generated from the following linear model:

$$\mathbf{x} = \mathbf{H}\boldsymbol{\theta} + \mathbf{w} \tag{3.198}$$

where $\mathbf{H}$ is a deterministic nonsingular matrix and $\mathbf{w}$ is a complex zero-mean Gaussian noise with positive definite covariance matrix $\mathbf{C}_w$. The log-likelihood function is given by

$$\ln f(\mathbf{x}; \boldsymbol{\theta}) = -\ln\left(\pi^N \cdot |\mathbf{C}_w|\right) - (\mathbf{x} - \mathbf{H}\boldsymbol{\theta})^H \mathbf{C}_w^{-1} (\mathbf{x} - \mathbf{H}\boldsymbol{\theta}), \tag{3.199}$$

which gives the gradient

$$\frac{\partial \ln f(\mathbf{x}; \boldsymbol{\theta})}{\partial \boldsymbol{\theta}^*} = \mathbf{H}^H \mathbf{C}_w^{-1} (\mathbf{x} - \mathbf{H}\boldsymbol{\theta}) \tag{3.200}$$

and the Hessian matrix

$$\mathbf{J}_2(\boldsymbol{\theta}) = \frac{\partial}{\partial \boldsymbol{\theta}^*} \left[\frac{\partial \ln f(\mathbf{x}; \boldsymbol{\theta})}{\partial \boldsymbol{\theta}^*}\right]^H = -\mathbf{H}^H \mathbf{C}_w^{-1} \mathbf{H}. \tag{3.201}$$

Since $\mathbf{C}_w^{-1}$ is positive definite (by Property 2.12), for any nonzero vector $\mathbf{q}$ we have $-\mathbf{q}^H \mathbf{H}^H \mathbf{C}_w^{-1} \mathbf{H}\mathbf{q} = -(\mathbf{H}\mathbf{q})^H \mathbf{C}_w^{-1} (\mathbf{H}\mathbf{q}) < 0$, and thus $\mathbf{J}_2(\boldsymbol{\theta})$ is negative definite. According to Theorem 2.44, the solution obtained by setting (3.200) to zero corresponds to the maximum of $f(\mathbf{x}; \boldsymbol{\theta})$. As a result, the ML estimator

$$\widehat{\boldsymbol{\theta}}_{\mathrm{ML}} = (\mathbf{H}^H \mathbf{C}_w^{-1} \mathbf{H})^{-1} \mathbf{H} \mathbf{C}_w^{-1} \mathbf{x}. \tag{3.202}$$

This indicates that for the linear data model given by (3.198) the ML estimator $\widehat{\boldsymbol{\theta}}_{\mathrm{ML}}$ is a linear estimator (a linear transformation of $\mathbf{x}$).

Properties of ML Estimators

Recall that any efficient estimator $\widehat{\boldsymbol{\theta}}$, if exists, must satisfy (3.186) for all $\boldsymbol{\theta}$. Accordingly, evaluating (3.186) at $\boldsymbol{\theta} = \widehat{\boldsymbol{\theta}}_{\mathrm{ML}}$ yields

$$\widehat{\boldsymbol{\theta}} - \widehat{\boldsymbol{\theta}}_{\mathrm{ML}} = \beta(\widehat{\boldsymbol{\theta}}_{\mathrm{ML}}) \cdot \mathbf{F}^{-1}(\widehat{\boldsymbol{\theta}}_{\mathrm{ML}}) \cdot \left.\frac{\partial \ln f(\mathbf{x}; \boldsymbol{\theta})}{\partial \boldsymbol{\theta}^*}\right|_{\boldsymbol{\theta} = \widehat{\boldsymbol{\theta}}_{\mathrm{ML}}}.$$

This, together with the fact that

$$\left.\frac{\partial \ln f(\mathbf{x};\boldsymbol{\theta})}{\partial \boldsymbol{\theta}^*}\right|_{\boldsymbol{\theta}=\widehat{\boldsymbol{\theta}}_{\mathrm{ML}}} = 0$$

is a necessary condition for the maximum of the log-likelihood function $\ln f(\mathbf{x};\boldsymbol{\theta})$, therefore leads to the following property of $\widehat{\boldsymbol{\theta}}_{\mathrm{ML}}$.

Property 3.59. *Any efficient estimator of unknown deterministic parameter vector $\boldsymbol{\theta}$, if it exists, is identical to the ML estimator $\widehat{\boldsymbol{\theta}}_{\mathrm{ML}}$.*

Another property of $\widehat{\boldsymbol{\theta}}_{\mathrm{ML}}$ is as follows.

Property 3.60. *For the linear data model given by (3.198), the ML estimator $\widehat{\boldsymbol{\theta}}_{\mathrm{ML}}$ given by (3.202) is (i) unbiased, (ii) efficient, (iii) UMVU, (iv) consistent, and (v) Gaussian distributed.*

The proof for the consistency of $\widehat{\boldsymbol{\theta}}_{\mathrm{ML}}$ can be found in [61, pp. 553–556], while the remaining parts of the proof are left as an exercise (Problem 3.23).

3.4.4 Method of Moments

Let $\widehat{\boldsymbol{\theta}}$ be an estimator of unknown parameter vector $\boldsymbol{\theta}$ based on a vector of N measurements $\mathbf{x} = (x[0], x[1], ..., x[N-1])^T$. The method of moments for obtaining $\widehat{\boldsymbol{\theta}}$ involves the following two steps.

(S1) Express $\boldsymbol{\theta}$ as a transformation of moment functions where the transformation is assumed to exist.
(S2) Obtain $\widehat{\boldsymbol{\theta}}$ by replacing the moment functions in (S1) with their respective estimators.

Several important examples of utilizing the method of moments are provided as follows.

Sample Correlations

Consider that the autocorrelation $r_x[l]$ is the unknown parameter to be estimated based on the set of N measurements $\{x[0], x[1], ..., x[N-1]\}$. Following Step (S1), we express

$$r_x[l] = E\{x[n]x^*[n-l]\} \tag{3.203}$$

and, following Step (S2), we consider the following estimator for $r_x[l]$:

$$\widehat{r}_x[l] = \frac{1}{N} \sum_{n=n_{\rm L}}^{n_{\rm U}} x[n]x^*[n-l] \tag{3.204}$$

where $n_{\rm L} = \max\{0, l\}$ and $n_{\rm U} = \min\{N-1, N-1+l\}$. Note that both $n_{\rm L}$ and $n_{\rm U}$ are functions of the lag l. Without loss of generality, let us focus on the case that $l \geq 0$, giving $n_{\rm L} = l$ and $n_{\rm U} = N-1$. It then follows from (3.204) that

$$E\{\widehat{r}_x[l]\} = \frac{1}{N} \sum_{n=l}^{N-1} E\{x[n]x^*[n-l]\} = \frac{N-l}{N} r_x[l]. \tag{3.205}$$

As a result, the estimator $\widehat{r}_x[0]$ is unbiased, whereas $\widehat{r}_x[l]$ for $l > 0$ is biased but asymptotically unbiased for finite l. As such, $\widehat{r}_x[l]$ is referred to as a *biased sample correlation* of $x[n]$. Furthermore, under certain conditions, $\widehat{r}_x[l]$ can be shown to be a consistent estimator [35, pp. 100–104]. Although there exist unbiased estimators for $r_x[l]$, it is still preferred to use the estimator $\widehat{r}_x[l]$ given by (3.204) because the resultant autocorrelation matrix is guaranteed to be positive definite (see Problem 3.24) and thus satisfies Properties 3.28 and 3.29.

For the cross-correlation $r_{xy}[l]$ of WSS processes $x[n]$ and $y[n]$, the corresponding *biased sample cross-correlation* is given by

$$\widehat{r}_{xy}[l] = \frac{1}{N} \sum_{n=n_{\rm L}}^{n_{\rm U}} x[n]y^*[n-l] \tag{3.206}$$

where $n_{\rm L} = \max\{0, l\}$ and $n_{\rm U} = \min\{N-1, N-1+l\}$. Similarly, the biased sample cross-correlation $\widehat{r}_{xy}[l]$ is consistent under certain conditions.

Sample Cumulants

Consider that the $(p+q)$th-order cumulant $C^x_{p,q}[l_1, l_2, ..., l_{p+q-1}]$ of a zero-mean stationary process $x[n]$ is the unknown parameter to be estimated based on the set of N measurements $\{x[0], x[1], ..., x[N-1]\}$. From (3.72), it follows that the *biased third-order sample cumulant* for $C^x_{2,1}[l_1, l_2]$ is given by

$$\begin{aligned}\widehat{C}^x_{2,1}[l_1, l_2] &= \widehat{E}\{x[n]x[n-l_1]x^*[n-l_2]\} \\ &= \frac{1}{N} \sum_{n=n_{\rm L}}^{n_{\rm U}} x[n]x[n-l_1]x^*[n-l_2] \end{aligned}\tag{3.207}$$

where $n_{\rm L} = \max\{0, l_1, l_2\}$ and $n_{\rm U} = \min\{N-1, N-1+l_1, N-1+l_2\}$. Similarly, the *biased fourth-order sample cumulant* for $C^x_{2,2}[l_1, l_2, l_3]$ is given by

$$\begin{aligned}\widehat{C}^x_{2,2}[l_1, l_2, l_3] = & \widehat{E}\{x[n]x[n-l_1]x^*[n-l_2]x^*[n-l_3]\} \\ & - \widehat{E}\{x[n]x[n-l_1]\}\widehat{E}\{x^*[n-l_2]x^*[n-l_3]\} \\ & - \widehat{E}\{x[n]x^*[n-l_2]\}\widehat{E}\{x[n-l_1]x^*[n-l_3]\} \\ & - \widehat{E}\{x[n]x^*[n-l_3]\}\widehat{E}\{x[n-l_1]x^*[n-l_2]\} \end{aligned} \tag{3.208}$$

where $\widehat{E}\{x_1x_2\}$ denotes the biased sample correlation of x_1 and x_2, and

$$\begin{aligned}&\widehat{E}\{x[n]x[n-l_1]x^*[n-l_2]x^*[n-l_3]\} \\ &\qquad = \frac{1}{N}\sum_{n=n_\mathrm{L}}^{n_\mathrm{U}} x[n]x[n-l_1]x^*[n-l_2]x^*[n-l_3]\end{aligned} \tag{3.209}$$

in which $n_\mathrm{L} = \max\{0, l_1, l_2, l_3\}$ and $n_\mathrm{U} = \min\{N-1, N-1+l_1, N-1+l_2, N-1+l_3\}$. Other higher-order cumulants can also be estimated in a similar fashion.

It has been shown that under certain conditions, the biased third-order sample cumulant given by (3.207) and the biased fourth-order sample cumulant given by (3.208) are asymptotically unbiased and consistent [35, p. 309]. Moreover, the larger the cumulant order, the higher the variance of the sample cumulant. This suggests that utilization of lower-order cumulants/polyspectra is preferable in practice. Fortunately, second-, third- and fourth-order cumulants are generally sufficient for most practical applications, and thus have been studied extensively for algorithm developments in many fields.

Spectral Estimation

Suppose $x[n]$ is a stationary process whose power spectra or polyspectra are the unknown parameters to be estimated. Basically, there are two types of methods for spectral estimation.

- *Parametric spectral estimation methods.* This type of method generally proceeds in three steps: (i) selecting a parametric model for $x[n]$, (ii) estimating the parameters of the selected model from measurements $\{x[0], x[1], ..., x[N-1]\}$, and (iii) obtaining the spectral estimator from the estimated parameters of the selected model [30, pp. 106, 107]. Some representative methods of this type are the AR, MA, and ARMA spectral estimation methods.
- *Nonparametric spectral estimation methods.* This type of method estimates the power spectrum of $x[n]$ directly from measurements $\{x[0], x[1], ..., x[N-1]\}$ (without assuming any parametric model). Some representative methods of this type are the periodogram spectral estimation method and the Blackman–Tukey spectral estimation method.

Generally speaking, parametric methods can provide accurate estimates when the presumed parametric model closely matches the true system, but will degrade severely when model mismatch occurs. On the other hand, nonparametric methods are robust against a wide variety of data structures because no assumption is made regarding data structure. In the following, two methods for power spectral estimation are described. The reader may consult [30] and [18] for excellent expositions on power spectral and polyspectral estimation methods, respectively.

Autoregressive Spectral Estimation

Let $x[n]$ be an AR process as shown in Fig. 3.32 where $u[n]$ is the driving input assumed to be a white WSS process with variance σ_u^2, and $h[n]$ is a causal stable system with transfer function

$$H(z) = \frac{1}{1 + \alpha_1 z^{-1} + \cdots + \alpha_p z^{-p}}. \tag{3.210}$$

The input–output relation of the AR(p) model $H(z)$ is given by

$$x[n] + a_1 x[n-1] + \cdots + a_p x[n-p] = u[n]. \tag{3.211}$$

Multiplying (3.211) by $x^*[n-l]$ and then taking expectation yields

$$r_x[l] + a_1 r_x[l-1] + \cdots + a_p r_x[l-p] = E\{u[n]x^*[n-l]\} \tag{3.212}$$

where $r_x[l]$ is the autocorrelation function of $x[n]$. Since $x[n] = h[n] \star u[n]$,

$$E\{u[n]x^*[n-l]\} = \sum_{k=-\infty}^{\infty} h^*[k] r_u[l+k] = \sigma_u^2 \cdot h^*[-l] \tag{3.213}$$

where $h[l] = 0$ for $l < 0$ (since $h[l]$ is causal) and $h[0] = \lim_{z\to\infty} H(z) = 1$ (by the *initial value theorem*). Formulating (3.212) into matrix form for $l = 0, 1, ..., p$ and using (3.213), we obtain

$$\begin{pmatrix} r_x[0] & r_x[-1] & \cdots & r_x[-p] \\ r_x[1] & r_x[0] & \cdots & r_x[1-p] \\ \vdots & \vdots & \ddots & \vdots \\ r_x[p] & r_x[p-1] & \cdots & r_x[0] \end{pmatrix} \begin{pmatrix} 1 \\ a_1 \\ \vdots \\ a_p \end{pmatrix} = \begin{pmatrix} \sigma_u^2 \\ 0 \\ \vdots \\ 0 \end{pmatrix}, \tag{3.214}$$

which are called the *Yule–Walker equations*. Accordingly, the AR coefficient vector $\mathbf{a}_p = (a_1, a_2, ..., a_p)^T$ can be estimated by solving the following set of equations:

$$\widehat{\mathbf{R}}_x \cdot \widehat{\mathbf{a}}_p = -\widehat{\mathbf{r}}_x \tag{3.215}$$

where

$$\widehat{\mathbf{R}}_x = \begin{pmatrix} \widehat{r}_x[0] & \widehat{r}_x^*[1] & \cdots & \widehat{r}_x^*[p-1] \\ \widehat{r}_x[1] & \widehat{r}_x[0] & \cdots & \widehat{r}_x^*[p-2] \\ \vdots & \vdots & \ddots & \vdots \\ \widehat{r}_x[p-1] & \widehat{r}_x[p-2] & \cdots & \widehat{r}_x[0] \end{pmatrix} \tag{3.216}$$

and $\widehat{\mathbf{r}}_x = (\widehat{r}_x[1], \widehat{r}_x[2], ..., \widehat{r}_x[p])^T$ in which $\widehat{r}_x[l]$ is the biased sample autocorrelation for $r_x[l]$. Furthermore, from (3.214), the variance σ_u^2 can be estimated via

$$\widehat{\sigma_u^2} = \widehat{r}_x[0] + \sum_{l=1}^{p} \widehat{a}_l \widehat{r}_x^*[l]. \tag{3.217}$$

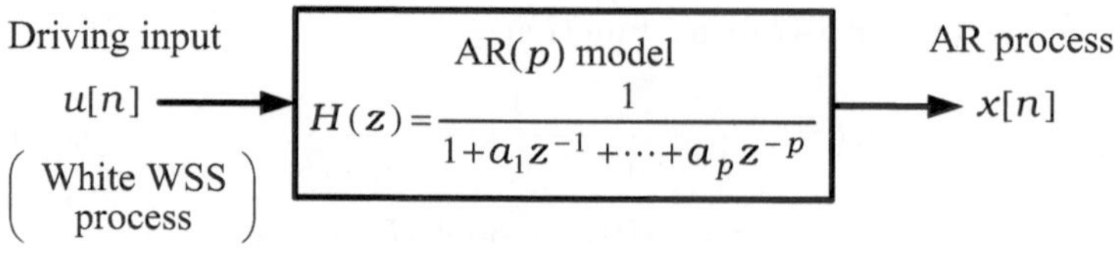

Fig. 3.32 Model for AR process $x[n]$

After $\widehat{a}_1, \widehat{a}_2, ..., \widehat{a}_p$ and $\widehat{\sigma_u^2}$ are obtained, the power spectrum $S_x(\omega)$ can be estimated based on (3.127). Specifically, the estimator of $S_x(\omega)$, referred to as the *AR spectral estimator*, is given by

$$\widehat{S}_x(\omega) = \frac{\widehat{\sigma_u^2}}{|1 + \widehat{a}_1 z^{-1} + \cdots + \widehat{a}_p z^{-p}|^2}. \tag{3.218}$$

When the AR process $x[n]$ is Gaussian, the AR spectral estimator can be shown to be a consistent estimator [35, p. 154].

Periodogram Spectral Estimation

By removing the ensemble average in (3.120) and taking the set of N measurements $\{x[0], x[1], ..., x[N-1]\}$ into account, we have the following estimator of power spectrum $S_x(\omega)$: [30, p. 65]

$$\widehat{S}_x(\omega) = \frac{1}{N} \left| \sum_{n=0}^{N-1} x[n] e^{-j\omega n} \right|^2. \tag{3.219}$$

The estimator is known as the *periodogram spectral estimator* and the corresponding estimate is known as the *periodogram.* It has been shown that the periodogram spectral estimator is an asymptotically unbiased estimator, but not a consistent estimator [35, pp. 107–109].

3.4.5 Minimum Mean-Square-Error Estimation

Suppose the parameter vector $\boldsymbol{\theta}$ to be estimated is random. Based on a vector of N measurements $\mathbf{x} = (x[0], x[1], ..., x[N-1])^T$, the MMSE estimator, denoted by $\widehat{\boldsymbol{\theta}}_{\text{MS}}$, is determined by minimizing the mean-square-error (MSE)

$$J_{\mathrm{MSE}}(\widehat{\boldsymbol{\theta}}) = E\left\{\|\boldsymbol{\varepsilon}\|^2\right\} = E\left\{\left\|\boldsymbol{\theta} - \widehat{\boldsymbol{\theta}}\right\|^2\right\} \tag{3.220}$$

where $\boldsymbol{\varepsilon} = \boldsymbol{\theta} - \widehat{\boldsymbol{\theta}}$ is the vector of estimation error, and $\widehat{\boldsymbol{\theta}} = \widehat{\boldsymbol{\theta}}_{\mathrm{MS}}$ as $J_{\mathrm{MSE}}(\widehat{\boldsymbol{\theta}})$ is minimum. Further express

$$J_{\mathrm{MSE}}(\widehat{\boldsymbol{\theta}}) = \int_{-\infty}^{\infty}\int_{-\infty}^{\infty} \|\boldsymbol{\varepsilon}\|^2 \cdot f(\mathbf{x}, \boldsymbol{\theta}) d\mathbf{x} d\boldsymbol{\theta} = \int_{-\infty}^{\infty} \widetilde{J}_{\mathrm{MSE}}(\widehat{\boldsymbol{\theta}}) \cdot f(\mathbf{x}) d\mathbf{x} \tag{3.221}$$

where

$$\widetilde{J}_{\mathrm{MSE}}(\widehat{\boldsymbol{\theta}}) = \int_{-\infty}^{\infty} \|\boldsymbol{\varepsilon}\|^2 \cdot f(\boldsymbol{\theta}|\mathbf{x}) d\boldsymbol{\theta} = E\left\{\|\boldsymbol{\varepsilon}\|^2|\mathbf{x}\right\}. \tag{3.222}$$

Since $f(\mathbf{x}) \geq 0$ for all $\mathbf{x}$, it follows that the minimum of $\widetilde{J}_{\mathrm{MSE}}(\widehat{\boldsymbol{\theta}})$ must also be the minimum of $J_{\mathrm{MSE}}(\widehat{\boldsymbol{\theta}})$, implying that $\widehat{\boldsymbol{\theta}}_{\mathrm{MS}}$ can be obtained from the minimum of $\widetilde{J}_{\mathrm{MSE}}(\widehat{\boldsymbol{\theta}})$.

Substituting $\boldsymbol{\varepsilon} = \boldsymbol{\theta} - \widehat{\boldsymbol{\theta}}$ into (3.222) yields

$$\begin{aligned}
\widetilde{J}_{\mathrm{MSE}}(\widehat{\boldsymbol{\theta}}) &= E\left\{(\boldsymbol{\theta} - \widehat{\boldsymbol{\theta}})^H(\boldsymbol{\theta} - \widehat{\boldsymbol{\theta}})|\mathbf{x}\right\} \\
&= E\left\{\|\boldsymbol{\theta}\|^2|\mathbf{x}\right\} - \widehat{\boldsymbol{\theta}}^H \cdot E\{\boldsymbol{\theta}|\mathbf{x}\} - E\{\boldsymbol{\theta}^H|\mathbf{x}\} \cdot \widehat{\boldsymbol{\theta}} + \|\widehat{\boldsymbol{\theta}}\|^2 \\
&= E\left\{\|\boldsymbol{\theta}\|^2|\mathbf{x}\right\} + \left\|\widehat{\boldsymbol{\theta}} - E\{\boldsymbol{\theta}|\mathbf{x}\}\right\|^2 - \left\|E\{\boldsymbol{\theta}|\mathbf{x}\}\right\|^2
\end{aligned} \tag{3.223}$$

where, in the second line, we have used the fact that given $\mathbf{x}$, the corresponding $\widehat{\boldsymbol{\theta}}$ is a constant vector. We therefore have the following result [17, pp. 175, 176].

Theorem 3.61 (Fundamental Theorem of Estimation Theory). *The MMSE estimator $\widehat{\boldsymbol{\theta}}_{\mathrm{MS}}$ of unknown random parameter vector $\boldsymbol{\theta}$ is the conditional mean given by*

$$\widehat{\boldsymbol{\theta}}_{\mathrm{MS}} = E\{\boldsymbol{\theta}|\mathbf{x}\} = \int_{-\infty}^{\infty} \boldsymbol{\theta} \cdot f(\boldsymbol{\theta}|\mathbf{x}) d\boldsymbol{\theta} \tag{3.224}$$

and the corresponding minimum value of MSE is given by

$$\widetilde{J}_{\mathrm{MSE}}(\widehat{\boldsymbol{\theta}}_{\mathrm{MS}}) = E\left\{\|\boldsymbol{\theta}\|^2|\mathbf{x}\right\} - \|\widehat{\boldsymbol{\theta}}_{\mathrm{MS}}\|^2. \tag{3.225}$$

In general, $\widehat{\boldsymbol{\theta}}_{\mathrm{MS}}$ given by (3.224) is a nonlinear estimator which may not be obtained easily. As such, in most practical applications, $\widehat{\boldsymbol{\theta}}_{\mathrm{MS}}$ is often confined to a linear estimator, called a *linear MMSE (LMMSE) estimator*. A typical representative of LMMSE estimators is the well-known *Wiener filter*, which will be detailed later on.

Properties of MMSE Estimators

From (3.224), it follows that $E\{\widehat{\boldsymbol{\theta}}_{\rm MS}\} = E\left\{E\{\boldsymbol{\theta}|\mathbf{x}\}\right\} = E\{\boldsymbol{\theta}\}$. Thus, we have the following property of $\widehat{\boldsymbol{\theta}}_{\rm MS}$.

Property 3.62. *The MMSE estimator* $\widehat{\boldsymbol{\theta}}_{\rm MS}$ *is unbiased.*

This property further leads to

$$J_{\rm MSE}(\widehat{\boldsymbol{\theta}}_{\rm MS}) = E\left\{\left\|\boldsymbol{\theta} - \widehat{\boldsymbol{\theta}}_{\rm MS}\right\|^2\right\} = \sum_k \mathrm{Var}(\widehat{\theta}_{\rm MS}[k]) \tag{3.226}$$

where $\widehat{\theta}_{\rm MS}[k]$ is the kth entry of $\widehat{\boldsymbol{\theta}}_{\rm MS}$. We thus have another property of $\widehat{\boldsymbol{\theta}}_{\rm MS}$.

Property 3.63. *Any UMVU estimator of unknown random parameter vector* $\boldsymbol{\theta}$*, if it exists, is identical to the MMSE estimator* $\widehat{\boldsymbol{\theta}}_{\rm MS}$.

3.4.6 Wiener Filtering

Wiener filtering is nothing but a special case of *linear optimum filtering* as depicted in Fig. 3.33. Given a set of N measurements $\{x[0], x[1], ..., x[N-1]\}$, the goal of linear optimum filtering is to design an optimum LTI filter $v[n]$ in some statistical sense such that the filter output

$$\widehat{d}[n] = v[n] \star x[n] \tag{3.227}$$

approximates the desired signal $d[n]$. This obviously assumes that $d[n]$ is related to $x[n]$ in some manner as illustrated in Table 3.10. Regarding Wiener filtering, the optimum LTI filter $v[n]$, called the *Wiener filter*, is designed by minimizing the MSE:

$$J_{\rm MSE}(v[n]) = E\{|\varepsilon[n]|^2\} \tag{3.228}$$

where

$$\varepsilon[n] = d[n] - \widehat{d}[n] \tag{3.229}$$

is the estimation error. Clearly, the Wiener filter $v[n]$ is an LMMSE estimator.

Orthogonality Principle

For simplicity, the Wiener filter $v[n]$ is assumed to be FIR with $v[n] = 0$ for $n < L_1$ and $n > L_2$, and its length $L = L_2 - L_1 + 1$. The estimation error given by (3.229) can be expressed as

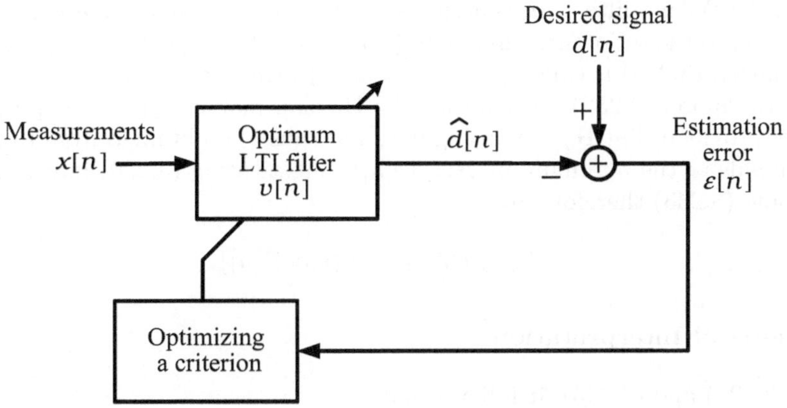

Fig. 3.33 Linear optimum filtering

Table 3.10 Typical problems of linear optimum filtering

Problem	Measurements	Desired Signal
Filtering of signal in noise	$x[n] = s[n] + w[n]$	$d[n] = s[n]$
Prediction of signal in noise	$x[n] = s[n] + w[n]$	$d[n] = s[n+\tau]$, $\tau > 0$
Smoothing of signal in noise	$x[n] = s[n] + w[n]$	$d[n] = s[n-\tau]$, $\tau > 0$

$$\varepsilon[n] = d[n] - \sum_{i=L_1}^{L_2} v[i]x[n-i] = d[n] - \mathbf{v}^T\mathbf{x}[n] \tag{3.230}$$

where $\mathbf{v} = (v[L_1], v[L_1+1], ..., v[L_2])^T$ and $\mathbf{x}[n] = (x[n-L_1], x[n-L_1-1], ..., x[n-L_2])^T$. From (3.228) and (3.230), it follows that the gradient

$$\frac{\partial J_{\text{MSE}}(v[n])}{\partial \mathbf{v}^*} = E\left\{\varepsilon[n]\frac{\partial \varepsilon^*[n]}{\partial \mathbf{v}^*}\right\} = -E\{\varepsilon[n]\mathbf{x}^*[n]\} \tag{3.231}$$

and the Hessian matrix

$$\mathbf{J}_2(v[n]) = \frac{\partial}{\partial \mathbf{v}^*}\left[\frac{\partial J_{\text{MSE}}(\mathbf{v})}{\partial \mathbf{v}^*}\right]^H = -E\left\{\frac{\partial \varepsilon^*[n]}{\partial \mathbf{v}^*}\mathbf{x}^T[n]\right\} = \mathbf{R}_x^* \tag{3.232}$$

where $\mathbf{R}_x = E\{\mathbf{x}[n]\mathbf{x}^H[n]\}$ is the $L \times L$ autocorrelation matrix. As a consequence of (3.231), the Wiener filter $v[n]$ satisfies

$$E\{\varepsilon[n]\mathbf{x}^*[n]\} = \mathbf{0}, \tag{3.233}$$

that is, the Wiener filter $v[n]$ is designed such that the estimation error $\varepsilon[n]$ is orthogonal to the set of measurements $\{x[n-L_1], x[n-L_1-1], ..., x[n-L_2]\}$. Accordingly, (3.233) is called the *orthogonality principle*.

Furthermore, (3.232) reveals that the Hessian matrix $\mathbf{J}_2(v[n])$ is positive definite provided that $\mathbf{R}_x$ is nonsingular, and thus $v[n]$ obtained from (3.233) corresponds to the minimum of $J_{\mathrm{MSE}}(v[n])$. Substituting (3.230) into (3.228) and using (3.233) therefore gives

$$\min\{J_{\mathrm{MSE}}(v[n])\} = E\{\varepsilon[n]d^*[n]\}. \tag{3.234}$$

Geometrical Interpretation

From (3.227) and (3.233), it follows that

$$E\{\varepsilon[n]\widehat{d}^*[n]\} = \sum_{k=L_1}^{L_2} v^*[k] \cdot E\{\varepsilon[n]x^*[n-k]\} = 0. \tag{3.235}$$

By viewing random signals and cross-correlations as vectors and inner products, respectively, we have the geometrical interpretation of (3.229) and (3.235) depicted in Fig. 3.34. The vector representing the estimation error $\varepsilon[n]$ is perpendicular to the vector representing an estimate of the desired signal, $\widehat{d}[n]$, which is a linear combination of measurements $x[n-L_1]$, $x[n-L_1-1]$, ..., $x[n-L_2]$. This geometry is referred to as the *statistician's Pythagorean theorem* [31]. In the same way, $\min\{J_{\mathrm{MSE}}(v[n])\}$ given by (3.234) can be interpreted as the projection of the vector representing the desired signal $d[n]$ onto the vector representing the estimation error $\varepsilon[n]$.

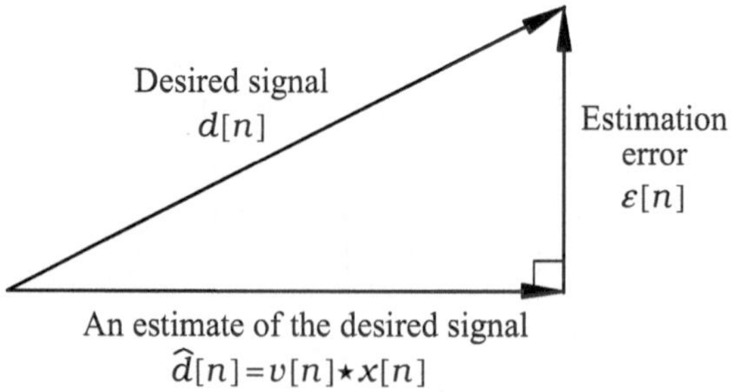

Fig. 3.34 Geometrical interpretation of Wiener filtering

Wiener–Hopf Equations

The coefficients of the Wiener filter $v[n]$ can be obtained by solving a set of linear equations derived from the orthogonality principle. Substituting (3.230) into (3.233) yields

$$E\{d[n]\mathbf{x}^*[n]\} = E\{\mathbf{v}^T\mathbf{x}[n]\mathbf{x}^*[n]\} = E\{\mathbf{x}^*[n]\mathbf{x}^T[n]\}\mathbf{v}$$

or

$$\mathbf{R}_x^*\mathbf{v} = \mathbf{r}_{dx} \tag{3.236}$$

where $\mathbf{r}_{dx} = E\{d[n]\mathbf{x}^*[n]\}$. The set of equations (3.236) is called the *Wiener–Hopf equations.* Moreover, substituting (3.230) into (3.234) yields

$$\min\{J_{\mathrm{MSE}}(v[n])\} = E\{|d[n]|^2\} - E\{d^*[n]\mathbf{x}^T[n]\}\mathbf{v} = \sigma_d^2 - \mathbf{r}_{dx}^H\mathbf{v} \tag{3.237}$$

where σ_d^2 is the variance of $d[n]$. Table 3.11 summarizes the Wiener filter $v[n]$.

Table 3.11 Summary of Wiener filter

	FIR Wiener Filter $v[n]$
Problem	Given measurements $x[n]$, find the Wiener filter $v[n]$, $n = L_1, L_1+1, ..., L_2$, to minimize the MSE $$J_{\mathrm{MSE}}(v[n]) = E\{|\varepsilon[n]|^2\}$$ where $\varepsilon[n] = d[n] - \mathbf{v}^T\mathbf{x}[n]$ is the estimation error, $d[n]$ is the desired signal, $\mathbf{v} = (v[L_1],\ v[L_1+1], \cdots,\ v[L_2])^T$, and $\mathbf{x}[n] = (x[n-L_1],\ x[n-L_1-1], \cdots,\ x[n-L_2])^T$.
Orthogonality principle	$E\{\varepsilon[n]\mathbf{x}^*[n]\} = \mathbf{0}$ $\min\{J_{\mathrm{MSE}}(v[n])\} = E\{\varepsilon[n]d^*[n]\}$
Wiener–Hopf equations	$\mathbf{R}_x^*\mathbf{v} = \mathbf{r}_{dx}$ where $\mathbf{R}_x = E\{\mathbf{x}[n]\mathbf{x}^H[n]\}$ and $\mathbf{r}_{dx} = E\{d[n]\mathbf{x}^*[n]\}$.
Minimum of the MSE	$\min\{J_{\mathrm{MSE}}(v[n])\} = \sigma_d^2 - \mathbf{r}_{dx}^H\mathbf{v}$ where $\sigma_d^2 = E\{|d[n]|^2\}$.

3.4.7 Least-Squares Estimation

Consider the data model shown in Fig. 3.35. The signal $s[n]$ is generated from a signal model governed by a set of unknown parameters $\{\theta[0], \theta[1], ..., \theta[L-1]\}$, and $x[n]$ is the measurement of $s[n]$ that may be perturbed by

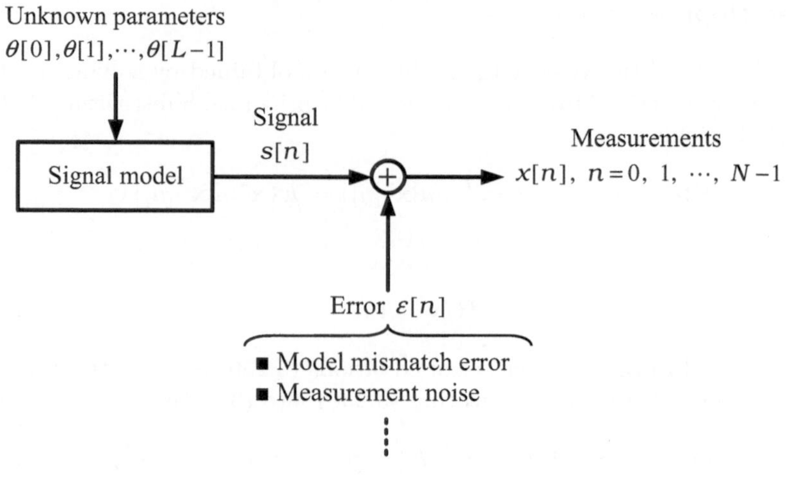

Fig. 3.35 Data model for the LS estimation problem

such errors as model mismatch error, measurement noise, etc. The resultant error due to these perturbations is denoted by $\varepsilon[n]$. With the collection of N measurements, the data model can be described as

$$\mathbf{x} = \mathbf{s} + \boldsymbol{\varepsilon} \tag{3.238}$$

where $\mathbf{x} = (x[0], x[1], ..., x[N-1])^T$, $\mathbf{s} = (s[0], s[1], ..., s[N-1])^T$, and $\boldsymbol{\varepsilon} = (\varepsilon[0], \varepsilon[1], ..., \varepsilon[N-1])^T$. Based on the data model (3.238), the LS estimation method[18] finds the unknown parameter vector $\boldsymbol{\theta} = (\theta[0], \theta[1], ..., \theta[L-1])^T$ by minimizing the sum of squared errors

$$J_{\mathrm{LS}}(\boldsymbol{\theta}) = \sum_{n=0}^{N-1} |\varepsilon[n]|^2 = \|\boldsymbol{\varepsilon}\|^2. \tag{3.239}$$

Note that $\boldsymbol{\theta}$ can be deterministic or random, and no probabilistic assumption needs to be made for $\mathbf{x}$. This therefore leads to broad applications of the LS estimation method.

Linear Signal Model

Consider the following linear signal model:

$$\mathbf{s} = \mathbf{H}\boldsymbol{\theta} \tag{3.240}$$

[18] The LS estimation method is due to the study of planetary motions by K. F. Gauss (1795), for whom a short biography can be found in [17, pp. 28–29].

where $\mathbf{H}$ is an $N \times L$ nonsingular matrix and $N \geq L$. With this signal model, the LS estimation problem is exactly identical to the full-rank overdetermined LS problem presented in Section 2.4, and thus the LS estimator

$$\widehat{\boldsymbol{\theta}}_{\mathrm{LS}} = (\mathbf{H}^H\mathbf{H})^{-1}\mathbf{H}^H\mathbf{x}. \tag{3.241}$$

From (3.238), (3.240) and (3.241), it follows that

$$\boldsymbol{\varepsilon} = \mathbf{x} - \mathbf{H}\widehat{\boldsymbol{\theta}}_{\mathrm{LS}} = \left[\mathbf{I} - \mathbf{H}(\mathbf{H}^H\mathbf{H})^{-1}\mathbf{H}^H\right]\mathbf{x} \tag{3.242}$$

and thereby

$$\min\{J_{\mathrm{LS}}(\boldsymbol{\theta})\} = \mathbf{x}^H\left[\mathbf{I} - \mathbf{H}(\mathbf{H}^H\mathbf{H})^{-1}\mathbf{H}^H\right]\mathbf{x}. \tag{3.243}$$

As a remark, the above-mentioned results can also be extended to any type of matrix $\mathbf{H}$ by means of the SVD as described in Section 2.4.

Weighted Least-Squares Estimation

When the relative reliability of each entry of $\mathbf{x}$ is known, one can improve the LS estimator by using the following *weighted least-squares (WLS) criterion*

$$J_{\mathrm{WLS}}(\boldsymbol{\theta}) = \boldsymbol{\varepsilon}^H\mathbf{W}\boldsymbol{\varepsilon} \tag{3.244}$$

where $\mathbf{W}$ is an $N \times N$ Hermitian, positive definite weighting matrix. Note that if $\mathbf{W} = \mathbf{I}$, then $J_{\mathrm{WLS}}(\boldsymbol{\theta})$ reduces to $J_{\mathrm{LS}}(\boldsymbol{\theta})$. Let us consider, again, the linear signal model (3.240). Then the gradient

$$\frac{\partial J_{\mathrm{WLS}}(\boldsymbol{\theta})}{\partial \boldsymbol{\theta}^*} = \mathbf{H}^H\mathbf{W}(\mathbf{x} - \mathbf{H}\boldsymbol{\theta}) \tag{3.245}$$

and the Hessian matrix

$$\mathbf{J}_2(\boldsymbol{\theta}) = \frac{\partial}{\partial \boldsymbol{\theta}^*}\left[\frac{\partial J_{\mathrm{WLS}}(\boldsymbol{\theta})}{\partial \boldsymbol{\theta}^*}\right]^H = \mathbf{H}^H\mathbf{W}\mathbf{H}. \tag{3.246}$$

Since $\mathbf{W}$ is positive definite, $\mathbf{q}^H\mathbf{H}^H\mathbf{W}\mathbf{H}\mathbf{q} = (\mathbf{H}\mathbf{q})^H\mathbf{W}(\mathbf{H}\mathbf{q}) > 0$ for any nonzero vector $\mathbf{q}$ and thus $\mathbf{J}_2(\boldsymbol{\theta})$ is positive definite. By Theorem 2.44, the solution obtained by setting (3.245) to zero corresponds to the minimum of $J_{\mathrm{WLS}}(\boldsymbol{\theta})$. As a result, the WLS estimator is obtained as

$$\widehat{\boldsymbol{\theta}}_{\mathrm{WLS}} = (\mathbf{H}^H\mathbf{W}\mathbf{H})^{-1}\mathbf{H}^H\mathbf{W}\mathbf{x}. \tag{3.247}$$

From (3.238), (3.240) and (3.247), it follows that

$$\boldsymbol{\varepsilon} = \left[\mathbf{I} - \mathbf{H}(\mathbf{H}^H\mathbf{W}\mathbf{H})^{-1}\mathbf{H}^H\mathbf{W}\right]\mathbf{x} \tag{3.248}$$

and accordingly

$$\min\{J_{\mathrm{WLS}}(\boldsymbol{\theta})\} = \mathbf{x}^H\left[\mathbf{W} - \mathbf{W}^H\mathbf{H}(\mathbf{H}^H\mathbf{W}\mathbf{H})^{-1}\mathbf{H}^H\mathbf{W}\right]\mathbf{x}. \tag{3.249}$$

Properties of WLS Estimators

The following property of the WLS estimator $\widehat{\boldsymbol{\theta}}_{\mathrm{WLS}}$ can be observed from (3.202) and (3.247) as well as (3.198), (3.238) and (3.240).

Property 3.64. *Suppose the error* $\boldsymbol{\varepsilon}$ *is a complex white Gaussian process with covariance matrix* $\mathbf{C}_{\boldsymbol{\varepsilon}}$*. With the weighting matrix* $\mathbf{W} = \mathbf{C}_{\boldsymbol{\varepsilon}}^{-1}$*, the WLS estimator* $\widehat{\boldsymbol{\theta}}_{\mathrm{WLS}}$ *given by (3.247) is identical to the ML estimator* $\widehat{\boldsymbol{\theta}}_{\mathrm{ML}}$ *given by (3.202).*

This property, together with Property 3.60, implies the following property.

Property 3.65. *Suppose the error* $\boldsymbol{\varepsilon}$ *is a complex white Gaussian process with covariance matrix* $\mathbf{C}_{\boldsymbol{\varepsilon}}$*. With the weighting matrix* $\mathbf{W} = \mathbf{C}_{\boldsymbol{\varepsilon}}^{-1}$*, the WLS estimator* $\widehat{\boldsymbol{\theta}}_{\mathrm{WLS}}$ *given by (3.247) is (i) unbiased, (ii) efficient, (iii) UMVU, (iv) consistent, and (v) Gaussian distributed.*

3.5 Summary

We have reviewed the fundamentals of discrete-time signals and systems, including the definitions, terminologies, transformation tools, and parametric models. We then reviewed random variables and random processes, including statistical characterization, second-order and higher-order statistics for both real and complex cases. Finally, estimation theory was introduced that includes the problem, properties of estimators and several representatives of estimation methods.

Appendix 3A
Relationship between Cumulants and Moments

Consider, first, that x_1, x_2, ..., x_k are real. From (3.67), it follows that the second characteristic function $\Psi(\boldsymbol{\omega})$ can be expanded as a Taylor series at $\boldsymbol{\omega} = \mathbf{0}$ with coefficients expressed in terms of cumulants. On the other hand, from (3.69) and (3.63), it follows that $\Psi(\boldsymbol{\omega})$ can also be expanded as a Taylor series at $\boldsymbol{\omega} = \mathbf{0}$ with coefficients expressed in terms of moments. Equating the two sets of coefficients of the Taylor series therefore leads to the following generic form of the relationship between cumulants and moments: [15–18]

$$\begin{aligned}&\mathrm{cum}\{x_1, x_2, ..., x_k\}\\&= \sum_{l=1}^{k} (-1)^{l-1}(l-1)! \cdot E\left\{\prod_{i\in I_1} x_i\right\} E\left\{\prod_{i\in I_2} x_i\right\} \cdots E\left\{\prod_{i\in I_l} x_i\right\}\end{aligned} \tag{3.250}$$

where the summation includes all possible partitions I_1, I_2, ..., I_l of the integer set $I = \{1, 2, ..., k\}$, i.e. $I_1 \cup I_2 \cup \cdots \cup I_l = I$. The use of (3.250) is illustrated in Table 3.12. From this table, we have

$$\begin{aligned}\mathrm{cum}\{x_1, x_2, x_3\} =& E\{x_1 x_2 x_3\} - E\{x_1\}E\{x_2 x_3\} - E\{x_2\}E\{x_1 x_3\} \\ & - E\{x_3\}E\{x_1 x_2\} + 2E\{x_1\}E\{x_2\}E\{x_3\}, \end{aligned} \tag{3.251}$$

which reduces to (3.72) as $E\{x_1\} = E\{x_2\} = E\{x_3\} = 0$.

Table 3.12 Illustration of Equation (3.250)

Expression of $\mathrm{cum}\{x_1, x_2, x_3\}$ in terms of Moments				
l	I_1	I_2	I_3	Terms in the summation
1	$\{1, 2, 3\}$			$E\{x_1 x_2 x_3\}$
2	$\{1\}$	$\{2, 3\}$		$-E\{x_1\}E\{x_2 x_3\}$
2	$\{2\}$	$\{1, 3\}$		$-E\{x_2\}E\{x_1 x_3\}$
2	$\{3\}$	$\{1, 2\}$		$-E\{x_3\}E\{x_1 x_2\}$
3	$\{1\}$	$\{2\}$	$\{3\}$	$2E\{x_1\}E\{x_2\}E\{x_3\}$

When x_1, x_2, ..., x_k are complex, the same form of relationship given by (3.250) can be obtained by treating $\{\omega_1, \omega_2, ..., \omega_k\}$ and $\{\omega_1^*, \omega_2^*, ..., \omega_k^*\}$ as two independent sets of variables and following the foregoing derivation with the Taylor series expanded for $\Psi(\boldsymbol{\omega}) = \Psi(\boldsymbol{\omega}, \boldsymbol{\omega}^*)$ at $(\boldsymbol{\omega}, \boldsymbol{\omega}^*) = (\mathbf{0}, \mathbf{0})$.

Appendix 3B Proof of Theorem 3.47

Since $\widehat{\theta}$ is unbiased,

$$E\{(\widehat{\theta} - \theta)^*\} = \int_{-\infty}^{\infty} f(\mathbf{x}; \theta) \cdot (\widehat{\theta} - \theta)^* d\mathbf{x} = 0. \tag{3.252}$$

Differentiating (3.252) with respect to θ^* yields

$$\int_{-\infty}^{\infty} \left[\frac{\partial f(\mathbf{x}; \theta)}{\partial \theta^*} (\widehat{\theta} - \theta)^* - f(\mathbf{x}; \theta) \right] d\mathbf{x} = 0. \tag{3.253}$$

Note that in deriving (3.253), we have used the assumption that $\partial f(\mathbf{x}; \theta)/\partial \theta^*$ exists and is absolutely integrable so that the order of differentiation and integration can be interchanged. Equation (3.253) further leads to

$$
\begin{aligned}
1 &= \left| \int_{-\infty}^{\infty} \frac{\partial \ln f(\mathbf{x};\theta)}{\partial \theta^*} f(\mathbf{x};\theta) \cdot (\widehat{\theta} - \theta)^* d\mathbf{x} \right|^2 = \left| E\left\{ \frac{\partial \ln f(\mathbf{x};\theta)}{\partial \theta^*} (\widehat{\theta} - \theta)^* \right\} \right|^2 \\
&\leq E\left\{ \left| \frac{\partial \ln f(\mathbf{x};\theta)}{\partial \theta^*} \right|^2 \right\} \cdot E\left\{ |\widehat{\theta} - \theta|^2 \right\} = F(\theta) \cdot \mathrm{Var}(\widehat{\theta}) \qquad (3.254)
\end{aligned}
$$

where the second line is obtained by using Theorem 3.9. Equation (3.254) therefore gives (3.176).

By Theorem 3.9, the equality of (3.254) holds if and only if (3.178) is satisfied where $\beta(\theta)$ is not a function of $\mathbf{x}$ or, equivalently, $\widehat{\theta}$ since the expectation operators in (3.254) are with respect to $\mathbf{x}$. This therefore completes the proof.

Q.E.D.

Appendix 3C
Proof of Theorem 3.52

Since $\widehat{\boldsymbol{\theta}}$ is unbiased,

$$
E\{(\widehat{\boldsymbol{\theta}} - \boldsymbol{\theta})^H\} = \int_{-\infty}^{\infty} f(\mathbf{x};\boldsymbol{\theta}) \cdot (\widehat{\boldsymbol{\theta}} - \boldsymbol{\theta})^H d\mathbf{x} = \mathbf{0}. \qquad (3.255)
$$

By the assumption that $\partial f(\mathbf{x};\boldsymbol{\theta})/\partial \boldsymbol{\theta}^*$ exists and is absolutely integrable, differentiating (3.255) with respect to $\boldsymbol{\theta}^*$ yields

$$
\int_{-\infty}^{\infty} \left[\frac{\partial f(\mathbf{x};\boldsymbol{\theta})}{\partial \boldsymbol{\theta}^*} (\widehat{\boldsymbol{\theta}} - \boldsymbol{\theta})^H - f(\mathbf{x};\boldsymbol{\theta}) \cdot \mathbf{I} \right] d\mathbf{x} = \mathbf{0}
$$

or

$$
\begin{aligned}
\mathbf{I} &= \int_{-\infty}^{\infty} \frac{\partial \ln f(\mathbf{x};\boldsymbol{\theta})}{\partial \boldsymbol{\theta}^*} f(\mathbf{x};\boldsymbol{\theta}) \cdot (\widehat{\boldsymbol{\theta}} - \boldsymbol{\theta})^H d\mathbf{x} \\
&= E\left\{ \frac{\partial \ln f(\mathbf{x};\boldsymbol{\theta})}{\partial \boldsymbol{\theta}^*} (\widehat{\boldsymbol{\theta}} - \boldsymbol{\theta})^H \right\} \qquad (3.256)
\end{aligned}
$$

Let $\mathbf{q}$ be any $L \times 1$ nonzero vector and assume that $\mathbf{F}(\boldsymbol{\theta})$ is nonsingular. Then $\mathbf{F}(\boldsymbol{\theta})$ is positive definite (by Property 2.13) and therefore $\mathbf{F}^{-1}(\boldsymbol{\theta})$ is also positive definite (by Property 2.12). By (3.256) and the Cauchy–Schwartz inequality (Theorem 3.9), we obtain

$$\left|\mathbf{q}^H\mathbf{F}^{-1}(\boldsymbol{\theta})\cdot\mathbf{I}\cdot\mathbf{q}\right|^2 = \left|E\left\{\left[\mathbf{q}^H\mathbf{F}^{-1}(\boldsymbol{\theta})\frac{\partial \ln f(\mathbf{x};\boldsymbol{\theta})}{\partial\boldsymbol{\theta}^*}\right]\cdot\left[(\widehat{\boldsymbol{\theta}}-\boldsymbol{\theta})^H\mathbf{q}\right]\right\}\right|^2$$
$$\leq E\left\{\left|\mathbf{q}^H\mathbf{F}^{-1}(\boldsymbol{\theta})\frac{\partial \ln f(\mathbf{x};\boldsymbol{\theta})}{\partial\boldsymbol{\theta}^*}\right|^2\right\}\cdot E\left\{\left|(\widehat{\boldsymbol{\theta}}-\boldsymbol{\theta})^H\mathbf{q}\right|^2\right\}$$
$$= \mathbf{q}^H\mathbf{F}^{-1}(\boldsymbol{\theta})\mathbf{q}\cdot\mathbf{q}^H\mathbf{C}(\widehat{\boldsymbol{\theta}})\mathbf{q}, \tag{3.257}$$

which implies that

$$\mathbf{q}^H\mathbf{F}^{-1}(\boldsymbol{\theta})\mathbf{q} \leq \mathbf{q}^H\mathbf{C}(\widehat{\boldsymbol{\theta}})\mathbf{q} \tag{3.258}$$

since $\mathbf{F}^{-1}(\boldsymbol{\theta})$ is positive definite. This therefore completes the proof of (3.184).

By Theorem 3.9, the equality of (3.258) holds if and only if

$$(\widehat{\boldsymbol{\theta}}-\boldsymbol{\theta})^T\mathbf{q}^* - \beta\mathbf{q}^H\mathbf{F}^{-1}(\boldsymbol{\theta})\frac{\partial \ln f(\mathbf{x};\boldsymbol{\theta})}{\partial\boldsymbol{\theta}^*} = 0$$

or

$$\mathbf{q}^H\left[(\widehat{\boldsymbol{\theta}}-\boldsymbol{\theta}) - \beta\mathbf{F}^{-1}(\boldsymbol{\theta})\frac{\partial \ln f(\mathbf{x};\boldsymbol{\theta})}{\partial\boldsymbol{\theta}^*}\right] = 0 \tag{3.259}$$

where $\beta \equiv \beta(\boldsymbol{\theta})$ is a nonzero function of $\boldsymbol{\theta}$ but not a function of $\mathbf{x}$ or, equivalently, $\widehat{\boldsymbol{\theta}}$ since the expectation operators in (3.257) are with respect to $\mathbf{x}$. Clearly, the condition (3.186) is equivalent to (3.259) since $\mathbf{q}$ is an arbitrary nonzero vector.

Q.E.D.

Problems

3.1. Compute the transfer functions and ROCs of the following LTI systems.

(a) A causal LTI system $h_1[n] = a^n u[n]$ (a right-sided sequence) where a is a constant and $u[n]$ is the *unit step function* defined as

$$u[n] = \begin{cases} 1, & n \geq 0, \\ 0, & n < 0. \end{cases}$$

(b) An anticausal LTI system $h_2[n] = -a^n u[-n-1]$ (a left-sided sequence) where a is a constant and $u[n]$ is the unit step function.

3.2. Prove Theorem 3.9.

3.3. Prove Theorem 3.12.

3.4. Prove Properties 3.13 to 3.18.

3.5. Consider that x is a zero-mean, real Gaussian random variable with variance σ_x^2. Show that the kth-order moment of x is given by

$$E\{x^k\} = \begin{cases} 0, & k = 1, 3, 5, ... \\ 1 \cdot 3 \cdot 5 \cdots (k-1)\sigma_x^k, & k = 2, 4, 6, ... \end{cases}$$

3.6. Show that if $\mathbf{x} = (x_1, x_2, ..., x_N)^T$ is a real Gaussian vector with mean $\mathbf{m}_x$ and covariance matrix $\mathbf{C}_x$ (a positive definite matrix), then its pdf is given by (3.91).

3.7. Prove Theorem 3.19 [24, Problems 2.14 and 2.15].

3.8. Prove Properties 3.20 to 3.23.

3.9. Compute the normalized skewness and normalized kurtosis of the following distributions.
(a) Uniform distribution given by (3.97).
(b) Laplace distribution given by (3.98).
(c) Exponential distribution given by (3.99).
(d) Bernoulli distribution given by (3.101).

3.10. Compute the normalized kurtosis of an M-PAM symbol, an M-PSK symbol and an M^2-QAM symbol where each symbol is assumed to be drawn from a set of M symbols with equal probability.

3.11. Let $r_x[l]$ be the autocorrelation function of a WSS process $x[n]$ and $\sum_{l=-\infty}^{\infty} |r_x[l]| < \infty$.
(a) Show that

$$\lim_{L\to\infty} \sum_{l=-L+1}^{L-1} \left(1 - \frac{|l|}{L}\right) r_x[l]e^{-j\omega l} = \sum_{l=-\infty}^{\infty} r_x[l]e^{-j\omega l}$$

(b) Show that the power spectrum $S_x(\omega)$ of $x[n]$ is given by (3.120).

3.12. Prove Property 3.29.

3.13. Let $x[n]$ be a WSS process with correlation matrix $\mathbf{R}_x$ given as (3.122) and power spectrum $S_x(\omega)$.
(a) Prove Property 3.31.
(b) Show that if $\mathbf{R}_x$ for $L \to \infty$ is positive definite, then $S_x(\omega) > 0$. (*Hint*: Use the result in part (a) of Problem 3.11.)

3.14. Consider the random signal $x[n]$ given by (3.123). Show that (3.140) holds under (HOS-1) and (HOS-2).

3.15. Prove Theorem 3.39.

3.16. Prove Theorem 3.41.

3.17. Consider the random signal $x[n]$ given by (3.158). Show that (3.162) holds under (CS-1) and (CS-2).

3.18. Derive (3.179) under the assumption that both $\partial f(\mathbf{x};\theta)/\partial\theta^*$ and $\partial^2 f(\mathbf{x};\theta)/\partial\theta\partial\theta^*$ exist and are absolutely integrable.

3.19. Prove Corollary 3.53.

3.20. Prove Theorem 3.55.

3.21. Prove Theorem 3.58.

3.22. Suppose $x[n]$ is a stationary process with mean m_x, variance σ_x^2 and autocovariance function $c_x[l] = E\{(x[n]-m_x)(x[n-l]-m_x)^*\}$. The sample-mean estimator for m_x is given by

$$\widehat{m}_x = \frac{1}{N}\sum_{n=0}^{N-1} x[n].$$

(a) Show that $\widehat{m}_x$ is unbiased.
(b) Show that $\widehat{m}_x$ is the UMVU estimator of m_x under the assumption that $x[n]$ is complex white Gaussian.
(c) Show that $\widehat{m}_x$ is a consistent estimator under the assumption that $\sum_{l=-\infty}^{\infty} |c_x[l]| < \infty$.

3.23. Show that based on the linear data model given by (3.198), the ML estimator $\widehat{\boldsymbol{\theta}}_{\mathrm{ML}}$ given by (3.202) is unbiased, efficient, UMVU, and Gaussian distributed.

3.24. Show that the autocorrelation matrix formed by the estimator $\widehat{r}_x[l]$ given by (3.204) is positive definite.

3.25. Suppose that both $\mathbf{x} = (x[0], x[1], ..., x[N-1])^T$ and $\boldsymbol{\theta} = (\theta[0], \theta[1], ..., \theta[L-1])^T$ are Gaussian with pdfs $\mathcal{N}(\mathbf{m}_x, \mathbf{C}_x)$ and $\mathcal{N}(\mathbf{m}_\theta, \mathbf{C}_\theta)$, respectively, and that $\mathbf{x}$ and $\boldsymbol{\theta}$ are jointly Gaussian. Show that the conditional pdf

$$f(\boldsymbol{\theta}|\mathbf{x}) = \frac{1}{\pi^L \cdot |\mathbf{C}_{\theta|x}|} \cdot \exp\left\{-(\boldsymbol{\theta} - \mathbf{m}_{\theta|x})^H \mathbf{C}_{\theta|\mathbf{x}}^{-1}(\boldsymbol{\theta} - \mathbf{m}_{\theta|x})\right\}$$

where the conditional mean

$$\mathbf{m}_{\theta|x} = E\{\boldsymbol{\theta}|\mathbf{x}\} = \mathbf{m}_\theta + \mathbf{C}_{\theta x}\mathbf{C}_x^{-1}(\mathbf{x} - \mathbf{m}_x)$$

and the conditional covariance matrix

$$\mathbf{C}_{\theta|x} = \mathbf{C}_\theta - \mathbf{C}_{\theta x}\mathbf{C}_x^{-1}\mathbf{C}_{x\theta}$$

in which $\mathbf{C}_{\theta x} = E\{(\boldsymbol{\theta} - \mathbf{m}_\theta)(\mathbf{x} - \mathbf{m}_x)^H\}$ and $\mathbf{C}_{x\theta} = E\{(\mathbf{x} - \mathbf{m}_x)(\boldsymbol{\theta} - \mathbf{m}_\theta)^H\}$.

3.26. Time Delay Estimation
Consider the following two sets of measurements

$$x_1[n] = s[n] + w_1[n]$$
$$x_2[n] = s[n - \tau] + w_2[n]$$

where $s[n]$ is the signal, $w_1[n]$ and $w_2[n]$ are the noise sources, and τ is the time delay to be estimated. Assume that $s[n]$, $w_1[n]$ and $w_2[n]$ are white WSS processes with variances σ_s^2, $\sigma_{w_1}^2$ and $\sigma_{w_2}^2$, respectively, and they are uncorrelated with each other. Based on $x_1[n]$ and $x_2[n]$, use a second-order causal FIR Wiener filter to obtain an estimate of the time delay τ for the case of the true $\tau = 1$.

Computer Assignments

3.1. Power Spectral Estimation
Consider that $x[n] = h[n] \star u[n]$ where $h[n]$ is a minimum-phase AR(2) system whose transfer function is given by

$$H(z) = \frac{1}{1 - 0.9z^{-1} + 0.2z^{-2}}$$

and $u[n]$ is a zero-mean, i.i.d., exponentially distributed stationary process with variance $\sigma_u^2 = 1$. Generate thirty independent sets of data $\{x[0], x[1], ..., x[511]\}$.

(a) Write a computer program to implement the AR spectral estimation method and obtain thirty AR spectral estimates with the thirty sets of data. Plot the thirty AR spectral estimates and their average, and explain what you observe.

(b) Write a computer program to implement the periodogram spectral estimation method and obtain thirty periodograms with the thirty sets of data. Plot the thirty periodograms and their average, and explain what you observe.

3.2. System Identification

As depicted in Fig. 3.36, consider the problem of identifying an LTI system $h[n]$ by means of the Wiener filter $v[n]$ with the sets of input and output measurements

$$x[n] = u[n] + w_1[n], \quad n = 0, 1, ..., N-1,$$
$$d[n] = h[n] \star u[n] + w_2[n], \quad n = 0, 1, ..., N-1$$

where $u[n]$ is the driving input of $h[n]$, and $w_1[n]$ and $w_2[n]$ are the sources of measurement noise. Assume that (i) the transfer function of $h[n]$ is given by

$$H(z) = h[0] + h[1]z^{-1} + h[2]z^{-2} = 1 - 1.8z^{-1} + 0.4z^{-2},$$

(ii) $u(n)$ is a zero-mean, exponentially distributed, i.i.d. stationary process with variance $\sigma_u^2 = 1$, skewness $C_3\{u[n]\} = 2$ and kurtosis $C_4\{u[n]\} = 6$, (iii) $w_1(n)$ and $w_2(n)$ are uncorrelated, zero-mean WSS Gaussian processes, and (iv) $u[n]$ is statistically independent of $w_1[n]$ and $w_2[n]$. Write a computer program to implement the system identification method, and perform thirty independent runs to obtain thirty estimates $\widehat{h}[n]$ for $N = 4000$ and SNR = 40, 10, 5 and 0 dB where the SNR for $x[n]$ is the same as that for $d[n]$. Show the mean ± one standard deviation of the thirty $\widehat{h}[n]$ obtained.

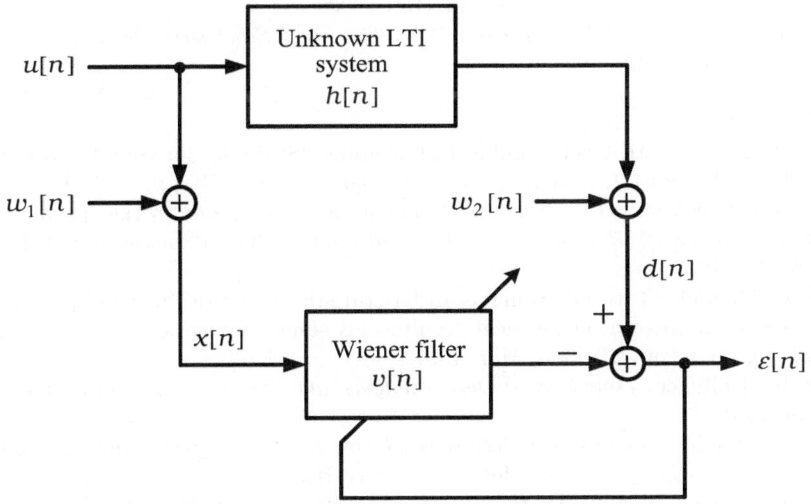

Fig. 3.36 Block diagram of system identification using Wiener filtering

References

1. A. V. Oppenheim and R. W. Schafer, *Discrete-time Signal Processing.* New Jersey: Prentice-Hall, 1999.
2. S. J. Orfanidis, *Introduction to Signal Processing.* New Jersey: Prentice-Hall, 1996.
3. O. K. Ersoy, *Fourier-related Transforms, Fast Algorithms and Applications.* New Jersey: Prentice-Hall, 1997.
4. S. A. Dianat and M. R. Raghuveer, "Fast algorithms for phase and magnitude reconstruction from bispectra," *Optical Engineering*, vol. 29, no. 5, pp. 504–512, May 1990.
5. H.-M. Chien, H.-L. Yang, and C.-Y. Chi, "Parametric cumulant based phase estimation of 1-D and 2-D nonminimum phase systems by allpass filtering," *IEEE Trans. Signal Processing*, vol. 45, no. 7, pp. 1742–1762, July 1997.
6. C.-Y. Chi, "Fourier series based nonminimum phase model for statistical signal processing," *IEEE Trans. Signal Processing*, vol. 47, no. 8, pp. 2228–2240, Aug. 1999.
7. C.-Y. Chen and C.-Y. Chi, "Nonminimum-phase complex Fourier series based model for statistical signal processing," in *Proc. IEEE Signal Processing Workshop on Higher-Order Statistics*, Caesarea, Israel, June 14–16, 1999, pp. 30–33.
8. C.-H. Chen, C.-Y. Chi, and C.-Y. Chen, "2-D Fourier series based model for nonminimum-phase linear shift-invariant systems and texture image classification," *IEEE Trans. Signal Processing*, vol. 50, no. 4, pp. 945–955, April 2002.
9. A. V. Oppenheim and R. W. Schafer, *Discrete-time Signal Processing.* New Jersey: Prentice-Hall, 1989.
10. W. A. Gardner, *Introduction to Random Processes with Applications to Signals and Systems.* New York: McGraw-Hill, 1990.
11. A. Papoulis, *Probability, Random Variables, and Stochastic Processes.* New York: McGraw-Hill, 1991.
12. Y. Dodge, Ed., *The Oxford Dictionary of Statistical Terms.* New York: Oxford University Press, 2003.
13. P. O. Amblard, M. Gaeta, and J. L. Lacoume, "Statistics for complex variables and signals - Part I: Variables," *Signal Processing*, vol. 53, pp. 1–13, 1996.
14. E. A. Cornish and R. A. Fisher, "Moments and cumulants in the specification of distributions," *Reviews of the International Statistical Institute*, vol. 5, pp. 307–320, 1937.
15. J. M. Mendel, "Tutorial on higher-order statistics (spectra) in signal processing and system theory: Theoretical results and some applications," *Proc. IEEE*, vol. 79, no. 3, pp. 278–305, Mar. 1991.
16. D. R. Brillinger, *Time Series: Data Analysis and Theory.* New York: McGraw-Hill, 1981.
17. J. M. Mendel, *Lessons in Estimation Theory for Signal Processing, Communications, and Control.* New Jersey: Prentice-Hall, 1995.
18. C. L. Nikias and A. P. Petropulu, *Higher-order Spectra Analysis: A Nonlinear Signal Processing Framework.* New Jersey: Prentice-Hall, 1993.
19. C. L. Nikias and J. M. Mendel, "Signal processing with higher-order spectra," *IEEE Signal Processing Magazine*, pp. 10–37, July 1993.
20. D. L. Harnett and J. L. Murphy, *Introductary Statistical Analysis.* California: Addison-Wesley, 1975.

21. L. Blank, *Statistical Procedures for Engineering, Management, and Science.* New York: McGraw-Hill, 1980.
22. S. Ghahramani, *Fundamentals of Probability.* New Jersey: Prentice-Hall, 2000.
23. O. Kempthorne and L. Folks, *Probability, Statistics, and Data Analysis.* Iowa: Iowa State Unitversity Press, 1971.
24. C. W. Therrien, *Discrete Random Signals and Statistical Signal Processing.* New Jersey: Prentice-Hall, 1992.
25. N. R. Goodman, "Statistical analysis based on a certain multivariate complex Gaussian distribution (an introduction)," *Annals of Math. Statistics*, pp. 152–177, 1963.
26. S. Haykin, *Blind Deconvolution.* New Jersey: Prentice-Hall, 1994.
27. T. Chonavel, *Statistical Signal Processing: Modelling and Estimation.* London: Springer-Verlag, 2002, translated by J. Ormrod.
28. G. L. Wise, "A cautionary aspect of stationary random processes," *IEEE Trans. Circuits and Systems*, vol. 38, no. 11, pp. 1409–1410, Nov. 1991.
29. J. G. Proakis, *Digital Communications*, 4th ed. New York: McGraw-Hill, 2001.
30. S. M. Kay, *Modern Spectral Estimation: Theory and Application.* New Jersey: Prentice-Hall, 1988.
31. S. Haykin, *Adaptive Filter Theory.* New Jersey: Prentice-Hall, 1996.
32. C. L. Nikias and M. R. Raghuveer, "Bispectrum estimation: A digital signal processing framework," *Proc. IEEE*, vol. 75, no. 7, pp. 869–891, July 1987.
33. D. Hatzinakos, Higher-order Spectral (H.O.S.) Analysis, in *Stochastic Techniques in Digital Signal Processing Systems*, Part 2 of 2, C.T. Leondes, ed. San Diego: Academic Press, 1994.
34. D. R. Brillinger, "Some history of the study of higher-order moments and spectra," in *Proc. IEEE Workshop on Higher-Order Spectral Analysis*, Vail, Colorado, June 28–30, 1989, pp. 41–45.
35. B. Porat, *Digital Processing of Random Signals: Theory and Methods.* New Jersey: Prentice-Hall, 1994.
36. G. B. Giannakis and M. K. Tsatsanis, "HOS or SOS for parametric modeling?" *Proc. IEEE*, pp. 3097–3100, 1991.
37. A. Swami, G. B. Giannakis, G. Zhou, and L. Srinivas, *Bibliography on Higher Order Statistics.* Virginia: University of Virginia, Jan. 1995.
38. P. A. Delaney and D. O. Walsh, "A bibliography of higher-order spectra and cumulants," *IEEE Signal Processing Magazine*, pp. 61–70, July 1994.
39. W. A. Gardner and L. E. Franks, "Characterization of cyclostationary random signal processes," *IEEE Trans. Information Theory*, vol. IT-21, no. 1, pp. 4–14, Jan. 1975.
40. W. A. Gardner, *Cyclostationarity in Communications and Signal Processing.* New York: IEEE Press, 1994.
41. H. Meyr, M. Moeneclaey, and S. A. Fechtel, *Digital Communication Receivers: Synchronization, Channel Estimation, and Signal Processing.* New York: John Wiley & Sons, 1998.
42. W. A. Gardner, "Exploitation of spectral redundancy in cyclostationary signals," *IEEE Signal Processing Magazine*, pp. 14–36, Apr. 1991.
43. G. B. Giannakis, "Linear cyclic correlation approaches for blind identification of FIR channels," in *Proc. IEEE 28th Asilomar Conference on Signals, Systems, and Computers*, 1995, pp. 420–424.

44. W. A. Gardner and C. M. Spooner, Cyclostationary Signal Processing, in *Stochastic Techniques in Digital Signal Processing Systems*, Part 2 of 2, C.T. Leondes, ed. San Diego: Academic Press, 1994.
45. C. M. Spooner, Higher-order Statistics for Nonlinear Processing of Cyclostationary Signals, A Chapter in *Cyclostationarity in Communications and Signal Processing*, W.A. Gardner, ed. New York: IEEE Press, 1994.
46. R. L. Freeman, "Bits, symbols, bauds, and bandwidth," *IEEE Communications Magazine*, pp. 96–99, Apr. 1998.
47. P. P. Vaidyanathan, *Multirate Systems and Filter Banks.* New Jersey: Prentice-Hall, 1993.
48. L. Tong, G. Xu, and T. Kailath, "Blind identification and equalization of multipath channels," in *Proc. IEEE International Conference on Communications*, Chicago, June 1992, pp. 1513–1517.
49. L. Tong, G. Xu, and T. Kailath, "A new approach to blind identification and equalization of multipath channels," in *Proc. IEEE 25th Asilomar Conference on Signals, Systems, and Computers*, Pacific Grove, California, Nov. 1991, pp. 856–860.
50. Z. Ding and Y. Li, "Channel identification using second order cyclic statistics," in *Proc. IEEE 26th Asilomar Conference on Signals, Systems, and Computers*, Pacific Grove, California, Oct. 1992, pp. 334–338.
51. L. Tong, G. Xu, and T. Kailath, "Fast blind equalization via antenna arrays," in *Proc. IEEE International Conference on Acoustics, Speech, and Signal Processing*, vol. 4, 1993, pp. 272–275.
52. L. Izzo, A. Napolitano, and L. Paura, Cyclostationary-exploiting Methods for Multipath-channel Identification, A Chapter in *Cyclostationarity in Communications and Signal Processing*, W. A. Gardner ed. New York: IEEE Press, 1994.
53. Z. Ding, Blind Channel Identification and Equalization Using Spectral Correlation Measurements, Part I: Frequency-domain Analysis, A Chapter in *Cyclostationarity in Communications and Signal Processing*, W. A. Gardner, ed. New York: IEEE Press, 1994.
54. Z. Ding and Y. Li, "On channel identification based on second-order cyclic spectra," *IEEE Trans. Signal Processing*, vol. 42, no. 5, pp. 1260–1264, May 1994.
55. L. Tong, G. Xu, and T. Kailath, "Blind identification and equalization based on second-order statistics: A time domain approach," *IEEE Trans. Information Theory*, vol. 40, no. 2, pp. 340–349, Mar. 1994.
56. S. M. Kay, *Fundamentals of Statistical Signal Processing: Estimation Theory.* New Jersey: Prentice-Hall, 1993.
57. M. D. Srinath and P. K. Rajasekaran, *An Introduction to Statistical Signal Processing with Applications.* New York: John Wiley & Sons, 1979.
58. H. W. Sorenson, *Parameter Estimation: Principles and Problems.* New York: Marcel Dekker, 1985.
59. H. Stark and J. W. Woods, *Probability and Random Processes with Applications to Signal Processing.* New Jersey: Prentice-Hall, 2002.
60. T. B. Fomby, R. C. Hill, and S. R. Johnson, *Advanced Econometric Methods.* New York: Springer-Verlag, 1984.
61. T. K. Moon and W. C. Stirling, *Mathematical Methods and Algorithms for Signal Processing.* New Jersey: Prentice-Hall, 2000.

4

SISO Blind Equalization Algorithms

In this chapter, we introduce some widely used SISO blind equalization algorithms, including an SOS based approach, namely a linear prediction approach, and two HOS based approaches, namely the maximum normalized cumulant equalization algorithm and the super-exponential equalization algorithm. We also present some simulation examples for testing these algorithms, as well as the applications of these algorithms to seismic exploration, speech signal processing and baud-spaced equalization in digital communications.

4.1 Linear Equalization

We start with an introduction to the problem of SISO blind equalization along with some performance indices, and then review two fundamental equalization criteria, namely peak distortion criterion and MMSE equalization criterion, which are often used to design nonblind equalizers.

4.1.1 Blind Equalization Problem

Problem Statement

Suppose that $u[n]$ is the source signal of interest and is distorted by an SISO LTI system $h[n]$ via the following data model: (see Fig. 4.1)

$$y[n] = x[n] + w[n] \tag{4.1}$$

where

$$x[n] = h[n] \star u[n] = \sum_{k=-\infty}^{\infty} h[k]u[n-k] \tag{4.2}$$

is the noise-free signal and $w[n]$ is the additive noise accounting for measurement noise as well as physical effects not explained by $x[n]$. From (4.1) and (4.2), it follows that

$$y[n] = h[0]u[n] + \underbrace{\sum_{k=-\infty,\ k\neq 0}^{\infty} h[k]u[n-k]}_{\text{ISI term}} + w[n]. \tag{4.3}$$

From (4.3), one can see that the desired sample (or symbol) $u[n]$, scaled by $h[0]$, in the first term is not only corrupted by the noise $w[n]$ but also interfered with by other samples (or symbols) in the second term. This latter effect is called *intersymbol interference (ISI)*.

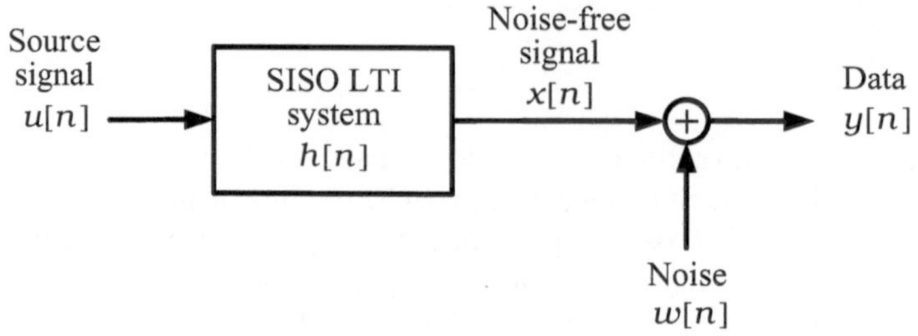

Fig. 4.1 Data model for SISO blind equalization

Blind equalization, also known as *blind deconvolution*, of the SISO LTI system $h[n]$ is a signal processing procedure to restore the source signal $u[n]$ with only the data $y[n]$ generated from (4.1) and (4.2). This problem arises in a variety of engineering and science areas such as seismic exploration, digital communications, speech signal processing, ultrasonic nondestructive evaluation, underwater acoustics, radio astronomy, and so on.

Blind Linear Equalization

To further explain the problem of SISO blind equalization, let us consider the *direct blind equalization approach* as shown in Fig. 4.2. Its goal is to design an SISO LTI filter, denoted by $v[n]$, using some presumed features of $u[n]$ such that the output of $v[n]$ given by

$$e[n] = v[n] \star y[n] = \sum_{k=-\infty}^{\infty} v[k]y[n-k] \tag{4.4}$$

approximates $u[n]$ as well as possible. The filter $v[n]$ is called the *deconvolution filter* or the *equalizer*, or the *blind equalizer* to emphasize the use of the blind approach, while its output $e[n]$ is called the *deconvolved signal* or the *equalized signal.* In practice, the equalizer $v[n]$ is commonly assumed to be an FIR filter with $v[n] = 0$ outside a preassigned domain of support $L_1 \leq n \leq L_2$ where

L_1 and L_2 are integers. As $L_1 = -\infty$ and $L_2 = \infty$, the equalizer $v[n]$ is said to be *doubly infinite*.

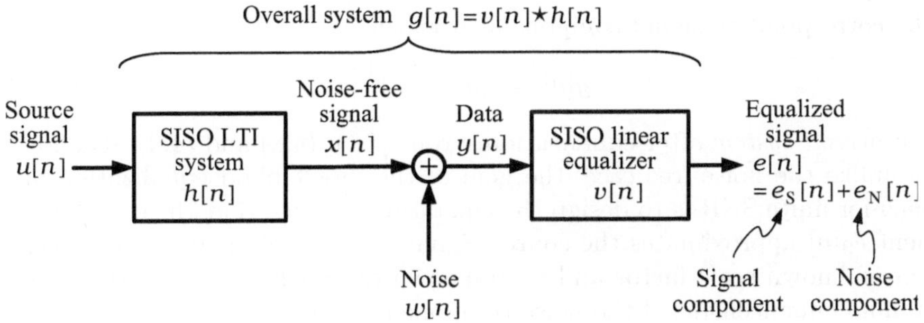

Fig. 4.2 Block diagram of SISO linear equalization

Define the signal-to-noise ratio (SNR) associated with the data $y[n]$ as

$$\text{SNR} \equiv \text{SNR}\{y[n]\} = \frac{E\{|x[n]|^2\}}{E\{|w[n]|^2\}} \tag{4.5}$$

where we have made use of the typical assumption that $u[n]$ and $w[n]$ are uncorrelated random processes. When SNR $= \infty$, i.e. the noise-free case, the goal of the direct blind equalization approach reduces to designing the equalizer $v[n]$ such that the equalized signal

$$e[n] = \alpha u[n - \tau] \tag{4.6}$$

where α is a real or complex constant and τ is an integer. Note that without further information about $u[n]$ or $h[n]$, the scale factor α and the time delay τ in (4.6) cannot be identified for the following reason. Let $\widetilde{u}[n] = \alpha u[n - \tau]$ be another source signal and $\widetilde{h}[n] = \alpha^{-1}h[n + \tau]$ another system. Then

$$\widetilde{h}[n] \star \widetilde{u}[n] = \sum_{k=-\infty}^{\infty} h[k + \tau]u[n - \tau - k] = \sum_{k=-\infty}^{\infty} h[k]u[n - k] = x[n],$$

indicating that both pairs $(u[n], h[n])$ and $(\widetilde{u}[n], \widetilde{h}[n])$ result in the same noise-free signal $x[n]$. Accordingly, provision of information about $x[n]$ (or $y[n]$) only is not sufficient to distinguish between them.

The equalized signal $e[n]$ given by (4.4) can be further expressed as

$$e[n] = e_S[n] + e_N[n] \tag{4.7}$$

where

$$e_N[n] = v[n] \star w[n] \quad \text{(by (4.1))} \tag{4.8}$$

corresponds to the noise component in $e[n]$ and

$$e_S[n] = v[n] \star x[n] = g[n] \star u[n] \quad \text{(by (4.1) and (4.2))} \tag{4.9}$$

is the corresponding signal component in which

$$g[n] = v[n] \star h[n] \tag{4.10}$$

is the *overall system* after equalization. From (4.7), (4.8) and (4.9), it follows that unlike the noise-free case, the goal of the direct blind equalization approach for finite SNR is to design the equalizer $v[n]$ such that the signal component $e_S[n]$ approximates the source signal $u[n]$ as well as possible (except for an unknown scale factor and an unknown time delay) while maintaining minimum enhancement of the noise component $e_N[n]$.

On the other hand, as mentioned in Chapter 1, one can also resort to the *indirect blind equalization approach* to restore the source $u[n]$ from the data $y[n]$. Its steps are as follows: (i) estimation of $h[n]$ by means of a *blind system identification (BSI) algorithm*, (ii) estimation, if needed, of other parameters such as the autocorrelation function of $y[n]$, and (iii) design of a nonblind equalizer with these estimated parameters for the retrieval of $u[n]$ (the goal of blind equalization).

Performance Indices

From (4.7) and (4.9), it follows that for SNR $= \infty$ the equalized signal

$$e[n] = g[\tau]u[n-\tau] + \underbrace{\sum_{k=-\infty,\ k\neq\tau}^{\infty} g[k]u[n-k]}_{\text{Residual ISI term}} . \tag{4.11}$$

If the overall system

$$g[n] = \alpha\delta[n-\tau] \tag{4.12}$$

where α is a constant and τ is an integer, then the residual ISI term in (4.11) disappears and $e[n] = \alpha u[n-\tau]$ which is exactly the objective for the noise-free case (see (4.6)). This fact suggests that $g[n]$ can serve to indicate the amount of residual ISI after equalization, thereby leading to the following commonly used performance index for the designed equalizer $v[n]$: [1, p. 124]

$$\mathrm{ISI}\{g[n]\} = \mathrm{ISI}\{\alpha g[n-\tau]\} = \frac{\sum_{n=-\infty}^{\infty} |g[n]|^2 - \max\{|g[n]|^2\}}{\max\{|g[n]|^2\}} \tag{4.13}$$

where α is any nonzero constant and τ is any integer. It is easy to see that $\mathrm{ISI}\{g[n]\} = 0$ if and only if $g[n] = \alpha\delta[n-\tau]$ for all $\alpha \neq 0$ and all τ. This implies that the smaller the value of $\mathrm{ISI}\{g[n]\}$, the closer the overall system $g[n]$ approaches a delta function.

On the other hand, to evaluate the degree of noise enhancement for the case of finite SNR, we may compare the SNR after equalization, defined as

$$\mathrm{SNR}\{e[n]\} = \frac{E\{|e_{\mathrm{S}}[n]|^2\}}{E\{|e_{\mathrm{N}}[n]|^2\}} \quad \text{(see (4.7))}, \tag{4.14}$$

with the SNR before equalization, i.e. $\mathrm{SNR}\{y[n]\}$ defined as (4.5). Alternatively, it may be more convenient to use the following performance index:

$$\rho\{v[n]\} \triangleq \frac{\mathrm{SNR}\{e[n]\}}{\mathrm{SNR}\{y[n]\}}, \tag{4.15}$$

which we refer to as the *SNR improvement-or-degradation ratio*. Note that $\rho\{v[n]\} > 1$ means SNR improvement after equalization, whereas $\rho\{v[n]\} < 1$ means SNR degradation after equalization.

4.1.2 Peak Distortion and MMSE Equalization Criteria

Peak Distortion Criterion

Referring to (4.11), we note that the magnitude of the residual ISI term

$$\left| \sum_{k=-\infty,\ k\neq\tau}^{\infty} g[k]u[n-k] \right| \leq \max\{|u[n]|\} \cdot \sum_{k=-\infty,\ k\neq\tau}^{\infty} |g[k]| \tag{4.16}$$

for any integer τ. This suggests that the worst-case residual ISI (the right-hand side of (4.16)) in the equalized signal can be reduced by finding the equalizer $v[n]$ which minimizes the following *peak distortion criterion*:[1] [3]

$$J_{\mathrm{PD}}(v[n]) = \sum_{n=-\infty,\ n\neq\tau}^{\infty} |g[n]| \tag{4.17}$$

where τ is an integer. Note that the choice of τ may significantly affect the minimum value of $J_{\mathrm{PD}}(v[n])$ when $v[n]$ is a finite-length equalizer.

From (4.17), it can be observed that $\min\{J_{\mathrm{PD}}(v[n])\} = 0$ if and only if $g[n]$ satisfies (4.12). This indicates that the equalizer $v[n]$ is designed to "force" $\mathrm{ISI}\{g[n]\}$ to zero. It is therefore called the *zero-forcing (ZF) equalizer*, and is denoted by $v_{\mathrm{ZF}}[n]$ for clarity. Accordingly, the frequency response of $g_{\mathrm{ZF}}[n]$ $(= v_{\mathrm{ZF}}[n] \star h[n])$ is given by

$$G_{\mathrm{ZF}}(\omega) = V_{\mathrm{ZF}}(\omega)H(\omega) = \alpha e^{-j\omega\tau}. \tag{4.18}$$

Let $h_{\mathrm{I}}[n]$ denote the inverse system of $h[n]$. Then (4.18) leads to

---

[1] Some varieties of peak distortion criterion along with their comparison can be found in [2].

$$V_{\mathrm{ZF}}(\omega) = \frac{\alpha e^{-j\omega\tau}}{H(\omega)} = \alpha e^{-j\omega\tau} H_{\mathrm{I}}(\omega), \tag{4.19}$$

provided that (i) $v_{\mathrm{ZF}}[n]$ is doubly infinite and (ii) $h_{\mathrm{I}}[n]$ is stable (i.e. $H(z)$ has no zeros on the unit circle). In other words, $v_{\mathrm{ZF}}[n]$ is equivalent to $h_{\mathrm{I}}[n]$ except for a scale factor and a time delay.

MMSE Equalization Criterion

Consider the Wiener filter $v[n]$ shown in Fig. 4.3 that minimizes the MSE

$$J_{\mathrm{MSE}}(v[n]) = E\left\{|\alpha u[n-\tau] - e[n]|^2\right\} \tag{4.20}$$

for a nonzero constant α and an integer τ where the scaled and delayed source signal $\alpha u[n-\tau]$ is regarded as the desired signal and the equalized signal $e[n]$ as an estimate of the desired signal. Note that the doubly infinite Wiener filter $v[n]$ for all $\alpha \neq 0$ and τ turns out to be the same except for a scale factor, whereas, like the ZF equalizer, the choice of τ may significantly affect the minimum value of $J_{\mathrm{MSE}}(v[n])$ when the length of $v[n]$ is finite. Owing to minimization of the MSE given by (4.20), the Wiener filter $v[n]$ is called an *LMMSE equalizer*, and is denoted by $v_{\mathrm{MS}}[n]$ for clarity.

Without loss of generality, let us consider only the case that $\alpha = 1$ and $\tau = 0$, and further make the following assumptions for MMSE equalization.

(WF-1) The SISO LTI system $h[n]$ is stable.
(WF-2) The source signal $u[n]$ is a WSS white process with variance σ_u^2.
(WF-3) The noise $w[n]$ is a WSS white process with variance σ_w^2.
(WF-4) The source signal $u[n]$ is statistically independent of the noise $w[n]$.

Under Assumptions (WF-1) through (WF-4) and the condition that $v_{\mathrm{MS}}[n]$ is doubly infinite, it can be shown by the orthogonality principle that the frequency response of $v_{\mathrm{MS}}[n]$ is given by (see Problem 4.1)

$$V_{\mathrm{MS}}(\omega) = \frac{\sigma_u^2 \cdot H^*(\omega)}{\sigma_u^2 \cdot |H(\omega)|^2 + \sigma_w^2}, \quad -\pi \leq \omega < \pi \tag{4.21}$$

and the corresponding overall system is given by

$$G_{\mathrm{MS}}(\omega) = V_{\mathrm{MS}}(\omega) \cdot H(\omega) = \frac{\sigma_u^2 \cdot |H(\omega)|^2}{\sigma_u^2 \cdot |H(\omega)|^2 + \sigma_w^2}. \tag{4.22}$$

Moreover, some properties of the LMMSE equalizer $v_{\mathrm{MS}}[n]$ are summarized as follows.[2]

---

[2] For the real case, the LMMSE equalizer $v_{\mathrm{MS}}[n]$ is identical to a (steady-state) minimum-variance deconvolution (MVD) filter reported in [4–6], and thus also shares the properties of the MVD filter [7,8].

Property 4.1. *The larger the SNR or the wider the bandwidth of the system* $h[n]$, *the closer the overall system* $g_{\mathrm{MS}}[n]$ $(= v_{\mathrm{MS}}[n] \star h[n])$ *is to* $\delta[n]$.

Property 4.2. *The LMMSE equalizer* $v_{\mathrm{MS}}[n]$ *reduces to a ZF equalizer* $v_{\mathrm{ZF}}[n]$ *when* $h[n]$ *is an allpass system.*

Property 4.3. *The LMMSE equalizer* $v_{\mathrm{MS}}[n]$ *is a perfect phase equalizer, i.e.* $\arg[V_{\mathrm{MS}}(\omega)] = -\arg[H(\omega)]$.

Property 4.4. *The overall system* $g_{\mathrm{MS}}[n]$ *is a legitimate autocorrelation function with (i)* $g_{\mathrm{MS}}[n] = g_{\mathrm{MS}}^*[-n]$ *and (ii)* $g_{\mathrm{MS}}[0] > |g_{\mathrm{MS}}[n]|$ *for all* $n \neq 0$.

Property 4.1 follows from (4.22) and Properties 4.2 and 4.3 can be observed directly from (4.21), while the proof of Property 4.4 is left as an exercise (Problem 4.2). Note that Property 4.2 is nothing but a special case of Property 4.1. Moreover, unlike the ZF equalizer, the LMMSE equalizer is always stable no matter whether $H(z)$ has zeros on the unit circle or not (see (4.21)).

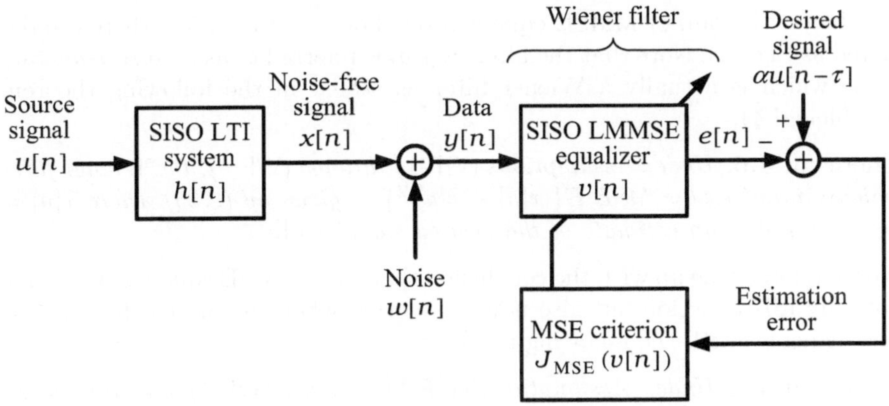

Fig. 4.3 Block diagram of SISO MMSE equalization

SNR Analysis for ZF and MMSE Equalization

Next, let us analyze the SNR improvement or degradation for both ZF and MMSE equalization under Assumptions (WF-1) through (WF-4). From (4.8) and (4.19), it follows that after ZF equalization, the power spectrum of the noise component $e_{\mathrm{N}}[n]$ is given by

$$S_{e_{\mathrm{N}}}(\omega) = \frac{\sigma_w^2}{|H(\omega)|^2}. \tag{4.23}$$

This implies that after ZF equalization, the noise spectrum will be enhanced for those frequencies at which $|H(\omega)|$ is low. A further result is given as follows.

Theorem 4.5. *Under Assumptions* (WF-1) *through* (WF-4) *and the condition of finite SNR, the SNR improvement-or-degradation ratio* $\rho\{v_{\rm ZF}[n]\} \leq 1$ *and the equality holds only when the system* $h[n]$ *is an allpass system.*

The proof is left as an exercise (Problem 4.3). This theorem states that if $h[n]$ is not an allpass system, then the ZF equalizer always results in SNR degradation after equalization, although it can completely eliminate the ISI induced by $h[n]$.

As for MMSE equalization, let us further express $V_{\rm MS}(\omega)$ given by (4.21) as follows:

$$V_{\rm MS}(\omega) = V_{\rm ZF}(\omega) V_{\rm NR}(\omega) \tag{4.24}$$

where $V_{\rm ZF}(\omega) = 1/H(\omega)$ is a ZF equalizer and

$$V_{\rm NR}(\omega) = \frac{\sigma_u^2 \cdot |H(\omega)|^2}{\sigma_u^2 \cdot |H(\omega)|^2 + \sigma_w^2}, \qquad -\pi \leq \omega < \pi. \tag{4.25}$$

The block diagram of MMSE equalization shown in Fig. 4.3 is therefore depicted as Fig. 4.4. Note that the filter $v_{\rm NR}[n]$ is referred to as a *noise-reduction filter*, which is actually a Wiener filter as stated in the following theorem (Problem 4.4).

Theorem 4.6. *Under Assumptions* (WF-1) *through* (WF-4), *the Wiener filter which minimizes the MSE* $E\{|x[n] - \widehat{x}[n]|^2\}$ *is given by (4.25) where* $\widehat{x}[n] = v_{\rm NR}[n] \star y[n]$ *is an estimate of the desired signal* $x[n]$.

We therefore come up with the conclusion that the LMMSE equalizer performs not only ISI reduction but also noise reduction when the SNR is finite. This also coincides with the following result.

Theorem 4.7. *Under Assumptions* (WF-1) *through* (WF-4) *and the condition of finite SNR, the SNR improvement-or-degradation ratio* $\rho\{v_{\rm MS}[n]\} \geq \rho\{v_{\rm ZF}[n]\}$ *and the equality holds only when the system* $h[n]$ *is an allpass system.*

The proof can be found in [9]. This theorem states that if $h[n]$ is not an allpass system, then the SNR in the equalized signal obtained by $v_{\rm MS}[n]$ is always higher than that obtained by $v_{\rm ZF}[n]$ for finite SNR.

4.2 SOS Based Blind Equalization Approach: Linear Prediction

To pave the way for the treatment of SOS based blind equalization, we start by reviewing the fundamentals of linear prediction, including forward and backward linear prediction error (LPE) filters, the well-known Levinson–Durbin

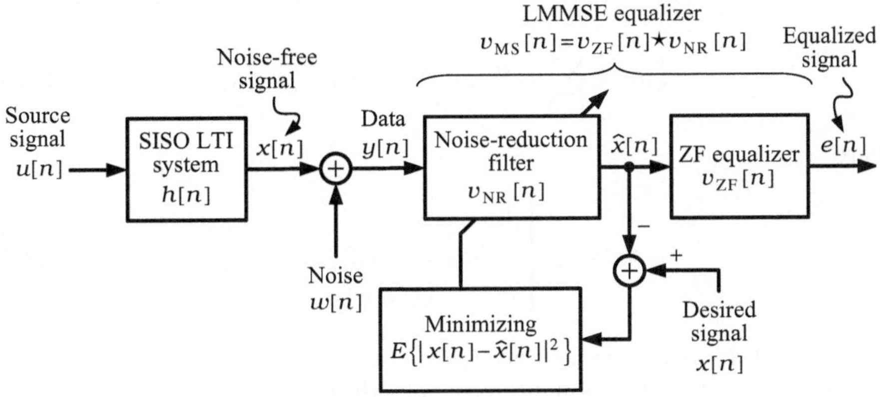

Fig. 4.4 MMSE equalization interpreted as noise reduction followed by ZF equalization

recursion for efficiently computing the coefficients of LPE filters, and the lattice structure of LPE filters. Then we focus on the topic of SISO blind equalization using LPE filters.

4.2.1 Forward and Backward Linear Prediction

Forward Linear Prediction

Suppose that we are given a finite set of samples (data) $\{y[0], y[1], ..., y[N-1]\}$ which were drawn from a zero-mean WSS process with positive definite autocorrelation matrix. As illustrated in Fig. 4.5a, *(one-step) forward linear prediction* is a signal processing procedure to estimate the "present" sample $y[n]$ by linearly combining the L "past" samples $y[n-1]$, $y[n-2]$, ..., $y[n-L]$ for $n = L, L+1, ..., N-1$. Although the so-called "present" sample $y[n]$ is already available (known), through this prediction procedure, we can obtain the parameters of interest that characterize $y[n]$. The goal can be simply achieved by means of Wiener filtering. Specifically, as shown in Fig. 4.5b, let $\widetilde{a}_1$, $\widetilde{a}_2$, ..., $\widetilde{a}_L$ be the L coefficients of an $(L-1)$th-order FIR Wiener filter to be designed such that the filter output

$$\widehat{y}_L[n] \triangleq \sum_{k=1}^{L} \widetilde{a}_k y[n-k] \tag{4.26}$$

approximates the desired sample $y[n]$ in the MMSE sense. This Wiener filter is called a *forward linear predictor* and the corresponding estimation error

$$e_L^f[n] \triangleq y[n] - \widehat{y}_L[n] \tag{4.27}$$

is called a *forward prediction error*. Substituting (4.26) into (4.27) yields

$$e_L^f[n] = \sum_{k=0}^{L} a_L[k] y[n-k] = a_L[n] \star y[n] \tag{4.28}$$

where

$$a_L[n] = \begin{cases} 1, & n = 0, \\ -\widetilde{a}_n, & n = 1, 2, ..., L, \\ 0, & \text{otherwise} \end{cases} \tag{4.29}$$

is called an *Lth-order forward linear prediction error (LPE) filter.*

According to the orthogonality principle, the coefficients $\widetilde{a}_1$, $\widetilde{a}_2$, ..., $\widetilde{a}_L$ can be obtained by solving the following Wiener–Hopf equations:

$$\begin{pmatrix} r_y[0] & r_y^*[1] & \cdots & r_y^*[L-1] \\ r_y[1] & r_y[0] & \cdots & r_y^*[L-2] \\ \vdots & \vdots & \ddots & \vdots \\ r_y[L-1] & r_y[L-2] & \cdots & r_y[0] \end{pmatrix} \begin{pmatrix} \widetilde{a}_1 \\ \widetilde{a}_2 \\ \vdots \\ \widetilde{a}_L \end{pmatrix} = \begin{pmatrix} r_y[1] \\ r_y[2] \\ \vdots \\ r_y[L] \end{pmatrix} \tag{4.30}$$

and the minimum value of $E\{|y[n] - \widehat{y}_L[n]|^2\}$ is given by

$$\sigma_L^2 \triangleq E\{|e_L^f[n]|^2\} = r_y[0] - \sum_{k=1}^{L} \widetilde{a}_k r_y^*[k] = \sum_{k=0}^{L} a_L[k] r_y^*[k]. \tag{4.31}$$

Combining (4.29), (4.30) and (4.31) forms

$$\begin{pmatrix} r_y[0] & r_y^*[1] & r_y^*[2] & \cdots & r_y^*[L] \\ r_y[1] & r_y[0] & r_y^*[1] & \cdots & r_y^*[L-1] \\ r_y[2] & r_y[1] & r_y[0] & \cdots & r_y^*[L-2] \\ \vdots & \vdots & \vdots & \ddots & \vdots \\ r_y[L] & r_y[L-1] & r_y[L-2] & \cdots & r_y[0] \end{pmatrix} \begin{pmatrix} 1 \\ a_L[1] \\ a_L[2] \\ \vdots \\ a_L[L] \end{pmatrix} = \begin{pmatrix} \sigma_L^2 \\ 0 \\ 0 \\ \vdots \\ 0 \end{pmatrix}, \tag{4.32}$$

which is called the *normal equations* or the *augmented Wiener–Hopf equations* associated with the Lth-order forward LPE filter $a_L[n]$.

Backward Linear Prediction

Similar to the forward linear prediction, *(one-step) backward linear prediction*, conceptually illustrated in Fig. 4.6a, is a signal processing procedure to estimate the sample $y[n-L]$ by linearly combining the L samples $y[n]$, $y[n-1]$,

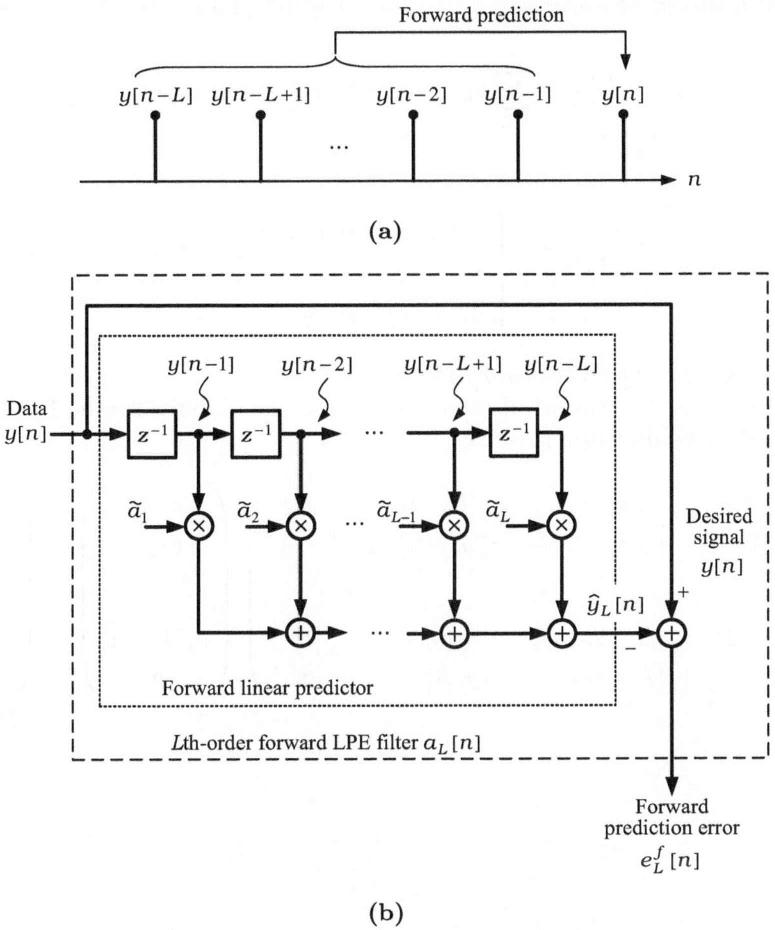

Fig. 4.5 (**a**) Conceptual illustration of the forward prediction and (**b**) tap-delay-line structure of an Lth-order forward LPE filter

..., $y[n-L+1]$. As shown in Fig. 4.6b, let $\widetilde{b}_0$, $\widetilde{b}_1$, ..., $\widetilde{b}_{L-1}$ be the L coefficients of an $(L-1)$th-order FIR Wiener filter to be designed such that the filter output

$$\widehat{y}_L[n-L] = \sum_{k=0}^{L-1} \widetilde{b}_k y[n-k] \tag{4.33}$$

approximates the desired sample $y[n-L]$ in the MMSE sense. This Wiener filter is called a *backward linear predictor* and the corresponding estimation error

$$e_L^b[n] = y[n-L] - \widehat{y}_L[n-L] \tag{4.34}$$

is called a *backward prediction error.* Substituting (4.33) into (4.34) yields

$$e_L^b[n] = \sum_{k=0}^{L} b_L[k]y[n-k] = b_L[n] \star y[n] \tag{4.35}$$

where

$$b_L[n] = \begin{cases} -\widetilde{b}_n, & n = 0, 1, ..., L-1, \\ 1, & n = L, \\ 0, & \text{otherwise} \end{cases} \tag{4.36}$$

is called an *Lth-order backward LPE filter.*

Similar to (4.32), the set of *normal equations* associated with the Lth-order backward LPE filter $b_L[n]$ is given by

$$\begin{pmatrix} r_y[0] & r_y^*[1] & \cdots & r_y^*[L-1] & r_y^*[L] \\ r_y[1] & r_y[0] & \cdots & r_y^*[L-2] & r_y^*[L-1] \\ \vdots & \vdots & \ddots & \vdots & \vdots \\ r_y[L-1] & r_y[L-2] & \cdots & r_y[0] & r_y^*[1] \\ r_y[L] & r_y[L-1] & \cdots & r_y[1] & r_y[0] \end{pmatrix} \begin{pmatrix} b_L[0] \\ b_L[1] \\ \vdots \\ b_L[L-1] \\ 1 \end{pmatrix} = \begin{pmatrix} 0 \\ 0 \\ \vdots \\ 0 \\ \sigma_L^2 \end{pmatrix} \tag{4.37}$$

where

$$\sigma_L^2 \triangleq E\{|e_L^b[n]|^2\} = \sum_{k=0}^{L} b_L[k] r_y[L-k] \tag{4.38}$$

is the minimum value of $E\{|y[n-L]-\widehat{y}_L[n-L]|^2\}$. Note that we have used the same notation 'σ_L^2' to stand for both the variance of the forward prediction error $e_L^f[n]$ and that of the backward prediction error $e_L^b[n]$ because they are identical, as explained below.

Relation between Forward and Backward LPE Filters

Reversing the equation order in (4.37) and taking complex conjugation yields

$$\begin{pmatrix} r_y[0] & r_y^*[1] & \cdots & r_y^*[L-1] & r_y^*[L] \\ r_y[1] & r_y[0] & \cdots & r_y^*[L-2] & r_y^*[L-1] \\ \vdots & \vdots & \ddots & \vdots & \vdots \\ r_y[L-1] & r_y[L-2] & \cdots & r_y[0] & r_y^*[1] \\ r_y[L] & r_y[L-1] & \cdots & r_y[1] & r_y[0] \end{pmatrix} \begin{pmatrix} 1 \\ b_L^*[L-1] \\ \vdots \\ b_L^*[1] \\ b_L^*[0] \end{pmatrix} = \begin{pmatrix} \sigma_L^2 \\ 0 \\ \vdots \\ 0 \\ 0 \end{pmatrix}.$$

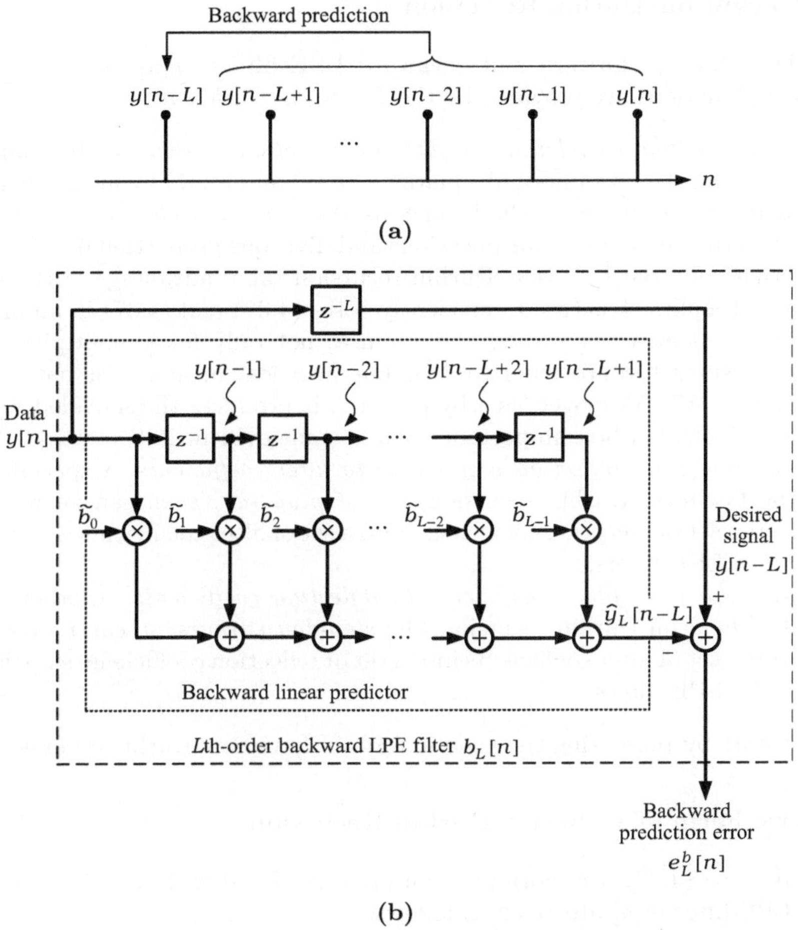

Fig. 4.6 (**a**) Conceptual illustration of the backward prediction and (**b**) tap-delay-line structure of an Lth-order backward LPE filter

Comparing this equation with (4.32) therefore gives the relation

$$b_L[n] = a_L^*[L-n] \quad \text{for } n = 0, 1, ..., L, \tag{4.39}$$

which further verifies that the variance of $e_L^f[n]$ is identical to that of $e_L^b[n]$.

4.2.2 Levinson–Durbin Recursion

For the Lth-order forward and backward LPE filters $a_L[n]$ and $b_L[n]$, the following functions are provided by the *Levinson–Durbin recursion.*[3]

- *Conversion from correlations to filter or reflection coefficients.* It is known that solving the set of normal equations (4.32) and (4.37) by means of standard matrix-inversion methods such as the *Gaussian elimination method* requires the amount of multiplications and divisions proportional to L^3 [12]. Alternatively, the Levinson–Durbin recursion takes advantage of the Hermitian Toeplitz structure to efficiently solve (4.32) and (4.37) in an order-recursive manner. Accordingly, it requires not only fewer multiplications and divisions (proportional to L^2), but also less memory for data storage [10, 11, 13]. Moreover, as a by-product, it produces the so-called *reflection coefficients* whose importance will become evident in Section 4.2.3.
- *Conversion from reflection coefficients to filter coefficients.* A special case of the Levinson–Durbin recursion, the *step-up recursion*, can be used to convert a set of reflection coefficients into a set of filter coefficients associated with the LPE filters.
- *Conversion from filter coefficients to reflection coefficients.* Another form of the Levinson-Durbin recursion, the *step-down recursion*, can be used to convert a set of filter coefficients into a set of reflection coefficients associated with the LPE filters.

Let us start by presenting the generic form of Levinson–Durbin recursion.

Generic Form of Levinson–Durbin Recursion

According to (4.32), the normal equations associated with an lth-order forward LPE filter $a_l[n]$ are given as follows:

$$\mathbf{R}_l \mathbf{a}_l = \begin{pmatrix} \sigma_l^2 \\ \mathbf{0}_l \end{pmatrix} \tag{4.40}$$

where $\mathbf{0}_l$ is an $l \times 1$ zero vector,

$$\mathbf{a}_l = (1, a_l[1], a_l[2], ..., a_l[l])^T, \tag{4.41}$$

and

[3] N. Levinson (1947) first proposed the so-called *Levinson recursion* to solve the Wiener–Hopf equations, which was then modified by J. Durbin (1960) to solve the normal equations — hence the name *Levinson–Durbin recursion* [10, 11].

$$\mathbf{R}_l = \left(\begin{array}{cccc|c} r_y[0] & r_y^*[1] & \cdots & r_y^*[l-1] & r_y^*[l] \\ r_y[1] & r_y[0] & \cdots & r_y^*[l-2] & r_y^*[l-1] \\ \vdots & \vdots & \ddots & \vdots & \vdots \\ r_y[l-1] & r_y[l-2] & \cdots & r_y[0] & r_y^*[1] \\ \hline r_y[l] & r_y[l-1] & \cdots & r_y[1] & r_y[0] \end{array} \right) = \begin{pmatrix} \mathbf{R}_{l-1} & \widetilde{\mathbf{r}}_{l-1}^* \\ \widetilde{\mathbf{r}}_{l-1}^T & r_y[0] \end{pmatrix} \tag{4.42}$$

in which $\widetilde{\mathbf{r}}_{l-1} = (r_y[l], r_y[l-1], ..., r_y[1])^T$ and $\mathbf{R}_0 = r_y[0]$. From (4.40) and (4.42), it follows that

$$\mathbf{R}_l \begin{pmatrix} \mathbf{a}_{l-1} \\ 0 \end{pmatrix} = \begin{pmatrix} \sigma_{l-1}^2 \\ \mathbf{0}_{l-1} \\ \Delta_{l-1} \end{pmatrix} \tag{4.43}$$

where

$$\Delta_{l-1} = \widetilde{\mathbf{r}}_{l-1}^T \mathbf{a}_{l-1} = r_y[l] + \sum_{k=1}^{l-1} r_y[l-k] a_{l-1}[k]. \tag{4.44}$$

Similarly, according to (4.37), the normal equations associated with an lth-order backward LPE filter $b_l[n]$ are as follows:

$$\mathbf{R}_l \mathbf{b}_l = \begin{pmatrix} \mathbf{0}_{l-1} \\ \sigma_l^2 \end{pmatrix} \tag{4.45}$$

where

$$\mathbf{b}_l = (b_l[0], b_l[1], ..., b_l[l-1], 1)^T \tag{4.46}$$

and

$$\mathbf{R}_l = \left(\begin{array}{c|cccc} r_y[0] & r_y^*[1] & \cdots & r_y^*[l-1] & r_y^*[l] \\ \hline r_y[1] & r_y[0] & \cdots & r_y^*[l-2] & r_y^*[l-1] \\ \vdots & \vdots & \ddots & \vdots & \vdots \\ r_y[l-1] & r_y[l-2] & \cdots & r_y[0] & r_y^*[1] \\ r_y[l] & r_y[l-1] & \cdots & r_y[1] & r_y[0] \end{array} \right) = \begin{pmatrix} r_y[0] & \mathbf{r}_{l-1}^H \\ \mathbf{r}_{l-1} & \mathbf{R}_{l-1} \end{pmatrix} \tag{4.47}$$

in which $\mathbf{r}_{l-1} = (r_y[1], r_y[2], ..., r_y[l])^T$. From (4.45) and (4.47), it follows that

$$\mathbf{R}_l \begin{pmatrix} 0 \\ \mathbf{b}_{l-1} \end{pmatrix} = \begin{pmatrix} \Delta_{l-1}' \\ \mathbf{0}_{l-1} \\ \sigma_{l-1}^2 \end{pmatrix} \tag{4.48}$$

where

$$\Delta'_{l-1} = \mathbf{r}_{l-1}^H \mathbf{b}_{l-1} = \sum_{k=0}^{l-2} r_y^*[k+1] b_{l-1}[k] + r_y^*[l]$$
$$= \Delta_{l-1}^* \quad \text{(see Problem 4.5)}. \tag{4.49}$$

By (4.43) and (4.48), we have

$$\mathbf{R}_l \left\{ \begin{pmatrix} \mathbf{a}_{l-1} \\ 0 \end{pmatrix} + \kappa_l \begin{pmatrix} 0 \\ \mathbf{b}_{l-1} \end{pmatrix} \right\} = \begin{pmatrix} \sigma_{l-1}^2 \\ \mathbf{0}_{l-1} \\ \Delta_{l-1} \end{pmatrix} + \kappa_l \begin{pmatrix} \Delta'_{l-1} \\ \mathbf{0}_{l-1} \\ \sigma_{l-1}^2 \end{pmatrix} \tag{4.50}$$

where κ_l is a real or complex constant. From (4.40) and (4.50), it follows that if κ_l is chosen such that

$$\begin{pmatrix} \sigma_{l-1}^2 \\ \mathbf{0}_{l-1} \\ \Delta_{l-1} \end{pmatrix} + \kappa_l \begin{pmatrix} \Delta'_{l-1} \\ \mathbf{0}_{l-1} \\ \sigma_{l-1}^2 \end{pmatrix} = \begin{pmatrix} \sigma_l^2 \\ \mathbf{0}_{l-1} \\ 0 \end{pmatrix}$$

or, equivalently,

$$\begin{cases} \sigma_{l-1}^2 + \kappa_l \Delta'_{l-1} = \sigma_l^2, \\ \Delta_{l-1} + \kappa_l \sigma_{l-1}^2 = 0, \end{cases} \tag{4.51}$$

then

$$\mathbf{a}_l = \begin{pmatrix} \mathbf{a}_{l-1} \\ 0 \end{pmatrix} + \kappa_l \begin{pmatrix} 0 \\ \mathbf{b}_{l-1} \end{pmatrix} \tag{4.52}$$

because the solution to (4.40) is unique ($\mathbf{R}_l$ is positive definite). In the same manner, by (4.43) and (4.48), we have

$$\mathbf{R}_l \left\{ \begin{pmatrix} 0 \\ \mathbf{b}_{l-1} \end{pmatrix} + \kappa'_l \begin{pmatrix} \mathbf{a}_{l-1} \\ 0 \end{pmatrix} \right\} = \begin{pmatrix} \Delta'_{l-1} \\ \mathbf{0}_{l-1} \\ \sigma_{l-1}^2 \end{pmatrix} + \kappa'_l \begin{pmatrix} \sigma_{l-1}^2 \\ \mathbf{0}_{l-1} \\ \Delta_{l-1} \end{pmatrix} \tag{4.53}$$

where κ'_l is a real or complex constant. From (4.45) and (4.53), it follows that if κ'_l is chosen such that

$$\begin{pmatrix} \Delta'_{l-1} \\ \mathbf{0}_{l-1} \\ \sigma_{l-1}^2 \end{pmatrix} + \kappa'_l \begin{pmatrix} \sigma_{l-1}^2 \\ \mathbf{0}_{l-1} \\ \Delta_{l-1} \end{pmatrix} = \begin{pmatrix} 0 \\ \mathbf{0}_{l-1} \\ \sigma_l^2 \end{pmatrix}$$

or, equivalently,

$$\begin{cases} \Delta'_{l-1} + \kappa'_l \sigma^2_{l-1} = 0, \\ \sigma^2_{l-1} + \kappa'_l \Delta_{l-1} = \sigma^2_l, \end{cases} \tag{4.54}$$

then

$$\mathbf{b}_l = \begin{pmatrix} 0 \\ \mathbf{b}_{l-1} \end{pmatrix} + \kappa'_l \begin{pmatrix} \mathbf{a}_{l-1} \\ 0 \end{pmatrix}. \tag{4.55}$$

For the forward LPE filter $a_l[n]$, (4.39) and (4.52) lead to

$$\begin{aligned} a_l[n] &= a_{l-1}[n] + \kappa_l b_{l-1}[n-1] \\ &= \begin{cases} a_{l-1}[n] + \kappa_l a^*_{l-1}[l-n], & \text{for } n = 0, 1, ..., l-1, \\ \kappa_l, & \text{for } n = l, \\ 0, & \text{otherwise} \end{cases} \end{aligned} \tag{4.56}$$

where $a_0[n] = \delta[n]$ (see (4.29)). Furthermore, by the second line of (4.51),

$$\kappa_l = -\frac{\Delta_{l-1}}{\sigma^2_{l-1}} \qquad \text{for } l = 1, 2, ..., L \tag{4.57}$$

and, by the first line of (4.51) and the second line of (4.49),

$$\begin{aligned} \sigma^2_l &= \sigma^2_{l-1} + \kappa_l \Delta^*_{l-1} = \sigma^2_{l-1} + \kappa_l(-\kappa_l \sigma^2_{l-1})^* \quad \text{(by (4.57))} \\ &= \sigma^2_{l-1}(1 - |\kappa_l|^2) \end{aligned} \tag{4.58}$$

where $\sigma^2_0 = r_y[0]$ (see (4.31)) and $\Delta_0 = r_y[1]$ (by (4.44)). The parameters κ_1, κ_2, ..., κ_L given by (4.57) are called the *reflection coefficients* because of their analogy to the reflection coefficients in transmission line models or acoustic tube models. By the orthogonality principle, κ_l can be further shown to be

$$\kappa_l = -\frac{E\left\{e^f_{l-1}[n]\left(e^b_{l-1}[n-1]\right)^*\right\}}{\sqrt{E\left\{\left|e^f_{l-1}[n]\right|^2\right\} \cdot E\left\{\left|e^b_{l-1}[n-1]\right|^2\right\}}} \tag{4.59}$$

(Problem 4.6), which is the negative of the correlation coefficient of $e^f_{l-1}[n]$ and $e^b_{l-1}[n-1]$ (see Definition 3.10).[4] From (4.59) and (3.54), it follows that

$$|\kappa_l| \leq 1 \qquad \text{for } l = 1, 2, ..., L, \tag{4.60}$$

[4] In the statistics literature, the reflection coefficients κ_1, κ_2, ..., κ_L are referred to as the *partial correlation (PARCOR) coefficients* [14].

which, together with (4.58), implies that

$$\sigma_0^2 \geq \sigma_1^2 \geq \cdots \geq \sigma_L^2 \geq 0, \tag{4.61}$$

i.e. the prediction-error variance would normally decrease as the filter order L increases. Also note, from (4.58), that if $\sigma_{l-1}^2 > 0$ and $\kappa_l = 1$, then $\sigma_l^2 = 0$, implying that the WSS process $y[n]$ is perfectly predictable. Table 4.1 summarizes Levinson–Durbin recursion in terms of the forward LPE filter.

Table 4.1 Levinson–Durbin recursion

Problem	Convert $\{r_y[0], r_y[1], ..., r_y[L]\}$ into $\{a_L[1], a_L[2], ..., a_L[L], \sigma_L^2\}$ or $\{\kappa_1, \kappa_2, ..., \kappa_L, \sigma_L^2\}$.
Initial conditions	Set $a_0[n] = \delta[n]$ and $\sigma_0^2 = r_y[0]$.

For filter order $l = 1, 2, ..., L$, compute

(i) $\Delta_{l-1} = r_y[l] + \sum_{k=1}^{l-1} r_y[l-k] a_{l-1}[k]$

(ii) $\kappa_l = -\dfrac{\Delta_{l-1}}{\sigma_{l-1}^2}$

(iii) $a_l[n] = \begin{cases} a_{l-1}[n] + \kappa_l a_{l-1}^*[l-n], & \text{for } n = 0, 1, ..., l-1, \\ \kappa_l, & \text{for } n = l, \\ 0, & \text{otherwise} \end{cases}$

(iv) $\sigma_l^2 = \sigma_{l-1}^2 (1 - |\kappa_l|^2)$

An example illustrating the Levinson–Durbin recursion is as follows.

Example 4.8

Given the correlations $r_y[0] = 4$ and $r_y[1] = r_y[2] = 1$, let us use the Levinson–Durbin recursion to compute the coefficients of the second-order forward LPE filter $a_2[n]$, the reflection coefficients κ_1 and κ_2, and the prediction-error variance σ_2^2. According to Table 4.1, we have the following steps.

1. Initial conditions: $a_0[0] = 1$ and $\sigma_0^2 = r_y[0] = 4$.
2. For filter order $l = 1$, compute
   (i) $\Delta_0 = r_y[1] = 1$.
   (ii) $\kappa_1 = -\Delta_0/\sigma_0^2 = -0.25$.
   (iii) $a_1[0] = 1$ and $a_1[1] = \kappa_1 = -0.25$.

(iv) $\sigma_1^2 = \sigma_0^2(1 - |\kappa_1|^2) = 3.75$.

3. For filter order $l = 2$, compute
   (i) $\Delta_1 = r_y[2] + r_y[1]a_1[1] = 0.75$.
   (ii) $\kappa_2 = -\Delta_1/\sigma_1^2 = -0.2$.
   (iii) $a_2[0] = 1$, $a_2[1] = a_1[1] + \kappa_2 a_1^*[1] = -0.2$, and $a_2[2] = \kappa_2 = -0.2$.
   (iv) $\sigma_2^2 = \sigma_1^2(1 - |\kappa_2|^2) = 3.6$.

One can see that $|\kappa_1| < 1$, $|\kappa_2| < 1$, and $\sigma_0^2 > \sigma_1^2 > \sigma_2^2 > 0$ for this case. □

As for the backward LPE filter $b_l[n]$, (4.39) and (4.55) lead to

$$b_l[n] = a_l^*[l-n] = b_{l-1}[n-1] + \kappa_l^* a_{l-1}[n]$$
$$= \begin{cases} \kappa_l^*, & \text{for } n = 0, \\ a_{l-1}^*[l-n] + \kappa_l^* a_{l-1}[n], & \text{for } n = 1, 2, ..., l-1, \\ 0, & \text{otherwise} \end{cases} \tag{4.62}$$

where we have used the fact that $\kappa_l' = \kappa_l^*$ (by (4.54), the second line of (4.49), and (4.57)). Description of Levinson–Durbin recursion in terms of the backward LPE filter is quite similar to that in Table 4.1 and thus is omitted here.

Step-Up Recursion

The problem considered here is to convert a set of reflection coefficients $\{\kappa_1, \kappa_2, ..., \kappa_L\}$ into a set of filter coefficients $\{a_L[1], a_L[2], ..., a_L[L]\}$ associated with the Lth-order forward LPE filter $a_L[n]$ by virtue of the generic form of the Levinson–Durbin recursion. More specifically, this problem is resolved by using the so-called *step-up recursion* [13], as shown in Table 4.2. Apparently, the step-up recursion is nothing but a degeneration of the Levinson–Durbin recursion.

Step-Down Recursion

Now consider the reverse problem that converts a set of filter coefficients $\{a_L[1], a_L[2], ..., a_L[L]\}$ into a set of reflection coefficients $\{\kappa_1, \kappa_2, ..., \kappa_L\}$ by virtue of the generic form of the Levinson–Durbin recursion. This problem is resolved by using the so-called *step-down recursion*[5] [13] derived as follows. From (4.56) and (4.62), it follows that

$$\begin{pmatrix} a_l[n] \\ a_l^*[l-n] \end{pmatrix} = \begin{pmatrix} 1 & \kappa_l \\ \kappa_l^* & 1 \end{pmatrix} \begin{pmatrix} a_{l-1}[n] \\ a_{l-1}^*[l-n] \end{pmatrix}. \tag{4.63}$$

[5] In some literature, the step-down recursion is referred to as the *inverse Levinson–Durbin recursion*, the *reverse-order Levinson recursion* or the *backward Levinson recursion* [10, 11].

Table 4.2 Step-up recursion

Problem	Convert $\{\kappa_1, \kappa_2, ..., \kappa_L, \sigma_L^2\}$ into $\{a_L[1], a_L[2], ..., a_L[L], \sigma_L^2\}$.
Initial condition	Set $a_0[n] = \delta[n]$.

For filter order $l = 1, 2, ..., L$, compute

$$a_l[n] = \begin{cases} a_{l-1}[n] + \kappa_l a_{l-1}^*[l-n], & \text{for } n = 0, 1, ..., l-1, \\ \kappa_l, & \text{for } n = l, \\ 0, & \text{otherwise} \end{cases}$$

By assuming $|\kappa_l| < 1$ for all l, we obtain

$$a_{l-1}[n] = \frac{a_l[n] - \kappa_l a_l^*[l-n]}{1 - |\kappa_l|^2} \quad \text{for } n = 0, 1, ..., l-1. \tag{4.64}$$

This therefore gives the step-down recursion shown in Table 4.3.

Table 4.3 Step-down recursion

Problem	Convert $\{a_L[1], a_L[2], ..., a_L[L], \sigma_L^2\}$ into $\{\kappa_1, \kappa_2, ..., \kappa_L, \sigma_L^2\}$.

For filter order $l = L, L-1, ..., 1$, compute

(i) $\kappa_l = a_l[l]$

(ii) $$a_{l-1}[n] = \begin{cases} \dfrac{a_l[n] - \kappa_l a_l^*[l-n]}{1 - |\kappa_l|^2}, & \text{for } n = 0, 1, ..., l-1, \\ 0, & \text{otherwise} \end{cases}$$

Figure 4.7 clarifies the use of the generic form of Levinson–Durbin recursion, the step-up recursion and the step-down recursion.

4.2.3 Lattice Linear Prediction Error Filters

For the lth-order forward and backward LPE filters $a_l[n]$ and $b_l[n]$, the associated forward prediction error $e_l^f[n]$ given by (4.28) and backward prediction

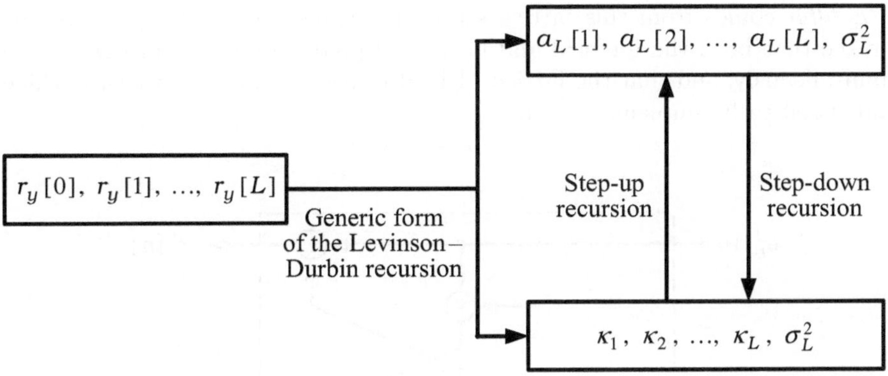

Fig. 4.7 Use of the Levinson–Durbin recursion

error $e_l^b[n]$ given by (4.35) can be further expressed as follows:

$$\begin{cases} e_l^f[n] = \mathbf{a}_l^T \mathbf{y}_l[n], \\ e_l^b[n] = \mathbf{b}_l^T \mathbf{y}_l[n] \end{cases} \tag{4.65}$$

where $\mathbf{a}_l$ and $\mathbf{b}_l$ have been defined by (4.41) and (4.46), respectively, and $\mathbf{y}_l[n] = (y[n], y[n-1], ..., y[n-l])^T$. By (4.52), (4.65) and the fact that

$$\mathbf{y}_l[n] = \begin{pmatrix} \mathbf{y}_{l-1}[n] \\ y[n-l] \end{pmatrix} = \begin{pmatrix} y[n] \\ \mathbf{y}_{l-1}[n-1] \end{pmatrix}, \tag{4.66}$$

we obtain

$$\begin{aligned} e_l^f[n] &= \left(\mathbf{a}_{l-1}^T \ \ 0\right) \begin{pmatrix} \mathbf{y}_{l-1}[n] \\ y[n-l] \end{pmatrix} + \kappa_l \left(0 \ \ \mathbf{b}_{l-1}^T\right) \begin{pmatrix} y[n] \\ \mathbf{y}_{l-1}[n-1] \end{pmatrix} \\ &= e_{l-1}^f[n] + \kappa_l \cdot e_{l-1}^b[n-1]. \end{aligned} \tag{4.67}$$

Similarly, by (4.55), (4.65) and (4.66), we obtain

$$\begin{aligned} e_l^b[n] &= \left(0 \ \ \mathbf{b}_{l-1}^T\right) \begin{pmatrix} y[n] \\ \mathbf{y}_{l-1}[n-1] \end{pmatrix} + \kappa_l^* \left(\mathbf{a}_{l-1}^T \ \ 0\right) \begin{pmatrix} \mathbf{y}_{l-1}[n] \\ y[n-l] \end{pmatrix} \\ &= e_{l-1}^b[n-1] + \kappa_l^* \cdot e_{l-1}^f[n]. \end{aligned} \tag{4.68}$$

As depicted in Fig. 4.8a, (4.67) and (4.68) form a stage (an elementary unit) of the so-called *lattice LPE filter*. Figure 4.8b depicts the cascaded connection of L stages for an Lth-order lattice LPE filter where $e_0^f[n] = e_0^b[n] = y[n]$

since $a_0[n] = b_0[n] = \delta[n]$ (see (4.28) and (4.35)). Obviously, the name *lattice LPE filter* comes from this lattice structure. Moreover, from Fig. 4.8b, one can see that both the forward and backward prediction errors are generated simultaneously, and that the lattice LPE filter has a modular structure which lends itself to IC implementation.

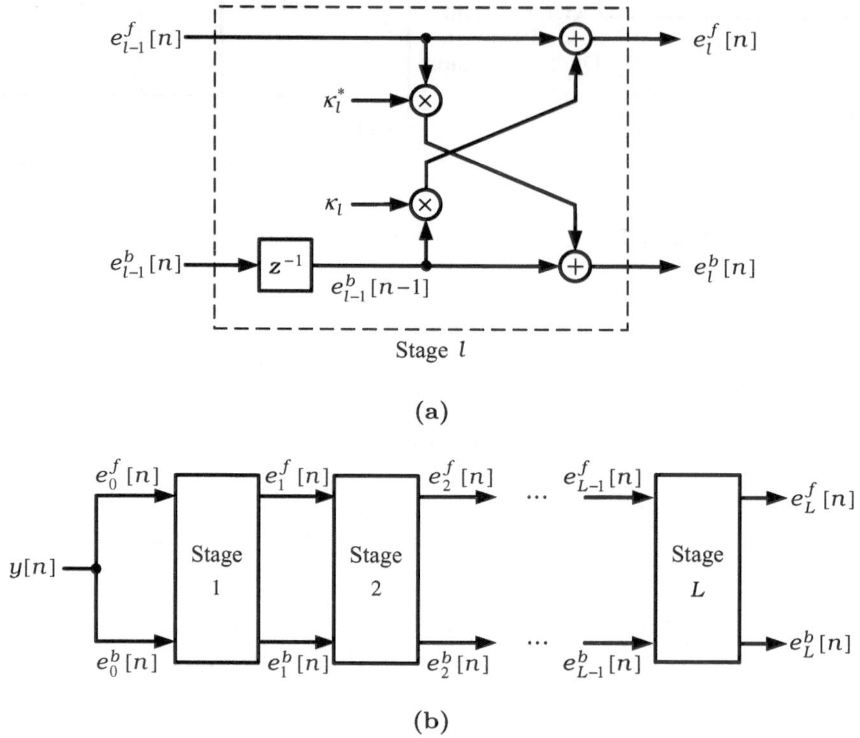

Fig. 4.8 (**a**) The lth stage and (**b**) block diagram of an Lth-order lattice LPE filter

Note that the tap-delay-line structure of the LPE filter is characterized by the filter coefficients, while the lattice structure of the LPE filter is characterized by the reflection coefficients, which can be efficiently estimated from data by means of the widely used *Burg algorithm*, to be described next.

Burg Algorithm

Suppose we are given a finite set of data $\{y[0], y[1], ..., y[N-1]\}$. For an Lth-order LPE filter, the Burg algorithm estimates the reflection coefficients κ_1, κ_2, ..., κ_L directly from the data set in a stage-recursive (an order-recursive)

manner. At stage l, the reflection coefficient κ_l is determined by minimizing the following objective function:

$$J_{\text{Burg}}^{[l]}(\kappa_l) = \sum_{n=l}^{N-1} \left\{ \left| e_l^f[n] \right|^2 + \left| e_l^b[n] \right|^2 \right\}. \tag{4.69}$$

From (4.67), (4.68) and (4.69), it follows that

$$\frac{\partial J_{\text{Burg}}^{[l]}(\kappa_l)}{\partial \kappa_l^*} = \sum_{n=l}^{N-1} \left\{ e_l^f[n] \left(e_{l-1}^b[n-1] \right)^* + e_{l-1}^f[n] \left(e_l^b[n] \right)^* \right\}. \tag{4.70}$$

Substituting (4.67) and (4.68) into (4.70) yields

$$\begin{aligned} \frac{\partial J_{\text{Burg}}^{[l]}(\kappa_l)}{\partial \kappa_l^*} &= \kappa_l \sum_{n=l}^{N-1} \left\{ \left| e_{l-1}^f[n] \right|^2 + \left| e_{l-1}^b[n-1] \right|^2 \right\} \\ &\quad + 2 \sum_{n=l}^{N-1} e_{l-1}^f[n] \left(e_{l-1}^b[n-1] \right)^* . \end{aligned}$$

Clearly, setting this result equal to zero gives the optimal estimator of κ_l associated with the minimum of $J_{\text{Burg}}^{[l]}(\kappa_l)$ as follows:

$$\widehat{\kappa}_l = \frac{-2 \sum_{n=l}^{N-1} e_{l-1}^f[n] \left(e_{l-1}^b[n-1] \right)^*}{\sum_{n=l}^{N-1} \left\{ \left| e_{l-1}^f[n] \right|^2 + \left| e_{l-1}^b[n-1] \right|^2 \right\}}. \tag{4.71}$$

The estimator $\widehat{\kappa}_l$ can be shown to satisfy $|\widehat{\kappa}_l| \leq 1$ (see [13, p. 317]), which is consistent with the theoretical bound given by (4.60). Table 4.4 summarizes the Burg algorithm.

As a remark, if our interest is to obtain the filter coefficients of the Lth-order forward LPE filter $a_L[n]$, we may, first, use the Burg algorithm to estimate the reflection coefficients $\kappa_1, \kappa_2, ..., \kappa_L$ from the set of data $\{y[0], y[1], ..., y[N-1]\}$ and then convert these reflection coefficients into filter coefficients $a_L[1], a_L[2], ..., a_L[L]$ by means of step-up recursion where $a_L[0] = 1$.

4.2.4 Linear Predictive Deconvolution

Having introduced the fundamentals of linear prediction, we now turn our attention to SISO blind equalization using LPE filters, referred to as *linear predictive deconvolution* or *linear predictive equalization.* Consider the finite set of data $\{y[0], y[1], ..., y[N-1]\}$ generated from the data model given by (4.1) and (4.2) with the following assumptions.

Table 4.4 Burg algorithm

Problem	Obtain the reflection coefficients $\kappa_1, \kappa_2, ..., \kappa_L$ from a set of data $\{y[0], y[1], ..., y[N-1]\}$.
Initial conditions	Set $e_0^f[n] = e_0^b[n] = y[n]$ for $n = 0, 1, ..., N-1$.

For stage $l = 1, 2, ..., L$, compute

$$\text{(i)}\ \kappa_l = \frac{-2\sum_{n=l}^{N-1} e_{l-1}^f[n](e_{l-1}^b[n-1])^*}{\sum_{n=l}^{N-1} |e_{l-1}^f[n]|^2 + |e_{l-1}^b[n-1]|^2}$$

$$\text{(ii)}\ \begin{cases} e_l^f[n] = e_{l-1}^f[n] + \kappa_l \cdot e_{l-1}^b[n-1] \\ e_l^b[n] = e_{l-1}^b[n-1] + \kappa_l^* \cdot e_{l-1}^f[n] \end{cases}$$

(A4-1) The SISO LTI system $h[n]$ is stable.
(A4-2) The source signal $u[n]$ is a WSS white process with variance σ_u^2.
(A4-3) The noise $w[n]$ is a zero-mean WSS (white or colored) process.
(A4-4) The source signal $u[n]$ is statistically independent of the noise $w[n]$.

Here, *linear predictive deconvolution* simply refers to use of an Lth-order forward LPE filter $a_L[n]$ as the blind equalizer whose coefficients are obtained from the data $y[n]$, $n = 0, 1, ..., N-1$. The corresponding forward prediction error $e_L^f[n]$ is the equalized signal, referred to as the *predictive deconvolved signal* or the *predictive equalized signal.*

Let $A_L(z)$ denote the transfer function of $a_L[n]$. For the noise-free case (SNR $= \infty$), the following property holds true when the transfer function of the LTI system $h[n]$ is given by

$$H(z) = \frac{1}{1 + \alpha_1 z^{-1} + \cdots + \alpha_p z^{-p}}, \tag{4.72}$$

i.e. the noise-free signal $x[n]$ is an AR process.

Property 4.9. *If the system $H(z)$ is an AR(p) model given by (4.72), then the Lth-order forward LPE filter $A_L(z) = 1/H(z)$ and*

$$a_L[n] = \begin{cases} 1, & for\ n = 0, \\ \alpha_n, & for\ n = 1, 2, ..., p, \\ 0, & otherwise \end{cases} \tag{4.73}$$

for $L \geq p$ *and* SNR $= \infty$. *Correspondingly, the predictive deconvolved signal* $e_L^f[n] = u[n]$.

This property follows directly from comparison between the normal equations given by (4.32) and the associated Yule–Walker equations as given by (3.214). In other words, for the noise-free case, the forward LPE filter $a_L[n]$ of sufficient order is exactly a ZF equalizer. In addition, Property 4.9 reveals that the forward LPE filter can also be used to provide the parameters of the AR spectral estimator (see (3.218)).

More generally, consider the case of finite SNR. According to Section 3.3.2, under Assumptions (A4-1) through (A4-4), $y[n]$ is a zero-mean WSS process and its power spectrum

$$S_y(\omega) = \sigma_u^2 |H(\omega)|^2 + S_w(\omega) \tag{4.74}$$

is a continuous function of ω on $[-\pi, \pi)$ where $S_w(\omega)$ is the power spectrum of $w[n]$. Note that any WSS process with continuous power spectrum may be represented as the output of a minimum-phase system driven by a WSS white noise [15]; such a representation is called the *canonical innovations representation*. Accordingly, we have the canonical innovations representation of the WSS process $y[n]$ as follows: (see Fig. 4.9)

$$y[n] = \widetilde{h}[n] \star \widetilde{u}[n] \tag{4.75}$$

where $\widetilde{h}[n]$ is a minimum-phase system with $\widetilde{h}[0] = 1$ and $\widetilde{u}[n]$ is a WSS white process with variance [15, p. 64]

$$\sigma_{\widetilde{u}}^2 = \exp\left\{ \frac{1}{2\pi} \int_{-\pi}^{\pi} \ln S_y(\omega) d\omega \right\}. \tag{4.76}$$

By definition, samples of the white process $\widetilde{u}[n]$ at different instants of time are uncorrelated with each other, meaning that each sample of $\widetilde{u}[n]$ brings "new information" or "innovations." For this reason, $\widetilde{u}[n]$ is also called an *innovations process*. Since any minimum-phase system can be approximated as an AR model of sufficient order, the canonical innovations representation (4.75) and Property 4.9 imply that as $L \to \infty$, $A_L(z) = 1/\widetilde{H}(z)$ and $e_L^f[n] = \widetilde{u}[n]$. Accordingly, we have the following property.

Property 4.10 (Whitening Property). *With sufficient filter order, the* L*th-order forward LPE filter* $a_L[n]$ *is a whitening filter and the predictive deconvolved signal* $e_L^f[n]$ *is a white process for any* SNR.

Moreover, for any SNR the power spectrum $S_y(\omega)$ can be estimated via the following relation:

$$S_y(\omega) = \sigma_{\tilde{u}}^2 \cdot \left|\widetilde{H}(\omega)\right|^2 = \frac{\sigma_{\tilde{u}}^2}{|A_L(\omega)|^2} \tag{4.77}$$

for sufficiently large L where, as mentioned above, $A_L(\omega)$ can be estimated by means of the Burg algorithm.

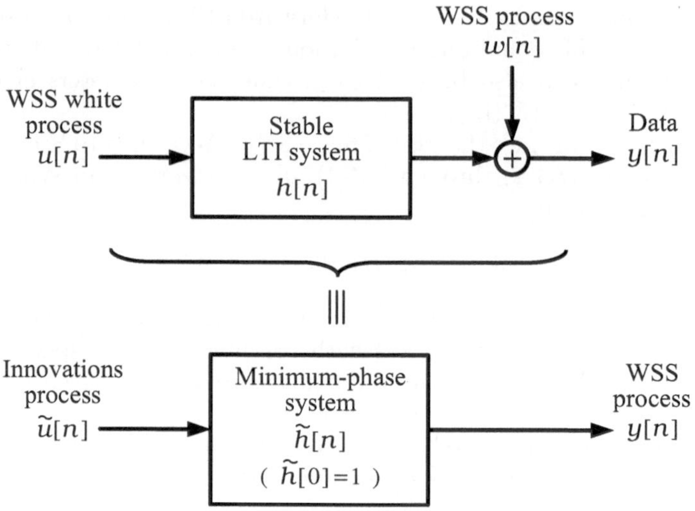

Fig. 4.9 Canonical innovations representation of the data $y[n]$

A further property which also holds true for any SNR is as follows.

Property 4.11 (Minimum-Phase Property). *The* L*th-order forward LPE filter* $a_L[n]$ *is minimum phase if and only if the associated reflection coefficients* $\kappa_1, \kappa_2, \ldots, \kappa_L$ *satisfy the condition that* $|\kappa_l| < 1$ *for all* l.

The proof can be found in the literature, e.g. [10, pp. 441–444] or [11, pp. 265–267]. Property 4.11, together with Property 4.9, implies that when SNR $= \infty$ and the system $h[n]$ is minimum phase, the forward LPE filter $a_L[n]$, which itself is also minimum phase, is sufficient for blind equalization of $h[n]$. However, when SNR $= \infty$ and

$$H(z) = H_{\rm MP}(z) \cdot H_{\rm AP}(z)$$

where $H_{\rm MP}(z)$ is a minimum-phase system and $H_{\rm AP}(z)$ is an allpass system, the Lth-order forward LPE filter $A_L(z)$ is identical to $1/H_{\rm MP}(z)$ except for a scale factor, provided that L is sufficiently large and $|\kappa_l| < 1$ for all l. In other words, the predictive deconvolved signal

$$e_L^f[n] = a_L[n] \star h[n] \star u[n] \simeq h_{\rm AP}[n] \star u[n] \quad (\text{SNR} = \infty)$$

is an *amplitude equalized signal* where $h_{\rm AP}[n]$ is the impulse response of the allpass system $H_{\rm AP}(z)$. As a consequence, when the system $h[n]$ is nonminimum phase, the (allpass) phase distortion will remain in the predictive deconvolved signal even for the noise-free case.

On the other hand, some properties regarding the Lth-order backward LPE filter $b_L[n]$ are, in turn, described as follows. The transfer function of the backward LPE filter $b_L[n]$ can be expressed as

$$\begin{aligned} B_L(z) &= \sum_{n=-\infty}^{\infty} b_L[n]z^{-n} = \sum_{n=-\infty}^{\infty} a_L^*[L-n]z^{-n} \quad \text{(by (4.39))} \\ &= z^{-L} \sum_{n=-\infty}^{\infty} a_L^*[n]z^{n} = z^{-L} \cdot A_L^*(1/z^*), \end{aligned} \tag{4.78}$$

which, together with Property 4.11, leads to the following property.

Property 4.12 (Maximum-Phase Property). *The Lth-order backward LPE filter $b_L[n]$ is maximum phase if and only if the associated reflection coefficients κ_1, κ_2, ..., κ_L satisfy the condition that $|\kappa_l| < 1$ for all l.*

Moreover, the following property can be proved by virtue of the orthogonality principle (Problem 4.7).

Property 4.13 (Orthogonality Property). *The backward prediction error $e_l^b[n]$ is uncorrelated with $e_m^b[n]$ for $m \neq l$, i.e.*

$$E\left\{e_l^b[n]\left(e_m^b[n]\right)^*\right\} = \begin{cases} \sigma_l^2, & \textit{for } m = l, \\ 0, & \textit{otherwise} \end{cases} \tag{4.79}$$

where σ_l^2 is the variance of $e_l^b[n]$.

LPE filters have been found useful in such applications as linear predictive deconvolution, power spectral estimation, linear predictive coding, and so forth. They also serve as amplitude equalizers (whitening filters) or preprocessing filters for other blind equalization algorithms, that will be detailed in the next section.

4.3 HOS Based Blind Equalization Approaches

Since the mid-1980s, the problem of SISO blind equalization has been tackled using HOS owing to the fact that HOS contains not only system magnitude information, but also system phase information (see Section 3.3.2). As mentioned in Section 4.1, we can divide HOS based blind equalization approaches into direct and indirect approaches. We will focus only on the direct approach with detailed treatment of two widely used HOS based algorithms, namely the maximum normalized cumulant (MNC) equalization algorithm and the super-exponential (SE) equalization algorithm. Other types of direct approach and indirect approach have been well reported in the literature; see [1, 16–21] (to name a few) and references therein.

According to the properties of HOS, the MNC and SE equalization algorithms are based on the following assumptions for the data $y[n]$ given by (4.1) and (4.2).

(A4-5) The SISO LTI system $h[n]$ is stable and its inverse system $h_{\mathrm{I}}[n]$ is also stable.[6]

(A4-6) The source signal $u[n]$ is a zero-mean, i.i.d., stationary non-Gaussian process with $(p+q)$th-order cumulant $C_{p,q}\{u[n]\} \neq 0$.

(A4-7) The noise $w[n]$ is a zero-mean, white or colored, WSS Gaussian process with autocorrelation function $r_w[l]$.

(A4-8) The source signal $u[n]$ is statistically independent of the noise $w[n]$.

Referring to Fig. 4.2, let us consider that the equalizer $v[n]$ to be designed is an FIR filter with $v[n] = 0$ outside the domain of support $L_1 \leq n \leq L_2$ where L_1 and L_2 are the preassigned integers. Let

$$\boldsymbol{v} = (v[L_1], v[L_1+1], ..., v[L_2])^T \tag{4.80}$$

be an $L \times 1$ parameter vector formed of the unknown equalizer coefficients where $L = L_2 - L_1 + 1$ is the equalizer length. Then the equalized signal $e[n]$ given by (4.4) can be expressed in vector form as

$$e[n] = \boldsymbol{v}^T \boldsymbol{y}[n] \tag{4.81}$$

where

$$\boldsymbol{y}[n] = (y[n-L_1], y[n-L_1-1], ..., y[n-L_2])^T \tag{4.82}$$

is an $L \times 1$ vector formed of the data $y[n]$. Next, let us present the MNC and SE equalization algorithms along with their analyses and improvements.

[6] We include the assumption of stable inverse system $h_{\mathrm{I}}[n]$ here for simplicity; relaxation of this assumption can be found in [22].

4.3.1 Maximum Normalized Cumulant Equalization Algorithm

MNC Equalization Criterion

For the MNC equalization algorithm, the equalizer $v[n]$ is designed by maximizing the following objective function:[7] [21, 23–34]

$$J_{p,q}(v[n]) = |\gamma_{p,q}\{e[n]\}| = \frac{|C_{p,q}\{e[n]\}|}{(\sigma_e^2)^{(p+q)/2}} \tag{4.83}$$

where p and q are nonnegative integers, $p+q \geq 3$, and $\gamma_{p,q}\{e[n]\}$ is the normalized $(p+q)$th-order cumulant of $e[n]$. For ease of subsequent descriptions, let us refer to the criterion of maximizing $J_{p,q}(v[n])$ as the *MNC equalization criterion*, the associated equalization as *MNC equalization*, and the associated equalizer $v[n]$ as an *MNC equalizer*, which is further denoted by $v_{\text{MNC}}[n]$ for clarity. When $y[n]$ is real (i.e. the real case), there is only one form of $J_{p,q}(v[n])$ corresponding to each particular value of $(p+q)$, irrespective of the possible combinations of p and q. As such, we also denote $J_{p,q}(v[n])$ by $J_{p+q}(v[n])$ for the real case. On the other hand, when $y[n]$ is complex, $J_{p,q}(v[n]) = J_{q,p}(v[n])$ by Theorem 3.12.

Under Assumptions (A4-5) through (A4-8) and by Theorem 3.39, the $(p+q)$th-order cumulant $C_{p,q}\{e[n]\}$ in the numerator of $J_{p,q}(v[n])$ can be expressed as

$$C_{p,q}\{e[n]\} = C_{p,q}\{u[n]\} \cdot \sum_{k=\infty}^{\infty} (g[k])^p (g^*[k])^q \quad \text{for } p+q \geq 3. \tag{4.84}$$

Furthermore, the variance σ_e^2 in the denominator of $J_{p,q}(v[n])$ can be shown to be

$$\sigma_e^2 = \sigma_u^2 \sum_{k=\infty}^{\infty} |g[k]|^2 + \sum_{k=-\infty}^{\infty} \sum_{i=-\infty}^{\infty} v[k]v^*[i]r_w^*[k-i]. \tag{4.85}$$

Substituting (4.84) and (4.85) into (4.83) yields

$$J_{p,q}(v[n]) = \frac{J_{p,q}^{(\infty)}(v[n])}{\left\{1 + \dfrac{\sum_{k=-\infty}^{\infty}\sum_{i=-\infty}^{\infty} v[k]v^*[i]r_w^*[k-i]}{\sigma_u^2 \sum_{k=\infty}^{\infty} |g[k]|^2}\right\}^{(p+q)/2}} \tag{4.86}$$

where

$$J_{p,q}^{(\infty)}(v[n]) = |\gamma_{p,q}\{u[n]\}| \cdot \frac{|\sum_{k=\infty}^{\infty} (g[k])^p (g^*[k])^q|}{\{\sum_{k=\infty}^{\infty} |g[k]|^2\}^{(p+q)/2}}, \tag{4.87}$$

[7] The criterion of maximizing $J_{p,q}(v[n])$ for blind equalization was due to Wiggins [23] and Donoho [24] for the real case, and due to Ulrych and Walker [25] and Shalvi and Weinstein [26] for the complex case.

which is identical to $J_{p,q}(v[n])$ as SNR $= \infty$. From (4.86) and (4.87), it follows that $J_{p,q}^{(\infty)}(v[n]) = J_{p,q}^{(\infty)}(\alpha v[n-\tau])$ and $J_{p,q}(v[n]) = J_{p,q}(\alpha v[n-\tau])$ for any nonzero constant α and any integer τ.

MNC equalization for the case of high SNR is supported by the following theorem.

Theorem 4.14 (Equalization Capability). *Under Assumptions* (A4-5) *and* (A4-6) *and as $L_1 \to -\infty$ and $L_2 \to \infty$, the objective function*

$$J_{p,q}(v[n]) = J_{p,q}^{(\infty)}(v[n]) \le |\gamma_{p,q}\{u[n]\}| \quad \textit{for } \text{SNR} = \infty \tag{4.88}$$

where $p+q \ge 3$. The equality of (4.88) holds if and only if

$$v_{\text{MNC}}[n] = \alpha h_{\text{I}}[n-\tau] \tag{4.89}$$

where $\alpha \neq 0$ is a real or complex constant and τ is an integer.

Proof: By Theorem 2.33, we have

$$\left| \sum_{k=\infty}^{\infty} (g[k])^p (g^*[k])^q \right| \le \sum_{k=\infty}^{\infty} |g[k]|^{p+q} \le \left\{ \sum_{k=\infty}^{\infty} |g[k]|^2 \right\}^{(p+q)/2} . \tag{4.90}$$

Equation (4.88) then follows from (4.87) and (4.90). Furthermore, by Theorem 2.33, the equality of (4.90) holds if and only if $g[k]$ is a delta function, that is equivalent to (4.89) under Assumption (A4-5) and as $L_1 \to -\infty$ and $L_2 \to \infty$.

Q.E.D.

Theorem 4.14 states that as SNR $= \infty$, the MNC equalizer $v_{\text{MNC}}[n]$ is equivalent to the inverse system (inverse filter) $h_{\text{I}}[n]$, except for a scale factor and a time delay. For this reason, the MNC criterion is often referred to as the *inverse filter criterion* [27, 29, 31, 32].

MNC Equalization Algorithm

It is easy to see, from (4.86) and (4.87), that $J_{p,q}(v[n]) \equiv J_{p,q}(\boldsymbol{v})$ is a highly nonlinear function of the parameter vector $\boldsymbol{v}$, indicating that derivation of closed-form solution for the optimum $\boldsymbol{v}$ is formidable. Therefore, we resort to gradient-type optimization methods to develop the MNC equalization algorithm for finding the (local) maximum of $J_{p,q}(\boldsymbol{v})$ and the relevant parameter vector $\boldsymbol{v}$ for the MNC equalizer $v_{\text{MNC}}[n]$.

Table 4.5 summarizes the generic steps of the MNC equalization algorithm. In Step (S3), we include a normalization operation for $\mathbf{d}^{[i]}$ (i.e. $\mathbf{d}^{[i]}/\|\mathbf{d}^{[i]}\|$) to confine the amount of update due to $\mathbf{d}^{[i]}$ to a limited range. This, in turn, reduces the search range, $[\mu_0/2^0, \mu_0/2^K]$, for finding the optimum step size $\mu^{[i]}$ for each iteration. Moreover, we also include a normalization operation

Table 4.5 MNC equalization algorithm

Parameter setting		Choose a cumulant order $(p+q) \geq 3$, an equalizer length $L = L_2 - L_1 + 1 > 1$, an initial parameter vector $\boldsymbol{v}^{[0]} \neq \mathbf{0}$, a positive real number μ_0 and a positive integer K for the step-size search range $[\mu_0/2^0, \mu_0/2^K]$, and a convergence tolerance $\zeta > 0$.
Steps	(S1)	Set the iteration number $i = 0$.
	(S2)	Compute the equalized signal $e^{[0]}[n] = (\boldsymbol{v}^{[0]})^T \boldsymbol{y}[n]$ and obtain the corresponding search direction $\mathbf{d}^{[0]}$ according to a certain gradient-type method.
	(S3)	Generate a new approximation to the parameter vector $\boldsymbol{v}$ via $$\begin{cases} \widetilde{\boldsymbol{v}}^{[i+1]} = \boldsymbol{v}^{[i]} + \mu^{[i]} \cdot \dfrac{\mathbf{d}^{[i]}}{\lVert \mathbf{d}^{[i]} \rVert} \\ \boldsymbol{v}^{[i+1]} = \dfrac{\widetilde{\boldsymbol{v}}^{[i+1]}}{\lVert \widetilde{\boldsymbol{v}}^{[i+1]} \rVert} \end{cases}$$ where an integer $k \in [0, K]$ is determined so that $\mu^{[i]} = \mu_0/2^k$ is the maximum step size leading to $J_{p,q}(\boldsymbol{v}^{[i+1]}) > J_{p,q}(\boldsymbol{v}^{[i]})$.
	(S4)	If $$\frac{\left\lvert J_{p,q}(\boldsymbol{v}^{[i+1]}) - J_{p,q}(\boldsymbol{v}^{[i]}) \right\rvert}{J_{p,q}(\boldsymbol{v}^{[i]})} \geq \zeta,$$ then go to Step (S5); otherwise, obtain a (local) maximum solution $\boldsymbol{v} = \boldsymbol{v}^{[i+1]}$ for the MNC equalizer $v_{\mathrm{MNC}}[n]$.
	(S5)	Compute the equalized signal $e^{[i+1]}[n] = (\boldsymbol{v}^{[i+1]})^T \boldsymbol{y}[n]$ and obtain the corresponding search direction $\mathbf{d}^{[i+1]}$.
	(S6)	Update the iteration number i by $(i+1)$ and go to Step (S3).

for $\boldsymbol{v}^{[i+1]}$ (i.e. $\widetilde{\boldsymbol{v}}^{[i+1]}/\lVert\widetilde{\boldsymbol{v}}^{[i+1]}\rVert$ in Step (S3)) to rule out the possibility of obtaining the trivial solution $\boldsymbol{v} = \mathbf{0}$. Note that the normalization operation for $\boldsymbol{v}^{[i+1]}$ causes no negative effect in obtaining the optimum solution because of the fact that $J_{p,q}(v[n])$ is invariant to any scaled version of $v[n]$. In Steps (S2) and (S5), the search direction $\mathbf{d}^{[i]}$ can be obtained by using the update equations of gradient-type optimization methods, that have been summarized in Table 2.2 for the steepest descent method, Table 2.3 for the Newton and approximate Newton methods, and Table 2.4 for the BFGS and approximate BFGS methods. No matter what gradient-type optimization method we use, computation of the first derivative $\partial J_{p,q}(\boldsymbol{v})/\partial \boldsymbol{v}^*$ is inevitable, that is therefore derived as follows. From (4.83), we have (see Problem 4.8)

$$\frac{\partial J_{p,q}(\boldsymbol{v})}{\partial \boldsymbol{v}^*} = \frac{J_{p,q}(\boldsymbol{v})}{2}\left\{\frac{1}{C_{p,q}\{e[n]\}}\cdot\frac{\partial C_{p,q}\{e[n]\}}{\partial \boldsymbol{v}^*} + \frac{1}{C_{q,p}\{e[n]\}}\cdot\frac{\partial C_{q,p}\{e[n]\}}{\partial \boldsymbol{v}^*} - \frac{p+q}{\sigma_e^2}\cdot\frac{\partial \sigma_e^2}{\partial \boldsymbol{v}^*}\right\} \quad (4.91)$$

in which $\partial C_{p,q}\{e[n]\}/\partial \boldsymbol{v}^*$, $\partial C_{q,p}\{e[n]\}/\partial \boldsymbol{v}^*$ and $\partial \sigma_e^2/\partial \boldsymbol{v}^*$ ($= \partial C_{1,1}\{e[n]\}/\partial \boldsymbol{v}^*$) can be derived using the following lemma (see also Problem 4.8).

Lemma 4.15. *For the equalized signal $e[n]$ given by (4.81),*

$$\frac{\partial C_{p,q}\{e[n]\}}{\partial v^*[k]} = \begin{cases} q\cdot \mathrm{cum}\{e[n]:p, e^*[n]:q-1, y^*[n-k]\}, & q\geq 1 \\ 0, & q=0 \end{cases} \quad (4.92)$$

where $p+q\geq 2$.

Applying Lemma 4.15 to (4.91) yields

$$\frac{\partial J_{p,q}(\boldsymbol{v})}{\partial \boldsymbol{v}^*} = \frac{J_{p,q}(\boldsymbol{v})}{2}\left\{\frac{q\cdot \mathrm{cum}\{e[n]:p, e^*[n]:q-1, \boldsymbol{y}^*[n]\}}{C_{p,q}\{e[n]\}} + \frac{p\cdot \mathrm{cum}\{e[n]:q, e^*[n]:p-1, \boldsymbol{y}^*[n]\}}{C_{q,p}\{e[n]\}} - \frac{p+q}{\sigma_e^2}\cdot E\{e[n]\boldsymbol{y}^*[n]\}\right\} \quad (4.93)$$

where

$$\mathrm{cum}\{e[n]:p, e^*[n]:q-1, \boldsymbol{y}^*[n]\} = \begin{pmatrix} \mathrm{cum}\{e[n]:p, e^*[n]:q-1, y^*[n-L_1]\} \\ \mathrm{cum}\{e[n]:p, e^*[n]:q-1, y^*[n-L_1+1]\} \\ \vdots \\ \mathrm{cum}\{e[n]:p, e^*[n]:q-1, y^*[n-L_2]\} \end{pmatrix}. \quad (4.94)$$

As described in Section 3.4.4, in practice, we need to replace all the correlations and cumulants in (4.93) by their respective sample correlations and sample cumulants, which are estimated from a finite set of data $\{y[0], y[1], ..., y[N-1]\}$. Note that the larger the cumulant order, the larger the variance of the sample cumulant. Accordingly, we usually consider $p+q=3$ or 4 for $J_{p,q}(\boldsymbol{v})$.

4.3.2 Super-Exponential Equalization Algorithm

The SE equalization algorithm, proposed by Shalvi and Weinstein [35], is an iterative algorithm for finding the blind equalizer $v[n]$. It originated from finding a set of update equations for $g[n]$ in the overall-system domain, that was then converted into the one for $v[n]$ in the equalizer domain. For ease of subsequent sections, we refer to the associated equalization as *SE equalization* and the associated equalizer $v[n]$ as an *SE equalizer*, which is further denoted by $v_{\mathrm{SE}}[n]$ for clarity.

Algorithm in Overall-System Domain

At iteration i, the overall system $g[n]$ is updated via

$$\begin{cases} \widetilde{g}^{[i+1]}[n] = (g^{[i]}[n])^p (g^{[i]}[n]^*)^{q-1} \\ g^{[i+1]}[n] = \widetilde{g}^{[i+1]}[n] / \|\widetilde{\boldsymbol{g}}^{[i+1]}\| \end{cases} \tag{4.95}$$

where both p and $(q-1)$ are nonnegative integers, $p+q \geq 3$, and

$$\widetilde{\boldsymbol{g}}^{[i+1]} = \left(..., \widetilde{g}^{[i+1]}[-1], \widetilde{g}^{[i+1]}[0], \widetilde{g}^{[i+1]}[1], ...\right)^T . \tag{4.96}$$

It then follows that

$$\left| \frac{g^{[i+1]}[n]}{g^{[i+1]}[m]} \right| = \left| \frac{g^{[i]}[n]}{g^{[i]}[m]} \right|^{p+q-1} = \cdots = \left| \frac{g^{[0]}[n]}{g^{[0]}[m]} \right|^{(p+q-1)^{i+1}} , \tag{4.97}$$

from which some observations are made as follows. Since $p+q \geq 3$, the coefficients of $g^{[i+1]}[n]$ with smaller amplitudes will decrease more rapidly than the ones with larger amplitudes. This, in turn, implies that $g^{[i+1]}[n]$ approaches a delta function more closely than $g^{[i]}[n]$, or equivalently, $\mathrm{ISI}\{g^{[i+1]}[n]\} < \mathrm{ISI}\{g^{[i]}[n]\}$. Moreover, the larger the value of $(p+q)$, the smaller the value of $\mathrm{ISI}\{g^{[i+1]}[n]\}$.

To analyze the convergence rate of $\mathrm{ISI}\{g^{[i+1]}[n]\}$, let us assume that

$$\left| g^{[0]}[n_0] \right| \geq \left| g^{[0]}[n_1] \right| \geq \left| g^{[0]}[n_2] \right| \geq \cdots , \tag{4.98}$$

which, together with (4.97), implies that

$$\left| g^{[i+1]}[n_0] \right| \geq \left| g^{[i+1]}[n_1] \right| \geq \left| g^{[i+1]}[n_2] \right| \geq \cdots , \tag{4.99}$$

where the coefficient $g^{[i+1]}[n_0]$ is also called the *leading coefficient.* That is, all the indices n_0, n_1, n_2, ..., are preserved along iterations. By (4.13), (4.97) and (4.99), we have

$$\begin{aligned} \mathrm{ISI}\{g^{[i+1]}[n]\} &= \sum_{n \neq n_0} \left| \frac{g^{[i+1]}[n]}{g^{[i+1]}[n_0]} \right|^2 = \sum_{n \neq n_0} \left| \frac{g^{[i]}[n]}{g^{[i]}[n_0]} \right|^{2(p+q-1)} \\ &\leq \left(\sum_{n \neq n_0} \left| \frac{g^{[i]}[n]}{g^{[i]}[n_0]} \right|^2 \right) \cdot \left| \frac{g^{[i]}[n_1]}{g^{[i]}[n_0]} \right|^{2(p+q-2)} \\ &= \mathrm{ISI}\{g^{[i]}[n]\} \cdot \left| \frac{g^{[i]}[n_1]}{g^{[i]}[n_0]} \right|^{2(p+q-2)} , \end{aligned} \tag{4.100}$$

which further gives rise to

$$\mathrm{ISI}\{g^{[i+1]}[n]\} \leq \mathrm{ISI}\{g^{[0]}[n]\} \cdot \left(\prod_{l=0}^{i} \left| \frac{g^{[l]}[n_1]}{g^{[l]}[n_0]} \right| \right)^{2(p+q-2)} . \tag{4.101}$$

By (4.97) and (4.101) and after some algebraic manipulations, we obtain

$$\mathrm{ISI}\{g^{[i+1]}[n]\} \leq \mathrm{ISI}\{g^{[0]}[n]\} \cdot \left| \frac{g^{[0]}[n_1]}{g^{[0]}[n_0]} \right|^{2\left[(p+q-1)^{i+1}-1\right]} . \tag{4.102}$$

According to (4.102), if $|g^{[0]}[n_1]|/|g^{[0]}[n_0]| < 1$, that is, the leading coefficient of $g^{[0]}[n]$ is unique, then $\mathrm{ISI}\{g^{[i+1]}[n]\}$ decreases to zero at least at a "super-exponential" (i.e. exponential in the power) rate, thereby leading to the name *super-exponential equalization algorithm.* Recall that $\mathrm{ISI}\{g[n]\} = 0$ implies $g[n] = \alpha\delta[n-\tau]$ where α is a scale factor and τ is a time delay. We accordingly come up with the following conclusion.

Theorem 4.16 (Equalization Capability). *Suppose the initial condition $g^{[0]}[n]$ has a unique leading coefficient. Then, under Assumptions* (A4-5) *and* (A4-6) *and as $L_1 \to -\infty$ and $L_2 \to \infty$, the SE equalizer*

$$v_{\mathrm{SE}}[n] = \alpha h_{\mathrm{I}}[n-\tau] \quad \textit{for } \mathrm{SNR} = \infty \tag{4.103}$$

where $\alpha \neq 0$ is a real or complex constant and τ is an integer.

On the other hand, in theory, if $g^{[0]}[n]$ has M (> 1) leading coefficients, then $g^{[i+1]}[n]$ after convergence will become a sequence composed of M nonzero components, rather than a delta function. In practice, any small deviations from this type of initial condition $g^{[0]}[n]$ are sufficient to ensure the convergence, however. Next, let us present how to convert the set of update equations given by (4.95) for $g[n]$ into the one for $v[n]$.

Algorithm in Equalizer Domain

Let $\widetilde{v}^{[i+1]}[n]$ and $v^{[i+1]}[n]$, $n = L_1, L_1+1, ..., L_2$, be the FIR equalizers corresponding to $\widetilde{g}^{[i+1]}[n]$ and $g^{[i+1]}[n]$, respectively. It is sensible to design $\widetilde{v}^{[i+1]}[n]$ by solving the following set of linear equations

$$\widetilde{v}^{[i+1]}[n] \star h[n] = \widetilde{g}^{[i+1]}[n], \quad n = ..., -1, 0, 1, ... \tag{4.104}$$

or, equivalently,

$$\mathbf{H}\widetilde{\boldsymbol{v}}^{[i+1]} = \widetilde{\boldsymbol{g}}^{[i+1]} \tag{4.105}$$

where

$$\mathbf{H} = \begin{pmatrix} & \vdots & & \\ h[-L_1-1] & h[-L_1-2] & \cdots & h[-L_2-1] \\ h[-L_1] & h[-L_1-1] & \cdots & h[-L_2] \\ h[-L_1+1] & h[-L_1] & \cdots & h[-L_2+1] \\ & \vdots & & \end{pmatrix}, \tag{4.106}$$

$$\widetilde{\boldsymbol{v}}^{[i+1]} = \left(\widetilde{v}^{[i+1]}[L_1], \widetilde{v}^{[i+1]}[L_1+1], ..., \widetilde{v}^{[i+1]}[L_2]\right)^T, \tag{4.107}$$

and $\widetilde{\boldsymbol{g}}^{[i+1]}$ is given by (4.96). Obviously, solving (4.105) for $\widetilde{\boldsymbol{v}}^{[i+1]}$ can be viewed as an LS problem and thus, by means of the LS method, we have the following LS solution:

$$\widetilde{\boldsymbol{v}}^{[i+1]} = (\mathbf{H}^H\mathbf{H})^{-1}\mathbf{H}^H\widetilde{\boldsymbol{g}}^{[i+1]} \tag{4.108}$$

where $\mathbf{H}$ is assumed to be of full rank for simplicity. This assumption, of course, can be relaxed by virtue of the SVD, as described in Section 2.4.

Similarly, we should design $v^{[i+1]}[n]$ to satisfy the set of linear equations

$$\mathbf{H}\boldsymbol{v}^{[i+1]} = \boldsymbol{g}^{[i+1]} \tag{4.109}$$

where

$$\boldsymbol{v}^{[i+1]} = \left(v^{[i+1]}[L_1], v^{[i+1]}[L_1+1], ..., v^{[i+1]}[L_2]\right)^T, \tag{4.110}$$

$$\boldsymbol{g}^{[i+1]} = \left(..., g^{[i+1]}[-1], g^{[i+1]}[0], g^{[i+1]}[1], ...\right)^T. \tag{4.111}$$

By the second line of (4.95), (4.105) and (4.109), we obtain

$$\boldsymbol{v}^{[i+1]} = \frac{\widetilde{\boldsymbol{v}}^{[i+1]}}{\sqrt{\left(\widetilde{\boldsymbol{v}}^{[i+1]}\right)^H \mathbf{H}^H\mathbf{H}\widetilde{\boldsymbol{v}}^{[i+1]}}}. \tag{4.112}$$

Since both $\mathbf{H}$ and $\widetilde{\boldsymbol{g}}^{[i+1]}$ are not available, the next step is to convert the set of update equations given by (4.108) and (4.112) into a realizable one in terms of correlations and cumulants. By (4.96), (4.106), (3.126) and Theorem 3.39, one can show that (see Problem 4.9)

$$\mathbf{H}^H\mathbf{H} = \frac{1}{\sigma_u^2}\mathbf{R}_y^* \quad \text{for SNR} = \infty \tag{4.113}$$

where $\mathbf{R}_y = E\{\mathbf{y}[n]\mathbf{y}^H[n]\}$ is the correlation matrix of $\mathbf{y}[n]$, and that

$$\mathbf{H}^H\widetilde{\boldsymbol{g}}^{[i+1]} = \frac{1}{C_{p,q}\{u[n]\}}\mathbf{d}_{ey}^{[i]} \quad \text{for any SNR} \tag{4.114}$$

where

$$\mathbf{d}_{ey}^{[i]} = \left(d_{ey}^{[i]}[L_1], d_{ey}^{[i]}[L_1+1], ..., d_{ey}^{[i]}[L_2] \right)^T \tag{4.115}$$

in which

$$d_{ey}^{[i]}[k] = \text{cum}\left\{ e^{[i]}[n] : p, \left(e^{[i]}[n]\right)^* : q-1, y^*[n-k] \right\} \tag{4.116}$$

and

$$e^{[i]}[n] = v^{[i]}[n] \star y[n] = \left(\boldsymbol{v}^{[i]}\right)^T \boldsymbol{y}[n] \tag{4.117}$$

is the equalized signal obtained at iteration $(i-1)$. Substituting (4.113) and (4.114) into (4.108) and (4.112) gives the set of update equations for the parameter vector $\boldsymbol{v}$ at iteration i as follows:

$$\begin{cases} \widetilde{\boldsymbol{v}}^{[i+1]} = \dfrac{\sigma_u^2}{C_{p,q}\{u[n]\}} \cdot \left(\mathbf{R}_y^*\right)^{-1} \mathbf{d}_{ey}^{[i]} \\ \boldsymbol{v}^{[i+1]} = \sqrt{\dfrac{\sigma_u^2}{\left(\widetilde{\boldsymbol{v}}^{[i+1]}\right)^H \mathbf{R}_y^* \widetilde{\boldsymbol{v}}^{[i+1]}}} \cdot \widetilde{\boldsymbol{v}}^{[i+1]} \end{cases} \tag{4.118}$$

To apply (4.118), we need (i) the autocorrelation function required by $\mathbf{R}_y$, (ii) the $(p+q)$th-order joint cumulant function required by $\mathbf{d}_{ey}^{[i]}$, (iii) the variance σ_u^2, and (iv) the $(p+q)$th-order cumulant $C_{p,q}\{u[n]\}$. Items (i) and (ii) can be estimated from the finite set of data $\{y[0], y[1], ..., y[N-1]\}$, whereas items (iii) and (iv) need be provided *a priori*. Moreover, if $\mathbf{H}$ is not of full column rank and SNR $= \infty$, then $\mathbf{R}_y$ is singular because $\text{rank}\{\mathbf{R}_y\} = \text{rank}\{\mathbf{H}^H\mathbf{H}\} = \text{rank}\{\mathbf{H}\}$ by (4.113). For this case, one can simply replace the inverse matrix $(\mathbf{R}_y^*)^{-1}$ in the first line of (4.118) by the pseudoinverse $(\mathbf{R}_y^*)^+$ obtained by virtue of the SVD (see Section 2.4).

Recall, from Section 4.1.1, that any scale factor is admissible for blind equalizers. We therefore can simplify the set of update equations (4.118) as follows:

$$\boldsymbol{v}^{[i+1]} = \frac{\left(\mathbf{R}_y^*\right)^{-1} \mathbf{d}_{ey}^{[i]}}{\left\| \left(\mathbf{R}_y^*\right)^{-1} \mathbf{d}_{ey}^{[i]} \right\|}, \tag{4.119}$$

which clearly does not require prior information about σ_u^2 and $C_{p,q}\{u[n]\}$. However, due to the lack of prior information about $C_{p,q}\{u[n]\}$, there may exist a phase ambiguity in $\boldsymbol{v}^{[i+1]}$ obtained at iteration i, or equivalently, a complex scale factor $e^{j\phi}$ between $\boldsymbol{v}^{[i+1]}$ and $\boldsymbol{v}^{[i]}$. Because of this, we use the following convergence rule to check the convergence properly:

$$\min_{\phi}\left\{\left\|\boldsymbol{v}^{[i+1]} - e^{j\phi}\boldsymbol{v}^{[i]}\right\|^2\right\} = 2 - 2\cdot\left|\left(\boldsymbol{v}^{[i+1]}\right)^H \boldsymbol{v}^{[i]}\right| < \zeta \tag{4.120}$$

or

$$\left|\left(\boldsymbol{v}^{[i+1]}\right)^H \boldsymbol{v}^{[i]}\right| > 1 - \frac{\zeta}{2} \tag{4.121}$$

where ζ is a positive real number preassigned for convergence tolerance. Table 4.6 summarizes the SE equalization algorithm based on (4.119) and (4.121). Note, from Table 4.6, that the inverse matrix $(\mathbf{R}_y^*)^{-1}$ in Step (S2) needs to be computed only once. This, together with the convergence analysis given by (4.102), indicates that the SE equalization algorithm is computationally efficient with a very fast convergence rate in ISI reduction in most circumstances. However, without an explicit objective function (of $y[n]$ or $e[n]$ or both), the SE equalization algorithm may sometimes diverge or have a slow convergence rate if the data length N is insufficient, the SNR is low, or the initial condition $\boldsymbol{v}^{[0]}$ is inappropriate.

Table 4.6 SE equalization algorithm

Parameter setting		Choose a cumulant order $(p+q) \geq 3$, an equalizer length $L = L_2 - L_1 + 1 > 1$, an initial parameter vector $\boldsymbol{v}^{[0]} \neq \mathbf{0}$, and a convergence tolerance $\zeta > 0$.
Steps	(S1)	Set the iteration number $i = 0$.
	(S2)	Estimate the correlation matrix $\mathbf{R}_y$ from the data $y[n]$, $n = 0, 1, ..., N-1$, and compute the inverse matrix $(\mathbf{R}_y^*)^{-1}$.
	(S3)	Compute the equalized signal $e^{[0]}[n] = (\boldsymbol{v}^{[0]})^T \boldsymbol{y}[n]$ and estimate the vector $\mathbf{d}_{ey}^{[0]}$ from $e^{[0]}[n]$ and $y[n]$, $n = 0, 1, ..., N-1$.
	(S4)	Update the parameter vector $\boldsymbol{v}$ at the ith iteration via $$\boldsymbol{v}^{[i+1]} = \frac{(\mathbf{R}_y^*)^{-1}\mathbf{d}_{ey}^{[i]}}{\left\|(\mathbf{R}_y^*)^{-1}\mathbf{d}_{ey}^{[i]}\right\|}.$$
	(S5)	If $$\left|\left(\boldsymbol{v}^{[i+1]}\right)^H \boldsymbol{v}^{[i]}\right| \leq 1 - \frac{\zeta}{2},$$ then go to Step (S6); otherwise, obtain the parameter vector $\boldsymbol{v} = \boldsymbol{v}^{[i+1]}$ for the SE equalizer $v_{\mathrm{SE}}[n]$.
	(S6)	Compute the equalized signal $e^{[i+1]}[n] = (\boldsymbol{v}^{[i+1]})^T \boldsymbol{y}[n]$ and estimate $\mathbf{d}_{ey}^{[i+1]}$ from $e^{[i+1]}[n]$ and $y[n]$, $n = 0, 1, ..., N-1$.
	(S7)	Update the iteration number i by $(i+1)$ and go to Step (S4).

Lattice Implementation

With the aid of a lattice LPE filter, we can implement the SE equalization algorithm in a lattice structure as depicted in Fig. 4.10.[8] The corresponding algorithm is called the *lattice super-exponential (lattice SE) equalization algorithm.*

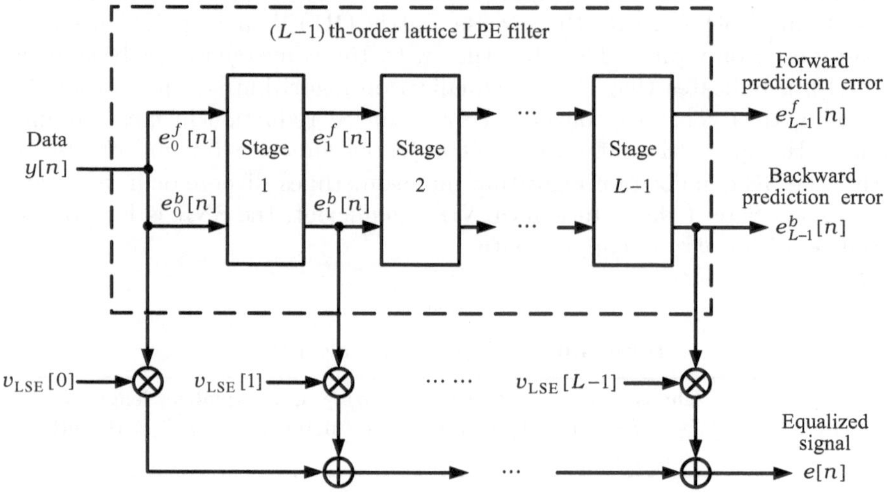

Fig. 4.10 Block diagram for the lattice SE equalization algorithm where each stage in the lattice LPE filter is the same as that in Fig. 4.8a

Specifically, we processing the data $y[n]$ by an $(L-1)$th-order backward LPE filter which is implemented with the lattice structure of $(L-1)$ stages. The resultant backward prediction errors $e_0^b[n], e_1^b[n], ..., e_{L-1}^b[n]$ are then processed by a multiple regression filter of length L. To find the coefficients of the multiple regression filter, denoted by $v_{\rm LSE}[n]$, $n = 0, 1, ..., L-1$, we apply the set of update equations (4.119) as follows. Let

$$\boldsymbol{v}_{\rm LSE} = (v_{\rm LSE}[0], v_{\rm LSE}[1], ..., v_{\rm LSE}[L-1])^T \tag{4.122}$$

denote the parameter vector to be found, and let

$$\mathbf{e}_b[n] = \left(e_0^b[n], e_1^b[n], ..., e_{L-1}^b[n]\right)^T \tag{4.123}$$

whose correlation matrix

$$\mathbf{R}_b \triangleq E\left\{\mathbf{e}_b[n]\mathbf{e}_b^H[n]\right\} = \mathrm{diag}\left\{\sigma_0^2, \sigma_1^2, ..., \sigma_{L-1}^2\right\} \tag{4.124}$$

[8] The structure shown in Fig. 4.10 is similar to that of a *joint-process estimator* [11].

(by Property 4.13) where σ_k^2 is the variance of $e_k^b[n]$. Using (4.119) with $y[n-k]$ replaced by $e_k^b[n]$, $\mathbf{R}_y$ replaced by $\mathbf{R}_b$ and $\boldsymbol{v}^{[i+1]}$ replaced by $\boldsymbol{v}_{\mathrm{LSE}}^{[i+1]}$, we have the set of update equations for finding $\boldsymbol{v}_{\mathrm{LSE}}$ at iteration i as follows:

$$\boldsymbol{v}_{\mathrm{LSE}}^{[i+1]} = \mathbf{d}_b^{[i]} / ||\mathbf{d}_b^{[i]}|| \tag{4.125}$$

where

$$\mathbf{d}_b^{[i]} = \left(d_b^{[i]}[0], d_b^{[i]}[1], ..., d_b^{[i]}[L-1] \right)^T \tag{4.126}$$

in which

$$d_b^{[i]}[k] = \frac{1}{\sigma_k^2} \cdot \mathrm{cum}\left\{ e^{[i]}[n] : p, \left(e^{[i]}[n] \right)^* : q-1, \left(e_k^b[n] \right)^* \right\} \tag{4.127}$$

and

$$e^{[i]}[n] = \sum_{k=0}^{L-1} v_{\mathrm{LSE}}^{[i]}[k] e_k^b[n] = \left(\boldsymbol{v}_{\mathrm{LSE}}^{[i]} \right)^T \mathbf{e}_b[n] \tag{4.128}$$

is the equalized signal obtained at iteration $(i-1)$. After convergence, we obtain the parameter vector $\boldsymbol{v}_{\mathrm{LSE}} = \boldsymbol{v}_{\mathrm{LSE}}^{[i+1]}$ for the multiple regression filter $v_{\mathrm{LSE}}[n]$. Table 4.7 gives the detailed steps of the lattice SE equalization algorithm. As a remark, without taking the parameter quantization effects into account, the lattice SE equalization algorithm has the same performance as the SE equalization algorithm [9, 36].

4.3.3 Algorithm Analyses

We have shown in Theorems 4.14 and 4.16 that, under the ideal condition of infinite SNR and doubly infinite equalizer, both the MNC equalization algorithm and the SE equalization algorithm (with an appropriate initial condition) give rise to the same solution $v[n] = \alpha h_{\mathrm{I}}[n-\tau]$, except for a scale factor and a time delay. Next, we further provide some insights into these algorithms, including their properties and relations, by considering the condition of finite SNR.[9] Again, for simplicity, we assume that the equalizer $v[n]$ is doubly infinite so that the analysis can be performed without taking the effect of finite-length truncation of $v[n]$ into account; for the finite-length effect, one can refer to [42].

[9] Analyses of the performance of the MNC and SE equalization algorithms and their relations can be found in a number of research works including [9, 37–41], to name a few.

Table 4.7 Lattice SE equalization algorithm

Parameter setting		Choose a cumulant order $(p+q) \geq 3$, an equalizer length $L > 1$, an initial parameter vector $\boldsymbol{v}_{\mathrm{LSE}}^{[0]} \neq \mathbf{0}$, and a convergence tolerance $\zeta > 0$.
Steps	(S1)	Set the iteration number $i = 0$.
	(S2)	Process the data $y[n]$, $n = 0, 1, ..., N-1$, by an $(L-1)$th-order lattice LPE filter to obtain the backward prediction error $e_k^b[n]$ and prediction-error variance σ_k^2 for $k = 0, 1, ..., L-1$.
	(S3)	Compute the equalized signal $e^{[0]}[n] = (\boldsymbol{v}_{\mathrm{LSE}}^{[0]})^T \mathbf{e}_b[n]$ and estimate the vector $\mathbf{d}_b^{[0]}$ from $e^{[0]}[n]$ and $e_k^b[n]$'s, $n = 0, 1, ..., N-1$.
	(S4)	Update the parameter vector $\boldsymbol{v}_{\mathrm{LSE}}$ at the ith iteration via $$\boldsymbol{v}_{\mathrm{LSE}}^{[i+1]} = \mathbf{d}_b^{[i]} / \|\mathbf{d}_b^{[i]}\|.$$
	(S5)	If $$\left| \left(\boldsymbol{v}_{\mathrm{LSE}}^{[i+1]}\right)^H \boldsymbol{v}_{\mathrm{LSE}}^{[i]} \right| \leq 1 - \frac{\zeta}{2},$$ then go to Step (S6); otherwise, obtain the parameter vector $\boldsymbol{v}_{\mathrm{LSE}} = \boldsymbol{v}_{\mathrm{LSE}}^{[i+1]}$ for the multiple regression filter $v_{\mathrm{LSE}}[n]$.
	(S6)	Compute the equalized signal $e^{[i+1]}[n] = (\boldsymbol{v}_{\mathrm{LSE}}^{[i+1]})^T \mathbf{e}_b[n]$ and estimate $\mathbf{d}_b^{[i+1]}$ from $e^{[i+1]}[n]$ and $e_k^b[n]$'s, $n = 0, 1, ..., N-1$.
	(S7)	Update the iteration number i by $(i+1)$ and go to Step (S4).

Properties of the MNC Equalizer

Like the LMMSE equalizer $v_{\mathrm{MS}}[n]$ (see Property 4.3), the MNC equalizer $v_{\mathrm{MNC}}[n]$ is also a perfect phase equalizer regardless of the value of SNR, as stated in the following property.

Property 4.17 (Linear Phase Property). *The phase response of the MNC equalizer $v_{\mathrm{MNC}}[n]$ is given by*

$$\arg[V_{\mathrm{MNC}}(\omega)] = -\arg[H(\omega)] + \omega\tau + \xi, \quad -\pi \leq \omega < \pi \tag{4.129}$$

where τ and ξ are real constants.

See Appendix 4A for the proof. This property states that the MNC equalizer $v_{\mathrm{MNC}}[n]$ completely cancels the phase distortion induced by the system $h[n]$ (up to a time delay τ and a constant phase shift ξ).

According to Property 4.17, let $g_{\mathrm{ZP}}[n]$ be a zero-phase system given by

$$g_{\mathrm{ZP}}[n] = e^{-j\xi} \cdot g_{\mathrm{MNC}}[n-\tau]. \tag{4.130}$$

Then, similar to $g_{\mathrm{MS}}[n]$ (see Property 4.4), the zero-phase system $g_{\mathrm{ZP}}[n]$ can be shown to possess the following property.

Property 4.18. *The zero-phase overall system $g_{\mathrm{ZP}}[n]$ defined as (4.130) is a legitimate autocorrelation function with (i) $g_{\mathrm{ZP}}[n] = g_{\mathrm{ZP}}^*[-n]$ and (ii) $g_{\mathrm{ZP}}[0] > |g_{\mathrm{ZP}}[n]|$ for all $n \neq 0$.*

This property reveals some insights into the signal component $e_{\mathrm{S}}[n]$ in the equalized signal (after MNC equalization). In particular, by expressing

$$e_{\mathrm{S}}[n] = e^{j\xi} \cdot \{g_{\mathrm{ZP}}[n] \star u[n+\tau]\} \qquad \text{(by (4.9) and (4.130)}, \tag{4.131}$$

we observe that if the source signal $u[n]$ consists of widely separated spikes, then the corresponding $e_{\mathrm{S}}[n]$ will be composed of symmetric wavelets (zero-phase patterns) with amplitudes proportional to $|u[n+\tau]|$. Meanwhile, the resolution of $e_{\mathrm{S}}[n]$ is determined by the width of the wavelet $g_{\mathrm{ZP}}[n]$. As we will see, these observations are of particular use to interpretation of equalized signals in seismic exploration.

To analyze SNR improvement or degradation after MNC equalization, we rewrite $J_{p,q}(v[n])$ given by (4.86) as a function of the equalizer $v[n]$ and the SNR improvement-or-degradation ratio $\rho\{v[n]\}$ as follows:

$$J_{p,q}(v[n]) = J_{p,q}^{(\infty)}(v[n]) \cdot \frac{1}{\left[1 + \dfrac{1}{\rho\{v[n]\} \cdot \mathrm{SNR}}\right]^{(p+q)/2}}. \tag{4.132}$$

From (4.132), it follows that $J_{p,q}(v[n])$ is always smaller than $J_{p,q}^{(\infty)}(v[n])$ for finite SNR since both SNR and $\rho\{v[n]\}$ are positive real numbers. One can also observe that the MNC equalizer $v_{\mathrm{MNC}}[n]$ associated with $J_{p,q}(v[n])$ partly maximizes $J_{p,q}^{(\infty)}(v[n])$ for the ISI reduction (as indicated by Theorem 4.14) and partly maximizes $\rho\{v[n]\}$ for noise reduction in the meantime. In other words, as the LMMSE equalizer $v_{\mathrm{MS}}[n]$, the MNC equalizer $v_{\mathrm{MNC}}[n]$ also performs noise reduction besides the ISI reduction. This, in turn, implies that as the former (see Theorem 4.7), the latter also performs noise reduction better than the ZF equalizer $v_{\mathrm{ZF}}[n]$, as exhibited by the following theorem.

Theorem 4.19. *Under Assumptions* (A4-5) *through* (A4-8) *and the condition of doubly infinite equalizer, the SNR improvement-or-degradation ratio $\rho\{v_{\mathrm{MNC}}[n]\} \geq \rho\{v_{\mathrm{ZF}}[n]\}$ for any SNR.*

We leave the proof as an exercise (Problem 4.10). Furthermore, a property of $\rho\{v_{\mathrm{MNC}}[n]\}$ is as follows.

Property 4.20. *Under Assumptions* (A4-5) *through* (A4-8) *and the condition of doubly infinite equalizer, the SNR improvement-or-degradation ratio $\rho\{v_{\mathrm{MNC}}[n]\}$ always increases as the SNR decreases.*

Refer to [9, 38] for inferences of this property. This property states that the lower the SNR, the more the MNC equalizer $v_{\mathrm{MNC}}[n]$ performs as a noise-reduction filter.

As for the properties of the SE equalizer $v_{\mathrm{SE}}[n]$, due to the lack of an explicit objective function, the analysis is far more difficult than that for the MNC equalizer $v_{\mathrm{MNC}}[n]$. Because of this, rather than giving the property analysis, we provide an analysis of the relationship between $v_{\mathrm{SE}}[n]$ and $v_{\mathrm{MNC}}[n]$, through which one can realize that $v_{\mathrm{SE}}[n]$ also shares the properties of $v_{\mathrm{MNC}}[n]$ under certain conditions.

Relationship between MNC and SE Equalizers

Let us, first, establish a connection between the MNC equalizer $v_{\mathrm{MNC}}[n]$ and the LMMSE equalizer $v_{\mathrm{MS}}[n]$. By setting the first derivative $\partial J_{p,q}(\boldsymbol{v})/\partial \boldsymbol{v}^*$ given by (4.93) to zero and using Theorem 3.39, all the stationary points of $J_{p,q}(v[n])$ can be shown to satisfy

$$q \cdot \frac{C_{p,q}\{u[n]\}}{C_{p,q}\{e[n]\}} \sum_{n=-\infty}^{\infty} \widetilde{g}_{p,q}[n] h^*[n-k] + p \cdot \frac{C_{q,p}\{u[n]\}}{C_{q,p}\{e[n]\}} \sum_{n=-\infty}^{\infty} \widetilde{g}_{q,p}[n] h^*[n-k]$$
$$= \frac{p+q}{\sigma_e^2} \left\{ \sigma_u^2 \sum_{n=-\infty}^{\infty} g[n] h^*[n-k] + \sum_{n=-\infty}^{\infty} v[n] r_w^*[n-k] \right\} \quad (4.133)$$

where

$$\widetilde{g}_{p,q}[n] \triangleq (g[n])^p \, (g^*[n])^{q-1} . \quad (4.134)$$

Taking the DTFT of (4.133) with respect to the index k yields

$$q \cdot \frac{C_{p,q}\{u[n]\}}{C_{p,q}\{e[n]\}} \cdot \widetilde{G}_{p,q}(\omega) H^*(\omega) + p \cdot \frac{C_{q,p}\{u[n]\}}{C_{q,p}\{e[n]\}} \cdot \widetilde{G}_{q,p}(\omega) H^*(\omega)$$
$$= \frac{p+q}{\sigma_e^2} \left[\sigma_u^2 \left| H(\omega) \right|^2 + S_w(\omega) \right] \cdot V(\omega). \quad (4.135)$$

The following theorem then follows from (4.135) and the condition that $|H(\omega)| \neq 0$ for all $\omega \in [-\pi, \pi)$ (by Assumption (A4-5)).

Theorem 4.21 (Relation between the MNC and LMMSE Equalizers). *Under Assumptions* (A4-5) *through* (A4-8) *and the condition of doubly infinite equalizer, the MNC equalizer $v_{\mathrm{MNC}}[n]$ associated with $J_{p,q}(v[n])$ is related to the LMMSE equalizer $v_{\mathrm{MS}}[n]$ via*

$$V_{\mathrm{MNC}}(\omega) = \left[\alpha_{p,q} \widetilde{G}_{p,q}(\omega) + \alpha_{q,p} \widetilde{G}_{q,p}(\omega) \right] \cdot V_{\mathrm{MS}}(\omega), \quad -\pi \leq \omega < \pi \quad (4.136)$$

where $\widetilde{G}_{p,q}(\omega) = \mathscr{F}\{\widetilde{g}_{p,q}[n]\}$ is given by (4.134),

$$\alpha_{p,q} \triangleq \frac{q}{p+q} \cdot \frac{\sigma_e^2}{\sigma_u^2} \cdot \frac{C_{p,q}\{u[n]\}}{C_{p,q}\{e[n]\}} \tag{4.137}$$

is a real or complex constant, and

$$V_{\mathrm{MS}}(\omega) = \frac{\sigma_u^2 \cdot H^*(\omega)}{\sigma_u^2 \cdot |H(\omega)|^2 + S_w(\omega)} \tag{4.138}$$

is the frequency response of the LMMSE equalizer $v_{\mathrm{MS}}[n]$.

Similarly, we establish the connection between the SE equalizer $v_{\mathrm{SE}}[n]$ and the LMMSE equalizer $v_{\mathrm{MS}}[n]$ as follows. Since the normalization operation in the second line of (4.118) is immaterial to this connection, we consider only the first line of (4.118) for simplicity. After convergence, the SE equalizer $v_{\mathrm{SE}}[n]$ obtained from (4.118) can be established to satisfy

$$E\{e[n]y^*[n-k]\} = \frac{\sigma_u^2}{C_{p,q}\{u[n]\}} \cdot \mathrm{cum}\{e[n]:p, e^*[n]:q-1, y^*[n-k]\}, \tag{4.139}$$

which can be further shown to be

$$\sigma_u^2 \sum_{n=-\infty}^{\infty} g[n]h^*[n-k] + \sum_{n=-\infty}^{\infty} v[n]r_w^*[n-k] = \sigma_u^2 \sum_{n=-\infty}^{\infty} \widetilde{g}_{p,q}[n]h^*[n-k] \tag{4.140}$$

where $\widetilde{g}_{p,q}[n]$ is given by (4.134). By taking the DTFT of (4.140) with respect to the index k and under the condition that $|H(\omega)| \neq 0$ for all $\omega \in [-\pi, \pi)$, we obtain the following theorem.

Theorem 4.22 (Relation between the SE and LMMSE Equalizers). *Under Assumptions* (A4-5) *through* (A4-8) *and the condition of doubly infinite equalizer, the SE equalizer $v_{\mathrm{SE}}[n]$ associated with the cumulant-order parameters (p,q) is related to the LMMSE equalizer $v_{\mathrm{MS}}[n]$ via*

$$V_{\mathrm{SE}}(\omega) = \widetilde{G}_{p,q}(\omega)V_{\mathrm{MS}}(\omega), \quad -\pi \leq \omega < \pi \tag{4.141}$$

where $\widetilde{G}_{p,q}(\omega) = \mathscr{F}\{\widetilde{g}_{p,q}[n]\}$ is given by (4.134) and $V_{\mathrm{MS}}(\omega)$ is given by (4.138).

For clarity, let us further denote the MNC equalizer $V_{\mathrm{MNC}}(\omega)$ associated with $J_{p,q}(v[n])$ by $V_{\mathrm{MNC}}^{[p,q]}(\omega)$, and the SE equalizer $V_{\mathrm{SE}}(\omega)$ associated with the cumulant-order parameters (p,q) by $V_{\mathrm{SE}}^{[p,q]}(\omega)$. Then, as a result of Theorems 4.21 and 4.22, we have the following theorem for the relation between the MNC and SE equalizers.

Theorem 4.23 (Relation between the MNC and SE Equalizers). *Under Assumptions* (A4-5) *through* (A4-8) *and the condition of doubly infinite equalizer, the MNC equalizer* $V_{\mathrm{MNC}}^{[p,q]}(\omega)$ *is related to the SE equalizer* $V_{\mathrm{SE}}^{[p,q]}(\omega)$ *via*

$$V_{\mathrm{MNC}}^{[p,q]}(\omega) = V_{\mathrm{MNC}}^{[q,p]}(\omega) = \alpha_{p,q} V_{\mathrm{SE}}^{[p,q]}(\omega) + \alpha_{q,p} V_{\mathrm{SE}}^{[q,p]}(\omega), \quad -\pi \leq \omega < \pi \tag{4.142}$$

where $\alpha_{p,q}$ *is a real or complex constant given by (4.137).*

One can observe, from Theorem 4.22 and (4.134), that as $y[n]$ is real, $\widetilde{g}_{p,q}[n] = \widetilde{g}^{p+q-1}[n] = \widetilde{g}_{q,p}[n]$ and accordingly $V_{\mathrm{SE}}^{[p,q]}(\omega) = V_{\mathrm{SE}}^{[q,p]}(\omega)$. As a result of this observation and Theorem 4.23, we have the following corollary.

Corollary 4.24 (Equivalence of the MNC and SE Equalizers). *Under Assumptions* (A4-5) *through* (A4-8) *and the condition of doubly infinite equalizer, the MNC equalizer* $V_{\mathrm{MNC}}^{[p,q]}(\omega)$ *is equivalent to the SE equalizer* $V_{\mathrm{SE}}^{[p,q]}(\omega)$ *if (i)* $y[n]$ *is real or (ii)* $y[n]$ *is complex and* $p = q$.

Let us emphasize that in addition to the optimum solution $v_{\mathrm{MNC}}[n]$, all the other stationary points of $J_{p,q}(v[n])$ also possess the relations given by Theorems 4.21 and 4.23, and the equivalence given by Corollary 4.24.

4.3.4 Algorithm Improvements

We may further improve the convergence rate and the reliability of the MNC and SE equalization algorithms by means of whitening preprocessing, better initial condition, and a hybrid framework, as described below.

Whitening Preprocessing

Recall, from Section 4.2.4, that when the data $y[n]$ are processed by the forward LPE filter $a_M[n]$ of sufficient order M, the resultant forward prediction error (the predictive deconvolved signal)

$$e_M^f[n] = a_M[n] \star y[n] \tag{4.143}$$

is basically an amplitude equalized signal. This fact is useful to blind equalization, as illustrated in Fig. 4.11. The whitening filter $a_M[n]$ simplifies the task of the blind equalizer $v[n]$ because only the phase distortion, together with some residual amplitude distortion, in $e_M^f[n]$ remains to be compensated by $v[n]$, thereby speeding up the blind equalization algorithm used.

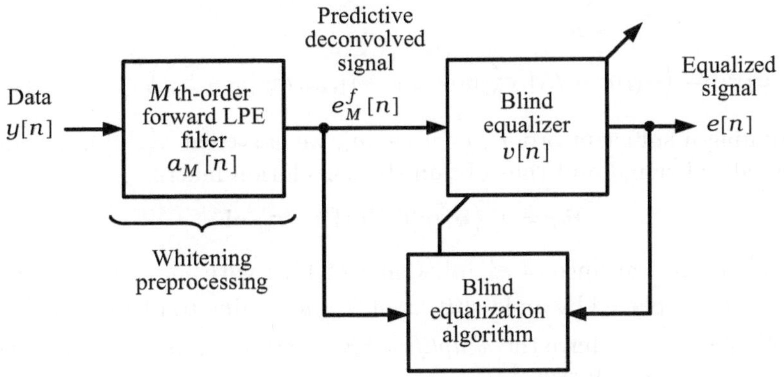

Fig. 4.11 Blind equalization with whitening preprocessing

Improved Initial Condition

Required by iterative blind equalization algorithms, the initial condition for $v[n]$ is usually chosen as $v^{[0]}[n] = \delta[n - L_C]$ for the sake of simplicity where $L_C = \lfloor (L_1 + L_2)/2 \rfloor$.[10] In some cases, a better initial condition may be needed to further improve the convergence rate as well as the reliability of the iterative blind equalization algorithm used. Next, let us present a low-complexity procedure for obtaining a better initial condition [9, 36].

The basic idea is to obtain the initial condition $v^{[0]}[n]$ by virtue of whitening preprocessing along with some simplification of the SE equalization algorithm, as depicted in Fig. 4.12. Specifically, the data $y[n]$ are preprocessed by an Mth-order forward LPE filter $a_M[n]$ where the length $M + 1 < L = L_2 - L_1 + 1$. The resultant predictive deconvolved signal $e_M^f[n]$, as given by (4.143), is then processed by the simplified blind equalizer $v_S[n]$, which has to be an FIR filter with $v_S[n] = 0$ outside the range $L_1 \le n \le \widetilde{L}_2$ where $\widetilde{L}_2 = L_2 - M$ such that the initial condition

$$v^{[0]}[n] = v_S[n] \star a_M[n] \tag{4.144}$$

to be obtained has the domain of support $L_1 \le n \le L_2$. We obtain the equalizer $v_S[n]$ by simplifying the set of update equations (4.119) as follows. Let

$$\boldsymbol{v}_S = \left(v_S[L_1], v_S[L_1 + 1], ..., v_S[\widetilde{L}_2] \right)^T \tag{4.145}$$

denote the parameter vector to be found, and the equalized signal

[10] The notation '$\lfloor x \rfloor$' represents the largest integer no larger than x.

$$e[n] = v_S[n] \star e_M^f[n] = \boldsymbol{v}_S^T \mathbf{e}_f[n] \tag{4.146}$$

where

$$\mathbf{e}_f[n] = \left(e_M^f[n-L_1], e_M^f[n-L_1-1], ..., e_M^f[n-\widetilde{L}_2]\right)^T. \tag{4.147}$$

By assuming a sufficient order M for $a_M[n]$, we can treat $e_M^f[n]$ as an amplitude equalized signal and thus obtain the correlation matrix

$$\mathbf{R}_f \triangleq E\left\{\mathbf{e}_f[n]\mathbf{e}_f^H[n]\right\} \approx \sigma_M^2 \cdot \mathbf{I} \tag{4.148}$$

where σ_M^2 is the variance of $e_M^f[n]$. Using (4.119) with $y[n-k]$ replaced by $e_M^f[n-k]$, $\mathbf{R}_y$ replaced by an identity matrix (according to (4.148)) and $\boldsymbol{v}^{[i+1]}$ replaced by $\boldsymbol{v}_S^{[i+1]}$, we have the *simplified SE equalization algorithm* for finding $\boldsymbol{v}_S$ at iteration i as follows:

$$\boldsymbol{v}_S^{[i+1]} = \mathbf{d}_f^{[i]} / ||\mathbf{d}_f^{[i]}|| \tag{4.149}$$

where

$$\mathbf{d}_f^{[i]} = \text{cum}\left\{e^{[i]}[n] : p, \left(e^{[i]}[n]\right)^* : q-1, \mathbf{e}_f^*[n]\right\} \tag{4.150}$$

and

$$e^{[i]}[n] = v_S^{[i]}[n] \star e_M^f[n] = \left(\boldsymbol{v}_S^{[i]}\right)^T \mathbf{e}_f[n] \tag{4.151}$$

is the equalized signal obtained at iteration $(i-1)$.

To further reduce the computational complexity, we perform the set of update equations (4.149) only once to obtain the parameter vector $\boldsymbol{v}_S$. In particular, with the initial condition $v_S^{[0]}[n] = \delta[n-\widetilde{L}_C]$ where

$$\widetilde{L}_C = \left\lfloor \frac{L_1 + \widetilde{L}_2}{2} \right\rfloor = \left\lfloor \frac{L_1 + L_2 - M}{2} \right\rfloor, \tag{4.152}$$

we have $e^{[0]}[n] = v_S^{[0]}[n] \star e_M^f[n] = e_M^f[n-\widetilde{L}_C]$, which further leads to

$$\mathbf{d}_f \triangleq \mathbf{d}_f^{[0]} = \text{cum}\left\{e_M^f[n-\widetilde{L}_C] : p, \left(e_M^f[n-\widetilde{L}_C]\right)^* : q-1, \mathbf{e}_f^*[n]\right\} \tag{4.153}$$

and

$$\boldsymbol{v}_S = \boldsymbol{v}_S^{[1]} = \mathbf{d}_f^{[0]} / ||\mathbf{d}_f^{[0]}|| = \mathbf{d}_f / ||\mathbf{d}_f||. \tag{4.154}$$

As a consequence, we obtain an improved initial condition $v^{[0]}[n]$ through whitening preprocessing followed by one-iteration operation of the simplified SE equalization algorithm. Table 4.8 summarizes the steps to obtain the improved initial condition $v^{[0]}[n]$. As a remark, if we apply the procedure of Table 4.8 to the lattice SE equalization algorithm, the forward LPE filter $a_M[n]$ needed by this procedure happens to be a by-product of the $(L-1)$th-order lattice LPE filter needed by the lattice SE equalization algorithm since $M < L-1$, thereby saving implementation complexity to obtain the improved initial condition $v^{[0]}[n]$.

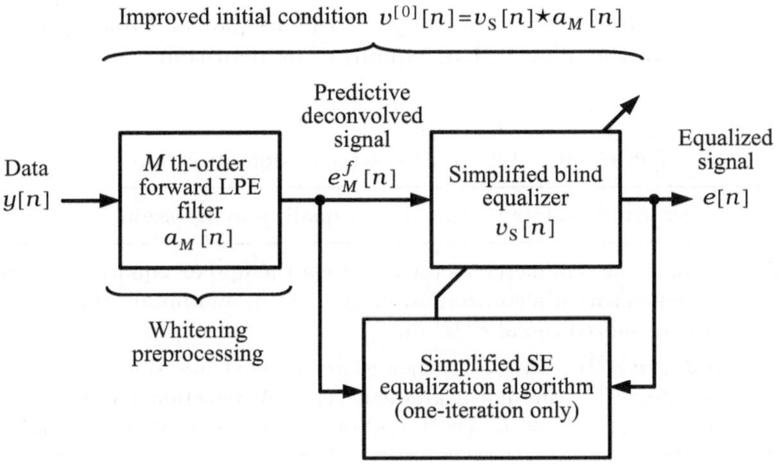

Fig. 4.12 Block diagram for obtaining the improved initial condition $v^{[0]}[n]$

Table 4.8 Steps for improved initial condition

Parameter setting		Choose a cumulant order $(p+q) \geq 3$, an equalizer length $L = L_2 - L_1 + 1 > 1$, and an LPE filter order $M < L - 1$.
Steps	(S1)	Preprocess the data $y[n]$, $n = 0, 1, ..., N-1$, by an Mth-order forward LPE filter $a_M[n]$ to obtain the predictive deconvolved signal $e_M^f[n]$, and form the vector $\mathbf{e}_f[n] = (e_M^f[n-L_1], e_M^f[n-L_1-1], ..., e_M^f[n-L_2+M])^T$.
	(S2)	Estimate the vector $\mathbf{d}_f$ from $e_M^f[n-\widetilde{L}_C]$ and $\mathbf{e}_f[n]$, $n = 0, 1, ..., N-1$, where $\widetilde{L}_C = \lfloor (L_1 + L_2 - M)/2 \rfloor$.
	(S3)	Obtain the parameter vector $\boldsymbol{v}_S = \mathbf{d}_f / \|\mathbf{d}_f\|$ for the simplified blind equalizer $v_S[n]$.
	(S4)	Obtain the initial condition $v^{[0]}[n] = v_S[n] \star a_M[n]$.

Hybrid Framework

Mboup and Regalia [43, 44] have reported that as SNR $= \infty$ and the data length $N = \infty$, the SE equalization algorithm for $p = q$ can be interpreted as a particular gradient-type optimization method for finding the maximum of the MNC criterion $J_{p,p}(v[n])$, and that the corresponding step size at each iteration is optimal for convergence speed. This interpretation, plus the equivalence stated in Corollary 4.24, therefore suggests a hybrid framework of MNC and SE equalization algorithms to obtain the MNC equalizer $v_{\text{MNC}}[n]$ in a collaborative and complementary manner, especially for the conditions of finite SNR and finite N [40]. The resultant algorithm, referred to as the *hybrid MNC equalization algorithm*, is summarized in Table 7.4 and illustrated in Fig. 4.13. Note that we can use the same convergence rule as that in Table 4.5 to check the convergence of the hybrid MNC equalization algorithm.

Table 4.9 Hybrid MNC equalization algorithm

Procedure to Obtain the MNC Equalizer at Iteration i
(T1) Update the parameter vector $\boldsymbol{v}^{[i+1]}$ via the update equations of the SE equalization algorithm given by (4.119), and obtain the associated equalized signal $e^{[i+1]}[n]$.
(T2) If $J_{p,q}(\boldsymbol{v}^{[i+1]}) > J_{p,q}(\boldsymbol{v}^{[i]})$, then go to the next iteration; otherwise, update $\boldsymbol{v}^{[i+1]}$ through a gradient-type optimization method such that $J_{p,q}(\boldsymbol{v}^{[i+1]}) > J_{p,q}(\boldsymbol{v}^{[i]})$, and obtain the associated $e^{[i+1]}[n]$.

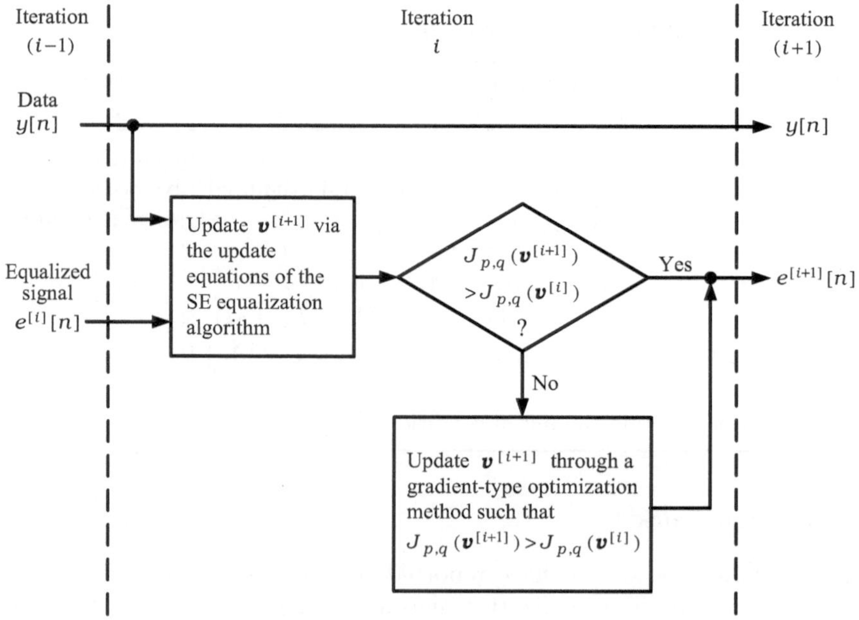

Fig. 4.13 Signal processing procedure for the hybrid MNC equalization algorithm

It is important to emphasize that according to Corollary 4.24, the hybrid MNC equalization algorithm is applicable only for the case of real $y[n]$ as well as the case of complex $y[n]$ and $p = q$. For these two cases, it follows, from

(4.93), (4.94), (4.115), (4.116) and (4.117), that the first derivative required by (T2) of Table 7.4 can be expressed as follows:

$$\left.\frac{\partial J_{p,q}(\boldsymbol{v})}{\partial \boldsymbol{v}^*}\right|_{\boldsymbol{v}\,=\,\boldsymbol{v}^{[i]}} = \frac{p+q}{2}\cdot J_{p,q}(\boldsymbol{v}^{[i]})\cdot\left\{\frac{\mathbf{d}_{ey}^{[i]}}{C_{p,q}\left\{e^{[i]}[n]\right\}} - \frac{\mathbf{R}_y^*\cdot\boldsymbol{v}^{[i]}}{E\left\{\left|e^{[i]}[n]\right|^2\right\}}\right\}. \tag{4.155}$$

Since both the correlation matrix $\mathbf{R}_y$, which only needs to be computed once (at the first iteration), and the vector $\mathbf{d}_{ey}^{[i]}$ needed in (T1) of Table 7.4 are already available for use in (T2), computation of (4.155) for (T2) is straightforward and simple. Moreover, in comparison with the MNC equalization algorithm, the hybrid MNC equalization algorithm exhibits fast convergence and significant computational saving by taking advantage of the SE equalization algorithm in (T1). Furthermore, it does not suffer from the divergence problem of the SE equalization algorithm due to the guaranteed convergence of (T2).

4.4 Simulation Examples for Algorithm Tests

Let us provide some simulation examples to show the efficacy of the linear prediction approach, the MNC, SE and hybrid MNC equalization algorithms, as well as their performance improvements (in terms of convergence speed and amount of ISI reduction) thanks to the whitening preprocessing and improved initial condition. For convenience, we use the following shorthand notation.

- "LPE" represents the simulation results for predictive deconvolution (equalization).
- "J3" and "J4" represent the simulation results for MNC equalization with $J_3(v[n])$ and $J_4(v[n])$, respectively.
- "SE3" and "SE4" represent the simulation results for SE equalization with $p+q=3$ and 4, respectively.
- "LPE-J4" represents the simulation results for MNC equalization with $J_4(v[n])$ and whitening preprocessing.
- "LPE-SE4" represents the simulation results for SE equalization with $p+q=4$ and whitening preprocessing.
- "ΔISI" represents the amount of ISI before equalization minus that after equalization, i.e. the ISI improvement or degradation.

Note that the larger the value of ΔISI, the better the performance of the blind equalization algorithm under test. Next, let us turn to the first example.

Example 4.25 (Efficacy of Predictive, MNC and SE Equalization)
In this example, the source signal $u[n]$ was assumed to be a real, zero-mean, exponentially distributed, i.i.d., stationary sequence with variance $\sigma_u^2 = 1$, skewness $C_3\{u[n]\} = 2$ and kurtosis $C_4\{u[n]\} = 6$. The following two cases were considered.

Case A. The system $h[n]$ was a minimum-phase causal FIR filter with coefficients $\{1, -0.7, 0.6, -0.4, 0.3, 0.1, 0.05\}$.

Case B. The system $h[n]$ was a nonminimum-phase ARMA(2,2) filter whose transfer function is given by

$$H(z) = \frac{1 - 1.6z^{-1} + 0.1z^{-2}}{1 + 0.1z^{-1} - 0.12z^{-2}}. \tag{4.156}$$

The data $y[n]$ were generated using (4.1) and (4.2) for data length $N = 4096$ and SNR = 20 dB where the noise $w[n]$ was assumed to be real white Gaussian. For predictive deconvolution, the Burg algorithm was used to obtain the forward LPE filter of order equal to 24. For MNC and SE equalization, the MNC equalization algorithm using the approximate BFGS method and the SE equalization algorithm were used to find the MNC equalizer $v_{\rm MNC}[n]$ and the SE equalizer $v_{\rm SE}[n]$, respectively, where $L_1 = 0$, $L_2 = 24$, and the typical initial condition $v^{[0]}[n] = \delta[n-12]$ were used. Thirty independent runs were performed for these algorithms.

Table 4.10 shows the averages of the thirty ΔISI's obtained (in dB) and the averaged number of iterations spent. Note that the amount of ISI before equalization is 0.4630 dB for Case A and −3.7536 dB for Case B. One can see, from Table 4.10, that the linear prediction approach outperforms the MNC and SE equalization algorithms for Case A (minimum-phase system), but failed in equalization for Case B (nonminimum-phase system). On the other hand, the MNC and SE equalization algorithms work well for both cases. Moreover, the MNC and SE equalization algorithms using third-order cumulants exhibit better performance than those using fourth-order cumulants since estimation of third-order cumulants is more accurate than that of fourth-order cumulants for finite data length. One can also observe that the convergence rate of the SE equalization algorithm is about three times faster than that of the MNC equalization algorithm.

□

Table 4.10 Simulation results of Example 4.25

		LPE	J3	J4	SE3	SE4
Case A	ΔISI (dB)	22.2788	20.1553	14.2476	20.1473	14.2467
	Number of iterations		28.1667	30.5333	9.0333	10.9667
Case B	ΔISI (dB)	–5.1241	17.0450	10.5202	17.0400	10.5229
	Number of iterations		29.6667	32.0333	8.0000	9.9333

Example 4.26 (Efficacy of Whitening Preprocessing)
The simulation conditions used in this example were the same as those in Example 4.25, except that the system $h[n]$ was a nonminimum-phase causal FIR filter with coefficients $\{0.4, 1, -0.7, 0.6, 0.3, -0.4, 0.1\}$. Algorithms used for predictive, MNC and SE equalization were also the same as those in Example 4.25. To maintain the same equalizer length for blind equalization in conjunction with whitening preprocessing, we used a forward LPE filter $a_M[n]$ of order $M = 7$ as the whitening filter followed by the MNC equalizer $v_{\text{MNC}}[n]$ or the SE equalizer $v_{\text{SE}}[n]$, both with $L_1 = 0$, $L_2 = 17$, and $v^{[0]}[n] = \delta[n-8]$. Thirty independent runs were performed for all these algorithms. Table 4.11 shows the averages of the thirty ΔISI's obtained (in dB) and the averaged number of iterations spent, where the amount of ISI before equalization is 1.0380 dB. It is clear, from Table 4.11, that whitening preprocessing not only improves the performance of the MNC and SE equalization algorithms with larger ΔISI, but also speeds up the two algorithms (with fewer iterations spent).

□

Table 4.11 Simulation results of Example 4.26

	LPE	J4	LPE-J4	SE4	LPE-SE4
ΔISI (dB)	7.0929	14.7092	16.2485	14.7111	16.2475
Number of iterations		25.0667	15.4667	11.1000	8.6667

Example 4.27 (Efficacy of Improved Initial Condition)
In this example, the source signal $u[n]$ was assumed to be a 4-QAM symbol sequence with unity variance and the system $h[n]$ was a nonminimum-phase ARMA(3,3) filter whose transfer function is given by

$$H(z) = \frac{1 + 0.1z^{-1} - 3.2725z^{-2} + 1.41125z^{-3}}{1 - 1.9z^{-1} + 1.1525z^{-2} - 0.1625z^{-3}}. \tag{4.157}$$

Figure 4.14 reveals the efficacy of the improved initial condition $v^{[0]}[n] = v_{\text{S}}[n] \star a_M[n]$ versus that of the typical initial condition $v^{[0]}[n] = \delta[n - L_{\text{C}}]$, where SNR = 20 dB, $L_1 = 0$, $L_2 = 20$, $M = 4$, and $L_{\text{C}} = 10$. From this figure, one can see that the values of ISI$\{g^{[0]}[n]\}$ associated with the improved initial condition are about 10 dB below those associated with the typical initial condition. Moreover, the SE equalization algorithm using the improved initial condition works well for all data lengths N with much faster convergence speed than that using the typical initial condition, and the latter converges more slowly for smaller N (1024, 2048 and 4096) and diverges for $N = 512$. These results demonstrate that the improved initial condition significantly improves the convergence speed and reliability of the SE equalization algorithm.

□

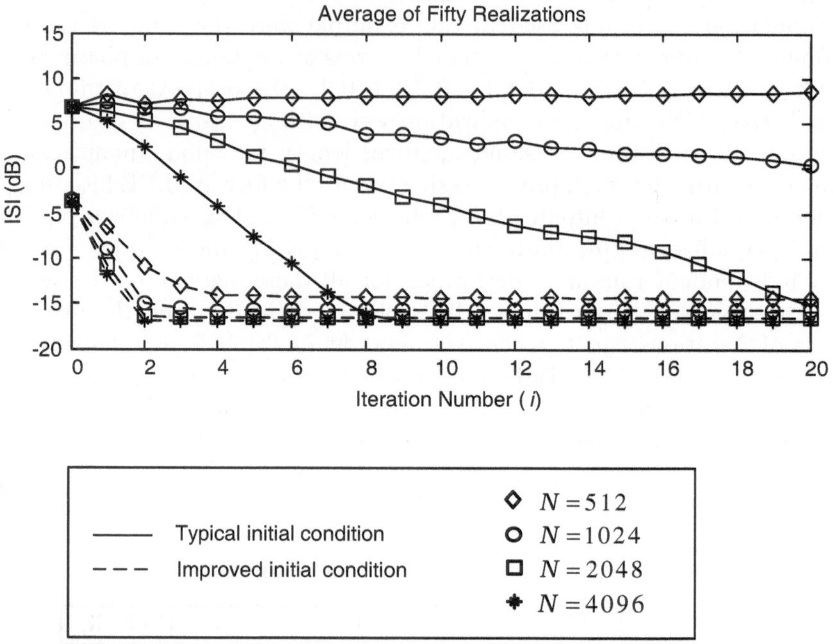

Fig. 4.14 Averages of fifty ISI$\{g^{[i]}[n]\}$s using the SE equalization algorithm

Example 4.28 (Efficacy of Hybrid MNC Equalization)
In this example, we considered the following two cases for generating the data $y[n]$.

Case A. The source signal $u[n]$ was a 16-QAM symbol sequence with unity variance. The system $h[n]$ was a fifth-order causal FIR filter whose coefficients were assumed to be uniformly distributed over the interval $[-0.5, 0.5]$ and were statistically independent of each other. Thirty independent runs with thirty independent realizations of $h[n]$ were performed to generate thirty sets of $y[n]$, $n = 0, 1, ..., N-1$, for $N = 10000$ and SNR $= \infty$.

Case B. The source signal $u[n]$ was a 4-QAM symbol sequence with unity variance and the system $h[n]$ was the same as that in Example 4.27. Thirty independent runs were performed to generate thirty sets of $y[n]$, $n = 0, 1, ..., N-1$, for $N = 2048$ and SNR $= 15$ dB where the noise $w[n]$ was real white Gaussian.

For Case A, the MNC equalization algorithm using the steepest descent method, the one using the approximate BFGS method, and the hybrid MNC equalization algorithm using the steepest descent method were used to obtain the MNC equalizer $v_{\mathrm{MNC}}[n]$ associated with $J_{2,2}(v[n])$, where $L_1 = 0$, $L_2 = 29$, and the typical initial condition $v^{[0]}[n] = \delta[n-14]$ were used. Figure 4.15 reveals the convergence speed of the hybrid MNC equalization algorithm versus that of the MNC equalization algorithm. Obviously, the former is much faster than the latter, thanks to utilization of the SE equalization algorithm in (T1) of Table 7.4. Figure 4.15 also verifies that the convergence speed of the MNC equalization algorithm using the approximate BFGS method is much faster than that using the steepest descent method.

For Case B, the SE equalization algorithm with $p = q = 2$ was used to obtain the SE equalizer $v_{\mathrm{SE}}[n]$, and the hybrid MNC equalization algorithm was used to obtain the MNC equalizer $v_{\mathrm{MNC}}[n]$ associated with $J_{2,2}(v[n])$, where $L_1 = 0$, $L_2 = 20$, and $v^{[0]}[n] = \delta[n-10]$. Figures 4.16a, b show thirty $J_{2,2}(v_{\mathrm{SE}}[n])$s and thirty $J_{2,2}(v_{\mathrm{MNC}}[n])$s, respectively. These two figures demonstrate the robustness of the hybrid MNC equalization algorithm against the divergence problem faced by the SE equalization algorithm, thanks to the explicit objective function $J_{p,q}(v[n])$ used in (T2) of Table 7.4. □

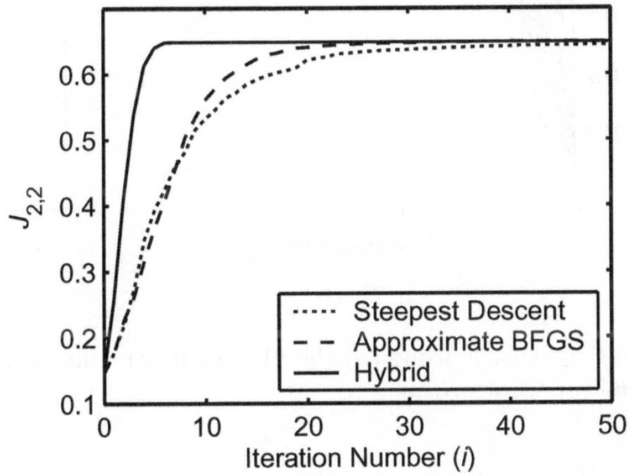

Fig. 4.15 Averages of thirty $J_{2,2}(v[n])$s using the MNC equalization algorithm (the dashed lines) and the hybrid MNC equalization algorithm (the solid line)

4.5 Some Applications

Having elaborated on the principles of SOS and HOS based blind equalization algorithms, we are now ready to introduce the applications of these algorithms.

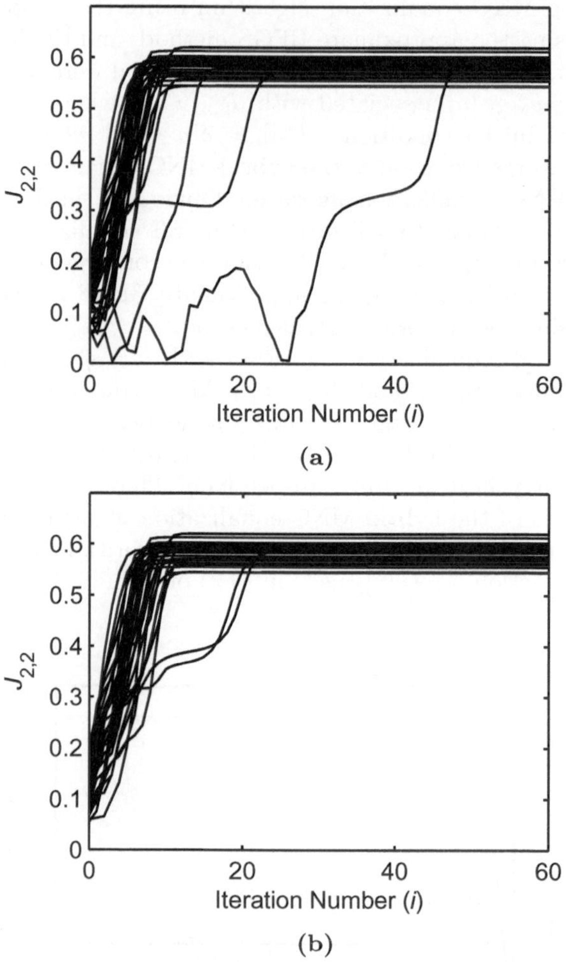

Fig. 4.16 Thirty $J_{2,2}(v[n])$s using (**a**) the SE equalization algorithm and (**b**) the hybrid MNC equalization algorithm

Our treatment will be confined to the following selected applications: seismic exploration, speech signal processing, and digital communications.

4.5.1 Seismic Exploration

Seismic exploration is concerned with identification of the earth's subsurface structure, and is often used when prospecting for potential reservoirs of petroleum and natural gas. We may divide geophysical techniques for seismic exploration into two branches: *reflection seismology* and *refraction seismology*. Reflection seismology is a traditional technique used to generate and interpret the earth's *reflection profile* which contains a certain amount of information about the earth's subsurface structure. Thus, reflection seismology has been widely applied in land and marine surveys. In comparison with reflection seismology, refraction seismology is a rather novel technique, especially in marine surveying, used to provide an improved reflection profile. However, the relevant technologies have been developed only to a limited extent [45,46]. Hence, we will focus only on the treatment of reflection seismology.

Reflection Seismology

Generally speaking, reflection seismology involves three stages of seismic signal processing, namely, seismic data acquisition, processing and interpretation. Figures 4.17a, b give schematic illustrations of land and marine seismic data acquisition, respectively. An acoustic source waveform, called a *seismic wavelet* or a *source signature*, is launched at a shot point, and then propagates through the earth's subsurface. Many types of energy sources have been employed to generate the seismic wavelet, including dynamite and other high-energy explosive sources, implosive air guns, electrical sparkers, vibrating chirp systems, etc. [47]. As the seismic wavelet encounters an interface between different geological layers, it is reflected and refracted at that interface due to impedance mismatch. The resultant reflected waves then propagate back to the earth's surface and are measured by an array of sensors (geophones for land survey and hydrophones for marine survey) that is collectively referred to as a *seismometer array*. At each sensor output, a set of measured seismic data is digitally recorded as a *seismic trace*. A collection of seismic traces are called a *seismogram*. By repeating this procedure at many source and sensor locations, one can produce a 2-D or 3-D image of the earth's reflection profile.

For modeling the earth's subsurface, it is traditional to utilize a layered system model as shown in Fig. 4.18, where $h[n]$ is the seismic wavelet, the interface between layer i and layer $(i+1)$ for $i = 1, 2, ...$, is characterized by the *reflection coefficient* κ_i (a real number) to account for the impendence mismatch, and τ_i is the travel time of the seismic wavelet from point A to point B via the ith layer. In addition, the layered earth system is usually accompanied by the following two assumptions [48, p. 31].

(A4-9) As the seismic wavelet $h[n]$ (a real sequence) propagates through each layer of the layered earth system, its waveform remains unchanged.

(A4-10) Each travel time τ_i can be represented by an integer multiple of the sampling period.

As a result of Assumptions (A4-9) and (A4-10), we have the convolutional model for the seismic trace, denoted by $y[n]$, as follows: (see Fig. 4.19)

$$y[n] = \sum_{i=1}^{\infty} \kappa_i h[n - \tau_i] + w[n] = h[n] \star u[n] + w[n] \tag{4.158}$$

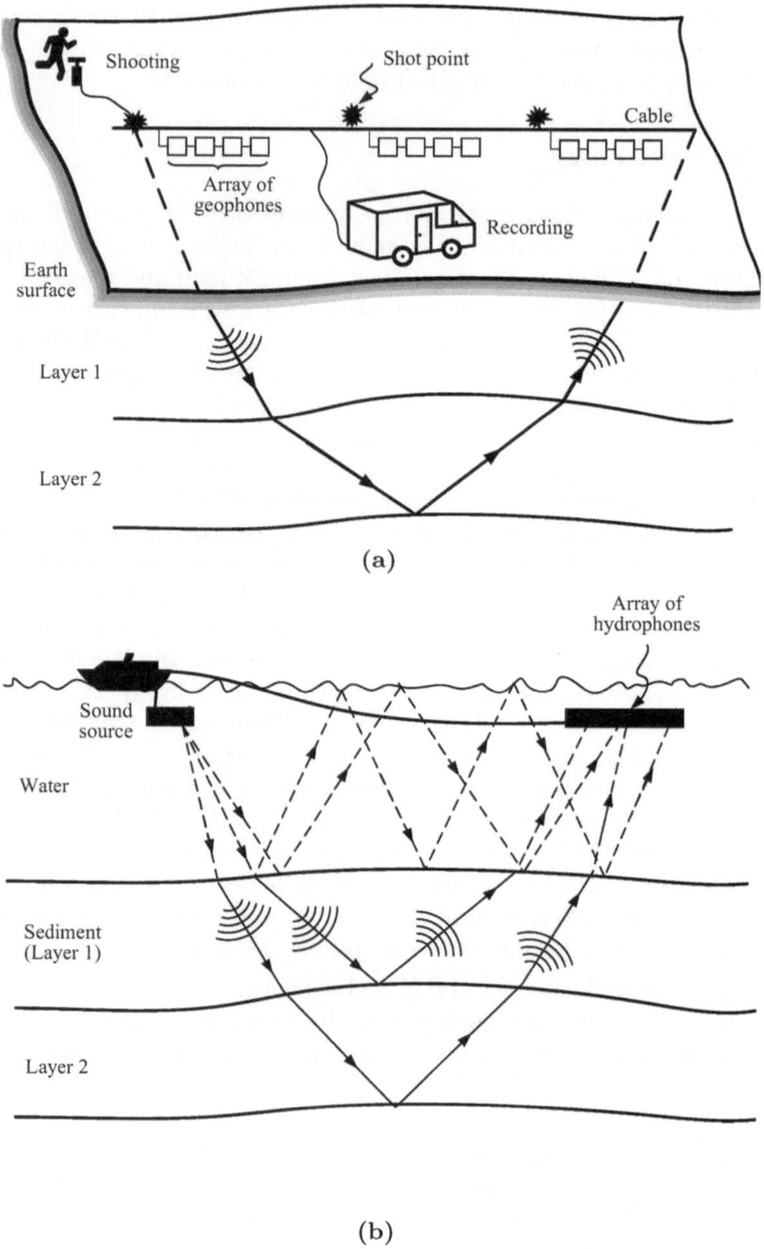

Fig. 4.17 Configurations of (**a**) land and (**b**) marine seismic data acquisitions

where the seismic wavelet $h[n]$ is treated as an LTI system,

$$u[n] = \sum_{i=1}^{\infty} \kappa_i \delta[n - \tau_i] \tag{4.159}$$

is called the *reflectivity sequence*, and $w[n]$ is the measurement noise. Note, from Fig. 4.19, that interpretation of the reflectivity sequence $u[n]$ as the system input and of the seismic wavelet $h[n]$ as the LTI system is counterintuitive from a physical point of view, but the counterintuitive approach is useful from a mathematical point of view due to the random nature of $u[n]$ [49]. As a remark, more realistic models for $y[n]$ can be constructed in terms of nonlinear and time-varying characteristics. Nevertheless, the LTI convolutional model (4.158), which basically holds within a limited area size or time gate for practical applications of real seismic data, can still be used as an approximation to a more realistic model.

From Fig. 4.18, we note that the reflectivity sequence $u[n]$, which reflects the physical properties of the earth's internal layers, contains certain information about the earth's subsurface structure below the center point C between point A and point B [50]. Accordingly, the objective here is to remove the effect of the source wavelet $h[n]$ from the seismic trace $y[n]$ so that an estimate of the reflectivity sequence $u[n]$ can be obtained. Obviously, this is a deconvolution problem, specifically referred to as *seismic deconvolution*. Since the early 1960s, seismic deconvolution has served as a routine computational data processing procedure in reflection seismology, and related topics have been an active field of geophysical research for nearly half a century [49, 51].

Approaches to Seismic Deconvolution

An intuitively direct approach to seismic deconvolution is the so-called *deterministic deconvolution*, which measures the seismic wavelet directly and then removes it from the seismic trace through a nonblind deconvolution procedure. Deterministic deconvolution has proved to be more effective for marine seismic exploration than land seismic exploration because seismic wavelet measuring is quite difficult in land environments [52]. In marine environments, the seismic wavelet can be measured by towing a deep hydrophone in the water, provided that the water is deep enough for the measurement not to be corrupted by reflected waves from the sea floor. However, it is still not an easy task to provide an accurate measurement of seismic wavelet due to the ghost resultant from the water–air interface (see Fig. 4.17b) and the filter effect of the towed hydrophone. As a consequence, the seismic wavelet $h[n]$ is usually not exactly known in practice, and accordingly we have to resort to blind approaches of seismic deconvolution.

Thus far, there have existed a multitude of blind approaches to seismic deconvolution, including predictive deconvolution, homomorphic deconvolution,

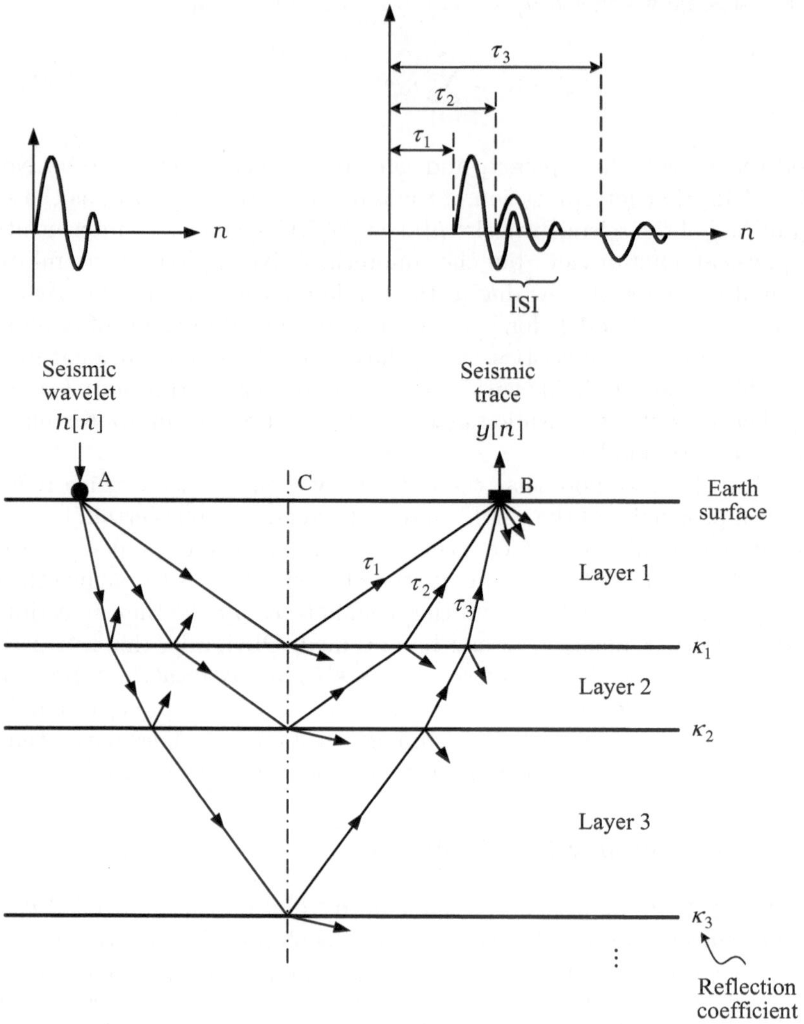

Fig. 4.18 Layered earth system

maximum-likelihood deconvolution, maximum or minimum entropy deconvolution, and so on. The volume of reprinted papers edited by Robinson and Osman [51] has collected a number of excellent papers on these blind approaches, in addition to the deterministic deconvolution. One can also refer to Arya and Holden's paper [52], which gives an overview of predictive deconvolution, homomorphic deconvolution, Kalman filtering and deterministic deconvolution for seismic exploration. For brevity, we will deal only with seis-

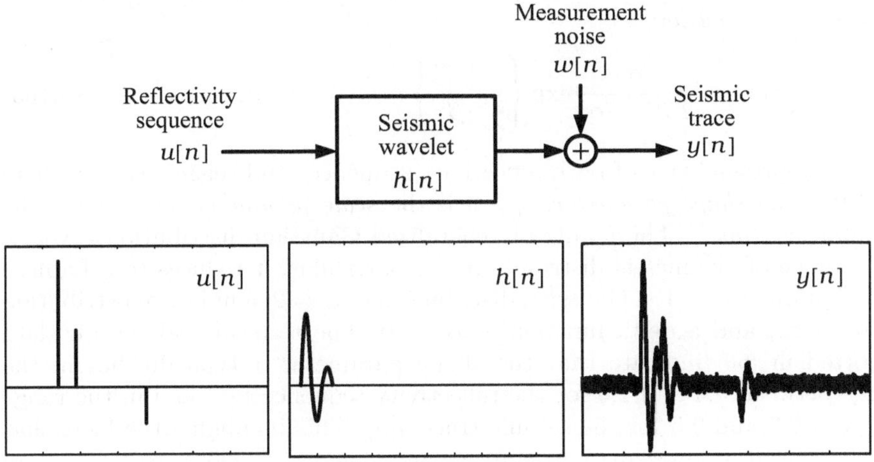

Fig. 4.19 Convolutional model for the seismic trace $y[n]$

mic deconvolution in terms of predictive deconvolution and MNC equalization, respectively.

Before applying predictive and MNC equalization to seismic deconvolution, let us first inspect the characteristics of the reflectivity sequence $u[n]$ as follows. In accordance with the physical properties of the earth's layered structure, it has been widely accepted that the reflectivity sequence can be thought of as a non-Gaussian sparse spike train with random amplitudes. Accordingly, Kormylo and Mendel [49, 53, 54] proposed a statistical model for $u[n]$, a *Bernoulli–Gaussian (B–G) model* given by

$$u[n] = u_{\rm B}[n] \cdot u_{\rm G}[n] \tag{4.160}$$

where $u_{\rm G}[n]$ is a real, zero-mean, WSS, white Gaussian process with variance $\sigma_{\rm G}^2$ and $u_{\rm B}[n]$ is a Bernoulli sequence with parameter λ, i.e.

$$\Pr\{u_{\rm B}[n]\} = \begin{cases} \lambda, & \text{for } u_{\rm B}[n] = 1, \\ 1-\lambda, & \text{for } u_{\rm B}[n] = 0. \end{cases} \tag{4.161}$$

Note that the parameter λ should be sufficiently small to account for the spikiness of $u[n]$. The B–G sequence $u[n]$ has variance $\sigma_u^2 = \lambda\sigma_{\rm G}^2$, skewness $C_3\{u[n]\} = 0$, and kurtosis $C_4\{u[n]\} = 3\lambda(1-\lambda)\sigma_{\rm G}^2$. A further model given by [55]

$$u[n] = u_{\rm LE}[n] + u_{\rm SE}[n] \tag{4.162}$$

is also used sometimes where $u_{\rm LE}[n]$ is a B–G sequence accounting for large events in $u[n]$, and $u_{\rm SE}[n]$ is a white Gaussian sequence accounting for smaller background events in $u[n]$.

In addition, it has been reported [56] that the family of *generalized Gaussian distribution*

$$f(x) = \frac{\alpha}{2\beta\Gamma(1/\alpha)} \exp\left\{-\frac{|x|^\alpha}{\beta}\right\}, \quad -\infty < x < \infty \tag{4.163}$$

is fairly representative of both reflectivity sequences and seismic traces where $\alpha > 0$ is the *shape parameter*, $\beta > 0$ is the *scale parameter*, and $\Gamma(\cdot)$ is the *gamma function*.[11] The family of generalized Gaussian distribution covers a wide range of symmetric distributions. In particular, it reduces to a Laplace distribution as $\alpha = 1$, a Gaussian distribution as $\alpha = 2$, a uniform distribution as $\alpha \to \infty$, and a delta function as $\alpha \to 0$. The tests of real seismic data reported in [56] indicate that the shape parameter α typically lies in the range between 0.4 and 1.5 for the reflectivity sequence $u[n]$ and in the range between 0.9 and 2.5 for the seismic trace $y[n]$. This highlights the facts that $u[n]$ is indeed a non-Gaussian sequence and that $y[n]$ is closer to a Gaussian process than $u[n]$. Therefore, it has been widely accepted that $u[n]$ satisfies both Assumptions (A4-2) and (A4-6).

As a matter of fact, the seismic wavelet $h[n]$ is usually nonminimum phase [49, p. 9]. Apparently, this poses a serious problem for predictive deconvolution due to Property 4.11, whereas this has no negative impact on the performance of MNC equalization. Next, let us present the original ideas behind the MNC equalization criterion $J_{p+q}(v[n])$ in seismic deconvolution.

Minimum Entropy Deconvolution

Since the reflectivity sequence $u[n]$ is a sparse spike train, it is visually clear that the correct equalizer $v[n]$ will lead to the equalized signal $e[n] = v[n] \star y[n]$ exhibiting a "simpler" appearance than the seismic trace $y[n]$. Based on this observation, Wiggins [23, 24] found the following objective function (to be maximized)

$$J_{\mathrm{MED}}(v[n]) = \frac{E\{e^4[n]\}}{(E\{e^2[n]\})^2} \tag{4.164}$$

which provides a visual judgement of "simplicity" (i.e. a performance index). Because a "simple" appearance, to a certain extent, corresponds to a small amount of information or *entropy*, $J_{\mathrm{MED}}(v[n])$ is referred to as a minimum entropy deconvolution (MED) criterion [24]. After Wiggins' work, several extensions and modifications of $J_{\mathrm{MED}}(v[n])$ have been proposed for seismic deconvolution. Among them, the MNC equalization criterion $J_{p+q}(v[n])$ was proposed by Donoho [24], who also proposed other types of MED criteria by virtue of the measures of Fisher's information and Shannon's information, in relation to the Wiggins' MED criterion $J_{\mathrm{MED}}(v[n])$ as follows:

[11] The *gamma function* $\Gamma(k+1) = k!$ for $k = 0, 1, 2, \ldots$

$$J_4(v[n]) = |J_{\mathrm{MED}}(v[n]) - 3| \,. \tag{4.165}$$

Clearly, $J_4(v[n])$ and $J_{\mathrm{MED}}(v[n])$ are equivalent when $u[n]$ has positive kurtosis.

It is worth mentioning that although the reflectivity sequence $u[n]$ is generally regarded as a real sequence, a constant phase shift may need to be further introduced to $u[n]$ for some cases, depending on the incident angle of the source wavelet, subsurface topography, etc. To estimate such a complex reflectivity sequence $u[n]$, Ulrych and Walker [25,51] proposed a complex version of the Wiggins' MED criterion as follows:

$$J_{\mathrm{MED}}(v[n]) = \frac{E\{|e[n]|^4\}}{(E\{|e[n]|^2\})^2}, \tag{4.166}$$

which relates to the MNC criterion, $J_{2,2}(v[n])$, via

$$J_{2,2}(v[n]) = |J_{\mathrm{MED}}(v[n]) - 2| \,, \tag{4.167}$$

provided that $E\{u^2[n]\} = 0$ (the typical case for complex stationary processes). Accordingly, (4.166) should be recognized as a first proposal for the complex version of the MNC criterion $J_{2,2}(v[n])$.

Simulation Example of Seismic Deconvolution

Let us show some simulation results of seismic deconvolution. In the simulation, we generated a B–G sequence with $\lambda = 0.05$ and $\sigma_{\mathrm{G}}^2 = 0.0225$ for the reflectivity sequence $u[n]$, and synthesized the noise-free seismic trace by convolving $u[n]$ with a nonminimum-phase ARMA(3,3) seismic wavelet $h[n]$ whose transfer function is given by (4.157). Then the noisy seismic trace $y[n]$ was obtained by adding a white Gaussian noise sequence $w[n]$ to the synthetic noise-free seismic trace. Figure 4.20a depicts the seismic wavelet $h[n]$, Fig. 4.20b a segment of the generated reflectivity sequence $u[n]$, and Fig. 4.20c a segment of the synthetic seismic trace $y[n]$ for SNR = 20 dB.

With the synthetic seismic trace $y[n]$, we used the Burg algorithm to obtain a forward LPE filter, $a_L[n]$, of order L equal to 16. We also considered $J_4(v[n])$ for MNC equalization since the skewness $C_3\{u[n]\} = 0$ and the kurtosis $C_4\{u[n]\} \approx 0.0032 \neq 0$ for the B–G sequence $u[n]$. The hybrid MNC algorithm was employed to find the MNC equalizer $v_{\mathrm{MNC}}[n]$, which was assumed to be a 16th-order causal FIR filter. A single run was performed for data length $N = 2048$. Figures 4.21a, b display a segment of the equalized signal (bars) using the LPE filter $a_L[n]$ and that using the MNC equalizer $v_{\mathrm{MNC}}[n]$, respectively, for SNR = 20 dB, where scale factors and time delays were artificially removed. One can see, from Figs. 4.21a, b, that the equalized signal using $v_{\mathrm{MNC}}[n]$ approximates $u[n]$ well, and exhibits a "simpler" appearance than the predictive deconvolved signal since an allpass distortion (due to nonminimum-phase $h[n]$) still remains in the latter.

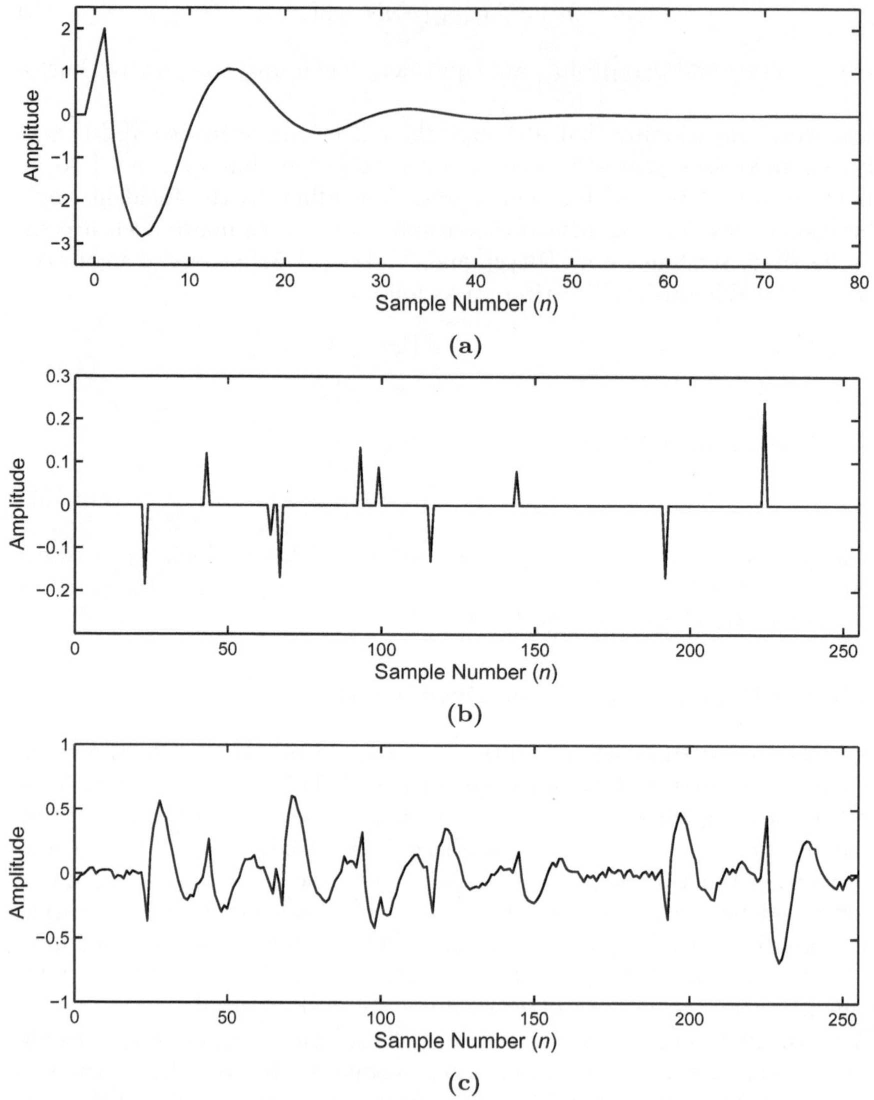

Fig. 4.20 (**a**) Seismic wavelet $h[n]$, (**b**) segment of the reflectivity sequence $u[n]$, and (**c**) segment of the synthetic seismic trace $y[n]$ for SNR = 20 dB

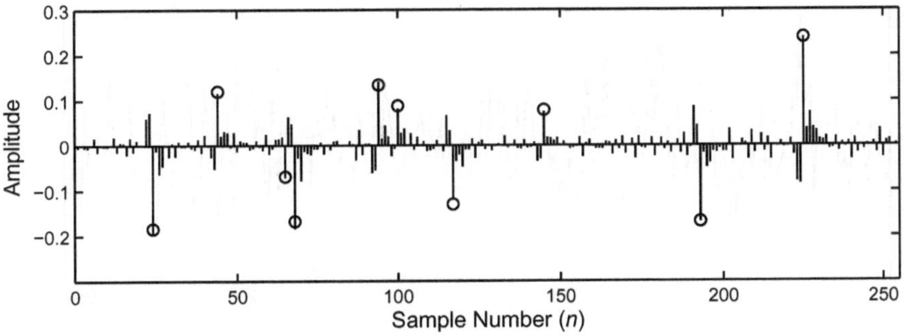

(a)

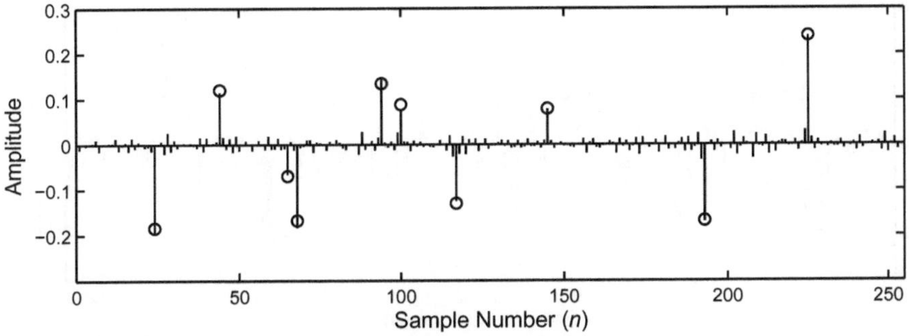

(b)

Fig. 4.21 Segment of the equalized signals (bars) using (**a**) a linear prediction approach and (**b**) the MNC equalization algorithm, along with the reflectivity sequence $u[n]$ (circles) for SNR $= 20$ dB

To provide further insight into seismic deconvolution, we repeated the above simulation using only the MNC equalization algorithm for SNR $= 0$ dB. The simulation results are displayed in Fig. 4.22. One can see, from Fig. 4.22c, that the signal component $e_S[n]$ consists of a sequence of pulses, each apparently being an approximately symmetric wavelet (corresponding to the zero-phase overall system $g[n]$) with amplitude proportional to the corresponding $u[n]$. These observations therefore agree with Property 4.18. Note, from Figs. 4.22b, c, the two close spikes at $n = 64$ and 67 are not discernible since the spacing between them is much narrower than the width of $g[n]$ for this case (SNR $= 0$ dB).

4.5.2 Speech Signal Processing

Blind deconvolution of speech data, specifically referred to as *speech deconvolution*, is an essential procedure of speech signal processing for a variety of applications, including speech analysis and synthesis, speech coding, speech

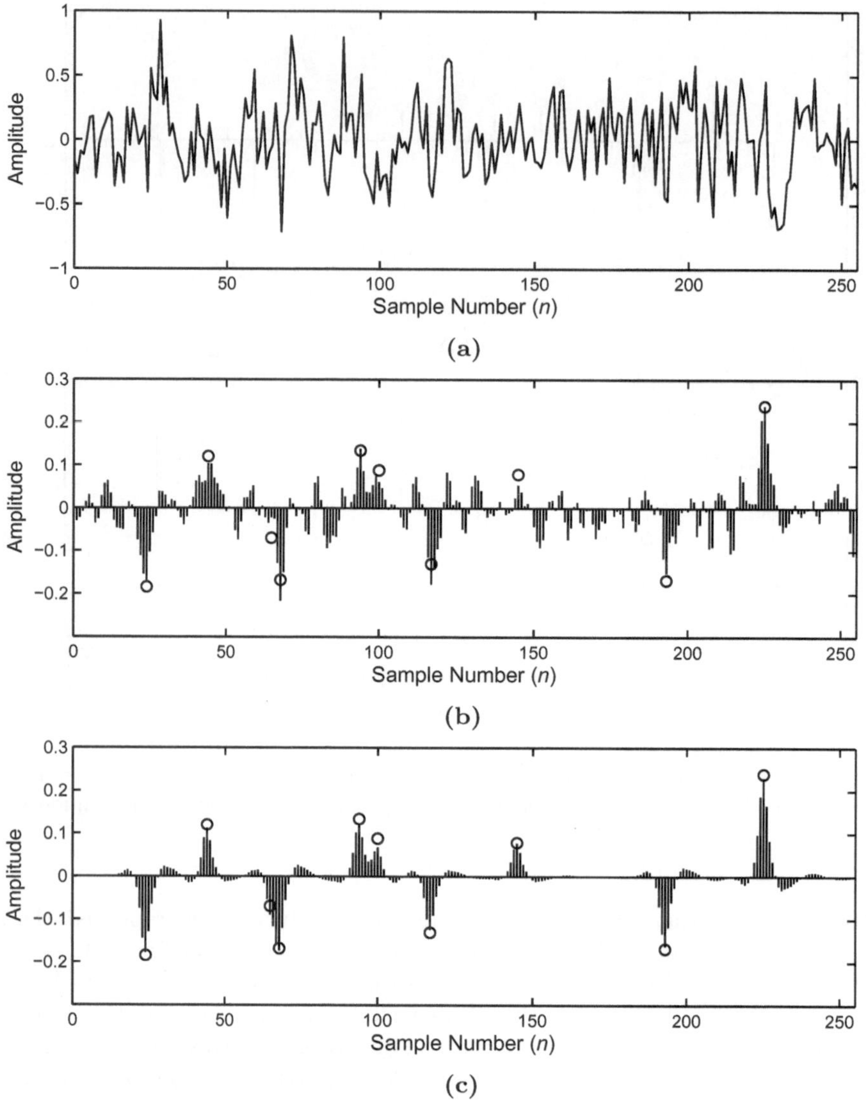

Fig. 4.22 Segment of (**a**) the synthetic seismic trace $y[n]$ for SNR = 0 dB, (**b**) the equalized signal $e[n]$ (bars) using MNC equalization, and (**c**) the signal component $e_S[n]$ (bars) of $e[n]$, along with the reflectivity sequence $u[n]$ (circles)

recognition, and speaker recognition (i.e. speaker verification and identification)[12] [57–64]. In the following, we will address mathematical models of speech production on which speech deconvolution algorithms often rely and, as an illustration, give the treatment of speech analysis and synthesis by means of predictive deconvolution and MNC equalization.

Speech Production Models

Figure 4.23 depicts a diagram of the simplified human vocal system. During normal breathing, the *vocal cords*, or the *vocal folds*, are held apart, allowing air to pass through the gap between the two vocal cords; this gap is called the *glottis*. As shown in Fig. 4.23, the *vocal tract* begins at the glottis and ends at the lips, and the air flow from the lungs serves as a source of energy to excite the vocal tract. The vocal tract then imposes a number of resonances upon the excitation to produce a speech sound. The frequencies of the resonances, called the *formant frequencies* or simply *formants*, depend on the shape and dimension of the vocal tract. Different sets of formants result in different sounds of speech.

According to the type of excitation, speech sounds can be divided into several classes, among which two elementary classes are as follows [58,60,63].

- *Voiced sounds.* Voiced sounds such as the phoneme /*a*/ in the utterance of "arm" are produced by forcing air through the glottis where the vocal cords are adjusted in terms of their tension so that they vibrate in a relaxation oscillation. This oscillation therefore generates a quasi-periodic pulse train that excites the vocal tract to produce voiced sounds.
- *Unvoiced sounds* or *fricative sounds.* Unvoiced sounds such as the phoneme /*f*/ in the utterance of "face" are produced by forming a constriction at some point within the vocal tract, and forcing air to pass through the constriction to produce turbulence. This therefore creates a broad-spectrum noise source that excites the vocal tract to produce unvoiced sounds.

Other classes of speech sounds such as plosive sounds and whispers merely result from the combinations of voiced source, unvoiced source and silence.

Owing to the different types of excitation and the different sets of formants, the characteristics of a speech signal are obviously time varying. As such, it is common to partition a speech signal into short segments for ease of further processing because each short segment can be thought of as a stationary process with certain fixed characteristics during that segment. This assumption typically holds true for segments with duration on the order of 30 or 40 ms [65, p. 724]. Note that the short segments are sometimes called the *analysis frames* or simply *frames*, and they often overlap frame to frame.

[12] *Speaker verification* is the process of verifying whether an unknown speaker is the person as claimed, while *speaker identification* is the process of associating an unknown speaker with a member in a population of known speakers [57].

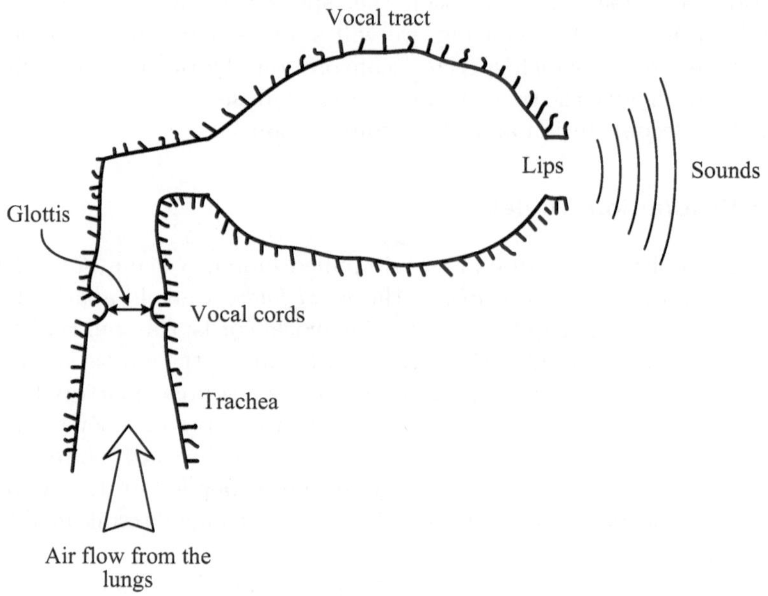

Fig. 4.23 Schematic diagram of the human vocal system

As a result of the foregoing discussions, it is reasonable to model each segment of a speech signal either as the output of an LTI system driven by a quasi-periodic impulse train for voiced speech or as the one driven by a random white noise for unvoiced speech. The period of the periodic impulse train is called the *pitch period*. Figure 4.24 depicts such a speech production model. One can see, from Fig. 4.24, that in addition to the effect of the vocal tract characterized by $h_{\rm V}[n]$, the speech production model also includes the effect of the vocal cords characterized by the glottal pulse model $h_{\rm G}[n]$ and the effect of the sound radiation at the lips characterized by the radiation model $h_{\rm R}[n]$. The glottal pulse model $h_{\rm G}[n]$ is of finite duration, while the radiation model $h_{\rm R}[n]$ will result in high frequency emphasis of the speech signal. Moreover, the resonances (formants) due to the vocal tract correspond to the poles of the vocal tract model $H_{\rm V}(z)$, and thus an AR model (an all-pole model) is a good representation of the vocal-tract effect for voiced speech. However, both poles and zeros (resonances and anti-resonances) also need be included in $H_{\rm V}(z)$ when unvoiced speech and nasals are to be considered. On the other hand, the quasi-periodic impulse train for voiced speech is characterized by the pitch period, while the probabilistic distribution of the white noise for unvoiced speech does not appear to be critical.

As shown in Fig. 4.25, it is convenient to combine the glottal pulse, vocal tract and radiation models to form the *speech production system*, denoted by

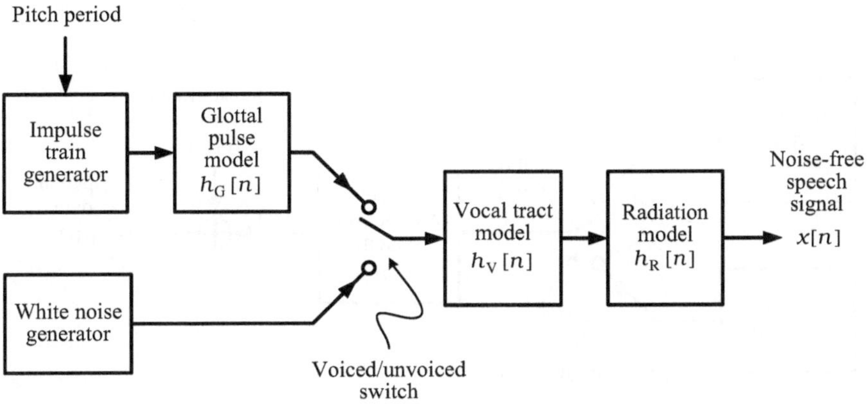

Fig. 4.24 Generic model for voiced and unvoiced speech production; after Rabiner and Schafer [58]

$h[n]$, as follows:

$$h[n] = \begin{cases} h_R[n] \star h_V[n] \star h_G[n], & \text{for voiced speech,} \\ h_R[n] \star h_V[n], & \text{for unvoiced speech.} \end{cases} \tag{4.168}$$

Accordingly, we have the convolutional model given by (4.1) and (4.2) for one segment of measured speech data, $y[n]$, where the source signal, $u[n]$, is either a quasi-periodic impulse train for voiced speech or a white noise for unvoiced speech.

Speech Analysis and Synthesis

The goal of speech analysis is to estimate the parameters of the speech production model such as the energy or gain of each segment, the type of speech sound (voiced or unvoiced), the pitch period, the speech production system $h[n]$, the formants, etc. Obviously, speech deconvolution is suited to separating $h[n]$ from the source signal $u[n]$. Correspondingly, speech can also be synthesized from these estimated parameters by taking the same parametric representation as that used in speech analysis. Consider, for example, the parametric representation in Fig. 4.25 for speech synthesis. Given the estimated pitch period, the impulse train generator for voiced sounds produces an impulse of unity amplitude at the beginning of each pitch period. On the other hand, due to a great many random effects involved in producing unvoiced speech, the white noise generated for unvoiced sounds can be assumed to be a white Gaussian process with unity variance [66]. The estimated gain is then applied to adjust the energy of the generated impulse train or white

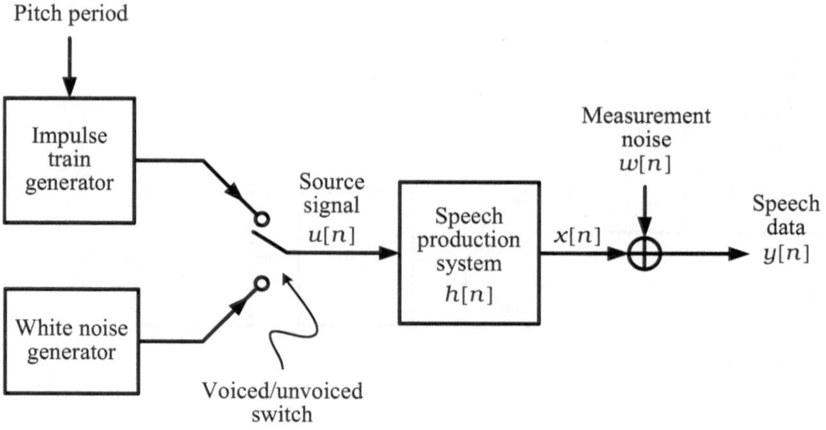

Fig. 4.25 Convolutional model for noisy speech data

Gaussian noise, that is then convolved by the estimate of $h[n]$ to generate the synthetic speech signal $x[n]$.

For speech analysis and synthesis, the most widely used approaches to speech deconvolution are linear predictive deconvolution and homomorphic speech deconvolution. The treatment of the latter is omitted here; the reader can find it in many textbooks such as [58]. In theory, the predictive deconvolved signal obtained by linear predictive deconvolution can be used to estimate the pitch period directly. In practice, however, a somewhat more sophisticated method such as the simple inverse filtering tracking (SIFT) method is generally used [67]. Furthermore, the formants can be estimated either directly from the coefficients of the LPE filter or the associated AR spectral estimate, by virtue of the fact that the poles of the speech production system $H(z)$ or the peaks in the spectrum of the speech signal basically correspond to the formants. In addition to linear predictive deconvolution, one can also apply MNC equalization to speech analysis and synthesis because each segment of voiced speech can reasonably be assumed to be a non-Gaussian stationary process with nonzero third-order and fourth-order cumulants; see [66, 68–71] for the details.

It is important to note that speech analysis and synthesis forms the basis of speech coding for such applications as speech communications. A representative of speech coding in speech communications is the so-called *linear predictive coding (LPC)* [58]. Specifically, the transmitter performs speech analysis by means of linear predictive deconvolution to obtain a set of estimated parameters, that is to be encoded in an efficient manner for reducing the bit rate of transmitting the encoded parameters. Correspondingly, the receiver decodes the encoded parameters and performs speech synthesis to generate a synthesized speech that approximates the original speech sent from the transmitter.

Note that some phase distortion often remains in the synthesized speech since the speech production system $H(z)$ is in general nonminimum phase. This phase distortion, however, has virtually no effect on speech perception because the human ear is fundamentally "phase deaf" [63, pp. 269–270]. Moreover, the parameters obtained by speech analysis can also serve as a set of features for speech recognition and speaker recognition. For such applications, SOS based speech analysis (by means of linear predictive deconvolution) can provide only the features related to the magnitude of $H(z)$. On the other hand, HOS based speech analysis such as the one using MNC equalization can provide not only the features related to the magnitude of $H(z)$ but also those related to the phase of $H(z)$, thereby leading to improved recognition performance over SOS based approaches. Examples of HOS based speech analysis for speech recognition can be found in [72–74].

Experimental Example of Speech Deconvolution

Let us present some experimental results for speech deconvolution with real voiced speech data. In the experiment, a set of real speech data for the sound /a/ uttered by a man was acquired through a 16-bit A/D converter with a sampling rate of 8 kHz. The set of measured speech data, shown in Fig. 4.26a, was then processed by a forward LPE filter $a_L[n]$ of order $L = 24$, and processed by an MNC equalizer $v_{\mathrm{MNC}}[n]$ with $L_1 = 0$ and $L_2 = 24$, where the coefficients of $a_L[n]$ were obtained with the Burg algorithm and those of $v_{\mathrm{MNC}}[n]$ obtained with the hybrid MNC equalization algorithm. Figures 4.26b–d display the predictive deconvolved signal using $a_L[n]$, the equalized signal using $v_{\mathrm{MNC}}[n]$ associated with $J_3(v[n])$ and that associated with $J_4(v[n])$, respectively. One can see, from Figs. 4.26b–d, that the equalized signal using $v_{\mathrm{MNC}}[n]$ approximates a pseudo-periodic impulse train much better than the predictive deconvolved signal since some phase distortion remains in the latter. This also verifies that the speech production system $h[n]$ may not be minimum phase.

With any of the equalized signals in Figs. 4.26b–d, the pitch period can easily be found to be 70 samples (i.e. 8.75 ms) from the two most significant impulses. Furthermore, the speech production system $h[n]$ can be estimated as $\widehat{H}(z) = 1/A_L(z)$ through the linear predictive deconvolution or $\widehat{H}(z) = 1/V_{\mathrm{MNC}}(z)$ through the MNC equalization, as shown in Fig. 4.27a. By convolving the system estimate, $\widehat{h}[n]$, with a periodic impulse train of period equal to 70 samples, we obtained the synthesized speeches displayed in Figs. 4.27b–d, where the synthesized speeches were adjusted to have the same energy as the original speech signal. One can see, from Fig. 4.27a, that the system estimate $\widehat{h}[n]$ associated with MNC equalization shows better resemblance to one pitch period of the original speech signal than that associated with linear predictive deconvolution. This, in turn, leads to the result that the synthesized speech signals associated with the former resemble the original speech signal better than those associated with the latter, as shown in Figs. 4.27b–d.

4.5.3 Baud-Spaced Equalization in Digital Communications

Digital communication systems can be divided into wired and wireless communication systems; the former utilize twisted pair, coaxial cable or optical fiber as the transmission media while the latter utilize air, water or free space as the transmission media. The fundamental principles for both systems are essentially similar. Thus, for brevity, we will deal with baud-spaced equalization only for wireless communication systems.

Wireless Communication Systems

Figure 4.28 depicts a typical wireless communication system, for which we assume that the information sequence to be transmitted is already in binary form. The components in the transmitter are explained as follows [3].

- *Source encoder and scrambler.* The source encoder, also called the *source compressor*, is used to remove the redundancy in the binary information sequence, thereby reducing the required transmission bandwidth. The compressed sequence may be further scrambled or randomized by the scrambler to ensure adequate binary transitions and energy dispersion. The scrambler is therefore also called the *randomizer* or the *energy disperser.*
- *Channel encoder and interleaver.* In a controlled manner, the channel encoder imposes some redundancy on the compressed/scrambled sequence to provide protection capability against the effects of noise and interference introduced by the transmission channel. The channel encoder is often accompanied by the interleaver to improve the protection capability against bursts of errors due to, for instance, the effect of channel fading.
- *Mapper and modulator.* The mapper transforms the coded sequence from the channel encoder into a sequence of PAM, PSK or QAM symbols, called a *symbol sequence.* The symbol sequence is real for PAM constellation, whereas it is complex for PSK and QAM constellations. The modulator then modulates the amplitude or phase, or both, of the symbol sequence onto a carrier to generate a (real) spectrally efficient signal waveform.
- *Radio-frequency (RF) section and antenna.* In the RF section, the center frequency of the signal waveform generated from the modulator is upconverted by a mixer to the specified frequency. The resultant RF waveform is then amplified by a power amplifier followed by a bandpass filter to reduce the effect of out-of-band radiation introduced by the nonlinearity of the power amplifier. Finally, the amplified and filtered RF waveform is radiated by the antenna.

As shown in Fig. 4.28, the receiver performs the reverse operations of the transmitter to recover the original information sequence.

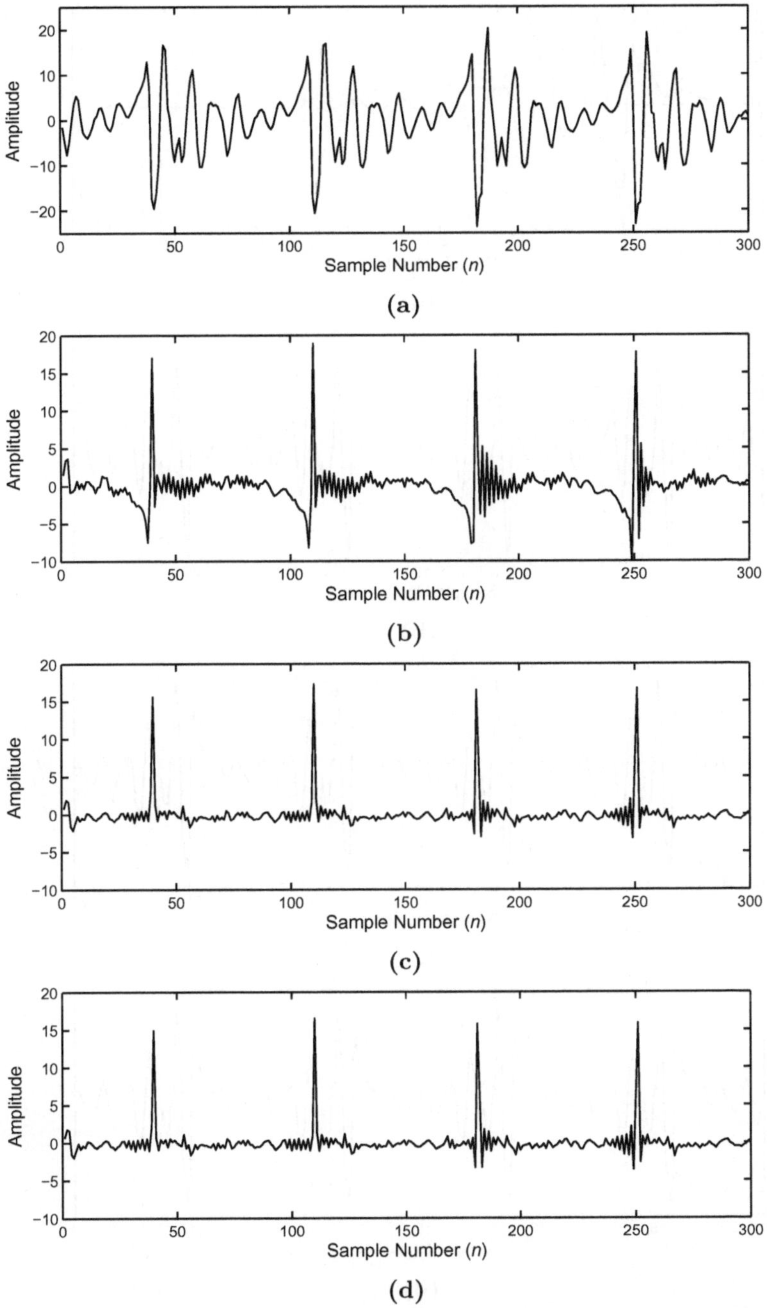

Fig. 4.26 (**a**) Segment of measured speech data for the sound $/a/$, (**b**) the predictive deconvolved signal, (**c**) the equalized signal using $J_3(v[n])$, and (**d**) the equalized signal using $J_4(v[n])$

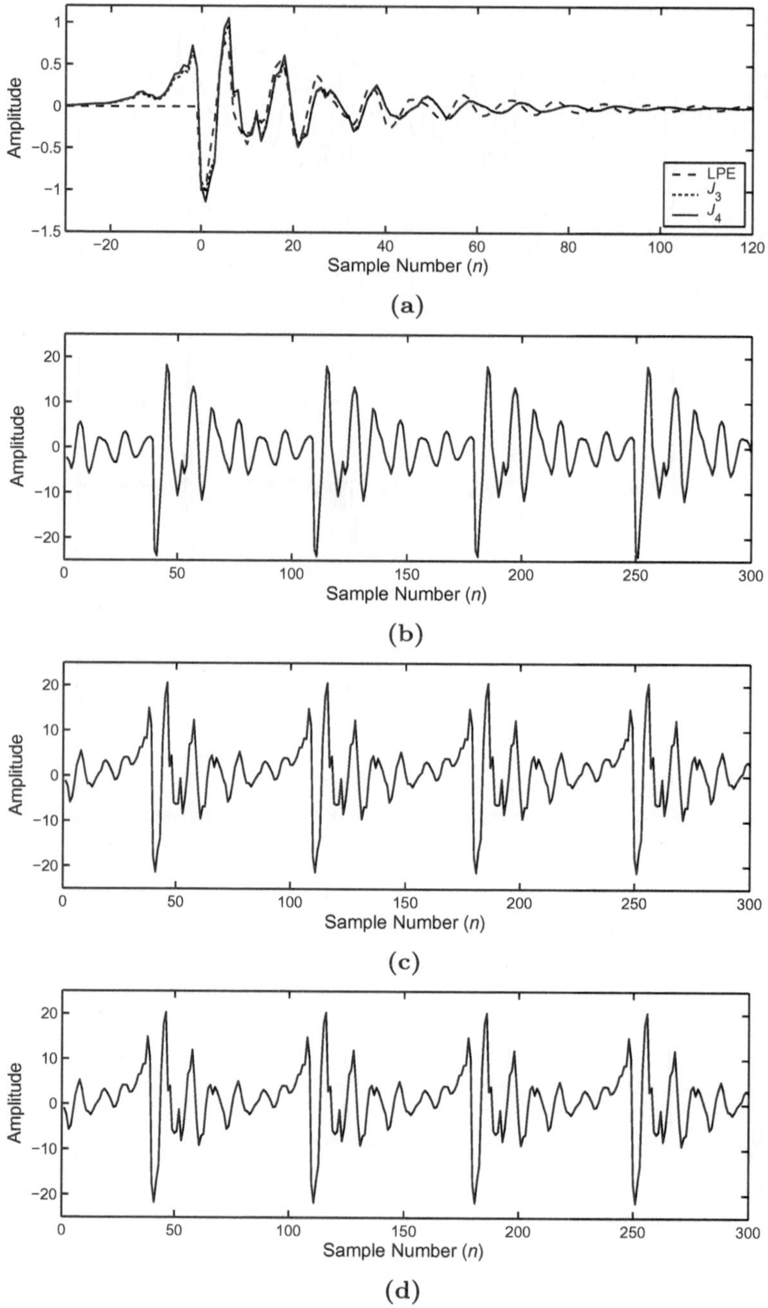

Fig. 4.27 (**a**) Estimates of the speech production system $h[n]$, and segment of the synthesized speech signals by using (**b**) $a_L[n]$, (**c**) $v_{\mathrm{MNC}}[n]$ associated with $J_3(v[n])$ and (**d**) $v_{\mathrm{MNC}}[n]$ associated with $J_4(v[n])$

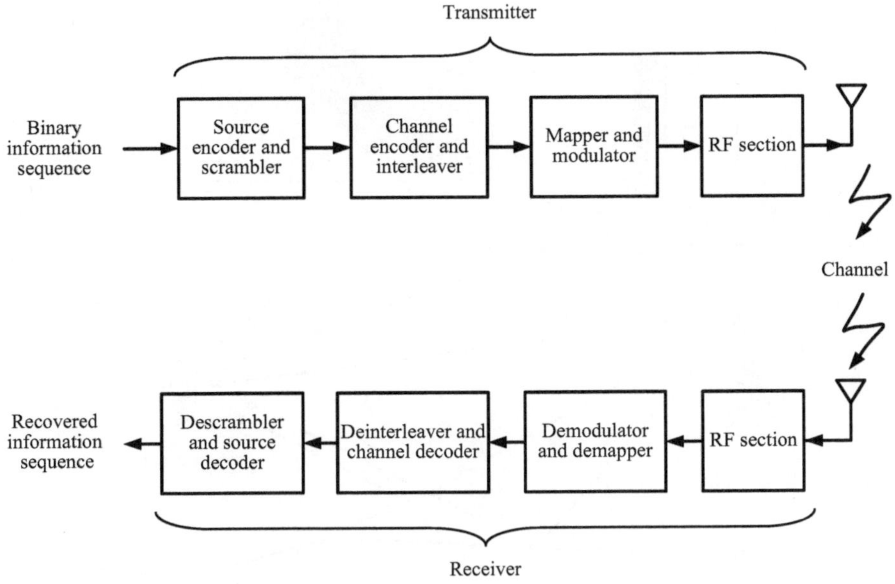

Fig. 4.28 Block diagram of a typical wireless communication system

Note that the transmission channel introduces not only the effects of noise and interference, but also the effects of frequency selectivity and time variation. Among these effects, the channel's frequency selectivity, which leads to the ISI in the received signal, is mainly due to limited bandwidth and multipath propagation of the transmission medium; see Fig. 4.29 for an example of multipath propagation. Next, let us establish the channel model on which baud-spaced equalizers are commonly based.

Channel Models

Consider the schematic diagram of a simplified wireless communication system in Fig. 4.30, which shows only the equivalent blocks for the modulator, RF section at the transmitting end, channel, RF section at the receiving end, and part of the demodulator. With linear modulation, the transmitted baseband signal, $s(t)$, is generated as follows:

$$s(t) = \sum_{k=-\infty}^{\infty} u[k] h_{\mathrm{T}}(t - kT) \tag{4.169}$$

where $u[k]$ is the (possibly complex-valued) symbol sequence generated from the mapper, T is the symbol period, and $h_{\mathrm{T}}(t)$ is the *transmitting filter*; further discussion on the transmitting filter will be given later. The baseband signal $s(t)$ is then upconverted to the transmitted bandpass signal, $s_{\mathrm{BP}}(t)$, as follows:

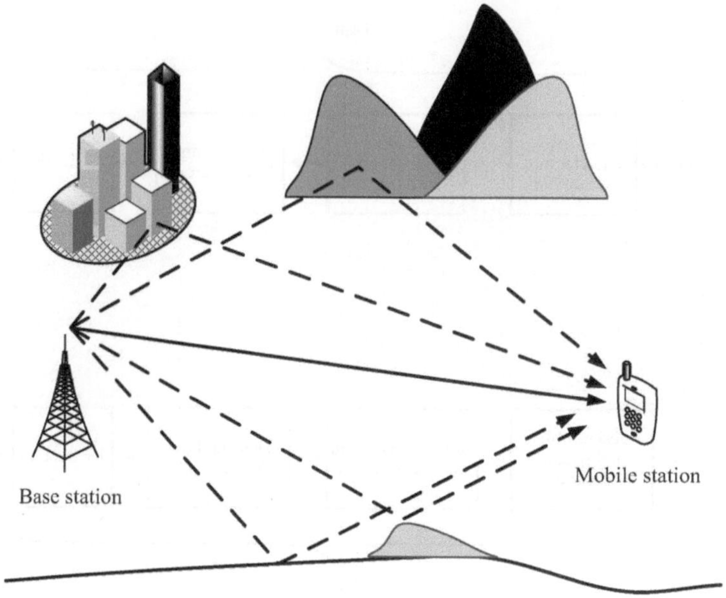

Fig. 4.29 Mobile communication system with multipath propagation

$$s_{\rm BP}(t) = \text{Re}\left\{s(t)e^{j[2\pi f_c t+\theta(t)]}\right\} = \text{Re}\left\{s_{\rm LP}(t)e^{j2\pi f_c t}\right\} \tag{4.170}$$

where f_c and $\theta(t)$ are the nominal carrier frequency and the phase jitter (fluctuation) of the oscillator at the transmitting end, respectively, and

$$s_{\rm LP}(t) = s(t)e^{j\theta(t)} \tag{4.171}$$

is the *equivalent lowpass signal* of $s_{\rm BP}(t)$.

At the receiver, we have the following received bandpass signal:

$$r_{\rm BP}(t) = h_{\rm BP}(t-\varepsilon) \star s_{\rm BP}(t) + w_{\rm BP}(t) \tag{4.172}$$

where $h_{\rm BP}(t)$ is the bandpass channel, $w_{\rm BP}(t)$ is the bandpass noise, and ε accounts for the propagation delay. In (4.172), we have made use of the assumption that the bandpass channel is an LTI system. After downconversion, we obtain the received baseband signal, $y(t)$, as follows:

$$y(t) = A \cdot h_{\rm R}(t) \star \left\{r_{\rm BP}(t)e^{-j[2\pi \widetilde{f}_c t+\widetilde{\theta}(t)]}\right\} \tag{4.173}$$

where A is the gain provided by the *automatic gain control (AGC)* circuit for gain adjustment, $\widetilde{f}_c$ and $\widetilde{\theta}(t)$ are the nominal carrier frequency and the phase jitter of the oscillator at the receiving end, respectively, and $h_{\rm R}(t)$ is the *receiving filter*; further discussion on the receiving filter will be given later.

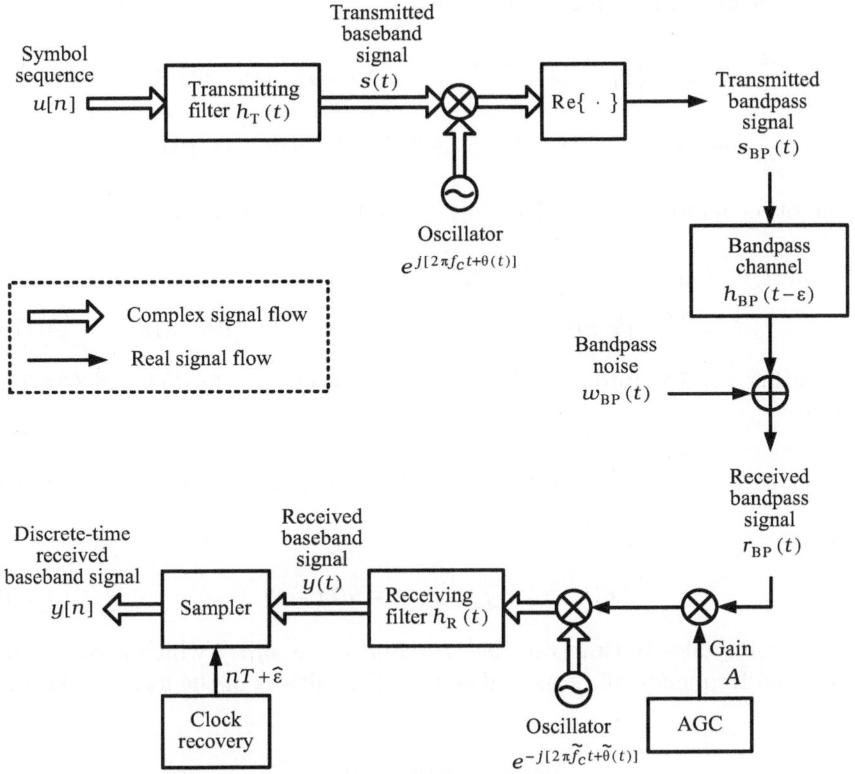

Fig. 4.30 Schematic diagram of a simplified wireless communication system

To simplify the channel model for $y(t)$, let us make the following reasonable assumptions.

(A4-11) The local carrier frequency at the receiving end, $\widetilde{f_c}$, well approximates the one at the transmitting end, f_c.

(A4-12) Both the phase fluctuations $\theta(t)$ and $\widetilde{\theta}(t)$ are slowly time varying.

(A4-13) The receiving filter $h_{\mathrm{R}}(t)$ is a lowpass filter.

Moreover, let

$$r_{\mathrm{BP}}(t) = \mathrm{Re}\left\{r_{\mathrm{LP}}(t)e^{j2\pi f_c t}\right\}, \tag{4.174}$$

$$w_{\mathrm{BP}}(t) = \mathrm{Re}\left\{w_{\mathrm{LP}}(t)e^{j2\pi f_c t}\right\}, \tag{4.175}$$

$$h_{\mathrm{BP}}(t) = 2 \cdot \mathrm{Re}\left\{h_{\mathrm{LP}}(t)e^{j2\pi f_c t}\right\}; \tag{4.176}$$

that is, $r_{\mathrm{LP}}(t)$ is the equivalent lowpass signal of $r_{\mathrm{BP}}(t)$, $w_{\mathrm{LP}}(t)$ the equivalent lowpass noise of $w_{\mathrm{BP}}(t)$, and $h_{\mathrm{LP}}(t)$ the equivalent lowpass channel of $h_{\mathrm{BP}}(t)$.

Substituting (4.174) into (4.173) yields

$$y(t) = \frac{A}{2} \cdot h_{\mathrm{R}}(t) \star \left\{ r_{\mathrm{LP}}(t) e^{-j[2\pi(\widetilde{f}_c - f_c)t + \widetilde{\theta}(t)]} + r^*_{\mathrm{LP}}(t) e^{-j[2\pi(\widetilde{f}_c + f_c)t + \widetilde{\theta}(t)]} \right\}$$
$$\approx \frac{A}{2} \cdot h_{\mathrm{R}}(t) \star \left\{ r_{\mathrm{LP}}(t) e^{-j[2\pi(\widetilde{f}_c - f_c)t + \widetilde{\theta}(t)]} \right\} \quad \text{(by (A4-13))}. \tag{4.177}$$

On the other hand, from (4.170), (4.171) and (4.172), one can obtain

$$r_{\mathrm{LP}}(t) = h_{\mathrm{LP}}(t - \varepsilon) \star \left[s(t) e^{j\theta(t)} \right] + w_{\mathrm{LP}}(t)$$
$$\approx e^{j\theta(t)} \cdot [h_{\mathrm{LP}}(t - \varepsilon) \star s(t)] + w_{\mathrm{LP}}(t) \quad \text{(by (A4-12))}. \tag{4.178}$$

Substituting (4.178) into (4.177) and by Assumptions (A4-11) and (A4-12), we have

$$y(t) \approx \frac{A}{2} e^{-j\Delta\theta(t)} \cdot [h_{\mathrm{R}}(t) \star h_{\mathrm{LP}}(t - \varepsilon) \star s(t)] + w(t) \tag{4.179}$$

where

$$\Delta\theta(t) \triangleq 2\pi(\widetilde{f}_c - f_c)t + \widetilde{\theta}(t) - \theta(t) \tag{4.180}$$

is the (possibly slowly time-varying) *residual phase offset* which accounts for the residual frequency offset as well as the phase jitters of the local oscillators, and

$$w(t) = \frac{A}{2} e^{-j[2\pi(\widetilde{f}_c - f_c)t + \widetilde{\theta}(t)]} \cdot [h_{\mathrm{R}}(t) \star w_{\mathrm{LP}}(t)] \tag{4.181}$$

corresponds to the equivalent baseband noise.

Substituting (4.169) into (4.179) yields

$$y(t) \approx e^{-j\Delta\theta(t)} \cdot \sum_{k=-\infty}^{\infty} u[k] h(t - kT - \varepsilon) + w(t) \tag{4.182}$$

where

$$h(t) = \frac{A}{2} \cdot h_{\mathrm{R}}(t) \star h_{\mathrm{LP}}(t) \star h_{\mathrm{T}}(t) \tag{4.183}$$

corresponds to the equivalent baseband channel. As shown in Fig. 4.30, the clock recovery block is used to provide an estimate of the symbol timing, $\widehat{\varepsilon}$, that is called the *timing phase*. With this timing phase $\widehat{\varepsilon}$, the discrete-time received baseband signal, $y[n]$, is acquired from $y(t)$ as follows:

$$y[n] = y(t = nT + \widehat{\varepsilon}) \approx e^{-j\Delta\theta_n} \cdot \sum_{k=-\infty}^{\infty} u[k] h[n-k] + w[n]$$
$$= e^{-j\Delta\theta_n} \cdot \{h[n] \star u[n]\} + w[n] \tag{4.184}$$

where $\Delta\theta_n = \Delta\theta(t = nT + \widehat{\varepsilon})$ is the discrete-time residual phase offset, $w[n] = w(t = nT + \widehat{\varepsilon})$ is the discrete-time baseband noise, and

$$h[n] = h(t = nT + \Delta\varepsilon) \tag{4.185}$$

is the discrete-time baseband channel in which $\Delta\varepsilon \triangleq \widehat{\varepsilon} - \varepsilon$ is called the *timing phase error.* Figure 4.31 depicts the equivalent discrete-time baseband channel model given by (4.184). From (4.183) and (4.185), it is clear that the baseband channel $h[n]$, and therefore the resultant ISI, is determined not only by the transmitting filter $h_{\rm T}(t)$, the equivalent lowpass channel $h_{\rm LP}(t)$ and the receiving filter $h_{\rm R}(t)$, but also by the timing phase $\widehat{\varepsilon}$.

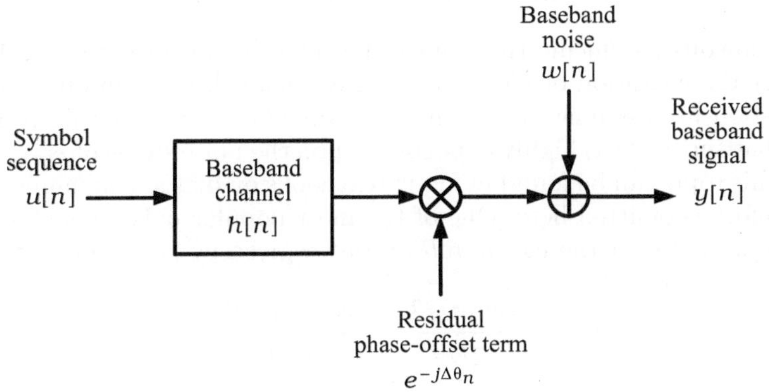

Fig. 4.31 Equivalent discrete-time baseband channel model

Having established the equivalent baseband channel model, we now explain the purposes of the transmitting and receiving filters. Consider the case of no pulse shaping, i.e. $h_{\rm T}(t) = {\rm rect}(t/T)$ where

$${\rm rect}(t) \triangleq \begin{cases} 1, & \text{for } -\dfrac{1}{2} \le t \le \dfrac{1}{2}, \\ 0, & \text{otherwise} \end{cases} \tag{4.186}$$

is a rectangular function. For this case, the transmitted baseband signal $s(t)$ consists of rectangular pulses whose bandwidth is actually unlimited. If these rectangular pulses pass through an ideal lowpass channel (an ideal bandlimited channel), then the received pulses will spread in time and therefore result in the ISI effect. It is for this reason that an appropriate pulse-shaping filter $h_{\rm T}(t)$ is needed to simultaneously reduce the transmitted pulse bandwidth and the ISI effect. On the other hand, the purpose of the receiving filter $h_{\rm R}(t)$ is two-fold. First, as stated in Assumption (A4-13), $h_{\rm R}(t)$ should perform as a lowpass filter so as to filter out the high-frequency term (i.e. the term with frequency $(\widetilde{f}_c + f_c)$) in the first line of (4.177) [75, p. 214]. The other purpose is that $h_R(t)$ should perform as a *matched filter*[13] so as to maximize the SNR in the received baseband signal $y[n]$ [75, p. 226]. From (4.183), it follows that the matched filter $h_R(t)$ should match the filter $h_{LP}(t) \star h_T(t)$. Unfortunately, information about the channel $h_{LP}(t)$ is generally not available to the receiver before proceeding with channel estimation or equalization. For this reason, a practical receiver typically replaces $h_R(t)$ by a filter matched to $h_T(t)$ only, i.e. $h_R(t) = h_T(T - t)$.

Based on the preceding discussions about the transmitting and receiving filters, what we need to consider is the design of the pulse-shaping filter $h_T(t)$ or, equivalently, the pulse

$$p(t) \triangleq h_R(t) \star h_T(t) = h_T(T - t) \star h_T(t) \tag{4.187}$$

for simultaneously reducing the transmitted pulse bandwidth and the ISI effect under the condition of ideal bandlimited channel. In addition, a further consideration in designing $p(t)$ is that the amount of ISI resulting from the timing phase error $\Delta\varepsilon$ is highly dependent upon the pulse shape of $p(t)$. Treatment of this topic can be found in many textbooks on digital communications, and therefore is omitted here. One of the most popular pulses for $p(t)$ is the *raised cosine pulse*, or the *cosine roll-off pulse*, given by [3, p. 560]

$$p(t) = \frac{\sin(\pi t/T)}{\pi t/T} \cdot \frac{\cos(\pi \alpha t/T)}{1 - 4\alpha^2 t^2/T^2} \tag{4.188}$$

where α is called the *roll-off factor* which ranges between 0 and 1. Other well-known pulses such as the *Gaussian pulse* can be found, for example, in [76, pp. 225–234].

Blind Approaches to Baud-Spaced Equalization

Since typical digital communication channels are not minimum phase, SOS based blind approaches cannot apply to baud-spaced equalization; hence, they will not be discussed here. On the other hand, HOS based blind approaches, which are applicable to baud-spaced equalization, can be divided into the following two classes [11].

- *Explicit methods.* Explicit methods utilizing HOS explicitly as the name indicates, include the MNC equalization algorithm, the SE equalization algorithm, the polyspectra based algorithms [17, 77, 78], etc.
- *Implicit methods.* Implicit methods utilizing higher-order moments implicitly, include the Sato algorithm [79], the constant modulus (CM) equalization algorithm [80–82], etc. They are also referred to as *Bussgang-type algo-*

[13] A filter whose impulse response is equal to $s(T - t)$ where $s(t)$ is confined to the time interval $0 \leq t \leq T$, is called the *matched filter* to the signal $s(t)$ [3, p. 237].

rithms because the equalized signal is approximately a *Bussgang process* [14] [11, p. 789].

Among these HOS based blind approaches, we will focus on the MNC and CM equalization algorithms for the following reasons. It is known that the MNC equalization algorithm is equivalent to the CM equalization algorithm under certain conditions [39,83], and that the CM equalization algorithm has been widely applied to alleviating the ISI effect induced in telephone, cable and radio channels in digital communications [18,19,84,85].

To apply the MNC equalization algorithm to baud-spaced equalization, we need, first, to inspect the basic assumptions for the symbol sequence $u[n]$ and the channel $h[n]$. For uncoded digital communication systems (i.e. no channel encoder included) with efficient source compression, each symbol in $u[n]$ can be reasonably assumed to be independently taken from a set of constellation points with equal probability [3, 75]. Accordingly, as mentioned in Section 3.2.4, $u[n]$ is a non-Gaussian process (satisfying Assumption (A4-6)) whose $(p+q)$th-order cumulants $C_{p,q}\{u[n]\} \neq 0$ for even $(p+q)$ and $C_{p,q}\{u[n]\} = 0$ for odd $(p+q)$. As a result, it is better to use the MNC equalization algorithm associated with the criterion $J_{2,2}(v[n])$. On the other hand, for coded digital communication systems, redundancy (memory) in $u[n]$ is introduced by the channel encoder in order to detect and/or correct erroneous bits at the receiving end. Different channel encoding schemes will result in different impacts on the validity of Assumption (A4-6) [86]. For such systems, Assumption (A4-6) may still be valid with a properly chosen channel encoding scheme or by adequately interleaving/scrambling the channel encoded sequence. As for the channel $h[n]$, it can only be treated as a short-term approximation to the real communication channel, which often exhibits a time-varying nature, especially for wireless communication systems. So the adaptive counterpart is more appropriate than the batch processing MNC equalization algorithm for baud-spaced equalization. The reader can find such an adaptive algorithm along with its derivation in [26, 30].

From (4.184) and by Assumptions (A4-11) and (A4-12), it follows that the equalized signal $e[n]$ associated with the MNC equalizer $v[n]$ is given by

$$\begin{aligned} e[n] &\approx \{v[n] \star h[n] \star u[n]\} \cdot e^{-j\Delta\theta_n} + \{v[n] \star w[n]\} \\ &= e_{\mathrm{S}}[n]e^{-j\Delta\theta_n} + e_{\mathrm{N}}[n] \quad \text{(by (4.9) and (4.8))} \end{aligned}$$

or

$$e[n]e^{j\Delta\theta_n} \approx e_{\mathrm{S}}[n] + e_{\mathrm{N}}[n]e^{j\Delta\theta_n} \tag{4.189}$$

where the noise term, $e_{\mathrm{N}}[n]e^{j\Delta\theta_n}$, is still Gaussian distributed by Assumption (A4-7). It can be seen that both $e[n]$ and $e[n]e^{j\Delta\theta_n}$ give rise to the same

[14] A stationary process $e[n]$ is said to be a *Bussgang process* if it satisfies the condition that $E\{e[n]e[n-k]\} = E\{e[n]f(e[n-k])\}$ where $f(\cdot)$ is a function of zero-memory nonlinearity [11, p. 788].

second-order moment $E\{|e[n]|^2\}$ and fourth-order moment $E\{|e[n]|^4\}$. As a result of this observation, the MNC criterion $J_{2,2}(v[n])$ is invariant to the residual phase offset $\Delta\theta_n$, and so is the MNC equalizer $v[n]$. This, in turn, allows the receiver to perform the carrier recovery after the baud-spaced MNC equalization, as shown in Fig. 4.32, thereby bringing some attractive features for carrier recovery (e.g. faster acquisition rate); refer to [85] for further details. Let us emphasize that the detection performance for the receiver equipped with a baud-spaced equalizer is quite sensitive to the timing phase error $\Delta\varepsilon$, that will be discussed in more detail in Chapter 6. For this reason, the receiver requires clock recovery preceding the baud-spaced equalizer.

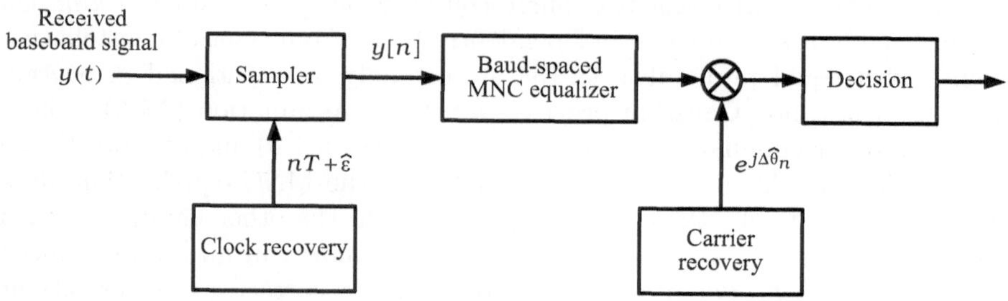

Fig. 4.32 Block diagram of the receiver with a baud-spaced MNC equalizer

Next, let us turn our attention to the CM equalization algorithm, which originated from the goal of finding a cost function that is independent of the carrier phase and symbol constellation. With this goal, Godard [80] proposed the so-called *Godard-p algorithm* for the design of the blind equalizer $v[n]$ by minimizing the function of *pth-order dispersion*:

$$D_p(v[n]) = E\left\{(|e[n]|^p - \Upsilon_p)^2\right\} \tag{4.190}$$

where p is a positive integer and

$$\Upsilon_p = \frac{E\{|u[n]|^{2p}\}}{E\{|u[n]|^p\}} \tag{4.191}$$

is referred to as constant modulus (CM). The special case, Godard-2 algorithm ($p = 2$), was also developed independently by Treichler and Agee [81], that is exactly the CM equalization algorithm. Note that in most literature, the CM equalization algorithm is simply called the *constant modulus algorithm (CMA)*. This name comes from the design philosophy that the equalized signal $e[n]$ must possess the same CM property as the symbol sequence $u[n]$. As a remark, if knowledge about the statistics of $u[n]$ is not available in advance, the value of Υ_2 in (4.190) can be replaced by an arbitrary positive number with a resultant scalar ambiguity in the equalized signal $e[n]$.

It should be noted that the CM equalization algorithm requires not only Assumption (A4-6), but also the condition that the kurtosis $C_{2,2}\{u[n]\} < 0$ (i.e. the sub-Gaussian distribution). Moreover, as the MNC criterion, the pth-order dispersion $D_p(v[n])$ is also invariant to the residual phase offset $\Delta\theta_n$, and thus the structure shown in Fig. 4.32 applies to baud-spaced CM equalization as well.

As a final remark, since both the MNC and CM equalization algorithms rely on the linear structure of the equalizer, their performance may be insufficient for the cases of high-order constellation (e.g. 256-QAM, 1024-QAM, etc.). It turns out that nonlinear blind equalizers such as the *decision-directed equalizer* or the *decision-feedback equalizer* are often needed in practical applications [87]. Generally speaking, compared to linear blind equalizers, nonlinear blind equalizers can provide better detection performance against the effects of ISI and noise, only when a good initial condition (e.g. a sufficiently low amount of ISI) is provided. Hence, practical QAM receivers generally employ a blind linear equalizer for initial ISI reduction and then switch to nonlinear equalization for further performance improvement after the convergence of the blind linear equalizer.

Simulation Example of Baud-Spaced Equalization

Let us show some simulation results of baud-spaced equalization using the hybrid MNC equalization algorithm with $J_{2,2}(v[n])$. In the simulation, we employed the channel model (4.184) to generate the received baseband signal $y[n]$. The residual phase offset was assumed to be $\Delta\theta_n = (2\pi n/100) + (5\pi/6)$ where the LTI channel $h[n]$ was a nonminimum-phase FIR system with three zeros at $z = 0.188 + j0.0075$, $-0.158 - j0.426$ and $6.5 - j6.35$. The symbol sequence $u[n]$ was assumed to be a QAM signal with unity variance and the noise sequence $w[n]$ was complex white Gaussian. A single run was performed to obtain the MNC equalizer $v_{\mathrm{MNC}}[n]$ with $L_1 = 0$ and $L_2 = 15$ for the following two cases.

Case A. 16-QAM constellation, SNR = 20 dB, and data length $N = 2048$.
Case B. 256-QAM constellation, SNR = 30 dB, and data length $N = 2048$.

For Case A, Fig. 4.33a displays the constellation diagram of the received signal $y[n]$ (i.e. $\mathrm{Im}\{y[n]\}$ versus $\mathrm{Re}\{y[n]\}$), Fig. 4.33b that of the equalized signal $e[n]$, and Fig. 4.33c that of the equalized signal after phase adjustment, namely, $e[n]e^{j\Delta\theta_n}$. One can observe that the constellation points in Fig. 4.33a are completely indiscernible, and that several circular patterns exist in Fig. 4.33b since the residual phase-offset term $e^{-j\Delta\theta_n}$ still remains in the equalized signal $e[n]$. In Fig. 4.33c, however, the constellation points are well discernible due to the perfect phase adjustment (accounting for the operation of carrier recovery). These results verify the fact that the MNC equalization algorithm is independent of the residual phase offset. Note that the amount of ISI is

−7.2268 dB before equalization and −22.4337 dB after the MNC equalization for Case A. That is, the MNC equalizer $v_{\mathrm{MNC}}[n]$ provides 15.2068 dB improvement in ISI.

On the other hand, the simulation results of Case B are displayed in Figs. 4.34a–c. Even though the MNC equalizer $v_{\mathrm{MNC}}[n]$ has provided around 14 dB ISI improvement for this case, circular patterns as seen in Fig. 4.33b are almost not visible in Fig. 4.34b, and the constellation points in Fig. 4.34c are hardly discernible in spite of the clear appearance of a square constellation pattern. This indicates that utilization of only blind linear equalizer is not sufficient for this case.

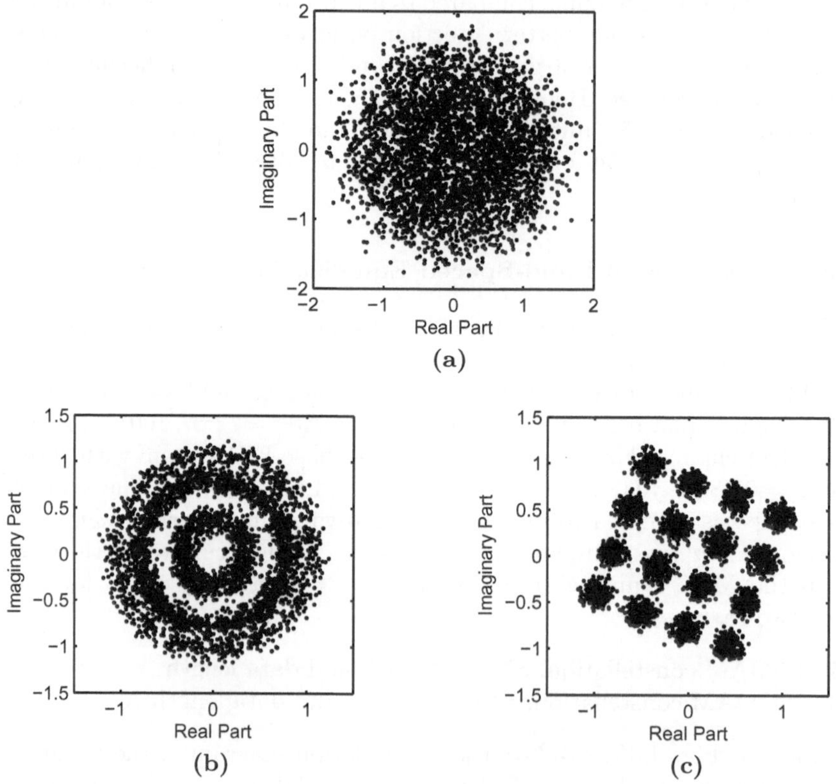

Fig. 4.33 Simulation results of Case A. Constellation diagram of (**a**) the received signal $y[n]$, (**b**) the equalized signal $e[n]$ and (**c**) the phase-adjusted equalized signal $e[n]e^{j\Delta\theta_n}$

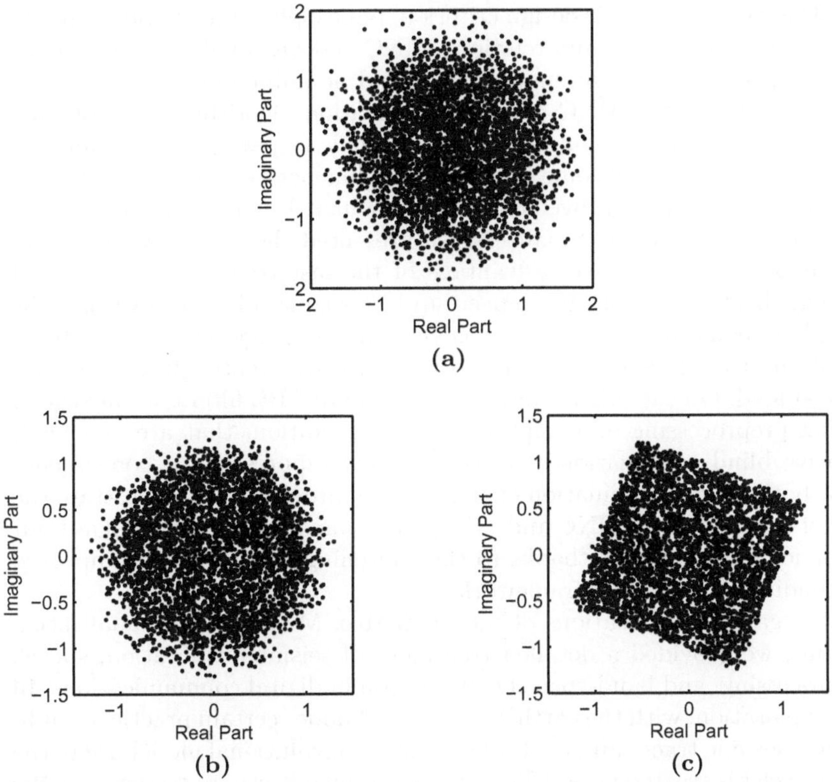

Fig. 4.34 Simulation results of Case B. Constellation diagram of (**a**) the received signal $y[n]$, (**b**) the equalized signal $e[n]$ and (**c**) the phase-adjusted equalized signal $e[n]e^{j\Delta\theta_n}$

4.6 Summary and Discussion

We have addressed the problem of SISO blind equalization along with two performance indices, namely the ISI and the SNR improvement-or-degradation ratio, for evaluation of designed blind equalizers, and have introduced some fundamentals of the well-known ZF and LMMSE equalizers, which are often used for nonblind equalization. We then introduced the fundamentals of linear prediction including forward and backward LPE filters along with the lattice structure and relevant algorithms, namely the Levinson–Durbin recursion and the Burg algorithm, for efficiently solving for and estimating the filter coefficients and reflection coefficients. After introducing the fundamentals of linear prediction, we dealt with SOS based blind equalization by means of forward LPE filters, that is known as (linear) predictive deconvolution and is capable of removing the ISI induced by minimum-phase systems only. As for HOS based blind equalization, we provided a detailed treatment of the MNC and

SE equalizers including their design criterion/philosophy, equalization capability, algorithms, properties, and relations. MNC and SE equalizers are capable of ISI mitigation for both minimum-phase and nonminimum-phase systems. Analytic results for the MNC and SE equalization algorithms are not only instrumental to interpretation of equalized signals (especially for seismic exploration), but also unveil some insights into their performance and relations that lead to algorithm improvements. In particular, by virtue of the relation between the MNC and SE equalizers, we presented the hybrid MNC equalization algorithm which takes advantage of the fast convergence feature of the SE equalization algorithm to speed up the original MNC algorithm. The hybrid MNC equalization algorithm is thought of as one of the best blind equalization algorithms based on performance, convergence speed and computational load. On the other hand, with the aid of LPE filters, we presented whitening preprocessing and improved initial condition, that are of benefit to iterative blind equalization algorithms in ISI suppression or convergence speed or both. Several simulation examples were provided to demonstrate the efficacy of the predictive, MNC and SE equalization algorithms as well as their performance improvements thanks to the whitening preprocessing, improved initial condition and hybrid framework.

With regard to applications of the predictive, MNC and SE equalization algorithms, we provided a detailed treatment of seismic exploration, speech signal processing, and baud-spaced equalization in digital communications. In seismic exploration, with the earth's layer model under certain practical conditions, each seismic trace can be established as a convolutional model where the seismic wavelet is treated as an LTI system and the reflectivity sequence as the source signal input to the LTI system. Due to the random nature of the reflectivity sequence, one can apply HOS based blind equalization approaches only using fourth-order or other even-order cumulants. Two representatives of blind seismic deconvolution approaches are the linear prediction approach and the minimum entropy deconvolution approach; the latter is also the predecessor of the MNC equalization algorithm. Let us emphasize that although seismic deconvolution was described via an SISO framework, the SIMO framework is also constantly considered in practice. In speech signal processing, the speech production model (a convolutional model) for voiced and unvoiced speeches was introduced, and speech analysis and synthesis by means of speech deconvolution was then introduced. Speech analysis and synthesis are essential to further speech processing applications such as speech coding, speech recognition and speaker recognition. Utilization of the linear prediction approach is adequate for speech coding since the phase distortion due to the nonminimum-phase speech production system has virtually no effect on speech perception. However, in addition to magnitude information, phase information may also provide useful features for performance improvements on speech recognition and speaker recognition, and so HOS based blind equalization approaches can be applied. In digital communications, we introduced a typical wireless communication system, and established the discrete-time baseband channel model based on baud-spaced sampling of the received signal. In addition to considering the physical channel effect, we also took into account the effects of the pulse shaping filter, matched filter, carrier frequency offset, phase jitter, and timing offset. For baud-spaced equalization, the MNC equalization algorithm as well as the widely used CM equalization algorithm were both introduced in the batch processing framework although their adaptive counterparts are usually applied in practice. It is important to note that blind linear equalizers are generally insufficient for baud-spaced equalization of high-order constellation systems such as 256-QAM and 1024-QAM systems, making them often used as the equalizer at the preliminary stage to provide an appropriate condition (a sufficiently low amount of ISI) for switching to a nonlinear equalization mode.

The material introduced in this chapter has covered almost all the essential concepts and design philosophies of blind equalization or deconvolution algorithms in terms of the SISO framework. In the following chapters, these continue to be the foundation of blind equalization and system identification for both the MIMO and 2-D cases.

Appendix 4A Proof of Property 4.17

By (3.119), (3.127) and (3.130), we have

$$\sigma_e^2 = \frac{1}{2\pi}\int_{-\pi}^{\pi} S_e(\omega)d\omega = \frac{1}{2\pi}\int_{-\pi}^{\pi} \left[\sigma_u^2|H(\omega)|^2 + S_w(\omega)\right]\cdot|V(\omega)|^2 d\omega. \quad (4.192)$$

This reveals that the denominator of $J_{p,q}(v[n])$ is dependent on the magnitude response $|V(\omega)|$, but independent of the phase response $\arg[V(\omega)]$.

Regarding the numerator of $J_{p,q}(v[n])$, let $\Phi(\omega) = \arg[G(\omega)] = \arg[H(\omega)] + \arg[V(\omega)]$ and $\Theta = \arg[C_{p,q}\{u[n]\}]$. By Property 3.17, Property 3.22 and (3.132), we have

$$C_{p,q}\{e[n]\} = C_{p,q}\{\mathrm{es}[n]\} = \left(\frac{1}{2\pi}\right)^{p+q-1}\int_{-\pi}^{\pi} S_{p,q}^{\mathrm{es}}(\omega_1,\omega_2,...,\omega_{p+q-1})d\boldsymbol{\omega} \quad (4.193)$$

where we have used the shorthand notation '$\int_{-\pi}^{\pi}$' and '$d\boldsymbol{\omega}$' to stand for '$\int_{-\pi}^{\pi}\cdots\int_{-\pi}^{\pi}$' and '$d\omega_1\cdots d\omega_{p+q-1}$,' respectively. This equation, together with (3.141), gives rise to

$$\begin{aligned}|C_{p,q}\{e[n]\}| &= \Bigg| \left(\frac{1}{2\pi}\right)^{p+q-1} \int_{-\pi}^{\pi} \left|S_{p,q}^{\rm es}(\omega_1,\omega_2,...,\omega_{p+q-1})\right| \\ &\quad \cdot \exp\left\{ j\left[\Theta + \Phi\left(\sum_{i=1}^{p+q-1}\omega_i\right) + \sum_{i=1}^{p-1}\Phi(-\omega_i) - \sum_{i=p}^{p+q-1}\Phi(\omega_i)\right]\right\} d\boldsymbol{\omega}\Bigg| \\ &\le \left(\frac{1}{2\pi}\right)^{p+q-1} \int_{-\pi}^{\pi} \left|S_{p,q}^{\rm es}(\omega_1,\omega_2,...,\omega_{p+q-1})\right| d\boldsymbol{\omega}. \end{aligned} \tag{4.194}$$

It is clear that the equality of (4.194) requires

$$\Phi\left(\sum_{i=1}^{p+q-1}\omega_i\right) + \sum_{i=1}^{p-1}\Phi(-\omega_i) - \sum_{i=p}^{p+q-1}\Phi(\omega_i) = \varphi, \quad \forall \omega_i \in [-\pi,\pi) \tag{4.195}$$

where φ is a real constant independent of ω_i. This requirement is equivalent to

$$\Phi\left(-\sum_{i=1}^{p-1}(-\omega_i) + \sum_{i=p}^{p+q-1}\omega_i\right) = -\sum_{i=1}^{p-1}\Phi(-\omega_i) + \sum_{i=p}^{p+q-1}\Phi(\omega_i) + \varphi, \tag{4.196}$$

implying that the optimum $\Phi(\omega)$ associated with the maximum of $J_{p,q}(v[n])$ is a linear function of ω for $\omega \in [-\pi,\pi)$. Therefore, we have finished the proof that $\arg[V_{\rm MNC}(\omega)]$ is given by (4.129) regardless of the value of $|V_{\rm MNC}(\omega)|$.

Q.E.D.

Problems

4.1. Show that under Assumptions (A4-1) through (A4-4), the frequency response of the doubly infinite LMMSE equalizer $v_{\rm MSE}[n]$ is given by (4.138).

4.2. Prove Property 4.4.

4.3. Prove Theorem 4.5. (*Hint*: Use the Cauchy–Schwartz inequality.)

4.4. Prove Theorem 4.6.

4.5. Prove the second line of (4.49).

4.6. Prove (4.59). (*Hint*: Use the orthogonality principle.)

4.7. Prove Property 4.13. (*Hint*: Use the orthogonality principle.)

4.8. Derive (4.91) and prove Lemma 4.15.

4.9. Prove (4.113) for SNR $= \infty$ and (4.114) for any SNR by virtue of (4.96), (4.106), (3.126) and Theorem 3.39.

4.10. Prove Theorem 4.19.

Computer Assignments

4.1. Consider the same simulation conditions as in Example 4.25, except that the system $H(z)$ is a nonminimum-phase ARMA(2,2) LTI filter given by

$$H(z) = \frac{1 - 2.7z^{-1} + 0.5z^{-2}}{1 + 0.1z^{-1} - 0.12z^{-2}}. \tag{4.197}$$

Write a computer program and run the simulation as in Example 4.25 for each of the following equalization algorithms.

(a) The linear prediction approach using the Burg algorithm.
(b) The MNC equalization algorithm using the approximate BFGS method.
(c) The SE equalization algorithm.
(d) The LSE equalization algorithm.
(e) The hybrid MNC equalization algorithm.

4.2. Consider the same simulation conditions as in Example 4.26, except that the system $H(z)$ is given by (4.157). Write a computer program and run the simulation as in Example 4.26 for each of the following equalization algorithms.

(a) The linear prediction approach using the Burg algorithm.
(b) The MNC equalization algorithm using the approximate BFGS method.
(c) The SE equalization algorithm.
(d) The MNC equalization algorithm preceded by whitening preprocessing.
(e) The SE equalization algorithm preceded by whitening preprocessing.

4.3. Consider the same simulation conditions as in Example 4.27, except that the system $H(z) = H_1(z)H_2(z)$ where $H_1(z)$ and $H_2(z)$ are causal FIR filters with coefficients $\{1, 0, -1\}$ and $\{0.04, -0.05, 0.07, -0.21, -0.5, 0.72, 0.36, 0, 0.21, 0.03, 0.07\}$, respectively. Write a computer program and run the simulation as in Example 4.27 for each of the following equalization algorithms.

(a) The SE equalization algorithm using the typical initial condition $v^{[0]}[n] = \delta[n - L_C]$.
(b) The SE equalization algorithm using the improved initial condition $v^{[0]}[n] = v_S[n] \star a_M[n]$.

References

1. S. Haykin, ed., *Blind Deconvolution.* New Jersey: Prentice-Hall, 1994.
2. R. D. Gitlin and J. E. Mazo, "Comparison of some cost functions for automatic equalization," *IEEE Trans. Communicatioins*, vol. 21, no. 3, pp. 233–237, Mar. 1973.
3. J. G. Proakis, *Digital Communications.* New York: McGraw-Hill, 2001.
4. J. M. Mendel, "White-noise estimators for seismic data processing in oil exploration," *IEEE Trans. Automatic Control*, vol. 22, no. 5, pp. 694–706, Oct. 1977.
5. J. M. Mendel, "Single-channel white-noise estimators for deconvolution," *Geophysics*, vol. 43, no. 1, pp. 102–124, Feb. 1978.
6. J. M. Mendel, "Minimum-variance deconvolution," *IEEE Trans. Geoscience and Remote Sensing*, vol. 19, pp. 161–171, 1981.
7. C.-Y. Chi and J. M. Mendel, "Performance of minimum-variance deconvolution filter," *IEEE Trans. Acoustics, Speech, and Signal Processing*, vol. 32, no. 6, pp. 1145–1153, Dec. 1984.
8. C.-Y. Chi, "A further analysis for the minimum-variance deconvolution filter performance," *IEEE Trans. Acoustics, Speech, and Signal Processing*, vol. 35, no. 6, pp. 888–889, June 1987.
9. C.-C. Feng, "Studies on Cumulant Based Inverse Filter Criteria for Blind Deconvolution," Ph.D. Dissertation, Department of Electrical Engineering, National Tsing Hua University, Taiwan, R.O.C., June 1999.
10. C. W. Therrien, *Discrete Random Signals and Statistical Signal Processing.* New Jersey: Prentice-Hall, 1992.
11. S. Haykin, *Adaptive Filter Theory.* New Jersey: Prentice-Hall, 1996.
12. G. H. Golub and C. F. van Loan, *Matrix Computations.* London: The Johns Hopkins University Press, 1989.
13. M. H. Hayes, *Statistical Digital Signal Processing and Modeling.* New York: John Wiley & Sons, 1996.
14. J. Makhoul, "Linear prediction: A tutorial review," *Proc. IEEE*, vol. 63, no. 4, pp. 561–580, Apr. 1975.
15. S. Haykin, *Modern Filters.* New York: Macmillan, 1989.
16. J. M. Mendel, "Tutorial on higher-order statistics (spectra) in signal processing and system theory: Theoretical results and some applications," *Proc. IEEE*, vol. 79, no. 3, pp. 278–305, Mar. 1991.
17. C. L. Nikias and A. P. Petropulu, *Higher-order Spectra Analysis: A Nonlinear Signal Processing Framework.* New Jersey: Prentice-Hall, 1993.
18. C. R. Johnson, Jr., P. Schniter, T. J. Endres, J. D. Behm, D. R. Brown, and R. A. Casas, "Blind equalization using the constant modulus criterion: A review," *Proc. IEEE*, vol. 86, no. 10, pp. 1927–1950, Oct. 1998.
19. J. K. Tugnait, L. Tong, and Z. Ding, "Single-user channel estimation and equalization," *IEEE Signal Processing Magazine*, vol. 17, no. 3, pp. 17–28, May 2000.
20. G. B. Giannakis, Y. Hua, P. Stoica, and L. Tong, ed., *Signal Processing Advances in Wireless and Mobile Communications.* Upper Saddle River: Prentice-Hall, 2001.
21. Z. Ding and Y. Li, ed., *Blind Equalization and Identification.* New York: Marcel Dekker, 2001.

22. C.-C. Feng and C.-Y. Chi, "Performance of Shalvi and Weinstein's deconvolution criteria for channels with/without zeros on the unit circle," *IEEE Trans. Signal Processing*, vol. 48, no. 2, pp. 571–575, Feb. 2000.
23. R. A. Wiggins, "Minimum entropy deconvolution," *Geoexploration*, vol. 16, pp. 21–35, 1978.
24. D. L. Donoho, On Minimum Entropy Deconvolution, in *Applied Time Series Analysis II*, D.F. Findly, ed. New York: Academic Press, 1981.
25. T. J. Ulrych and C. Walker, "Analytic minimum entropy deconvolution," *Geophysics*, vol. 47, no. 9, pp. 1295–1302, Sept. 1982.
26. O. Shalvi and E. Weinstein, "New criteria for blind deconvolution of nonminimum phase systems (channels)," *IEEE Trans. Information Theory*, vol. 36, no. 2, pp. 312–321, Mar. 1990.
27. J. K. Tugnait, "Inverse filter criteria for estimation of linear parametric models using higher order statistics," in *Proc. IEEE International Conference on Acoustics, Speech, and Signal Processing*, vol. 5, Toronto, Ontario, Canada, Apr. 14–17, 1991, pp. 3101–3104.
28. J. K. Tugnait, "Comments on 'New criteria for blind deconvolution of nonminimum phase systems (channels)'," *IEEE Trans. Inform. Theory*, vol. 38, no. 1, pp. 210–213, Jan. 1992.
29. J. K. Tugnait, "Estimation of linear parametric models using inverse filter criteria and higher order statistics," *IEEE Trans. Signal Processing*, vol. 41, no. 11, pp. 3196–3199, Nov. 1993.
30. O. Shalvi and E. Weinstein, Universal Methods for Blind Deconvolution, A Chapter in *Blind Deconvolution*, S. Haykin, ed. New Jersey: Prentice-Hall, 1994.
31. C.-Y. Chi and M.-C. Wu, "A unified class of inverse filter criteria using two cumulants for blind deconvolution and equalization," in *Proc. IEEE International Conference on Acoustics, Speech, and Signal Processing*, Detroit, Michigan, May 9–12, 1995, pp. 1960–1963.
32. C.-Y. Chi and M.-C. Wu, "Inverse filter criteria for blind deconvolution and equalization using two cumulants," *Signal Processing*, vol. 43, no. 1, pp. 55–63, Apr. 1995.
33. J. A. Cadzow and X. Li, "Blind deconvolution," *Digital Signal Processing Journal*, vol. 5, no. 1, pp. 3–20, Jan. 1995.
34. J. A. Cadzow, "Blind deconvolution via cumulant extrema," *IEEE Signal Processing Magazine*, vol. 13, no. 3, pp. 24–42, May 1996.
35. O. Shalvi and E. Weinstein, "Super-exponential methods for blind deconvolution," *IEEE Trans. Informtion Theory*, vol. 39, no. 2, pp. 504–519, Mar. 1993.
36. C.-C. Feng and C.-Y. Chi, "A two-step lattice super-exponential algorithm for blind equalization," in *Proc. Fourth Computer and Communications Symposium*, Taoyuan, Taiwan, Oct. 7–8, 1998, pp. 329–335.
37. P. A. Regalia and M. Mboup, "Undermodeled equalization: a characterization of stationary points for a family of blind criteria," *IEEE Trans. Signal Processing*, vol. 47, no. 3, pp. 760–770, Mar. 1999.
38. C.-C. Feng and C.-Y. Chi, "Performance of cumulant based inverse filters for blind deconvolution," *IEEE Trans. Signal Processing*, vol. 47, no. 7, pp. 1922–1935, July 1999.
39. C.-Y. Chi, C.-Y. Chen, and B.-W. Li, "On super-exponential algorithm, constant modulus algorithm and inverse filter cirteria for blind equalization," in

Proc. IEEE Signal Processing Workshop on Statistical Signal and Array Processing, Pocono Manor, Pennsylvania, Aug. 14–16, 2000, pp. 216–220.
40. C.-Y. Chi, C.-C. Feng, and C.-Y. Chen, "Performance of super-exponential algorithm for blind equalization," in *Proc. IEEE 51st Vehicular Technology Conference*, Tokyo, Japan, May 15–18, 2000, pp. 1864–1868.
41. M. Gu and L. Tong, "Domains of attraction of Shalvi-Weinstein receivers," *IEEE Trans. Signal Processing*, vol. 49, no. 7, pp. 1397–1408, July 2001.
42. Z. Ding, C. R. Johnson, Jr., and R. A. Kennedy, Global Convergence Issues with Linear Blind Adaptive Equalizers, A Chapter in *Blind Deconvolution*, S. Haykin, ed. New Jersey: Prentice-Hall, 1994.
43. M. Mboup and P. A. Regalia, "On the equivalence between the super-exponential algorithm and a gradient search method," in *Proc. IEEE International Conference on Acoustics, Speech, and Signal Processing*, Paris, France, Mar. 15–19, 1999, pp. 2643–2646.
44. M. Mboup and P. A. Regalia, "A gradient search interpretation of the super-exponential algorithm," *IEEE Trans. Information Theory*, vol. 46, no. 7, pp. 2731–2734, Nov. 2000.
45. I. Rodríguez, A. Mánuel, A. Carlosena, A. Bermúdez, J. del Río, and S. Shariat, "Signal processing in ocean bottom seismographs for refraction seismology," in *Proc. IEEE Instrumentation and Measurement Technology Conference*, Como, Italy, May 18–20, 2004, pp. 2086–2089.
46. D. D. Sternlicht, "Looking back: A history of marine seismic exploration," *IEEE Potentials*, vol. 18, no. 1, pp. 36–38, 1999.
47. L. C. Wood and S. Treitel, "Seismic signal processing," *Proc. IEEE*, vol. 63, no. 4, pp. 649–661, Apr. 1975.
48. M. T. Silvia and E. A. Robinson, *Deconvolution of Geophysical Time Series in the Exploration for Oil and Natural Gas.* Amsterdam, Netherlands: Elsevier, 1979.
49. J. M. Mendel, *Optimal Seismic Deconvolution: An Estimation-based Approach.* New York: Academic Press, 1983.
50. H. Luo and Y. Li, "The application of blind channel identification techniques to prestack seismic deconvolution," *Proc. IEEE*, vol. 86, no. 10, pp. 2082–2089, Oct. 1998.
51. E. A. Robinson and O. M. Osman, ed., *Deconvolution 2*, ser. Geophysics Reprint Series, no. 17. Oklahoma: Society of Exploration Geophysicists, 1996.
52. V. K. Arya and H. D. Holden, "Deconvolution of seismic data – An overview," *IEEE Trans. Geoscience Electronics*, vol. GE-16, no. 2, pp. 95–98, Apr. 1978.
53. J. Kormylo and J. M. Mendel, "Maximum-likelihood deconvolution," *IEEE Trans. Geoscience and Remote Sensing*, vol. GE-21, pp. 72–82, 1983.
54. J. M. Mendel, *Maximum-likelihood Deconvolution: A Journey into Model Based Signal Processing.* New York: Springer-Verlag, 1990.
55. J. M. Mendel, "Some modeling problems in reflection seismology," *IEEE ASSP Magazine*, pp. 4–17, Apr. 1986.
56. W. C. Gray, "Variable norm deconvolution," Ph.D. Dissertation, Dept. Geophysics, Stanford University, Aug. 1979.
57. Q. Li, B.-H. Juang, C.-H. Lee, Q. Zhou, and F. K. Soong, "Recent advancements in automatic speaker authentication," *IEEE Robotics and Automation Magazine*, vol. 6, no. 1, pp. 24–34, Mar. 1999.
58. L. R. Rabiner and R. W. Schafer, *Digital Signal Processing of Speech Signals.* New Jersey: Prentice-Hall, 1978.

59. D. O'Shaughnessy, "Speaker recognition," *IEEE ASSP Magazine*, pp. 4–17, Oct. 1986.
60. T. W. Parsons, *Voice and Speech Processing.* New York: McGraw-Hill, 1987.
61. J. M. Naik, "Speaker verification: A tutorial," *IEEE Communications Magazine*, vol. 28, no. 1, pp. 42–48, Jan. 1990.
62. A. S. Spanias and F. H. Wu, "Speech coding and speech recognition technologies: A review," in *Proc. IEEE International Sympoisum on Circuits and Systems*, vol. 1, June 11–14, 1991, pp. 572–577.
63. J. R. Deller, Jr., J. G. Proakis, and J. H. L. Hansen, *Discrete-time Processing of Speech Signals.* New York: Macmillan, 1993.
64. J. W. Picone, "Signal modeling techniques in speech recognition," *Proc. IEEE*, vol. 81, no. 9, pp. 1215–1247, Sept. 1993.
65. A. V. Oppenheim and R. W. Schafer, *Discrete-time Signal Processing.* New Jersey: Prentice-Hall, 1989.
66. M. C. Doğan and J. M. Mendel, "Real-time robust pitch detector," in *Proc. IEEE International Conference on Acoustics, Speech, and Signal Processing*, vol. 1, San Francisco, California, Mar. 23–26, 1992, pp. 129–132.
67. J. D. Markel, "The SIFT algorithm for fundamental frequency estimation," *IEEE Trans. Audio and Electroacoustics*, vol. AU-20, no. 5, pp. 367–377, Dec. 1972.
68. C.-Y. Chi and W.-T. Chen, The Deconvolution of Speech Signals by the Utlization of Higher-order Statistics Based Algorithms, A Chapter in *Control and Dynamic Systems*, part 2 of 2, C. T. Leondes, ed. New York: Academic Press, 1994.
69. H.-M. Chien, H.-L. Yang, and C.-Y. Chi, "Parametric cumulant based phase estimation of 1-D and 2-D nonminimum phase systems by allpass filtering," *IEEE Trans. Signal Processing*, vol. 45, no. 7, pp. 1742–1762, July 1997.
70. E. Nemer, R. Goubran, and S. Mahmoud, "The third-order cumulant of speech signals with application to reliable pitch estimation," in *Proc. IEEE Signal Processing Workshop on Statistical Signal and Array Processing*, Portland, Oregon, Sept. 14–16, 1998, pp. 427–430.
71. C.-Y. Chi, "Fourier series based nonminimum phase model for statistical signal processing," *IEEE Trans. Signal Processing*, vol. 47, no. 8, pp. 2228–2240, Aug. 1999.
72. K. K. Paliwal and M. M. Sondhi, "Recognition of noisy speech using cumulant-based linear prediction analysis," in *Proc. IEEE International Conference on Acoustics, Speech, and Signal Processing*, vol. 1, Toronto, Ontario, Apr. 14–17, 1991, pp. 429–432.
73. A. Moreno, S. Tortola, J. Vidal, and J. A. R. Fonollosa, "New HOS-based parameter estimation methods for speech recognition in noisy environments," in *Proc. IEEE International Conference on Acoustics, Speech, and Signal Processing*, vol. 1, Detroit, Michigan, May 9–12, 1995, pp. 429–432.
74. M. C. Orr, B. J. Lithgow, and R. Mahony, "Speech features found in a continuous high order statistical analysis of speech," in *Proc. the Second Joint EMBS/BMES Conference*, vol. 1, Houston, Texas, Oct. 23–26, 2002, pp. 180–181.
75. H. Meyr, M. Moeneclaey, and S. A. Fechtel, *Digital Communication Receivers: Synchronization, Channel Estimation, and Signal Processing.* New York: John Wiley & Sons, 1998.

76. T. S. Rappaport, *Wireless Communications: Principles and Practice.* New Jersey: Prentice-Hall, 1996.
77. D. Hatzinakos and C. L. Nikias, "Blind equalization using a tricepstrum-based algorithm," *IEEE Trans. Communicatioins*, vol. 39, no. 5, pp. 669–682, May 1991.
78. D. Hatzinakos and C. L. Nikias, Blind Equalization Based on Higher-Order Statistics (H.O.S.), A Chapter in *Blind Deconvolution*, S. Haykin, ed. New Jersey: Prentice-Hall, 1994.
79. Y. Sato, "A method of self-recovering equalization for multi-level amplitude modulation," *IEEE Trans. Communicatioins*, vol. 23, pp. 679–682, June 1975.
80. D. Godard, "Self recovering equalization and carrier tracking in two-dimensional data communication systems," *IEEE Trans. Communicatioins*, vol. 28, no. 11, pp. 1867–1875, Nov. 1980.
81. J. R. Treichler and B. G. Agee, "A new approach to multipath correction of constant modulus signals," *IEEE Trans. Acoustics, Speech, and Signal Processing*, vol. 31, no. 2, pp. 349–472, Apr. 1983.
82. J. R. Treichler and M. G. Larimore, "New processing techniques based on the constant modulus adaptive algorithm," *IEEE Trans. Acoustics, Speech, and Signal Processing*, vol. 33, no. 2, pp. 420–431, Apr. 1985.
83. P. A. Regalia, "On the equivalence between the Godard and Shalvi-Weinstein schemes of blind equalization," *Signal Processing*, vol. 73, pp. 185–190, 1999.
84. J. R. Treichler, I. Fijalkow, and C. R. Johnson, Jr., "Fractionally spaced equalizers: how long should they really be?" *IEEE Signal Processing Magazine*, pp. 65–81, May 1996.
85. J. R. Treichler, M. G. Larimore, and J. C. Harp, "Practical blind demodulators for high-order QAM signals," *Proc. IEEE*, vol. 86, no. 10, pp. 1907–1926, 1998.
86. J. Mannerkoski and V. Koivunen, "Autocorrelation properties of channel encoded sequences – Applicability to blind equalization," *IEEE Trans. Signal Processing*, vol. 48, no. 12, pp. 3501–3507, Dec. 2000.
87. S. U. H. Qureshi, "Adaptive equalization," *Proc. IEEE*, vol. 73, no. 9, pp. 1349–1387, Sept. 1985.

5

MIMO Blind Equalization Algorithms

This chapter introduces some widely used fundamental MIMO blind equalization algorithms using either SOS or HOS of MIMO system outputs. A subspace approach and a linear prediction approach using SOS are introduced for the SIMO case, and a matrix pencil method is introduced for the MIMO case. As to HOS based approaches, the MNC and SE equalization algorithms are introduced for an MIMO system with temporally i.i.d. inputs followed by their properties and relations which lead to an improved MNC equalization algorithm with much faster convergence speed and lower computational load than the MNC equalization algorithm. Then an equalization-GCD equalization algorithm is introduced for an MIMO system with temporally colored inputs, which makes use of the improved MNC equalization algorithm and a greatest common divisor computation algorithm. Finally, some simulation results for the introduced SIMO and MIMO blind equalization algorithms are presented. The chapter concludes with a summary and discussion.

5.1 MIMO Linear Time-Invariant Systems

Discrete-time SISO LTI systems (channels) have been introduced in Section 3.1. Discrete-time MIMO LTI systems are basically the extension of discrete-time SISO LTI systems, but some of their basic properties and definitions are rather different. Moreover, an MIMO LTI system itself is a multi-variable LTI system for which some of its properties must be considered in the design of MIMO equalization algorithms. These properties comprise poles/zeros, *normal rank* of MIMO LTI systems that are addressed next. For simplicity, an MIMO system also refers to an MIMO LTI system hereafter.

5.1.1 Definitions and Properties

Consider a K-input M-output discrete-time LTI system defining the relation between K inputs $u_1[n], u_2[n], ..., u_K[n]$ and M outputs $x_1[n], x_2[n], ..., x_M[n]$.

Let

$$\mathbf{u}[n] = (u_1[n], u_2[n], ..., u_K[n])^T$$

and

$$\mathbf{x}[n] = (x_1[n], x_2[n], ..., x_M[n])^T$$

be the input and output vectors of this MIMO discrete-time LTI system, respectively. Then, for any input vector $\mathbf{u}[n]$, the output vector $\mathbf{x}[n]$ is completely determined by the discrete-time convolutional model

$$\mathbf{x}[n] = \mathbf{H}[n] \star \mathbf{u}[n] = \sum_{k=-\infty}^{\infty} \mathbf{H}[k]\mathbf{u}[n-k] \tag{5.1}$$

where the same notation '$\star$' as for the SISO case represents the operation of MIMO *convolution* (*linear convolution*),[1] and $\mathbf{H}[n]$ is an $M \times K$ matrix defined as

$$\mathbf{H}[n] = \begin{pmatrix} h_{11}[n] & h_{12}[n] & \cdots & h_{1K}[n] \\ h_{21}[n] & h_{22}[n] & \cdots & h_{2K}[n] \\ \vdots & \vdots & \ddots & \vdots \\ h_{M1}[n] & h_{M2}[n] & \cdots & h_{MK}[n] \end{pmatrix} \tag{5.2}$$

in which the (i, j)th element $h_{ij}[n]$ is the impulse response of the SISO system from the jth input $u_j[n]$ to the ith output $x_i[n]$.

The $M \times K$ matrix sequence $\mathbf{H}[n]$ defined by (5.2) is also called the impulse response of the MIMO LTI system, with the number of rows equal to the number of system outputs and the number of columns equal to the number of system inputs. Specifically, let

$$\mathbf{h}_j[n] = (h_{1j}[n], h_{2j}[n], ..., h_{Mj}[n])^T$$

(an $M \times 1$ vector) be the jth column of $\mathbf{H}[n]$, i.e.

$$\mathbf{H}[n] = (\mathbf{h}_1[n], \mathbf{h}_2[n], ..., \mathbf{h}_K[n]) .$$

Then, the system output vector $\mathbf{x}[n]$ given by (5.1) can also be expressed as

$$\mathbf{x}[n] = \sum_{j=1}^{K} \mathbf{x}_j[n] = \sum_{j=1}^{K} \mathbf{h}_j[n] \star u_j[n] \tag{5.3}$$

where

$$\mathbf{x}_j[n] = \mathbf{h}_j[n] \star u_j[n] = \sum_{k=-\infty}^{\infty} \mathbf{h}_j[k]u_j[n-k] \tag{5.4}$$

[1] Note that the convolution operation defined in the MIMO case is associative and distributive but not commutative because matrix multiplication is not commutative, i.e. $\mathbf{AB} \neq \mathbf{BA}$ in general.

is the contribution from the jth input $u_j[n]$ to the M outputs of $\mathbf{x}[n]$. Moreover, the ith output $x_i[n]$ of the system, from (5.1) and (5.2), can be seen to be

$$x_i[n] = \sum_{j=1}^{K} \sum_{k=-\infty}^{\infty} h_{ij}[k]u_j[n-k]$$

which is also a mapping from all the K system inputs to the output $x_i[n]$.

The properties of the MIMO LTI system can be introduced either directly from the properties of each individual SISO system $h_{ij}[n]$ or through transform-domain analysis of the impulse response $\mathbf{H}[n]$. Some definitions and properties of this system are presented next.

Definition 5.1 (Causality). *An MIMO system $\mathbf{H}[n]$ is causal if for every choice of n_0 the output vector sequence $\mathbf{x}[n]$ at $n = n_0$ depends only on the input vector sequence $\mathbf{u}[n]$ for $n \leq n_0$.*

A property directly related to the above definition is as follows.

Property 5.2. *An MIMO system $\mathbf{H}[n]$ is causal if and only if each component $h_{ij}[n]$ of $\mathbf{H}[n]$ is causal; otherwise, $\mathbf{H}[n]$ is noncausal.*

Definition 5.3 (Stability). *An MIMO system $\mathbf{H}[n]$ is stable in BIBO sense if any bounded input vector sequence $\mathbf{u}[n]$ produces a bounded output vector sequence $\mathbf{x}[n]$.*[2]

A property directly related to the stability of an MIMO system is as follows.

Property 5.4. *An MIMO system $\mathbf{H}[n]$ is BIBO stable if and only if each component $h_{ij}[n]$ of $\mathbf{H}[n]$ is BIBO stable.*

Definition 5.5 (FIR). *An MIMO system $\mathbf{H}[n]$ is FIR if there exist finite integers n_L and n_R such that $\mathbf{H}[n] = \mathbf{0}_{M\times K}$ ($M \times K$ zero matrix) for $n < n_L$ and $n > n_R$.*

A property of FIR MIMO systems is as follows.

Property 5.6. *An MIMO system $\mathbf{H}[n]$ is FIR if and only if each component $h_{ij}[n]$ of $\mathbf{H}[n]$ is FIR; otherwise, $\mathbf{H}[n]$ is IIR.*

---

[2] A vector valued sequence $\mathbf{u}[n]$ is bounded if the *maximal amplitude* or *peak*

$$\|\mathbf{u}\|_\infty = \sup_n \|\mathbf{u}[n]\|$$

is finite, where the symbol $\|\mathbf{u}\|_\infty$ denotes the ℓ^∞ norm of $\mathbf{u}[n]$.

As in the SISO case, the transform-domain analysis is advisable for illuminating some important properties of an MIMO system such as poles and zeros related properties. The definitions of the frequency response and the transfer function of MIMO systems are defined below, respectively, by analogy with the definitions for SISO systems.

Definition 5.7 (Frequency Response). *The frequency response* $\boldsymbol{\mathcal{H}}(\omega)$ *of the MIMO system is defined as the DTFT of the impulse response* $\mathbf{H}[n]$, *i.e.*

$$\boldsymbol{\mathcal{H}}(\omega) = \sum_{n=-\infty}^{\infty} \mathbf{H}[n]e^{-j\omega n} = \begin{pmatrix} H_{11}(\omega) & H_{12}(\omega) & \cdots & H_{1K}(\omega) \\ H_{21}(\omega) & H_{22}(\omega) & \cdots & H_{2K}(\omega) \\ \vdots & \vdots & \ddots & \vdots \\ H_{M1}(\omega) & H_{M2}(\omega) & \cdots & H_{MK}(\omega) \end{pmatrix}$$

where

$$H_{ij}(\omega) = \sum_{n=-\infty}^{\infty} h_{ij}[n]e^{-j\omega n}.$$

The physical meaning of the *frequency response* for the MIMO case is pretty much the same as for the SISO case. Specifically, a $K \times 1$ *complex multichannel sinusoidal* is defined as

$$\mathbf{u}_s[n] = \begin{pmatrix} A_1 \exp(j\omega_0 n + \phi_1) \\ A_2 \exp(j\omega_0 n + \phi_2) \\ \vdots \\ A_K \exp(j\omega_0 n + \phi_K) \end{pmatrix} = \mathbf{a} \exp(j\omega_0 n) \tag{5.5}$$

where $A_i \geq 0, \forall i$ and $\mathbf{a} = (A_1 \exp(j\phi_1), A_2 \exp(j\phi_2), ..., A_K \exp(j\phi_K))^T$. Each element of the multichannel sinusoidal $\mathbf{u}_s[n]$ is characterized by a complex sinusoidal of the same frequency but with different magnitude and phase in general. If this multichannel sinusoidal $\mathbf{u}_s[n]$ is input to an MIMO LTI system characterized by the impulse response $\mathbf{H}[n]$, the output of this system can be obtained from (5.1) as

$$\begin{aligned} \mathbf{x}[n] &= \mathbf{H}[k] \star \mathbf{u}_s[n] = \sum_{k=-\infty}^{\infty} \mathbf{H}[k]\mathbf{u}_s[n-k] \\ &= \sum_{k=-\infty}^{\infty} \mathbf{H}[k]\mathbf{a} \exp(j\omega_0(n-k)) = \left\{ \sum_{k=-\infty}^{\infty} \mathbf{H}[k] \exp(-j\omega_0 k) \right\} \mathbf{a} \exp(j\omega_0 n) \\ &= \boldsymbol{\mathcal{H}}(\omega_0)\mathbf{u}_s[n] \quad \text{(by Definition 5.7).} \end{aligned}$$

As for the SISO case, the response of an MIMO system to a complex multichannel sinusoidal is also a complex multichannel sinusoidal of the same frequency but modified in magnitude and phase by the associated frequency response of this system.

Definition 5.8 (Transfer Function). *The transfer function $\mathcal{H}(z)$ of the MIMO system is defined as the z-transform of the impulse response $\mathbf{H}[n]$, i.e.*

$$\mathcal{H}(z) = \sum_{n=-\infty}^{\infty} \mathbf{H}[n]z^{-n} = \begin{pmatrix} H_{11}(z) & H_{12}(z) & \cdots & H_{1K}(z) \\ H_{21}(z) & H_{22}(z) & \cdots & H_{2K}(z) \\ \vdots & \vdots & \ddots & \vdots \\ H_{M1}(z) & H_{M2}(z) & \cdots & H_{MK}(z) \end{pmatrix}$$

where

$$H_{ij}(z) = \sum_{n=-\infty}^{\infty} h_{ij}[n]z^{-n}.$$

The normal rank of the transfer function matrix is described as follows.

Definition 5.9 (Normal Rank). *The normal rank of $\mathcal{H}(z)$, denoted as r, is defined as the maximum rank of $\mathcal{H}(z)$ over all z (including $z = 0$ and $z = \infty$).*

It is easy to see that $r \leq \min(M, K)$.[3]

The definition of poles of a transfer function $\mathcal{H}(z)$ given below is similar to that for SISO systems.

Definition 5.10 (Poles of MIMO Systems). *The poles of a transfer function $\mathcal{H}(z)$ are defined as those values of complex variable z for which at least one component $H_{ij}(z)$ of $\mathcal{H}(z)$ becomes infinity.*

A property related to poles of an MIMO system is as follows.

Property 5.11. *An MIMO system $\mathbf{H}[n]$ is stable if and only if it has no poles on the unit-circle. Thus, any MIMO FIR system is stable.*

Unlike the SISO system, the zeros of an MIMO system are defined as the values of the complex variable z for which $\mathcal{H}(z)$ *loses rank*, as defined as follows.

Definition 5.12 (Zeros of MIMO Systems). *The zeros of an $M \times K$ transfer function $\mathcal{H}(z)$ are defined as those values of complex variable z for which* $\text{rank}(\mathcal{H}(z)) < r$ *(the normal rank of $\mathcal{H}(z)$).*

The zeros defined above are also called *transmission zeros* due to the following reasons. Assume that the transfer function $\mathcal{H}(z)$ is an $M \times K$ matrix with $M \geq K$. If $\mathcal{H}(z)$ loses rank at $z = z_o$, i.e. $\text{rank}(\mathcal{H}(z_o)) < r \leq$

[3] If $r = \min(M, K)$, the matrix function $\mathcal{H}(z)$ is said to be *full* normal rank.

$min(M, K) = K$, it can be shown that there exist at least one input vector $\mathbf{u}_o \neq \mathbf{0}$ (zero right direction)[4], such that

$$\mathcal{H}(z_o)\mathbf{u}_o = \mathbf{0} \tag{5.6}$$

implying that the vector $\mathbf{u}_o$ belongs to the null space of $\mathcal{H}(z_o)$.[5] Hence, the *transmission* from certain inputs is *blocked* at $z = z_o$, thereby giving rise to the name "transmission zeros" for the zeros defined above. An example for illuminating the zeros of a 2×2 system is given next.

Example 5.13 (Transmission Zeros)
Consider the following 2×2 transfer function

$$\mathcal{H}(z) = \begin{pmatrix} (z+0.5)(z^{-1}+0.5) & z-0.2 \\ 0 & 1 \end{pmatrix}.$$

Obviously, the rank of $\mathcal{H}(z)$ is equal to unity at $z = -0.5$. According to Definition 5.12, $z = -0.5$ is a zero of $\mathcal{H}(z)$. One can easily find a nonzero vector $\mathbf{u}_o$ (such as $\mathbf{u}_o = (1,\ 0)^T$) such that

$$\mathcal{H}(z = -0.5)\mathbf{u}_o = \mathbf{0}. \qquad \square$$

It seems to be possible to find zeros of some transfer functions by inspection as presented in the above example. Note that the zeros of each individual $H_{ij}(z)$ of the transfer function $\mathcal{H}(z)$ may not be the zeros of $\mathcal{H}(z)$. If all the components of a column of $\mathcal{H}(z)$ have a common zero, then it is also a zero of $\mathcal{H}(z)$. In general, it is nontrivial to find the zeros of $\mathcal{H}(z)$. Fortunately, given the transfer function $\mathcal{H}(z)$ of an MIMO system, the exact locations as well as the multiplicity of poles and zeros of the system can be effectively determined through a transformation of $\mathcal{H}(z)$ into a pseudo-diagonal form, called the *Smith–McMillan Form* [1], which will be introduced later in Section 5.1.2.

With the above definitions of poles and zeros, we can now define the *minimum-phase system* for the MIMO case as follows.

Definition 5.14 (Minimum-Phase System). *An MIMO system* $\mathbf{H}[n]$ *is said to be a minimum-phase system if all its poles and (transmission) zeros are strictly inside the unit circle.*

So, a minimum-phase MIMO FIR system possesses the following property.

---

4 If $z = z_o$ is a zero of $\mathcal{H}(z)$, there also exist at least one nonzero vector $\mathbf{w}_o$ (zero left direction) such that $\mathbf{w}_o^T\mathcal{H}(z_o) = \mathbf{0}^T$. The vector $\mathbf{w}_o^T$ belongs to the null space generated by the rows of $\mathcal{H}(z_o)$.

5 The number of linearly independent vectors that satisfy (5.6) depends on the rank loss of $\mathcal{H}(z_o)$. This number is known as the *geometric multiplicity* of the zero.

Property 5.15. *An MIMO FIR system* $\mathbf{H}[n]$ *is minimum-phase if all its (transmission) zeros lie inside the unit circle.*

Finally, for a given $M \times K$ transfer function $\boldsymbol{\mathcal{H}}(z)$ with $M \geq K$, the inverse system of $\boldsymbol{\mathcal{H}}(z)$ is defined as follows.

Definition 5.16 (Inverse System). *A* $K \times M$ *LTI system* $\mathbf{H}_{\mathrm{I}}[n]$ *is called the (left) inverse system of the MIMO system* $\mathbf{H}[n]$ *if*

$$\mathbf{H}_{\mathrm{I}}[n] \star \mathbf{H}[n] = \delta[n]\mathbf{I} \tag{5.7}$$

or equivalently,

$$\boldsymbol{\mathcal{H}}_{\mathrm{I}}(\omega)\boldsymbol{\mathcal{H}}(\omega) = \mathbf{I}. \tag{5.8}$$

Some of the existing algorithms for MIMO blind equalization are actually for the optimum design of the left inverse system $\mathbf{H}_{\mathrm{I}}[n]$ in some statistical sense.

5.1.2 Smith–McMillan Form

First of all, let us introduce some definitions for polynomials and polynomial matrices, and some elementary operations through which the *Smith–McMillan form* of a transfer function can be achieved systematically.

Definition 5.17 (Monic Polynomial). *A polynomial* $P(z)$ *is said to be monic if it has the coefficient* 1 *for the highest power of* z*.*

Definition 5.18 (Coprime Polynomials). *Two polynomials are said to be coprime if they have no common factors, or equivalently no common roots.*

Definition 5.19 (Polynomial Matrix). *An* $M \times K$ *matrix* $\boldsymbol{\mathcal{P}}(z) = [P_{ij}(z)]$ *is a polynomial matrix if* $P_{ij}(z)$ *is a polynomial in* z*, for all* $i = 1, 2, ..., M$ *and* $j = 1, 2, ..., K$*.*

Definition 5.20 (Unimodular Matrix). *A square polynomial matrix* $\boldsymbol{\mathcal{U}}(z)$ *is said to be a unimodular matrix if its determinant* $\det(\boldsymbol{\mathcal{U}}(z))$ *is a nonzero constant (independent of* z*). So, the inverse of a unimodular matrix is also a unimodular matrix.*

Definition 5.21 (Elementary Operations). *An elementary operation on a polynomial matrix is one of the following three operations:*

(EO-1) *interchange of two rows (columns);*
(EO-2) *multiplication of one row (column) by a constant;*
(EO-3) *addition of one row (column) times a polynomial to another row (column).*

A common property of these elementary operations is that they do not change the rank of the polynomial matrix on which the operations are performed. Each of these elementary operations can be represented as a pre- or post-multiplication of the target polynomial matrix by a suitable *elementary matrix* as defined below.

Definition 5.22 (Elementary Matrix). *A left (right) elementary matrix is a matrix such that, when it multiplies from the left (right) a polynomial matrix, then it performs a row (column) elementary operation on the polynomial matrix. All elementary matrices are unimodular.*

Definition 5.23 (Equivalent Matrices). *Two polynomial matrices $\mathcal{P}_1(z)$ and $\mathcal{P}_2(z)$ are equivalent matrices, denoted by $\mathcal{P}_1(z) \sim \mathcal{P}_2(z)$, if there exist a set of left elementary matrices $\{\mathbf{L}_1(z), \mathbf{L}_2(z), ..., \mathbf{L}_{n_L}(z)\}$ and a set of right elementary matrices $\{\mathbf{R}_1(z), \mathbf{R}_2(z), ..., \mathbf{R}_{n_R}(z)\}$, such that*

$$\mathcal{P}_1(z) = \mathbf{L}_1(z)\mathbf{L}_2(z)\cdots\mathbf{L}_{n_L}(z)\mathcal{P}_2(z)\mathbf{R}_1(z)\mathbf{R}_2(z)\cdots\mathbf{R}_{n_R}(z). \tag{5.9}$$

Smith Form of Polynomial Matrices

Consider an $M \times K$ polynomial matrix $\mathcal{P}(z)$. By a suitable choice of the elementary matrices $\mathbf{L}_i(z)$ and $\mathbf{R}_j(z)$, the polynomial matrix $\mathcal{P}(z)$ can be *transformed* into an $M \times K$ pseudo-diagonal matrix $\mathcal{S}(z)$ through a sequence of elementary operations as stated in Definition 5.23 and the following theorem:

Theorem 5.24 (Smith Form). *Let $\mathcal{P}(z)$ be an $M \times K$ polynomial matrix of normal rank r. Then $\mathcal{P}(z) \sim \mathcal{S}(z)$ where*

$$\mathcal{S}(z) = \begin{cases} \left(\overline{\mathcal{E}}(z)\ \mathbf{0}_{M\times(K-M)}\right), & \text{for } M < K, \\ \overline{\mathcal{E}}(z), & \text{for } M = K, \\ \begin{pmatrix} \overline{\mathcal{E}}(z) \\ \mathbf{0}_{(M-K)\times K} \end{pmatrix}, & \text{for } M > K \end{cases} \tag{5.10}$$

in which

$$\overline{\mathcal{E}}(z) = \mathrm{diag}(\overline{\varepsilon}_1(z), \overline{\varepsilon}_2(z), ..., \overline{\varepsilon}_r(z), 0, 0, ..., 0)$$

is a $\min(M,K) \times \min(M,K)$ *diagonal matrix. Furthermore, $\overline{\varepsilon}_i(z)$, $i = 1, 2, ...,$ r, are monic polynomials, and $\overline{\varepsilon}_i(z)$ is a factor of $\overline{\varepsilon}_{i+1}(z)$ (or $\overline{\varepsilon}_i(z)|\overline{\varepsilon}_{i+1}(z)$), i.e. $\overline{\varepsilon}_i(z)$ divides $\overline{\varepsilon}_{i+1}(z)$.*

See [1] for the proof of Theorem 5.24. The pseudo-diagonal matrix $\mathcal{S}(z)$ given by (5.10) is called the *Smith form* of $\mathcal{P}(z)$.

Instead of finding the polynomials $\overline{\varepsilon}_i(z)$'s from $\mathcal{P}(z)$ via a sequence of elementary operations, one can obtain them from the *determinant divisors* of $\mathcal{P}(z)$:

$$D_0(z) \triangleq 1$$
$$D_i(z) = \text{greatest common divisor (GCD) for}$$
$$\text{all } i \times i \text{ sub-determinants (i.e. minors) of } \boldsymbol{\mathcal{P}}(z), \quad i = 1, 2, ..., r,$$

where each GCD[6] is normalized as a monic polynomial. The polynomials $\overline{\varepsilon}_i(z)$ can be shown to be given by [2]:

$$\overline{\varepsilon}_i(z) = \frac{D_i(z)}{D_{i-1}(z)}, \qquad i = 1, 2, ..., r.$$

Smith–McMillan Form of Rational Matrices

Consider an $M \times K$ rational matrix $\boldsymbol{\mathcal{G}}(z) = [G_{ij}(z)]$. Let $d(z)$ be the *least common multiple* (LCM) of the denominators of all elements $G_{ij}(z)$'s of $\boldsymbol{\mathcal{G}}(z)$. Then, $\boldsymbol{\mathcal{G}}(z)$ can be written as

$$\boldsymbol{\mathcal{G}}(z) = \frac{1}{d(z)} \boldsymbol{\mathcal{P}}(z) \tag{5.11}$$

where $\boldsymbol{\mathcal{P}}(z)$ is a polynomial matrix.

A straightforward application of Theorem 5.24 to the polynomial matrix $\boldsymbol{\mathcal{P}}(z)$ in (5.11) leads to the following result, which gives a pseudo-diagonal form for a rational matrix:

Theorem 5.25 (Smith–McMillan Form). *Let $\boldsymbol{\mathcal{G}}(z) = [G_{ij}(z)]_{M \times K}$ be an $M \times K$ matrix of normal rank r, where each $G_{ij}(z)$ is a rational function of z. Then, $\boldsymbol{\mathcal{G}}(z) \sim \boldsymbol{\mathcal{M}}(z)$, where*

$$\boldsymbol{\mathcal{M}}(z) = \begin{cases} \left(\boldsymbol{\mathcal{E}}(z) \ \mathbf{0}_{M \times (K-M)} \right), & \text{for } M < K, \\ \boldsymbol{\mathcal{E}}(z), & \text{for } M = K, \\ \begin{pmatrix} \boldsymbol{\mathcal{E}}(z) \\ \mathbf{0}_{(M-K) \times K} \end{pmatrix}, & \text{for } M > K \end{cases} \tag{5.12}$$

where

$$\boldsymbol{\mathcal{E}}(z) = \text{diag}\left(\frac{\varepsilon_1(z)}{\psi_1(z)}, \frac{\varepsilon_2(z)}{\psi_2(z)}, ..., \frac{\varepsilon_r(z)}{\psi_r(z)}, 0, 0, ..., 0 \right) \tag{5.13}$$

is a $\min(M, K) \times \min(M, K)$ diagonal rational matrix in which $\{\varepsilon_i(z), \psi_i(z)\}$ is a pair of monic and coprime polynomials. Furthermore, $\varepsilon_i(z)$ is a factor of $\varepsilon_{i+1}(z)$ and $\psi_i(z)$ is a factor of $\psi_{i-1}(z)$, i.e. $\varepsilon_i(z)|\varepsilon_{i+1}(z)$ and $\psi_i(z)|\psi_{i-1}(z)$.

[6] If $a(z)$ and $b(z)$ are polynomials, the GCD of $a(z)$ and $b(z)$ is the polynomial $c(z)$ that has the two properties: (i) $c(z)|a(z)$ and $c(z)|b(z)$, and (ii) for any $\overline{c}(z)$ such that $\overline{c}(z)|a(z)$ and $\overline{c}(z)|b(z)$, then $\overline{c}(z)|c(z)$.

The pseudo-diagonal matrix $\boldsymbol{\mathcal{M}}(z)$ given by (5.12) is called the Smith–McMillan form of $\boldsymbol{\mathcal{G}}(z)$.

Proof: We first write $\boldsymbol{\mathcal{G}}(z)$ as in (5.11). By Theorem 5.24, one can find the Smith form, i.e. $\boldsymbol{\mathcal{S}}(z)$, of $\boldsymbol{\mathcal{P}}(z)$ through elementary operations. Thus, we have

$$\boldsymbol{\mathcal{G}}(z) \sim \frac{1}{d(z)}\boldsymbol{\mathcal{S}}(z) = \boldsymbol{\mathcal{M}}(z),$$

which leads to the form given by (5.12) and (5.13) after all possible cancellations between $\overline{\varepsilon}_i(z)$'s and $d(z)$.

Example 5.26 (Smith–McMillan Form)
To manifest the computations involved in obtaining the Smith–McMillan form of a transfer function, consider the 3×2 transfer function $\boldsymbol{\mathcal{G}}(z)$ below:

$$\boldsymbol{\mathcal{G}}(z) = \begin{pmatrix} \dfrac{1}{(z+1)(z+2)} & \dfrac{-1}{(z+1)(z+2)} \\ \dfrac{z^2+z-4}{(z+1)(z+2)} & \dfrac{2z^2-z-8}{(z+1)(z+2)} \\ \dfrac{z-2}{z+1} & \dfrac{2z-4}{z+1} \end{pmatrix}.$$

Note that the normal rank of $\boldsymbol{\mathcal{G}}(z)$ is $r = 2$. The Smith–McMillan form of $\boldsymbol{\mathcal{G}}(z)$ can be achieved through the following steps.

Step 1. Determine $d(z)$. The LCM of all denominators in $\boldsymbol{\mathcal{G}}(z)$ is

$$d(z) = (z+1)(z+2).$$

Step 2. Determine $\boldsymbol{\mathcal{P}}(z)$.

$$\boldsymbol{\mathcal{P}}(z) = d(z)\boldsymbol{\mathcal{G}}(z) = \begin{pmatrix} 1 & -1 \\ z^2+z-4 & 2z^2-z-8 \\ (z-2)(z+2) & (2z-4)(z+2) \end{pmatrix}.$$

Step 3. Find the set of polynomials $D_i(z)$. $D_0(z) = 1$,

$$D_1(z) = \text{GCD}\{1, -1, z^2+z-4, 2z^2-z-8, z^2-4, 2z^2-8\} = 1,$$

and

$$D_2(z) = \text{GCD}\left\{\begin{vmatrix} 1 & -1 \\ z^2+z-4 & 2z^2-z-8 \end{vmatrix}, \begin{vmatrix} 1 & -1 \\ z^2-4 & 2z^2-8 \end{vmatrix}, \begin{vmatrix} z^2+z-4 & 2z^2-z-8 \\ z^2-4 & 2z^2-8 \end{vmatrix}\right\}$$
$$= \text{GCD}\left\{(z+2)(z-2), (z+2)(z-2), z(z+2)(z-2)\right\}$$
$$= (z+2)(z-2).$$

Step 4. Compute $\bar{\varepsilon}_i(z)$ and determine $\boldsymbol{\mathcal{S}}(z)$. They are

$$\bar{\varepsilon}_1(z) = \frac{D_1(z)}{D_0(z)} = \frac{1}{1} = 1.$$
$$\bar{\varepsilon}_2(z) = \frac{D_2(z)}{D_1(z)} = \frac{(z+2)(z-2)}{1} = (z+2)(z-2).$$

The Smith form of $\boldsymbol{\mathcal{P}}(z)$ is thus

$$\boldsymbol{\mathcal{S}}(z) = \begin{pmatrix} 1 & 0 \\ 0 & (z+2)(z-2) \\ 0 & 0 \end{pmatrix}.$$

Step 5. Compute $\varepsilon_i(z)$ and $\psi_i(z)$, and then determine $\boldsymbol{\mathcal{M}}(z)$. The elements in $\boldsymbol{\mathcal{M}}(z)$ are

$$\frac{\varepsilon_1(z)}{\psi_1(z)} = \frac{\bar{\varepsilon}_1(z)}{d(z)} = \frac{1}{(z+1)(z+2)}.$$
$$\frac{\varepsilon_2(z)}{\psi_2(z)} = \frac{\bar{\varepsilon}_2(z)}{d(z)} = \frac{(z+2)(z-2)}{(z+1)(z+2)} = \frac{z-2}{z+1}.$$

Therefore, the Smith–McMillan form of $\boldsymbol{\mathcal{G}}(z)$ is

$$\boldsymbol{\mathcal{G}}(z) \sim \boldsymbol{\mathcal{M}}(z) = \begin{pmatrix} \dfrac{1}{(z+1)(z+2)} & 0 \\ 0 & \dfrac{z-2}{z+1} \\ 0 & 0 \end{pmatrix}.$$

□

Poles and Zeros

Given an $M \times K$ transfer function $\boldsymbol{\mathcal{G}}(z)$, the poles and zeros of $\boldsymbol{\mathcal{G}}(z)$ can be found from the elements of $\boldsymbol{\mathcal{M}}(z)$, which is the Smith–McMillan form of $\boldsymbol{\mathcal{G}}(z)$. The *pole polynomial* associated with $\boldsymbol{\mathcal{G}}(z)$ is defined as

$$P_{\text{pole}}(z) = \prod_{i=1}^{r} \psi_i(z) = \psi_1(z)\psi_2(z)\cdots\psi_r(z). \tag{5.14}$$

The roots of $P_{\text{pole}}(z) = 0$ are the poles of $\boldsymbol{\mathcal{G}}(z)$. Let $B(z)$ be a polynomial either of z or of z^{-1} and

$$\deg(B(z)) = \text{degree of } B(z).$$

Then $\deg(P_{\text{pole}}(z))$ is known as the *McMillan degree*, which is also the total number of poles of the system. Repeated poles can also be identified from $P_{\text{pole}}(z)$ by inspection.

In the same fashion, the *zero polynomial* associated with $\boldsymbol{\mathcal{G}}(z)$ is defined as

$$P_{\text{zero}}(z) = \prod_{i=1}^{r} \varepsilon_i(z) = \varepsilon_1(z)\varepsilon_2(z)\cdots\varepsilon_r(z). \tag{5.15}$$

The roots of $P_{\text{zero}}(z) = 0$ are the transmission zeros of the transfer function $\boldsymbol{\mathcal{G}}(z)$. It can be seen, from (5.15), that any transmission zero of the system must be a factor of at least one of the polynomials $\varepsilon_i(z)$. The normal rank of both $\boldsymbol{\mathcal{G}}(z)$ and $\boldsymbol{\mathcal{M}}(z)$ is r. It is obvious from (5.12) that if any $\varepsilon_i(z) = 0$ for a certain value $z = z_o$, then the rank of $\boldsymbol{\mathcal{M}}(z)$ drops below r for $z = z_o$. Since the ranks of $\boldsymbol{\mathcal{G}}(z)$ and $\boldsymbol{\mathcal{M}}(z)$ are the same, the rank of $\boldsymbol{\mathcal{G}}(z)$ also drops below r for $z = z_o$.

Example 5.27 (Poles and Zeros from Smith–McMillan Form)
This example continues Example 5.26. The pole and zero polynomials of $\boldsymbol{\mathcal{G}}(z)$ can be identified by applying (5.14) and (5.15) to $\boldsymbol{\mathcal{M}}(z)$ obtained in Example 5.26. As a result, the pole polynomial is given by

$$P_{\text{pole}}(z) = [(z+1)(z+2)](z+1) = (z+1)^2(z+2),$$

so the McMillan degree for this system is three. Two poles are located at the $z = -1$ (one in $[\boldsymbol{\mathcal{M}}(z)]_{1,1}$ and the other in $[\boldsymbol{\mathcal{M}}(z)]_{2,2}$), and thus the system is unstable. The other pole is located at $z = -2$. Similarly, the zero polynomial is given by

$$P_{\text{zero}}(z) = z - 2,$$

so there is a (transmission) zero located at $z = 2$. It can easily be checked that the rank of $\boldsymbol{\mathcal{G}}(z)$ drops from 2 to 1 for $z = 2$.

□

5.2 Linear Equalization

In this section we start with an introduction to the MIMO blind equalization problem followed by two fundamental and widely-used equalization criteria, i.e. the peak distortion and MMSE equalization criteria.

5.2.1 Blind Equalization Problem

Problem Statement

Suppose that $u_1[n]$, $u_2[n]$, ..., $u_K[n]$ are the source signals of interest and distorted by an $M \times K$ system $\mathbf{H}[n]$ as shown in Fig. 5.1. The noisy $M \times 1$ output vector $\mathbf{y}[n]$ can be expressed as

$$\mathbf{y}[n] = (y_1[n], y_2[n], ..., y_M[n])^T = \mathbf{x}[n] + \mathbf{w}[n] \tag{5.16}$$

where

$$\mathbf{x}[n] = \mathbf{H}[n] \star \mathbf{u}[n] = \sum_{i=-\infty}^{\infty} \mathbf{H}[i]\mathbf{u}[n-i] \tag{5.17}$$

is the *noise-free* output vector, $\mathbf{u}[n] = (u_1[n], u_2[n], ..., u_K[n])^T$ consists of the K system inputs, and $\mathbf{w}[n] = (w_1[n], w_2[n], ..., w_M[n])^T$ is an $M \times 1$ vector noise.

In general, the system input vector $\mathbf{u}[n]$ and noise vector $\mathbf{w}[n]$ are mutually independent stationary vector random processes, and therefore both the noise-free output vector $\mathbf{x}[n]$ and the noisy output vector $\mathbf{y}[n]$ are stationary vector random processes as well. Then the correlation function and power spectral matrix of the noisy output vector $\mathbf{y}[n]$ are related to those of the noise-free output vector $\mathbf{x}[n]$ as follows:

$$\mathbf{R}_{\mathbf{y}}[l] = E\{\mathbf{y}[n]\mathbf{y}^H[n-l]\} = \mathbf{R}_{\mathbf{x}}[l] + \mathbf{R}_{\mathbf{w}}[l]$$
$$\boldsymbol{\mathcal{S}}_{\mathbf{y}}(\omega) = \mathscr{F}\{\mathbf{R}_{\mathbf{y}}[l]\} = \boldsymbol{\mathcal{S}}_{\mathbf{x}}(\omega) + \boldsymbol{\mathcal{S}}_{\mathbf{w}}(\omega).$$

The signal quality of the received signal $\mathbf{y}[n]$ given by (5.16) can be quantified by the total SNR defined as

$$\mathrm{SNR} = \frac{E\{||\mathbf{x}[n]||^2\}}{E\{||\mathbf{w}[n]||^2\}} = \frac{\mathrm{tr}\{\mathbf{R}_{\mathbf{x}}[0]\}}{\mathrm{tr}\{\mathbf{R}_{\mathbf{w}}[0]\}}. \tag{5.18}$$

In addition to the additive noise, it can be seen from (5.16) and (5.17), that the system $\mathbf{H}[n]$ simultaneously introduces not only "*temporal distortion*" when $\mathbf{H}[n] \neq \mathbf{A}\delta[n-\tau]$ but also "*spatial distortion*" when $\mathbf{H}[n]$ is not diagonal for all n. In wireless communications, the temporal distortion is called the intersymbol interference (ISI) and the spatial distortion is called the multiple access interference (MAI). The blind equalization problem is to extract a desired system input $u_k[n]$ (or multiple system inputs) with a given set of measurements $\mathbf{y}[n]$ without information of the system $\mathbf{H}[n]$. In other words, it is a problem to eliminate both the ISI (temporal distortion) and MAI (spatial distortion) for recovery of the desired system input $u_k[n]$ (or multiple system inputs) from the received $\mathbf{y}[n]$ without information on $\mathbf{H}[n]$.

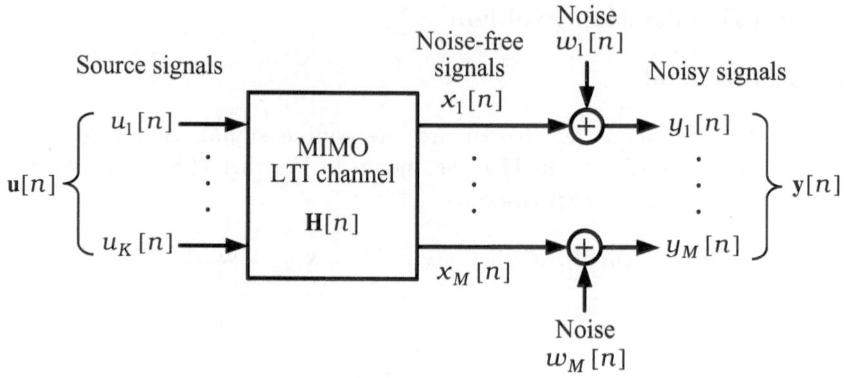

Fig. 5.1 The MIMO system model

MIMO Linear Equalization

Let $\mathbf{v}[n] = (v_1[n], v_2[n], ..., v_M[n])^T$ denote a multiple–input single–output (MISO) linear equalizer to be designed as shown in Fig. 5.2, that consists of a bank of linear FIR filters, with $\mathbf{v}[n] \neq \mathbf{0}$ for $n = L_1, L_1 + 1, ..., L_2$ and length $L = L_2 - L_1 + 1$.

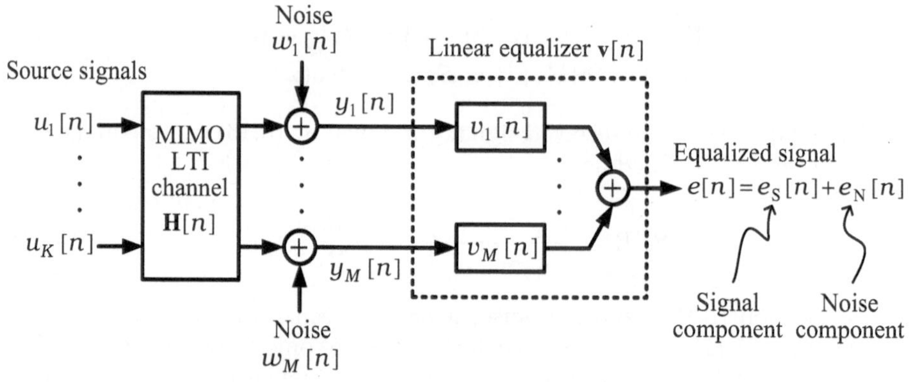

Fig. 5.2 Block diagram of MISO linear equalization

The *equalized signal* (or extracted system input) $e[n]$ can be expressed as

$$e[n] = \mathbf{v}^T[n] \star \mathbf{y}[n] = \sum_{j=1}^{M} v_j[n] \star y_j[n] = e_S[n] + e_N[n] \qquad (5.19)$$

where

$$e_{\mathrm{N}}[n] = \mathbf{v}^T[n] \star \mathbf{w}[n] \quad \text{(by (5.16))} \tag{5.20}$$

corresponds to the noise component in $e[n]$ and

$$\begin{aligned} e_{\mathrm{S}}[n] &= \mathbf{v}^T[n] \star \mathbf{H}[n] \star \mathbf{u}[n] \quad \text{(by (5.16))} \\ &= \mathbf{g}^T[n] \star \mathbf{u}[n] = \sum_{j=1}^{K} g_j[n] \star u_j[n] \end{aligned} \tag{5.21}$$

is the corresponding signal component in which

$$\mathbf{g}[n] = (g_1[n], g_2[n], ..., g_K[n])^T = \mathbf{H}^T[n] \star \mathbf{v}[n] \tag{5.22}$$

is the *overall system* after equalization. The signal component $e_{\mathrm{S}}[n]$ given by (5.21) is depicted in Fig. 5.3 in terms of the overall system $\mathbf{g}[n]$. The goal of MIMO blind equalization is to design an "optimum" equalizer $\mathbf{v}[n]$ such that the signal component $e_{\mathrm{S}}[n]$ approximates one input signal $u_\ell[n]$ (up to a scale factor and a time delay) where $\ell \in \{1, 2, ..., K\}$ is unknown. Note that the determination of ℓ usually needs prior information about the inputs and the associated subchannel $\mathbf{h}_k[n]$, depending on the application. Furthermore, all the K system inputs can be estimated through a multistage successive cancellation (MSC) procedure [3], which will be introduced in Section 5.4.1 later. Finding the optimum equalizer $\mathbf{v}[n]$ is the so-called direct approach for blind equalization as introduced above. On the other hand, extraction of the desired system input (or multiple system inputs) from the received $\mathbf{y}[n]$ can also be achieved indirectly by blind system identification (also called blind channel estimation) and then by utilization of a nonblind equalizer with the use of the estimated system.

It is also possible to design a $K \times M$ blind equalizer $\boldsymbol{\mathcal{V}}(\omega)$ to extract all the K inputs simultaneously. Because the system output $\mathbf{y}[n]$ is invariant to any permutation of the system inputs, a "perfect" blind MIMO equalizer $\boldsymbol{\mathcal{V}}(\omega)$ should satisfy

$$\boldsymbol{\mathcal{V}}(\omega)\boldsymbol{\mathcal{H}}(\omega) = \mathbf{PA}\boldsymbol{\mathcal{D}}(\omega) \tag{5.23}$$

where $\mathbf{P}$ is a $K \times K$ permutation matrix,

$$\mathbf{A} = \mathrm{diag}(\alpha_1, \alpha_2, ..., \alpha_K)$$

is a *multiple-scaling matrix* and

$$\boldsymbol{\mathcal{D}}(\omega) = \mathrm{diag}(e^{-j\omega\tau_1}, e^{-j\omega\tau_2}, ..., e^{-j\omega\tau_K})$$

is a *multiple-delay matrix.*

Evaluation of how an equalized signal $e[n]$ accurately approximates the associated input $\alpha u_\ell[n - \tau]$ is necessary in the MIMO equalizer design. Next, Let us introduce two widely used performance indices.

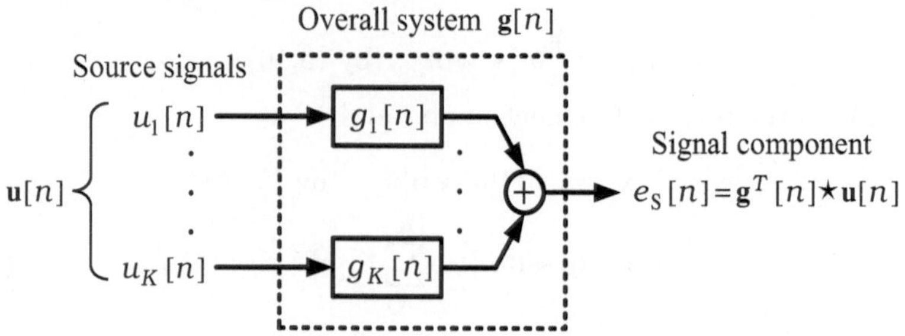

Fig. 5.3 Equivalent block diagram without noise components in Fig. 5.2

Performance Indices

Assuming that $\mathbf{w}[n] = \mathbf{0}$ and thus $e_{\mathrm{N}}[n] = 0$, one can easily see, from (5.21), that the better the equalized signal $e[n] = e_{\mathrm{S}}[n]$ approximates $\alpha u_\ell[n-\tau]$, the better the resultant overall system $\mathbf{g}[n]$ approximates $\alpha\delta[n-\tau]\boldsymbol{\eta}_\ell$. The first commonly used performance index for the designed equalizer is *the amount of ISI* which is defined as:

$$\mathrm{ISI}(\mathbf{g}[n]) = \frac{\sum_{n=-\infty}^{\infty} \|\mathbf{g}[n]\|^2 - \max_{i,n}\{|g_i[n]|^2\}}{\max_{i,n}\{|g_i[n]|^2\}} \geq 0. \tag{5.24}$$

Note that $\mathrm{ISI}(\mathbf{g}[n]) = 0$ if and only if $\mathbf{g}[n] = \alpha\delta[n-\tau]\boldsymbol{\eta}_\ell$ (for some values of α, τ, and ℓ). Therefore, the smaller the value of $\mathrm{ISI}(\mathbf{g}[n])$, the closer the overall system $\mathbf{g}[n]$ to $\alpha\delta[n-\tau]\boldsymbol{\eta}_\ell$.

Another widely used performance index with noise effects taken into account is "*signal-to-interference-plus-noise ratio (SINR)*" of the equalized signal $e[n]$ defined as

$$\mathrm{SINR}(e[n]) = \frac{\sigma_{u_\ell}^2 \cdot |g_\ell[n_0]|^2}{E\left\{|e_{\mathrm{S}}[n]|^2\right\} + E\left\{|e_{\mathrm{N}}[n]|^2\right\} - \sigma_{u_\ell}^2 \cdot |g_\ell[n_0]|^2} \tag{5.25}$$

where $\sigma_{u_\ell}^2 = E\{|u_\ell[n]|^2\}$ and $|g_\ell[n_0]| = \max_{k,n}\{|g_k[n]|\}$.

5.2.2 Peak Distortion and MMSE Equalization Criteria

ZF Equalization

An equalizer $\mathbf{v}[n]$ designed by minimizing the following *peak distortion equalization criterion*

$$J_{\mathrm{PD}}(e[n]) = \mathrm{ISI}(\mathbf{g}[n]) \tag{5.26}$$

is called a ZF equalizer $\mathbf{v}_{\rm ZF}[n]$ due to the fact that the associated $\mathrm{ISI}(\mathbf{g}[n]) = 0$ as $J_{\rm PD}(e[n])$ is minimized and the resultant equalized signal $e[n] = \alpha_\ell u_\ell[n-\tau_\ell]$ (perfect equalization) under the noise-free condition.

Let $\boldsymbol{V}_{\rm ZF}^{[k]}(\omega)$ be an $M \times 1$ MISO ZF equalizer associated with $u_k[n]$. Then the $K \times M$ MIMO ZF equalizer $\boldsymbol{\mathcal{V}}_{\rm ZF}(\omega)$ is therefore

$$\boldsymbol{\mathcal{V}}_{\rm ZF}^T(\omega) = \left(\boldsymbol{V}_{\rm ZF}^{[1]}(\omega), \boldsymbol{V}_{\rm ZF}^{[2]}(\omega), ..., \boldsymbol{V}_{\rm ZF}^{[K]}(\omega)\right),$$

which, according to the peak distortion equalization criterion, satisfies

$$\boldsymbol{\mathcal{V}}_{\rm ZF}(\omega)\boldsymbol{\mathcal{H}}(\omega) = \mathrm{diag}(\alpha_1 e^{-j\omega\tau_1}, \alpha_2 e^{-j\omega\tau_2}, ..., \alpha_K e^{-j\omega\tau_K}) \tag{5.27}$$

or equivalently,

$$\mathbf{V}_{\rm ZF}[n] \star \mathbf{H}[n] = \mathrm{diag}(\alpha_1\delta[n-\tau_1], \alpha_2\delta[n-\tau_2], ..., \alpha_K\delta[n-\tau_K]).$$

It may happen that some of the $\boldsymbol{V}_{\rm ZF}^{[k]}(\omega)$ are unstable. It has been known that a stable ZF equalizer for an SISO system exists if the system has no zeros on the unit circle. For an MIMO system, the condition of full column rank for $\boldsymbol{\mathcal{H}}(z)$ is necessary and sufficient for the existence of a stable ZF equalizer based on the following theorem:

Theorem 5.28 (Existence of Stable ZF Equalizer). *There exists a stable ZF equalizer for an MIMO system* $\mathbf{H}[n]$ *if and only if the following ZF-condition (or linear equalizability condition) is satisfied:*

(C-ZF) $\boldsymbol{\mathcal{H}}(z)$ *is of full column rank for all* $|z| = 1$.

See [4, pp. 295,296] for the proof of Theorem 5.28.

Note that the ZF equalizer $\mathbf{v}_{\rm ZF}[n]$ theoretically needs to be doubly infinite for a general $\mathbf{H}[n]$. However, if the $\mathbf{H}[n]$ is an FIR system satisfying (C-ZF) with $M > K$, then a finite-length equalizer may suffice to achieve ZF equalization [5].

In practical applications, the SNR given by (5.18) is always finite and a ZF equalizer may suffer from the noise enhancement problem (see (5.19) and (5.20)). Therefore, in addition to the suppression of the spatial distortion (or MAI) and temporal distortion (or ISI), the noise reduction should be taken into account in the equalizer design. The MMSE equalization criterion, which serves for these design considerations, is generally preferable to the peak distortion equalization.

MMSE Equalization

An equalizer $\mathbf{v}[n]$ designed by minimizing the following MSE criterion

$$J_{\rm MSE}(e[n]) = E\{|e[n] - u_k[n-\tau_k]|^2\} \tag{5.28}$$

where $k \in \{1, 2, ..., K\}$ and τ_k is an integer, is called an LMMSE equalizer $\mathbf{v}_{\mathrm{MS}}[n]$. It should be noted that the LMMSE equalizer is designed for a specific delay τ_k such that $e[n]$ is a good approximation to $u_k[n - \tau_k]$, and thus a different LMMSE equalizer will be obtained for different choices of τ, especially when the equalizer length is finite.[7]

Let $\boldsymbol{V}_{\mathrm{MS}}^{[k]}(\omega)$ be an $M \times 1$ MISO LMMSE equalizer associated with $u_k[n]$ and $\tau_k = 0$. Assume that $\mathbf{v}_{\mathrm{MS}}^{[k]}[n]$, $k = 1, 2, ..., K$, are doubly infinite. Then, the $K \times M$ LMMSE equalizer $\boldsymbol{\mathcal{V}}_{\mathrm{MS}}(\omega)$ can be obtained by the orthogonality principle as follows (Problem 5.1):

$$\begin{aligned}\boldsymbol{\mathcal{V}}_{\mathrm{MS}}^T(\omega) &= \left(\boldsymbol{V}_{\mathrm{MS}}^{[1]}(\omega), \boldsymbol{V}_{\mathrm{MS}}^{[2]}(\omega), ..., \boldsymbol{V}_{\mathrm{MS}}^{[K]}(\omega)\right) \\ &= \left(\boldsymbol{\mathcal{S}}_{\mathbf{y}}^T(\omega)\right)^{-1} \cdot \boldsymbol{\mathcal{S}}_{\mathbf{yu}}^T(\omega) = \left(\boldsymbol{\mathcal{S}}_{\mathbf{y}}^T(\omega)\right)^{-1} \cdot \boldsymbol{\mathcal{H}}^*(\omega) \cdot \boldsymbol{\mathcal{S}}_{\mathbf{u}}^T(\omega) \qquad (5.29)\end{aligned}$$

where

$$\boldsymbol{\mathcal{S}}_{\mathbf{yu}}(\omega) = \mathscr{F}\{\mathbf{R}_{\mathbf{yu}}[l]\} = \mathscr{F}\{E\{\mathbf{y}[n]\mathbf{u}^H[n-l]\}\}$$

is the cross-power spectrum matrix of $\mathbf{y}[n]$ and $\mathbf{u}[n]$. According to (5.29), for finite SNR, the LMMSE equalizer performs ISI as well as MAI suppression and noise reduction simultaneously.

5.3 SOS Based Blind Equalization Approaches

5.3.1 Blind SIMO Equalization

As with SISO blind equalization, some statistical assumptions about the system inputs and some conditions about the unknown system are required by most algorithms using SOS or HOS. In 1991, Tong, Xu and Kailath [6] proposed the blind identifiability and equalizability of SIMO ($K = 1$) LTI systems using only SOS of the system outputs. Their work led to a number of SOS based SIMO blind system estimation and equalization algorithms as reported in [7–10] under certain assumptions. One may refer to [11] for an overview of these techniques as well as the associated assumptions and conditions. These methods have the attractive characteristic that system estimates can always be obtained by optimizing a quadratic cost function. As mentioned previously, these SOS based equalization approaches usually obtain a system estimate followed by the use of the associated LMMSE equalizer.

Consider the following SIMO model ($M > K = 1$):

$$\mathbf{y}[n] = \mathbf{h}[n] \star u[n] + \mathbf{w}[n] = \sum_{k=0}^{L_h} \mathbf{h}[k] u[n-k] + \mathbf{w}[n] \qquad (5.30)$$

[7] Practically, the *best* delay τ_k with the *minimum* $J_{\mathrm{MSE}}(e[n])$ is suggested for the designed finite-length equalizer.

where $\mathbf{h}[n] = (h_1[n], h_2[n], ..., h_M[n])^T$ is an $M \times 1$ FIR system of order equal to L_h, $u[n]$ is a single scalar input and $\mathbf{w}[n] = (w_1[n], w_2[n], ..., w_M[n])^T$ is an $M \times 1$ vector noise. The SIMO system can be formed when either multiple receiving antennas are used, or when oversampling of the system output (as the baud rate is less than the Nyquist rate, i.e. excess bandwidth exists) is performed in communication systems. By stacking $(N_w + 1)$ successive observations, we define the following vectors:

$$\boldsymbol{y}[n] = \left(\mathbf{y}^T[n], \mathbf{y}^T[n-1], ..., \mathbf{y}^T[n-N_w]\right)^T \tag{5.31}$$

$$\boldsymbol{u}[n] = (u[n], u[n-1], ..., u[n-L_h-N_w])^T \tag{5.32}$$

$$\boldsymbol{w}[n] = \left(\mathbf{w}^T[n], \mathbf{w}^T[n-1], ..., \mathbf{w}^T[n-N_w]\right)^T . \tag{5.33}$$

It can be obtained, from (5.30), that

$$\boldsymbol{y}[n] = \boldsymbol{\Psi}\boldsymbol{u}[n] + \boldsymbol{w}[n] = \boldsymbol{x}[n] + \boldsymbol{w}[n] \tag{5.34}$$

where $\boldsymbol{x}[n] = \boldsymbol{\Psi}\boldsymbol{u}[n]$ and

$$\boldsymbol{\Psi} = \begin{pmatrix} \mathbf{h}[0] & \mathbf{h}[1] & \cdots & \mathbf{h}[L_h] & \mathbf{0} & \cdots & \mathbf{0} \\ \mathbf{0} & \mathbf{h}[0] & \mathbf{h}[1] & \cdots & \mathbf{h}[L_h] & \ddots & \mathbf{0} \\ \vdots & \ddots & \ddots & \ddots & \ddots & \ddots & \mathbf{0} \\ \mathbf{0} & \cdots & \mathbf{0} & \mathbf{h}[0] & \mathbf{h}[1] & \cdots & \mathbf{h}[L_h] \end{pmatrix}$$

is an $M(N_w + 1) \times (N_w + L_h + 1)$ convolution matrix with block Toeplitz structure. Moreover, the system output correlation $\mathbf{R}_{\boldsymbol{y}}[0]$ can be seen to be

$$\mathbf{R}_{\boldsymbol{y}}[0] = \mathbf{R}_{\boldsymbol{x}}[0] + \mathbf{R}_{\boldsymbol{w}}[0] \tag{5.35}$$

where

$$\mathbf{R}_{\boldsymbol{x}}[0] = \boldsymbol{\Psi}\mathbf{R}_{\boldsymbol{u}}[0]\boldsymbol{\Psi}^H \tag{5.36}$$

is an $M(N_w + 1) \times M(N_w + 1)$ nonnegative definite matrix.

Noise Subspace Approach

Subspace approaches primarily exploit the separability of noise and signal subspaces for both system estimation and equalization, provided that the system order is known in advance. Among these approaches, those for SIMO blind equalization are basically developed under the following assumptions about the SIMO model $\mathbf{y}[n]$ given by (5.30).

(A5-1) The $M \times 1$ LTI system $\mathbf{h}[n]$ is BIBO stable.

(A5-2) *Channel disparity condition*: The M polynomials $H_i(z), i = 1, 2, ..., M$ of the transfer function $\boldsymbol{H}(z)$ have no common factors (i.e. no common subchannel zeros).

(A5-3) The system input $u[n]$ is a stationary zero-mean random process with $\mathbf{R}_{\boldsymbol{u}}[0]$ being of full rank, i.e. the system input is *persistently exciting.*[8]

(A5-4) The noise vector sequence $\mathbf{w}[n]$ is a zero-mean spatially uncorrelated and temporally white (s.u.t.w.) vector random process with correlation function $\mathbf{R}_{\mathbf{w}}[l] = \sigma_w^2 \delta[l]\mathbf{I}$.

(A5-5) The system input $u[n]$ is statistically independent of the noise $\mathbf{w}[n]$.

Accordingly, by Assumption (A5-4), it can easily be seen that

$$\mathbf{R}_{\boldsymbol{w}}[0] = \sigma_w^2 \mathbf{I}.$$

Therefore, the system output covariance given by (5.35) becomes

$$\mathbf{R}_{\boldsymbol{y}}[0] = \mathbf{R}_{\boldsymbol{x}}[0] + \sigma_w^2 \mathbf{I} \tag{5.37}$$

where $\mathbf{R}_{\boldsymbol{x}}[0]$ is defined by (5.36). With the choice of

$$N_w \geq L_h,$$

one can show that all the columns of $\boldsymbol{\Psi}$ are linearly independent if and only if the transfer function $\boldsymbol{H}(z) \neq \mathbf{0}$ for all $z \neq 0$ [7]. That is, the convolution matrix $\boldsymbol{\Psi}$ will be of full column rank if and only if Assumption (A5-2) is satisfied.

Notice that, as $N_w \geq L_h$, $\boldsymbol{\Psi}$ is a "tall" matrix with $M(N_w + 1) > (N_w + L_h + 1)$ and thus $\mathbf{R}_{\boldsymbol{x}}[0]$ given by (5.36) is rank deficient with

$$\text{rank}\{\mathbf{R}_{\boldsymbol{x}}[0]\} = \text{rank}\{\mathbf{R}_{\boldsymbol{u}}[0]\} = N_w + L_h + 1$$

under the channel disparity condition. The range space of $\mathbf{R}_{\boldsymbol{x}}[0]$ (which is the column space of $\boldsymbol{\Psi}$) is commonly called the *signal subspace*, while its orthogonal complement, i.e. the null space of $\mathbf{R}_{\boldsymbol{x}}[0]$, is called the *noise subspace.* Hence, the dimension of the signal subspace is

$$d_s = \text{rank}\{\boldsymbol{\Psi}\} = N_w + L_h + 1$$

and the dimension of the noise subspace is

$$d_n = M(N_w + 1) - d_s = M(N_w + 1) - (N_w + L_h + 1). \tag{5.38}$$

By eigendecomposition, $\mathbf{R}_{\boldsymbol{y}}[0]$ can be expressed as

$$\mathbf{R}_{\boldsymbol{y}}[0] = \mathbf{Q}_y \boldsymbol{\Lambda}_y \mathbf{Q}_y^H \tag{5.39}$$

[8] Obviously, a temporally white input does make $\mathbf{R}_{\boldsymbol{u}}[0]$ full rank. However, it should be noted that subspace methods do not require $u[n]$ to be temporally white.

where

$$\mathbf{Q}_y = (\mathbf{q}_{y,1}, \mathbf{q}_{y,2}, ..., \mathbf{q}_{y,M(N_w+1)}) \tag{5.40}$$

is a matrix whose columns are the orthonormal eigenvectors of $\mathbf{R}_{\boldsymbol{y}}[0]$ and

$$\mathbf{\Lambda}_y = \text{diag}(\lambda_{y,1}, \lambda_{y,2}, ..., \lambda_{y,M(N_w+1)}) \tag{5.41}$$

is a diagonal matrix consisting of the associated *nonnegative* eigenvalues with $\lambda_{y,1} \geq \lambda_{y,2} \geq \cdots \geq \lambda_{y,M(N_w+1)} \geq 0$. Let $\mathbf{q}_{x,1}, \mathbf{q}_{x,2}, ..., \mathbf{q}_{x,d_s}$ be the d_s eigenvectors of the nonnegative definite (but rank deficient) matrix $\mathbf{R}_{\boldsymbol{x}}[0]$ associated with the largest d_s eigenvalues $\lambda_{x,1} \geq \lambda_{x,2} \geq \cdots \geq \lambda_{x,d_s} > 0$ (and the other d_n eigenvalues equal to zero). Then $\mathbf{R}_{\boldsymbol{x}}[0]$ can be diagonalized as

$$\mathbf{Q}_x^H \mathbf{R}_{\boldsymbol{x}}[0] \mathbf{Q}_x = \mathbf{\Lambda}_x \tag{5.42}$$

where

$$\mathbf{Q}_x = (\mathbf{q}_{x,1}, \mathbf{q}_{x,2}, ..., \mathbf{q}_{x,d_s}, \mathbf{0}, \mathbf{0}, ..., \mathbf{0}) \tag{5.43}$$

and

$$\mathbf{\Lambda}_x = \text{diag}(\lambda_{x,1}, \lambda_{x,2}, ..., \lambda_{x,d_s}, 0, 0, ..., 0). \tag{5.44}$$

From (5.35) and (5.42) through (5.44), it is easy to infer that

$$\mathbf{q}_{x,i}^H \mathbf{R}_{\boldsymbol{y}}[0] \mathbf{q}_{x,i} = \lambda_{x,i} + \sigma_w^2, \quad i = 1, 2, ..., d_s.$$

Furthermore, $\mathbf{Q}_y$ given by (5.40) can be partitioned into

$$\mathbf{Q}_y = (\mathbf{Q}_s, \mathbf{Q}_n) \tag{5.45}$$

where

$$\mathbf{Q}_s = (\mathbf{q}_{y,1}, \mathbf{q}_{y,2}, ..., \mathbf{q}_{y,d_s}) = (\mathbf{q}_{x,1}, \mathbf{q}_{x,2}, ..., \mathbf{q}_{x,d_s}) \tag{5.46}$$

$$\mathbf{Q}_n = (\mathbf{q}_{y,d_s+1}, \mathbf{q}_{y,d_s+2}, ..., \mathbf{q}_{y,M(N_w+1)}) \tag{5.47}$$

and all the eigenvalues of $\mathbf{R}_{\boldsymbol{y}}[0]$ are given by

$$\lambda_{y,i} = \begin{cases} \lambda_{x,i} + \sigma_w^2, & i = 1, 2, ..., d_s, \\ \sigma_w^2, & i = d_s + 1, d_s + 2, ..., d_s + d_n. \end{cases}$$

Let us emphasize that the range spaces of $\mathbf{Q}_s$ and $\mathbf{Q}_n$ are exactly the signal subspace and the noise subspace, respectively.

By the facts that the signal subspace (the range space of $\mathbf{Q}_s$) is orthogonal to the noise subspace (the range space of $\mathbf{Q}_n$) and that $\mathbf{Q}_s$, $\mathbf{R}_{\boldsymbol{x}}[0]$ and $\mathbf{\Psi}$ have the same range space (by (5.36) and (5.46)), it can easily be proved that

$$\mathbf{Q}_n^H \boldsymbol{\Psi} = \mathbf{0}_{d_n \times d_s}. \tag{5.48}$$

At first glance, the above relation provides a set of $d_n \times d_s$ equations which seem not to be sufficient for solving for the $M(N_w+1)d_s$ unknowns of $\boldsymbol{\Psi}$ because $d_n < M(N_w+1)$ (by (5.38)). However, the number of unknowns in $\boldsymbol{\Psi}$ is much smaller than $M(N_w+1) \cdot d_s$ due to its block Toeplitz structure. In fact, it has been shown that $\boldsymbol{\Psi}$ can be uniquely solved from (5.48) subject to the constraint that $\boldsymbol{\Psi}$ is a nonzero block Toeplitz matrix [8]. Consequently, the *stacked* system impulse-response vector defined as

$$\boldsymbol{h} = \begin{pmatrix} \mathbf{h}[0] \\ \mathbf{h}[1] \\ \vdots \\ \mathbf{h}[L_h] \end{pmatrix}_{M(L_h+1)\times 1} \tag{5.49}$$

can be uniquely identified (up to a nonzero scale factor) by minimizing a quadratic cost function stated in the following theorem.

Theorem 5.29. *Given the system order L_h, the stacked system impulse-response vector $\mathbf{h}$ given by (5.49) can be uniquely identified (up to a nonzero scalar ambiguity) by minimizing*

$$J_{\mathrm{SS}}(\boldsymbol{h}) = \boldsymbol{h}^H \left(\sum_{i=d_s+1}^{M(N_w+1)} \boldsymbol{\mathcal{Q}}_i \boldsymbol{\mathcal{Q}}_i^H \right) \boldsymbol{h} \tag{5.50}$$

subject to $\|\boldsymbol{h}\|^2 = 1$, where

$$\boldsymbol{\mathcal{Q}}_i = \begin{pmatrix} \mathbf{q}_{y,i}^{(0)} & \mathbf{q}_{y,i}^{(1)} & \cdots & \mathbf{q}_{y,i}^{(N_w)} & \mathbf{0} & \cdots & \mathbf{0} \\ \mathbf{0} & \mathbf{q}_{y,i}^{(0)} & \mathbf{q}_{y,i}^{(1)} & \cdots & \mathbf{q}_{y,i}^{(N_w)} & \ddots & \mathbf{0} \\ \vdots & \ddots & \ddots & \ddots & \ddots & \ddots & \vdots \\ \mathbf{0} & \cdots & \mathbf{0} & \mathbf{q}_{y,i}^{(0)} & \mathbf{q}_{y,i}^{(1)} & \cdots\cdots & \mathbf{q}_{y,i}^{(N_w)} \end{pmatrix}$$

in which the $M \times 1$ vectors $\mathbf{q}_{y,i}^{(j)}, j = 0, 1, ..., N_w$ are the (N_w+1) partitions of

$$\mathbf{q}_{y,i}^T = \left(\left(\mathbf{q}_{y,i}^{(0)}\right)^T, \left(\mathbf{q}_{y,i}^{(1)}\right)^T, ..., \left(\mathbf{q}_{y,i}^{(N_w)}\right)^T \right).$$

The proof of Theorem 5.29 can be found in [8]. By this theorem, the solution to minimizing $J_{\mathrm{SS}}(\boldsymbol{h})$ given by (5.50) is equivalent to finding the eigenvector associated with the minimum eigenvalue of the matrix inside the parentheses on the right-hand side of (5.50). The resultant noise subspace approach is summarized in Table 5.1. Then the system input $u[n]$ can be estimated by

$$\widehat{u}[n] = \mathbf{v}^T[n] \star \mathbf{y}[n] \tag{5.51}$$

where $\mathbf{v}[n] = \mathbf{v}_{\mathrm{ZF}}[n]$ (the ZF equalizer) or $\mathbf{v}[n] = \mathbf{v}_{\mathrm{MS}}[n]$ (the LMMSE equalizer) given as follows

$$\boldsymbol{V}_{\mathrm{ZF}}(\omega) = \widehat{\boldsymbol{H}}_{\mathrm{I}}^T(\omega) = \left(\widehat{\boldsymbol{H}}^+(\omega)\right)^T = \widehat{\boldsymbol{H}}^*(\omega)\left(\widehat{\boldsymbol{H}}^T(\omega)\widehat{\boldsymbol{H}}^*(\omega)\right)^{-1} \tag{5.52}$$

$$\boldsymbol{V}_{\mathrm{MS}}(\omega) = \sigma_u^2\left(\boldsymbol{\mathcal{S}}_{\mathbf{y}}^T(\omega)\right)^{-1} \cdot \widehat{\boldsymbol{H}}^*(\omega) \tag{5.53}$$

where $\boldsymbol{\mathcal{S}}_{\mathbf{y}}(\omega)$ can be obtained by taking the Fourier transform of the sample correlation function $\widehat{\mathbf{R}}_{\mathbf{y}}[l]$.

Table 5.1 Noise subspace approach

Parameters setting	Given the system order L_h and the dimension M of the observation vector $\mathbf{y}[n]$, choose $N_w \geq L_h$ and concatenate (N_w+1) successive observations into the vector $\boldsymbol{y}[n]$ defined by (5.31).
EVD	Find the d_n eigenvectors $\mathbf{q}_{y,i}, i = d_s+1, d_s+2, ..., M(N_w+1)$ associated with the d_n smallest eigenvalues of $\mathbf{R}_{\boldsymbol{y}}[0]$.
System estimation	Find the stacked system impulse response vector $\boldsymbol{h}$ defined by (5.49) by minimizing $J_{\mathrm{SS}}(\boldsymbol{h})$ given by (5.50). Obtain the system estimate $\widehat{\mathbf{h}}[n]$ by partitioning $\boldsymbol{h}$ into subvectors of length equal to M.
Equalization	Design a ZF or LMMSE equalizer with the obtained system estimate $\widehat{\mathbf{h}}[n]$, by (5.52) or (5.53), respectively.

In summary, the noise subspace approach, a typical subspace approach, can also be viewed as a method of moments. Its advantage is the closed-form solution for system estimation but it has some shortcomings. This approach requires the system order to be given or estimated correctly in advance, and its performance is quite sensitive to system order mismatch. Additionally, this approach tends to fail when the channel disparity condition is nearly violated [12,13].[9] Moreover, it is computationally expensive due to the eigendecomposition of the data correlation matrix of large dimension.

[9] Recently, the noise subspace method was modified by Ali, Manton and Hua [14] and was shown to be robust in the presence of common subchannel zeros and system order overestimation errors by virtue of exploiting the transmitter redundancy through using a trailing zero precoder. Nevertheless, the use of a precoding procedure limits applications of the modified subspace method.

Single-Stage Linear Prediction Approach

The fundamental concept of linear prediction approaches for system estimation arises from the observation that as the SIMO system is an FIR system, its output can also be modeled as a vector AR process under certain conditions.[10] This important observation allows blind multichannel equalization/estimation by linear least-squares estimation with a closed-form solution. These approaches for SIMO blind equalization are basically developed with the following assumptions about the SIMO model $\mathbf{y}[n]$ given by (5.30).

(A5-6) The $M \times 1$ LTI system $\mathbf{h}[n]$ is BIBO stable.

(A5-7) The M polynomials $H_i(z), i = 1, 2, ..., M$ of the transfer function $\boldsymbol{H}(z)$ have no common factors (i.e. no common subchannel zeros).

(A5-8) The system input $u[n]$ is a stationary zero-mean temporally white random process with variance σ_u^2.

According to the generalized Bezout Identity [15], if the channel disparity condition is satisfied by $\mathbf{h}[n]$, there exists an $M \times 1$ causal FIR filter $\mathbf{v}_{\mathrm{BZ}}[n]$ such that

$$\boldsymbol{V}_{\mathrm{BZ}}^T(z)\boldsymbol{H}(z) = 1,$$

indicating that $\mathbf{v}_{\mathrm{BZ}}[n]$ is actually a causal FIR ZF equalizer. In other words, *in the absence of noise* ($\mathbf{w}[n] = \mathbf{0}$),

$$u[n] = \mathbf{v}_{\mathrm{BZ}}^T[n] \star \mathbf{y}[n] = \sum_{i=0}^{L_{\mathrm{BZ}}} \mathbf{v}_{\mathrm{BZ}}^T[i]\mathbf{y}[n-i], \tag{5.54}$$

or equivalently,

$$\mathbf{v}_{\mathrm{BZ}}^T[0]\mathbf{y}[n] = -\sum_{i=1}^{L_{\mathrm{BZ}}} \mathbf{v}_{\mathrm{BZ}}^T[i]\mathbf{y}[n-i] + u[n],$$

which implies that $\mathbf{y}[n]$ is also a finite-order vector AR process.

Define the linear prediction error (or innovations process) *vector* associated with the finite-order vector AR process $\mathbf{y}[n]$ as

$$\begin{aligned}\boldsymbol{\varepsilon}[n] &= \mathbf{V}_{\mathrm{LP}}[n] \star \mathbf{y}[n] = \mathbf{y}[n] - \sum_{k=1}^{N_w+1} \mathbf{V}_{\mathrm{LP}}[k]\mathbf{y}[n-k] \\ &= \mathbf{y}[n] - \boldsymbol{\mathcal{V}}_{\mathrm{LP}}\boldsymbol{y}[n-1]\end{aligned} \tag{5.55}$$

where $\mathbf{V}_{\mathrm{LP}}[n]$ is an $M \times M$ finite-length linear prediction filter with $\mathbf{V}_{\mathrm{LP}}[0] = \mathbf{I}$, $\boldsymbol{y}[n]$ is defined as (5.31), and

$$\boldsymbol{\mathcal{V}}_{\mathrm{LP}} = (\mathbf{V}_{\mathrm{LP}}[1], \mathbf{V}_{\mathrm{LP}}[2], ..., \mathbf{V}_{\mathrm{LP}}[N_w+1]).$$

[10] This property is related to the generalized Bezout identity.

Because of the assumption that $u[n]$ is temporally white and because of the fact that the resultant innovation process $\boldsymbol{\varepsilon}[n]$ is also temporally white, it can be inferred that

$$\begin{aligned}\boldsymbol{\varepsilon}[n] &= \sum_{k=0}^{L_h} \mathbf{h}[k]u[n-k] - \boldsymbol{\mathcal{V}}_{\mathrm{LP}}\boldsymbol{\Psi}\boldsymbol{u}[n-1] \\ &= \mathbf{h}[0]u[n] + \left(\sum_{k=1}^{L_h} \mathbf{h}[k]u[n-k] - \boldsymbol{\mathcal{V}}_{\mathrm{LP}}\boldsymbol{\Psi}\boldsymbol{u}[n-1]\right) = \mathbf{h}[0]u[n] \end{aligned} \tag{5.56}$$

implying the term inside the parentheses in (5.56) must equal $\mathbf{0}$.

The linear prediction filter $\mathbf{V}_{\mathrm{LP}}[n]$ can be determined by minimizing the mean square error $E\{\|\boldsymbol{\varepsilon}[n]\|^2\}$ where $\boldsymbol{\varepsilon}[n]$ is given by (5.55). The orthogonality principle directly leads to (Problem 5.2)

$$\boldsymbol{\mathcal{V}}_{\mathrm{LP}} = E\{\mathbf{y}[n]\boldsymbol{y}^H[n-1]\}\mathbf{R}_{\boldsymbol{y}}^{+}[0]. \tag{5.57}$$

Therefore, one can obtain the linear prediction filter $\mathbf{V}_{\mathrm{LP}}[n]$ from the SOS of $\mathbf{y}[n]$ according to (5.57) and then process $\mathbf{y}[n]$ by the $\mathbf{V}_{\mathrm{LP}}[n]$ obtained to get the linear prediction error vector $\boldsymbol{\varepsilon}[n]$ given by (5.55).

From (5.56), the correlation function of $\boldsymbol{\varepsilon}[n]$ can be seen to be

$$\begin{aligned}\mathbf{R}_{\boldsymbol{\varepsilon}}[0] &= E\{\boldsymbol{\varepsilon}[n]\boldsymbol{\varepsilon}^H[n]\} \\ &= \mathbf{h}[0]E\{u[n]u^*[n]\}\mathbf{h}^H[0] = \sigma_u^2\mathbf{h}[0]\mathbf{h}^H[0]. \end{aligned} \tag{5.58}$$

By virtue of the fact that $\mathbf{h}[0]\mathbf{h}^H[0]$ is a rank-one matrix, $\mathbf{h}[0]$ can be estimated (up to a scalar ambiguity) as the eigenvector of $\mathbf{R}_{\boldsymbol{\varepsilon}}[0]$ associated with the largest eigenvalue. Once an estimate of $\mathbf{h}[0]$, denoted as $\widehat{\mathbf{h}}[0]$, is obtained, the system input $u[n]$ can be restored either directly from the innovations process $\boldsymbol{\varepsilon}[n]$ as

$$\widehat{u}[n] = \widehat{\mathbf{h}}^H[0]\boldsymbol{\varepsilon}[n] \quad \text{(by (5.56))} \tag{5.59}$$

or from the system output vector $\mathbf{y}[n]$ as

$$\widehat{u}[n] = \left(\widehat{\mathbf{h}}^H[0]\mathbf{V}_{\mathrm{LP}}[n]\right) \star \mathbf{y}[n] = \widehat{\mathbf{v}}_{\mathrm{ZF}}^T[n] \star \mathbf{y}[n] \tag{5.60}$$

up to a constant scale factor. The above approach for estimation of $u[n]$ is called the "single-stage" linear prediction approach, which is summarized in Table 5.2. Additionally, with the input estimate $\widehat{u}[n]$ obtained, one can estimate the system using input–output system identification methods such as the least-squares method.

In summary, the single-stage linear prediction approach does not need knowledge of the system order and thus is robust against system order overestimation error. However, its performance depends critically on the accuracy of

Table 5.2 Single-stage linear prediction approach

Parameters setting	Given the system order L_h and the dimension M of the observation vector $\mathbf{y}[n]$, choose $N_w \geq L_h$ and concatenate $(N_w + 1)$ successive observations into the vector $\boldsymbol{y}[n]$ defined by (5.31).
Linear prediction	Find the linear prediction filter coefficients $\boldsymbol{\mathcal{V}}_{\mathrm{LP}}$ using (5.57). Obtain the linear prediction error vector $\boldsymbol{\varepsilon}[n]$ by (5.55).
System estimation	Obtain $\widehat{\mathbf{h}}[0]$ as the eigenvector of $\mathbf{R}_{\boldsymbol{\varepsilon}}[0]$ associated with the largest eigenvalue.
Equalization	Obtain the system input estimate $\widehat{u}[n]$ using either (5.59) or (5.60).

the estimated $\widehat{\mathbf{h}}[0]$ (the leading coefficient vector of the SIMO system). However, $\|\mathbf{h}[0]\|$ may be much smaller than $\max_n\{\|\mathbf{h}[n]\|\}$ and thus the estimate $\widehat{u}[n]$ given by (5.59) may not be a very accurate estimate of the system input $u[n]$. To improve the performance degradation due to small $\|\mathbf{h}[0]\|$, some other prediction error based approaches have been reported such as the outer product decomposition algorithm [16] and *multi-step* linear prediction algorithm [17].

5.3.2 Blind MIMO Equalization

Temporally White Inputs

The SOS based blind equalization algorithms introduced in the previous subsection for the SIMO case seem to be extendable to the MIMO case. Unfortunately, when the system inputs are s.u.t.w., this is not possible due to some inherent ambiguities of the SOS of the MIMO system outputs, even though some subspace methods can estimate the MIMO system up to an upper triangular unimodular matrix ambiguity. To resolve this ambiguity needs further information of the system such as the structure of the system, which, however, may always be unknown in practice. The following example illuminates the system ambiguities existent in SOS as the system inputs are s.u.t.w.

Example 5.30 (SOS Based MIMO System Estimation)
Consider an MIMO system with s.u.t.w. system input vector $\mathbf{u}[n]$ and $\boldsymbol{\mathcal{S}}_{\mathbf{u}}(\omega) = \sigma_u^2\mathbf{I}$. Then the power spectral matrix of $\mathbf{y}[n]$ can be expressed as

$$\begin{aligned}\boldsymbol{\mathcal{S}}_{\mathbf{y}}(\omega) &= \sigma_u^2 \boldsymbol{\mathcal{H}}(\omega)\boldsymbol{\mathcal{H}}^H(\omega) + \boldsymbol{\mathcal{S}}_{\mathbf{w}}(\omega) \\ &= \sigma_u^2 \boldsymbol{\mathcal{H}}(\omega)\mathbf{U}\mathbf{U}^H\boldsymbol{\mathcal{H}}^H(\omega) + \boldsymbol{\mathcal{S}}_{\mathbf{w}}(\omega) \\ &= \sigma_u^2 \widetilde{\boldsymbol{\mathcal{H}}}(\omega)\widetilde{\boldsymbol{\mathcal{H}}}^H(\omega) + \boldsymbol{\mathcal{S}}_{\mathbf{w}}(\omega)\end{aligned}$$

where $\mathbf{U}$ is an arbitrary unitary matrix and $\widetilde{\boldsymbol{\mathcal{H}}}(\omega) = \boldsymbol{\mathcal{H}}(\omega)\mathbf{U}$. As the unitary matrix is a nonpermutation matrix, $\boldsymbol{\mathcal{H}}(\omega)$ and $\widetilde{\boldsymbol{\mathcal{H}}}(\omega)$ are two different MIMO systems resulting in ambiguities in system estimation using SOS.

□

Temporally Colored Inputs

For the case of spatially uncorrelated and temporally colored (s.u.t.c.) inputs with *distinct* power spectra, the identifiability of an *irreducible* (see Assumption (A5-10) below) MIMO FIR system $\mathbf{H}[n]$ using SOS of the system output vector $\mathbf{y}[n]$ has been proven by Hua and Tugnait [18]. Meanwhile, some SOS based blind system identification and equalization methods have been reported, such as Gorokhov and Loubaton's subspace method [19], Abed-Meraim and Hua's minimum noise subspace method [20], the matrix pencil (MP) method proposed by Ma *et al.* [21], the blind identification via decorrelating subchannels (BIVDS) approach proposed by Hua *et al.* [22], An and Hua's blind identification via decorrelating the whole channel (BIVDW) approach [23], and so forth. All of these SOS based methods require $M > K$ [19–23]. Moreover, both the BIVDS and BIVDW methods further make the assumption that the power spectra of the driving inputs are sufficiently diverse [22, 23]. This assumption may always be invalid in some practical applications (such as wireless communications). Therefore, let us introduce only the MP method in this subsection.

Matrix Pencil Method

The MP method for MIMO blind equalization of s.u.t.c. inputs is developed basically with the assumption that $SNR = \infty$ and the following assumptions about the MIMO model $\mathbf{y}[n]$ given by (5.16) and (5.17).

(A5-9) $\boldsymbol{\mathcal{H}}(z)$ is an FIR system of length $L_H + 1$ and $M > K$.
(A5-10) $\boldsymbol{\mathcal{H}}(z)$ is irreducible, i.e. $\boldsymbol{\mathcal{H}}(z)$ is of full column rank for all $z \neq 0$.
(A5-11) $\boldsymbol{\mathcal{H}}(z)$ is column-reduced, i.e. the $M \times K$ matrix

$$(\mathbf{h}_1[L_1], \mathbf{h}_2[L_2], ..., \mathbf{h}_K[L_K])$$

is of full column rank where L_k is the degree of the polynomial vector $\boldsymbol{H}_k(z) = \mathcal{Z}\{\mathbf{h}_k[n]\}$ (in z^{-1}) defined as

$$L_k = \max_i \deg\{H_{i,k}(z)\}.$$

(Note that $L_H = \max_k\{L_k\}$, the degree of the MIMO system $\boldsymbol{\mathcal{H}}(z)$).

(A5-12) The channel inputs $u_k[n]$, $k = 1, 2, ..., K$ are persistently exciting, and mutually uncorrelated with distinct nonwhite power spectra.

Let us define the following vectors

$$\begin{aligned}
\boldsymbol{y}_m[n] &= (y_m[n], y_m[n-1], ..., y_m[n-N_w])^T \\
\boldsymbol{u}_k[n] &= (u_k[n], u_k[n-1], ..., u_k[n-L_H-N_w])^T \\
\boldsymbol{y}[n] &= \left(\boldsymbol{y}_1^T[n], \boldsymbol{y}_2^T[n], ..., \boldsymbol{y}_M^T[n]\right)^T \\
\boldsymbol{u}[n] &= \left(\boldsymbol{u}_1^T[n], \boldsymbol{u}_2^T[n], ..., \boldsymbol{u}_K^T[n]\right)^T
\end{aligned}$$

and the following $(N_w+1) \times (L_H+N_w+1)$ Toeplitz matrix

$$\boldsymbol{\Psi}_{mk} = \begin{pmatrix} h_{mk}[0] & h_{mk}[1] & \cdots & h_{mk}[L_H] & 0 & \cdots & 0 \\ 0 & h_{mk}[0] & h_{mk}[1] & \cdots & h_{mk}[L_H] & \ddots & 0 \\ \vdots & \ddots & \ddots & \ddots & \ddots & \ddots & 0 \\ 0 & \cdots & 0 & h_{mk}[0] & h_{mk}[1] & \cdots & h_{mk}[L_H] \end{pmatrix}$$

where N_w is chosen such that $M(N_w+1) \geq K(L_H+N_w+1)$.

Under the noise-free assumption and Assumptions (A5-10) and (A5-11) [24], it can be shown, by (5.16) and Assumption (A5-9), that

$$\boldsymbol{y}[n] = \boldsymbol{\Psi}\boldsymbol{u}[n] \tag{5.61}$$

where

$$\boldsymbol{\Psi} = \begin{pmatrix} \boldsymbol{\Psi}_{11} & \boldsymbol{\Psi}_{12} & \cdots & \boldsymbol{\Psi}_{1K} \\ \boldsymbol{\Psi}_{21} & \boldsymbol{\Psi}_{22} & \cdots & \boldsymbol{\Psi}_{2K} \\ \vdots & \vdots & \ddots & \vdots \\ \boldsymbol{\Psi}_{M1} & \boldsymbol{\Psi}_{M2} & \cdots & \boldsymbol{\Psi}_{MK} \end{pmatrix} \tag{5.62}$$

is of full column rank [24]. Moreover, the system output correlation matrix $\mathbf{R}_{\boldsymbol{y}}[l]$ can be seen to be

$$\mathbf{R}_{\boldsymbol{y}}[l] = \boldsymbol{\Psi}\mathbf{R}_{\boldsymbol{u}}[l]\boldsymbol{\Psi}^H \tag{5.63}$$

where

$$\mathbf{R}_{\boldsymbol{u}}[l] = \mathrm{diag}\{\mathbf{R}_{\boldsymbol{u}_1}[l], \mathbf{R}_{\boldsymbol{u}_2}[l], ..., \mathbf{R}_{\boldsymbol{u}_K}[l]\}$$

is a block diagonal matrix (by Assumption (A5-12)).

Consider the generalized eigenvalue problem of solving

$$\mathbf{R}_{\boldsymbol{y}}[l_1]\mathbf{q} = \lambda\mathbf{R}_{\boldsymbol{y}}[l_2]\mathbf{q}, \qquad l_1 \neq l_2. \tag{5.64}$$

By (5.63), the generalized eigenvectors $\mathbf{q}$ in (5.64) must satisfy

$$\mathbf{\Psi}\text{diag}\{\mathbf{R}_{\boldsymbol{u}_1}[l_1] - \lambda\mathbf{R}_{\boldsymbol{u}_1}[l_2], ..., \mathbf{R}_{\boldsymbol{u}_K}[l_1] - \lambda\mathbf{R}_{\boldsymbol{u}_K}[l_2]\}\mathbf{\Psi}^H\mathbf{q} = \mathbf{0}, \tag{5.65}$$

or equivalently,[11]

$$\mathbf{R}_{\boldsymbol{u}_k}[l_1]\mathbf{b}_k = \lambda\mathbf{R}_{\boldsymbol{u}_k}[l_2]\mathbf{b}_k, \quad k = 1, 2, ..., K \tag{5.66}$$

where

$$(\mathbf{b}_1^T, \mathbf{b}_2^T, ..., \mathbf{b}_K^T)^T = \mathbf{\Psi}^H\mathbf{q}.$$

By Assumption (A5-12), matrices $\mathbf{R}_{\boldsymbol{u}_k}[l_1]$ and $\mathbf{R}_{\boldsymbol{u}_k}[l_2]$ generally form a distinct pair for each k. Therefore, any of the K sets of generalized eigenvalues of (5.66) usually include some nonzero distinct generalized eigenvalues which do not belong to all the other sets, and some other generalized eigenvalues. The generalized eigenvectors associated with these distinct generalized eigenvalues can be used for inputs extraction from $\mathbf{y}[n]$ based on the following theorem.

Theorem 5.31. *Assume that there exist κ nonzero distinct generalized eigenvalues $\lambda_1, \lambda_2, ..., \lambda_\kappa$ of (5.64) with the κ associated generalized eigenvectors $\mathbf{q}_1, \mathbf{q}_2, ..., \mathbf{q}_\kappa$, called essential generalized eigenvectors, and that $\mathbf{\Psi}^H\mathbf{q}_i = (\mathbf{b}_{i1}^T, \mathbf{b}_{i2}^T, ..., \mathbf{b}_{iK}^T)^T \neq \mathbf{0}, i = 1, 2, ..., \kappa$. Then, for each essential generalized eigenvector $\mathbf{q}_i$, $\mathbf{b}_{i\ell} \neq \mathbf{0}$ for one and only one $\ell \in \{1, 2, ..., K\}$.*

The proof of Theorem 5.31 is left as an exercise (Problem 5.3).

First of all, those generalized eigenvectors $\mathbf{q}_j$ satisfying $\mathbf{\Psi}^H\mathbf{q}_j = \mathbf{0}$ can easily be identified by comparing the value of $\varepsilon_j \triangleq \mathbf{q}_j^H\mathbf{R}_{\boldsymbol{y}}[0]\mathbf{q}_j$ with a threshold ξ (a small positive number). If

$$\varepsilon_j = \mathbf{q}_j^H\mathbf{\Psi}\left(\text{diag}\{\mathbf{R}_{\boldsymbol{u}_1}[0], \mathbf{R}_{\boldsymbol{u}_2}[0], ..., \mathbf{R}_{\boldsymbol{u}_K}[0]\}\right)\mathbf{\Psi}^H\mathbf{q}_j < \xi, \tag{5.67}$$

then $\mathbf{\Psi}^H\mathbf{q}_j = \mathbf{0}$ is acceptable because $\mathbf{R}_{\boldsymbol{u}_k}[0]$ has full rank by Assumption (A5-12). These generalized eigenvectors are not useful and will be discarded. Using an essential generalized eigenvector $\mathbf{q}_i$ (by Theorem 5.31), one can obtain a filtered (distorted) version of one input $u_\ell[n]$ as

$$s_i[n] = \mathbf{q}_i^H\boldsymbol{y}[n] = \mathbf{q}_i^H\mathbf{\Psi}\boldsymbol{u}[n] = \mathbf{b}_{i\ell}^H\boldsymbol{u}_\ell[n]. \tag{5.68}$$

Therefore, κ filtered inputs $s_i[n]$ can be obtained by the κ essential generalized eigenvectors. A systematic classification method such as the simple hierarchical dimensionality reduction approach [25] can be utilized to categorize the κ extracted inputs $s_i[n]$ into K groups, each associated with a distinct input $u_k[n]$. After classification, each group can then be formulated as an SIMO system model with those $s_i[n]$ in the group as the system outputs. Thus the existing SIMO blind equalization/identification approaches such as the subspace and linear prediction approaches introduced in the previous subsection

[11] Matrices of the form $\mathbf{A} - b\mathbf{C}$ are known as *matrix pencils*.

can be employed to obtain an input estimate using all the $s_i[n]$ of the associated group. The above equalization approach constitutes the matrix pencil approach which is summarized in Table 5.3.

In summary, the matrix pencil approach is a two-stage method in which stage 1 utilizes a matrix pencil between the two chosen system output correlation matrices at different lags (l_1, l_2) for extracting filtered inputs,[12] whereas stage 2 employs an SIMO blind equalization/identification approach for estimating some inputs from the filtered inputs. It should be noted that the number of essential generalized eigenvectors for the chosen (l_1, l_2) is usually uncertain in stage 1. In other words, some of the K inputs may never be extracted in stage 1, and therefore a deflation strategy may be adopted to extract the remaining inputs as reported in [21].

Table 5.3 Matrix pencil approach

Parameters setting	Choose l_1, l_2, a positive number ξ and N_w such that $M(N_w + 1) \geq K(L_H + N_w + 1)$.
Solving for generalized eigenvalues	Solve the generalized eigenvalue problem given by (5.64) for obtaining the κ essential generalized eigenvectors $\mathbf{q}_1, \mathbf{q}_2, ..., \mathbf{q}_\kappa$.
Filtered inputs extraction	Extract the filtered inputs using the essential generalized eigenvectors $\mathbf{q}_i$ obtained using (5.68).
Classification	Divide the κ filtered inputs into K groups using a systematic classification method.
Equalization	Obtain one input estimate for each group using an SIMO blind equalization approach.

5.4 HOS Based Blind Equalization Approaches

Blind equalization of MIMO systems using HOS have been investigated extensively [3, 27–36] since 1990, basically under the following assumptions for the MIMO system model $\mathbf{y}[n]$ given by (5.16) and (5.17):

12 Choi *et al.* [26] have suggested that the symmetry of the system output correlation matrices can be ensured by forcing $\mathbf{R}_{\boldsymbol{y}}[l] = (\mathbf{R}_{\boldsymbol{y}}[l] + \mathbf{R}_{\boldsymbol{y}}^H[l])/2$ to avoid the numerical problem in calculating the generalized eigenvectors of pencil matrices.

(A5-13) The $M \times K$ LTI system $\mathbf{H}[n]$ is BIBO stable.

(A5-14) $M \geq K$, i.e. the number of system outputs is no less than the number of system inputs.

(A5-15) The system inputs $u_k[n]$, $k = 1, 2, ..., K$, are stationary zero-mean, nonGaussian mutually independent (or equivalently, spatially independent) random processes with variance $\sigma_{u_k}^2 = E\{|u_k[n]|^2\}$ and $(p+q)$th-order cumulant $C_{p,q}\{u_k[n]\} \neq 0, \forall k$ where p and q are nonnegative integers and $(p+q) \geq 3$.

(A5-16) The noise $\mathbf{w}[n]$ is a zero-mean, Gaussian vector random process, which can be spatially correlated and temporally colored with correlation function $\mathbf{R}_{\mathbf{w}}[l]$.

(A5-17) The system input vector $\mathbf{u}[n]$ is statistically independent of the noise vector $\mathbf{w}[n]$.

In Section 4.3 we have introduced two typical SISO blind equalization algorithms, i.e. MNC and SE equalization algorithms, using HOS. Now we introduce their MIMO counterparts, referred to as MIMO-MNC and MIMO-SE equalization algorithms, respectively. Let us consider the case that each of the system inputs is temporally i.i.d. (i.e. spatially independent and temporally independent (s.i.t.i.)) followed by the case that each of the system inputs can be either temporally i.i.d. or colored (i.e. spatially independent and temporally colored (s.i.t.c.)).

5.4.1 Temporally IID Inputs

Consider that the MISO linear equalizer $\mathbf{v}[n]$ to be designed is an FIR system for which $\mathbf{v}[n] \neq \mathbf{0}$, $n = L_1, L_1+1, ..., L_2$. Let $\boldsymbol{v}_m = (v_m[L_1], v_m[L_1+1], ..., v_m[L_2])^T$, and $\boldsymbol{v} = (\boldsymbol{v}_1^T, \boldsymbol{v}_2^T, ..., \boldsymbol{v}_M^T)^T$ denote an $(ML) \times 1$ vector consisting of equalizer coefficients where $L = L_2 - L_1 + 1$ is the equalizer length. The equalized signal $e[n]$ given by (5.19), the output of the MISO equalizer $\mathbf{v}[n]$, can be expressed as

$$e[n] = \boldsymbol{v}^T \boldsymbol{y}[n] \tag{5.69}$$

where $\boldsymbol{y}[n]$ is an $(ML) \times 1$ vector formed of the system output vector $\mathbf{y}[n]$ (see (5.16)) as follows

$$\boldsymbol{y}[n] = \left(\boldsymbol{y}_1^T[n], \boldsymbol{y}_2^T[n], ..., \boldsymbol{y}_M^T[n]\right)^T$$

in which

$$\boldsymbol{y}_m[n] = (y_m[n-L_1], y_m[n-L_1-1], ..., y_m[n-L_2])^T .$$

With Assumptions (A5-14) through (A5-17) and the assumption that system inputs are temporally i.i.d., it can be shown, from (5.20), (5.19) and

(5.22), that the correlation function $r_{e_N}[k]$ of the Gaussian noise term $e_N[n]$ in (5.19) can be expressed as

$$\begin{aligned} r_{e_N}[k] &= E\{e_N[n]e_N^*[n-k]\} \\ &= \sum_{n_1=-\infty}^{\infty}\sum_{n_2=-\infty}^{\infty} \mathbf{v}^T[n_1]\mathbf{R_w}[k+n_2-n_1]\mathbf{v}^*[n_2] \\ &= \sum_{l=1}^{M}\sum_{i=1}^{M}[\mathbf{R_w}[k]]_{il} \star v_i[k] \star v_l^*[-k], \end{aligned} \tag{5.70}$$

and that

$$\sigma_e^2 = E\{|e[n]|^2\} = \sum_{i=1}^{K}\sigma_{u_i}^2\left(\sum_{n=-\infty}^{\infty}|g_i[n]|^2\right) + r_{e_N}[0] \tag{5.71}$$

$$C_{p,q}\{e[n]\} = \sum_{i=1}^{K} C_{p,q}\{u_i[n]\}\left(\sum_{n=-\infty}^{\infty} g_i^p[n](g_i^*[n])^q\right), \quad p+q \ge 3 \tag{5.72}$$

since $C_{p,q}\{e_N[n]\} = 0$ for all $p+q \ge 3$.

Maximum Normalized Cumulant Equalization Algorithm

The MNC criterion for blind equalization presented in Section 4.3.1 is applicable to both of SISO and MIMO systems, and is repeated here for convenience.

$$\begin{aligned} \text{MNC Criterion:} \quad J_{p,q}(\boldsymbol{v}) &= J_{p,q}(e[n]) \\ &= \frac{|C_{p,q}\{e[n]\}|}{E\{|e[n]|^2\}^{(p+q)/2}} = \frac{|C_{p,q}\{e[n]\}|}{\sigma_e^{(p+q)}} \end{aligned} \tag{5.73}$$

where $e[n] = \boldsymbol{v}^T\boldsymbol{y}[n]$ and p and q are nonnegative integers with $p+q \ge 3$. The MNC criterion for the MIMO system is supported by the following theorem [3,37]:

Theorem 5.32 (MIMO-MNC). *Suppose that $\mathbf{y}[n]$ consists of the outputs of an MIMO system given by (5.16) satisfying Assumptions* (A5-13) *through* (A5-17). *Further suppose that $SNR = \infty$ and the MIMO system $\mathbf{H}[n]$ satisfies the ZF-condition* (C-ZF) *(given in Theorem 5.28). Then as $L_1 \to -\infty$ and $L_2 \to \infty$, $J_{p,q}(\boldsymbol{v})$ with $p+q \ge 3$ is maximum if and only if $\mathbf{g}[n] = \alpha\delta[n-\tau]\boldsymbol{\eta}_\ell$, i.e.*

$$e[n] = \alpha u_\ell[n-\tau] \tag{5.74}$$

and

$$\max\{J_{p,q}(\boldsymbol{v})\} = J_{p,q}(u_\ell[n]) = \max_k\{J_{p,q}(u_k[n])\}, \tag{5.75}$$

where $\alpha \neq 0$ and τ (integer) are an unknown scale factor and an unknown time delay, respectively.

The proof of Theorem 5.32 (through using Theorem 2.33) is left as an exercise (Problems 5.4). Theorem 5.32 also implies that as the values of $J_{p,q}(u_k[n])$ are the same for all k, the optimum $e[n]$ can be an estimate of any one of the K inputs. In addition to the global optimum equalizer presented in Theorem 5.32, among local maxima and minima (stationary points) of $J_{p,q}(\boldsymbol{v})$, there are K stable local maxima as presented in the following theorem.

Theorem 5.33 (Local Maxima of MIMO-MNC). *Under the same assumptions made in Theorem 5.32, as $L_1 \to -\infty$ and $L_2 \to \infty$, there are K stable local maxima for $J_{p,q}(\boldsymbol{v})$ each associated with a $\mathbf{g}[n] = \alpha_\ell \delta[n-\tau_\ell]\boldsymbol{\eta}_\ell$, $\ell \in \{1, 2, ..., K\}$ and the other local maxima are unstable equilibria for the following cases:*

(C1) $p + q > 2$ as $\mathbf{y}[n]$ is real.
(C2) $p = q \geq 2$ as $\mathbf{y}[n]$ is complex.
(C3) $p + q > 2$, $p \neq q$, $\sigma_{u_k}^2 = \sigma_u^2$ and $C_{p,q}\{u_k[n]\} = \gamma \neq 0$, $k = 1, 2, ..., K$ as $\mathbf{y}[n]$ is complex.

Tugnait [3] proves that Theorem 5.33 is true for $(p, q) = (2, 1)$ and $(p, q) = (2, 2)$ in Case (C1), and for $(p, q) = (2, 2)$ in Case (C2). Chi and Chen [37] prove the theorem basically following a procedure similar to that given in [3]. The proof of the theorem is algebraically lengthy and omitted here.

Because of the lack of a closed-form solution for the optimum $\boldsymbol{v}$ of the nonlinear objective function $J_{p,q}(\boldsymbol{v})$, the efficient gradient-type BFGS method given in Table 2.4 can be employed to obtain the optimum $\boldsymbol{v}$, a local maximum rather than a global maximum of $J_{p,q}(\boldsymbol{v})$. The resultant MIMO-MNC equalization algorithm which provides the optimum $\mathbf{v}_{\text{MNC}}[n]$ or $\boldsymbol{v}_{\text{MNC}}$ is summarized in Table 5.4.

Table 5.4 MIMO-MNC equalization algorithm

Parameters setting	Choose equalizer length L, cumulant order (p, q) and convergence tolerance $\zeta > 0$.
Initial condition	Set $\boldsymbol{v}^{[0]}$.
Iteration $i = 0, 1, ...$	Update $\boldsymbol{v}^{[i+1]}$ using the BFGS method summarized in Table 2.4 such that $J_{p,q}(\boldsymbol{v}^{[i+1]}) > J_{p,q}(\boldsymbol{v}^{[i]})$.
Convergence check	If $\lvert J_{p,q}(\boldsymbol{v}^{[i+1]}) - J_{p,q}(\boldsymbol{v}^{[i]})\rvert / J_{p,q}(\boldsymbol{v}^{[i]}) \geq \zeta$, then go to the next iteration; otherwise the optimum $\boldsymbol{v}_{\text{MNC}} = \boldsymbol{v}^{[i]}$ has been obtained.

Super-Exponential Equalization Algorithm

The MIMO-SE equalization algorithm reported by Yeung and Yau [30] and Inouye and Tanebe [31] is a straightforward extension of the corresponding algorithm for the SISO case presented in Section 4.3.2. At the ith iteration, the equalizer coefficient vector $\boldsymbol{v}$ for the MIMO case is updated by the following linear equations: [38]

$$\begin{cases} \widetilde{\boldsymbol{v}}^{[i+1]} = (\mathbf{R}_{\boldsymbol{y}}^*[0])^{-1} \cdot \mathbf{d}_{e\boldsymbol{y}}^{[i]}, \\ \boldsymbol{v}^{[i+1]} = \dfrac{\widetilde{\boldsymbol{v}}^{[i+1]}}{\left\|\widetilde{\boldsymbol{v}}^{[i+1]}\right\|} \end{cases} \tag{5.76}$$

where

$$\mathbf{d}_{e\boldsymbol{y}}^{[i]} = \mathrm{cum}\{e^{[i]}[n] : p, (e^{[i]}[n])^* : q-1, \boldsymbol{y}^*[n]\} \tag{5.77}$$

in which p and q are nonnegative integers and $p+q \geq 3$, and $e^{[i]}[n] = (\boldsymbol{v}^{[i]})^T \boldsymbol{y}[n]$ is the equalized signal obtained at the ith iteration. To avoid possible numerical problems caused by rank deficiency of $\mathbf{R}_{\boldsymbol{y}}[0]$, it is better to obtain inverse matrix $(\mathbf{R}_{\boldsymbol{y}}[0])^{-1}$ by SVD.

Again, under the same assumptions made in Theorem 5.32, the MIMO-SE algorithm will converge at the super-exponential rate (i.e. $\mathrm{ISI}(\mathbf{g}^{[i]}[n] = \mathbf{H}^T[n] \star \mathbf{v}^{[i]}[n])$ decreases to zero at a super-exponential rate), and end up with the equalized signal

$$e[n] = \alpha u_\ell[n-\tau] \tag{5.78}$$

where α is a real/complex constant, τ is an integer, and $\ell \in \{1, 2, ..., K\}$. Similarly, the convergence rule for the MIMO-SE equalization algorithm can be chosen as that given by (4.121) used in the SISO-SE equalization algorithm. The resultant MIMO-SE equalization algorithm for obtaining $\mathbf{v}_{\mathrm{SE}}[n]$ or $\boldsymbol{v}_{\mathrm{SE}}$ is summarized in Table 5.5.

The computational efficiency of the SISO-SE algorithm mentioned in Section 4.3.2 also applies to the MIMO-SE equalization algorithm; namely, the MIMO-SE equalization algorithm is computationally efficient with faster convergence than the MIMO-MNC equalization algorithm, but may diverge for finite data length N and finite SNR.

Properties and Relations

For finite SNR and sufficient equalizer length, the MIMO-MNC equalizer and MIMO-SE equalizer introduced above are closely related as stated in the following property [39]:

Property 5.34. *With sufficient equalizer length L, the MIMO-MNC equalizer $\mathbf{v}_{\mathrm{MNC}}[n]$ and the MIMO-SE equalizer $\mathbf{v}_{\mathrm{SE}}[n]$ are basically the same (up to a*

Table 5.5 MIMO-SE equalization algorithm

Parameters setting	Choose equalizer length L, cumulant order (p, q) and convergence tolerance $\zeta > 0$.
Inverse matrix calculation	Estimate $\mathbf{R}_{\boldsymbol{y}}[0]$ with sample correlations of $\boldsymbol{y}[n]$. Obtain $(\mathbf{R}_{\boldsymbol{y}}[0])^{-1}$ using SVD.
Initial condition	Set $\boldsymbol{v}^{[0]}$.
Iteration $i = 0, 1, ...$	Update $\boldsymbol{v}^{[i+1]}$ by (5.76).
Convergence check	If $\lvert(\boldsymbol{v}^{[i+1]})^H \boldsymbol{v}^{[i]}\rvert > 1 - \zeta/2$, then go to the next iteration; otherwise, $\boldsymbol{v}_{\mathrm{SE}} = \boldsymbol{v}^{[i]}$ has been obtained.

scale factor) for $p+q \geq 3$ as $\mathbf{y}[n]$ is real and for $p = q \geq 2$ as $\mathbf{y}[n]$ is complex.

The proof of Property 5.34 is presented in Appendix 5A. This property reveals that with *finite data length*, the MIMO-MNC equalizer $\mathbf{v}_{\mathrm{MNC}}[n]$ and MIMO-SE equalizer $\mathbf{v}_{\mathrm{SE}}[n]$ should exhibit similar performance and behavior for $p + q \geq 3$ as $\mathbf{y}[n]$ is real and for $p = q \geq 2$ as $\mathbf{y}[n]$ is complex.

Next, let us introduce two interesting properties of the MIMO-MNC equalizer. Define

$$\mathbf{D}_{p,q} = \mathrm{diag}\left\{C_{p,q}\{u_1[n]\}, C_{p,q}\{u_2[n]\}, ..., C_{p,q}\{u_K[n]\}\right\} \tag{5.79}$$

$$\widetilde{g}_{p,q}[n; k] = (g_k[n])^p (g_k^*[n])^{q-1} \tag{5.80}$$

$$\widetilde{\boldsymbol{G}}_{p,q}(\omega) = \left(\mathcal{F}\{\widetilde{g}_{p,q}[n; 1]\}, \mathcal{F}\{\widetilde{g}_{p,q}[n; 2]\}, ..., \mathcal{F}\{\widetilde{g}_{p,q}[n; K]\}\right)^T. \tag{5.81}$$

Note that $\mathbf{D}_{1,1} = \mathrm{diag}\{\sigma_{u_1}^2, \sigma_{u_2}^2, ..., \sigma_{u_K}^2\}$ by (5.79). Under Assumptions (A6-13) through (A6-17), it can be shown, from (5.29), that the $K \times M$ LMMSE equalizer $\boldsymbol{\mathcal{V}}_{\mathrm{MS}}(\omega)$ can be expressed as

$$\boldsymbol{\mathcal{V}}_{\mathrm{MS}}^T(\omega) = \left[\boldsymbol{\mathcal{S}}_{\mathbf{y}}^T(\omega)\right]^{-1} \cdot \boldsymbol{\mathcal{H}}^*(\omega) \cdot \mathbf{D}_{1,1} \tag{5.82}$$

where

$$\boldsymbol{\mathcal{S}}_{\mathbf{y}}(\omega) = \boldsymbol{\mathcal{H}}(\omega) \cdot \mathbf{D}_{1,1} \cdot \boldsymbol{\mathcal{H}}^H(\omega) + \boldsymbol{\mathcal{S}}_{\mathbf{w}}(\omega). \tag{5.83}$$

The two properties of the MIMO-MNC equalizer for any SNR are as follows [39]:

Property 5.35. *With sufficient equalizer length, the MNC equalizer $\mathbf{v}_{\mathrm{MNC}}[n]$ is related to the MIMO LMMSE equalizer $\boldsymbol{\mathcal{V}}_{\mathrm{MS}}(\omega)$ via*

$$\boldsymbol{V}_{\mathrm{MNC}}(\omega) = \boldsymbol{\mathcal{V}}_{\mathrm{MS}}^T(\omega) \cdot \mathbf{Q}(\omega) \tag{5.84}$$

where[13]

$$\mathbf{Q}(\omega) = \mathbf{D}_{1,1}^{-1} \left\{ \alpha_{p,q} \mathbf{D}_{p,q} \widetilde{\boldsymbol{G}}_{p,q}(\omega) + \alpha_{q,p} \mathbf{D}_{q,p} \widetilde{\boldsymbol{G}}_{q,p}(\omega) \right\} \tag{5.85}$$

in which

$$\alpha_{p,q} = \frac{q \cdot \sigma_e^2}{(p+q) \cdot C_{p,q}\{e[n]\}}. \tag{5.86}$$

Property 5.36. *As the noise* $\mathbf{w}[n]$ *is spatially uncorrelated (i.e.* $[\mathbf{R}_\mathbf{w}[k]]_{il} = 0$ *for* $i \neq l$*), each component of the overall system* $\mathbf{g}[n]$ *associated with the* $\mathbf{v}_{\text{MNC}}[n]$ *with sufficient equalizer length is linear phase, i.e.*

$$\arg[G_k(\omega)] = \omega \tau_k + \varphi_k, \qquad -\pi \leq \omega < \pi \tag{5.87}$$

where τ_k *and* φ_k *are real constants.*

The proof of Property 5.35 is given in Appendix 5B, and the proof of Property 5.36 is left as an exercise (Problem 5.5). These relations and properties motivate that the theoretical MIMO-MNC equalizer without a closed-form solution can be obtained analytically, and that the MIMO-MNC equalization algorithm can be implemented efficiently (fast convergence speed and guaranteed convergence) with the use of the same update equation for the equalizer used by the MIMO-SE equalization algorithm.

The theoretical solution for $\mathbf{v}_{\text{MNC}}[n]$ is necessary in the simulation and algorithm tests during the algorithm design. According to Property 5.35, the *theoretical* MIMO-MNC equalizer $\mathbf{v}_{\text{MNC}}[n]$ can also be efficiently obtained from the nonblind $\boldsymbol{\mathcal{V}}_{\text{MS}}(\omega)$ given by (5.82) using the following FFT-based iterative algorithm [39], called Algorithm MNC-LMMSE, where $\boldsymbol{V}[k]$, $\boldsymbol{H}[k]$, $\boldsymbol{G}[k]$, $\widetilde{\boldsymbol{G}}_{p,q}[k]$ and $\boldsymbol{\mathcal{V}}_{\text{MS}}[k]$ denote $\mathcal{N}$-point FFTs of $\mathbf{v}[n]$, $\mathbf{H}[n]$, $\mathbf{g}[n]$, $\widetilde{\mathbf{g}}_{p,q}[n]$ and $\mathbf{v}_{\text{MS}}[n]$, respectively.

Algorithm MNC-LMMSE

(S1) Set $i = 0$. Choose an initial condition $\mathbf{v}^{[0]}[n]$ for $\mathbf{v}[n]$ and a convergence tolerance $\zeta > 0$.

(S2) Set $i = i+1$. Compute the $\mathcal{N}$-point DFT $\boldsymbol{V}^{[i-1]}[k]$.[14] Compute $\boldsymbol{G}^{[i-1]}[k] = \boldsymbol{\mathcal{H}}^T[k]\boldsymbol{V}^{[i-1]}[k]$ by (5.22) and then obtain its $\mathcal{N}$-point inverse DFT $\mathbf{g}^{[i-1]}[n]$.

---

[13] As $p = q \geq 2$ is considered, $\mathbf{Q}(\omega)$ given by (5.85) can be simplified as

$$\mathbf{Q}(\omega) = \frac{\sigma_e^2}{C_{p,q}\{e[n]\}} \cdot \mathbf{D}_{1,1}^{-1} \mathbf{D}_{p,p} \widetilde{\boldsymbol{G}}_{p,p}(\omega) \quad \text{(as } p = q \geq 2\text{)}.$$

[14] It is never limited by the length of $\mathbf{v}[n]$ as long as the DFT length $\mathcal{N}$ is chosen sufficiently large such that aliasing effects on the resultant $\mathbf{v}[n]$ are negligible.

(S3) Compute $\widetilde{\boldsymbol{G}}_{p,q}[k]$ and $\widetilde{\boldsymbol{G}}_{q,p}[k]$ using (5.80) and (5.81) with $\mathbf{g}[n] = \mathbf{g}^{[i-1]}[n]$. Then compute $\mathbf{Q}[k]$ using (5.85).

(S4) Compute $\widetilde{\boldsymbol{V}}[k] = \boldsymbol{\mathcal{V}}_{\mathrm{MS}}^{T}[k]\mathbf{Q}[k]$ by (5.84) followed by its $\mathcal{N}$-point inverse DFT $\widetilde{\mathbf{v}}[n]$. Then obtain $\mathbf{v}^{[i]}[n] = \widetilde{\mathbf{v}}[n]/(\sum_n \|\widetilde{\mathbf{v}}[n]\|^2)^{1/2}$. If $J_{p,q}(\mathbf{v}^{[i]}[n]) > J_{p,q}(\mathbf{v}^{[i-1]}[n])$ (where the theoretical $J_{p,q}(\mathbf{v}^{[i]}[n])$ is computed by using (5.73), (5.70), (5.71) and (5.72)), go to (S6).

(S5) Compute $\Delta\boldsymbol{V}[k] = \boldsymbol{V}^{[i]}[k] - \boldsymbol{V}^{[i-1]}[k]$, and update $\boldsymbol{V}^{[i]}[k]$ via

$$\boldsymbol{V}^{[i]}[k] = \boldsymbol{V}^{[i-1]}[k] + \mu \cdot \Delta\boldsymbol{V}[k]$$

where the step size μ is chosen such that $J_{p,q}(\mathbf{v}^{[i]}[n]) > J_{p,q}(\mathbf{v}^{[i-1]}[n])$. Then normalize $\mathbf{v}^{[i]}[n]$ by $\mathbf{v}^{[i]}[n]/(\sum_n \|\mathbf{v}^{[i]}[n]\|^2)^{1/2}$.

(S6) If $\sum_n \|\mathbf{v}^{[i]}[n] - \mathbf{v}^{[i-1]}[n]\|^2 > \zeta$, then go to (S2); otherwise, the theoretical (true) $\mathbf{v}_{\mathrm{MNC}}[n] = \mathbf{v}^{[i]}[n]$ is obtained.

Note that the convergence of Algorithm MNC-LMMSE can be guaranteed because $J_{p,q}(\mathbf{v}^{[i]}[n])$ (which is bounded) increases at each iteration, and (S5) is rarely performed. Let us emphasize that Algorithm MNC-LMMSE is never an MIMO blind deconvolution algorithm, and that it is merely an iterative algorithm that requires the system response $\boldsymbol{\mathcal{H}}(\omega)$, variances $\sigma_{u_k}^2$ and $(p+q)$th-order cumulants $C_{p,q}\{u_k[n]\}$ of system inputs and noise correlation function $\mathbf{R}_{\mathbf{w}}[k]$ to compute the true MIMO-MNC equalizer $\mathbf{v}_{\mathrm{MNC}}[n]$.

MIMO Hybrid MNC Equalization Algorithm

Recall that the MIMO-SE equalization algorithm is more computationally efficient and faster than the MIMO-MNC equalization algorithm but the former faces a potential divergence issue for finite SNR and limited data length. In view of Property 5.34, the MIMO-SE equalization algorithm is actually searching for the same equalizer as the MIMO-MNC equalization algorithm, whereas the latter is guaranteed convergent. These facts suggest an iterative hybrid MNC equalization algorithm as illustrated in Fig. 4.13 for the SISO case with $y[n]$ being replaced by $\mathbf{y}[n]$. The iterative hybrid MNC equalization algorithm is summarized below.

MIMO Hybrid MNC Equalization Algorithm:

(S1) Like the MIMO-SE equalization algorithm, obtain $\boldsymbol{v}^{[i+1]}$ by (5.76), and obtain the associated $e^{[i+1]}[n] = \left(\boldsymbol{v}^{[i]}\right)^T \mathbf{y}[n]$.

(S2) If $J_{p,q}(\boldsymbol{v}^{[i+1]}) < J_{p,q}(\boldsymbol{v}^{[i]})$, update $\boldsymbol{v}^{[i+1]}$ through a gradient-type optimization method such that $J_{p,q}(\boldsymbol{v}^{[i+1]}) > J_{p,q}(\boldsymbol{v}^{[i]})$ and obtain the associated $e^{[i+1]}[n]$.

(S3) If $|J_{p,q}(\boldsymbol{v}^{[i+1]}) - J_{p,q}(\boldsymbol{v}^{[i]})|/J_{p,q}(\boldsymbol{v}^{[i]}) \geq \zeta$, then go to the next iteration; otherwise $\boldsymbol{v}_{\mathrm{MNC}} = \boldsymbol{v}^{[i+1]}$ has been obtained.

It is important to emphasize that according to Property 5.34, the hybrid MNC equalization algorithm is applicable only for the case of real $\mathbf{y}[n]$ and for the case of complex $\mathbf{y}[n]$ and $p = q$. The MIMO hybrid MNC equalization algorithm uses the MIMO SE equalization algorithm in (S1) for fast convergence (basically with super-exponential rate) which usually happens in most of iterations before convergence, and a gradient-type optimization method in (S2) for the guaranteed convergence. Specifically, the first derivative $\partial J_{p,q}(\boldsymbol{v})/\partial \boldsymbol{v}^*|_{\boldsymbol{v}=\boldsymbol{v}^{[i]}}$ required by (S2) can be expressed as (4.155) with $\mathbf{R}_y$ and $\mathbf{d}_{ey}^{[i]}$ being replaced by $\mathbf{R}_{\boldsymbol{y}}$ and $\mathbf{d}_{e\boldsymbol{y}}^{[i]}$, respectively.

Estimation of All the System Inputs

One can express the system output vector $\mathbf{y}[n]$ given by (5.16) as

$$\begin{aligned}\mathbf{y}[n] &= (\mathbf{h}_1[n], \mathbf{h}_2[n], ..., \mathbf{h}_K[n]) \star \mathbf{u}[n] + \mathbf{w}[n] \\ &= \sum_{k=1}^{K} \mathbf{h}_k[n] \star u_k[n] + \mathbf{w}[n] = \sum_{k=1}^{K} \mathbf{x}_k[n] + \mathbf{w}[n] \end{aligned} \tag{5.88}$$

where

$$\mathbf{x}_k[n] = \mathbf{h}_k[n] \star u_k[n] \tag{5.89}$$

is the contribution in $\mathbf{y}[n]$ from the kth system input. With the system input estimate $\widehat{u}_\ell[n] = e[n]$, $\ell \in \{1, 2, ..., K\}$ obtained by an MIMO blind equalization algorithm such as the MIMO-MNC algorithm, the system $\mathbf{h}_\ell[n]$ can be estimated by the so-called *input–output cross-correlation (IOCC) method* [3], as follows:

$$\widehat{\mathbf{h}}_\ell[n] = \frac{E\{\mathbf{y}[m+n]\widehat{u}_\ell^*[m]\}}{E\{|\widehat{u}_\ell[m]|^2\}}. \tag{5.90}$$

Note that $\widehat{u}_\ell[n] = \alpha u_\ell[n-\tau]$ as $\mathbf{w}[n] = \mathbf{0}$ (by Theorem 5.32), implying $\widehat{\mathbf{h}}_\ell[n] = \mathbf{h}_\ell[n+\tau]/\alpha$ by (5.90). Therefore, the contribution in $\mathbf{y}[n]$ due to $u_\ell[n]$ can be estimated as

$$\begin{aligned}\widehat{\mathbf{x}}_\ell[n] &= \widehat{\mathbf{h}}_\ell[n] \star \widehat{u}_\ell[n] \quad \text{(by (5.89))} \\ &= \mathbf{h}_\ell[n] \star u_\ell[n] \quad \text{as } \mathbf{w}[n] = \mathbf{0}.\end{aligned}$$

Cancelling $\widehat{\mathbf{x}}_\ell[n]$ from the data $\mathbf{y}[n]$ yields

$$\begin{aligned}\mathbf{y}[n] - \widehat{\mathbf{x}}_\ell[n] &= \mathbf{y}[n] - \widehat{\mathbf{h}}_\ell[n] \star \widehat{u}_\ell[n] \quad \text{(by (5.88))} \\ &= \mathbf{y}[n] - \widehat{\mathbf{h}}_\ell[n] \star \widehat{\mathbf{v}}^T[n] \star \mathbf{y}[n] \end{aligned} \tag{5.91}$$

which corresponds to the outputs of an $M \times (K-1)$ system driven by $(K-1)$ inputs $u_k[n]$, $k = 1, ..., \ell-1, \ell+1, ..., K$.

The widely used multistage successive cancellation (MSC) procedure [3] for the restoration of all system inputs is shown in Fig. 5.4 and is summarized as follows.

MSC Procedure:
Given measurements $\mathbf{y}[n], n = 0, 1, ..., N-1$, obtain the system input estimates $\widehat{u}_1[n]$, $\widehat{u}_2[n]$, ..., $\widehat{u}_K[n]$ through K stages (usually in a nonsequential order) that includes the following two steps at each stage.

(S1) Find a system input estimate, denoted as $\widehat{u}_\ell[n]$ (where ℓ is unknown), by any MIMO blind equalization algorithm such as the MIMO hybrid MNC algorithm, and then obtain the associated system estimate $\widehat{\mathbf{h}}_\ell[n]$ by (5.90).

(S2) Update $\mathbf{y}[n]$ by $\mathbf{y}[n] - \widehat{\mathbf{h}}_\ell[n] \star \widehat{u}_\ell[n]$, namely, cancel the contribution of $\widehat{u}_\ell[n]$ from $\mathbf{y}[n]$.

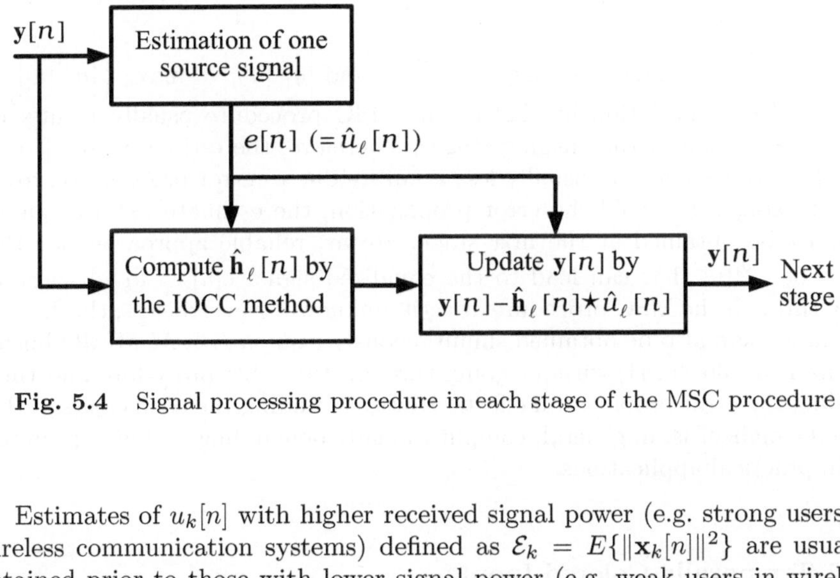

Fig. 5.4 Signal processing procedure in each stage of the MSC procedure

Estimates of $u_k[n]$ with higher received signal power (e.g. strong users in wireless communication systems) defined as $\mathcal{E}_k = E\{\|\mathbf{x}_k[n]\|^2\}$ are usually obtained prior to those with lower signal power (e.g. weak users in wireless communication systems). Assume that the system input $u_1[n]$ is of interest and obtained through the MSC procedure. Then the associated output SINR defined by (5.25), denoted by SINR_1, which is needed for performance evaluation during the algorithm design, is presented next.

Assume that $u_1[n]$ is estimated at the lth stage of the MSC procedure and $e^{[i]}[n]$, $\mathbf{h}^{[i]}[n]$ and $\mathbf{v}^{[i]}[n]$ are the equalizer output, system estimate and optimum equalizer, respectively, obtained at the ith stage for $i = 1, ..., l$. Define

$$\mathcal{T}^{[l]}(z) = \mathcal{Z}\{\mathbf{T}^{[l]}[n]\}$$
$$= \begin{cases} \mathbf{I}, & l = 1, \\ \mathcal{T}^{[l-1]}(z)\left(\mathbf{I} - \boldsymbol{H}^{[l-1]}(z)\cdot\left(\boldsymbol{V}^{[l-1]}(z)\right)^T\right), & l \geq 2. \end{cases}$$

Letting $\boldsymbol{Y}^{[l]}(z) = \mathcal{T}^{[l]}(z) \cdot \boldsymbol{Y}(z)$ and $\boldsymbol{W}^{[l]}(z) = \mathcal{T}^{[l]}(z) \cdot \boldsymbol{W}(z)$, then the equalizer output $e^{[l]}[n]$ at the lth stage can be shown, from (5.91), to be

$$e^{[l]}[n] = \left(\mathbf{v}^{[l]}[n]\right)^T \star \mathbf{y}^{[l]}[n] = \left(\mathbf{g}^{[l]}[n]\right)^T \star \mathbf{u}[n] + e_{\mathrm{N}}^{[l]}[n]$$

where $\left(\mathbf{g}^{[l]}[n]\right)^T = \left(\mathbf{v}^{[l]}[n]\right)^T \star \mathbf{T}^{[l]}[n] \star \mathbf{H}[n]$ and $e_{\mathrm{N}}^{[l]}[n] = \left(\mathbf{v}^{[l]}[n]\right)^T \star \mathbf{w}^{[l]}[n]$. Therefore, SINR_1 can be calculated from

$$\mathrm{SINR}_1 = \frac{\sigma_{u_1}^2 \cdot \left|g_1^{[l]}[n_0]\right|^2}{\sum_{k=1}^{K} \sigma_{u_k}^2 \left(\sum_n \left|g_k^{[l]}[n]\right|^2\right) + E\left\{\left|e_{\mathrm{N}}^{[l]}[n]\right|^2\right\} - \sigma_{u_1}^2 \cdot \left|g_1^{[l]}[n_0]\right|^2} \tag{5.92}$$

where $g_1^{[l]}[n]$ is the first component of $\mathbf{g}^{[l]}[n]$ and $\left|g_1^{[l]}[n_0]\right| = \max_n \left\{\left|g_1^{[l]}[n]\right|\right\}$.

Imperfect cancelation in (S2) of the MSC procedure usually results in error propagation in the ensuing stages. Therefore, the estimates $\widehat{u}_k[n]$s obtained at later stages are usually less accurate due to error propagation from stage to stage. To avoid the error propagation, the estimate $\widehat{u}_k[n]$ of interest must be obtained at the first stage. So far, reliable approaches for the choice of $\mathbf{v}_k^{[0]}[n]$ that can lead to the equalized signal $e[n] = \widehat{u}_k[n]$ without going through the MSC procedure are still unknown. Nevertheless, the K system inputs can also be obtained simultaneously using a $K \times M$ MIMO linear equalizer [34, 36, 40, 41] without going through the MSC procedure and thus avoiding the error propagation effects. However, finding the coefficients of the MIMO equalizer is, in general, computationally demanding and may prohibit use in practical applications.

5.4.2 Temporally Colored Inputs

Let us consider the case that the driving inputs of an FIR MIMO system are s.i.t.c., and introduce an *equalization-GCD* algorithm for blind equalization and system identification for this case. This blind equalization algorithm comprises the MIMO blind equalization algorithms for s.i.t.i. inputs introduced in Section 5.4.1, and a GCD computation algorithm, which is introduced first for the equalization-GCD algorithm without notational confusion.

GCD Computation

Let $\boldsymbol{F}(z) = (F_1(z), F_2(z), ..., F_M(z))^T$ be an $M \times 1$ polynomial vector of z^{-1} with a GCD $B(z)$ among the M components of $\boldsymbol{F}(z)$, i.e.

$$\boldsymbol{F}(z) = \boldsymbol{H}(z) \cdot B(z)$$

where $\boldsymbol{H}(z) = (H_1(z), H_2(z), ..., H_M(z))^T$ with no common factor among the M polynomials $H_k(z)$'s. Qiu *et al.* [42] proposed a subspace method (summarized in Appendix 5C) for estimating $\mathbf{h}[n]$ and $b[n]$ with the given $\boldsymbol{F}(z)$ (or equivalently $\mathbf{f}[n]$). Their GCD computation method, which requires the degree of $\boldsymbol{H}(z)$ given in advance, not only provides closed-form solutions for the estimates of $\mathbf{h}[n]$ and $b[n]$ but also is robust against noise (due to quantization error, observation error, modeling error, etc.) in $\boldsymbol{F}(z)$.

Model Assumptions

In addition to Assumptions (A5-13) through (A5-17) made in Section 5.4 on the signal model $\mathbf{y}[n]$ given by (5.16) and (5.17), let us further make the following assumptions:

(A5-18) The $M \times K$ system $\boldsymbol{\mathcal{H}}(z) = (\boldsymbol{H}_1(z), \boldsymbol{H}_2(z), ..., \boldsymbol{H}_K(z))$ is FIR with $M > 1$. No common factor exists among the M components of $\boldsymbol{H}_k(z) = (H_{1,k}(z), H_{2,k}(z), ..., H_{M,k}(z))^T$, $\forall k$, and their degrees L_k are known *a priori*.

(A5-19) Each of the system inputs $u_k[n]$, $k \in \{1, 2, ..., K\}$, can be modeled by an MA process as

$$u_k[n] = b_k[n] \star s_k[n] = s_k[n] + \sum_{i=1}^{M_k} b_k[i] s_k[n-i] \tag{5.93}$$

where $b_k[n]$, $k \in \{1, 2, ..., K\}$, is an LTI FIR system of unknown order $M_k \geq 0$ with $b_k[0] = 1$ and $B_k(z) = \mathcal{Z}\{b_k[n]\} \neq 0$ for $|z| = 1$, and $s_k[n]$ are stationary zero-mean s.i.t.i. nonGaussian random processes with $C_{p,q}\{s_k[n]\} \neq 0, \forall k$ and $p + q \geq 3$.

Under Assumption (A5-19), $\mathbf{y}[n]$ given by (5.16) and (5.17) can also be expressed as the following equivalent MIMO model (Fig. 5.5):

$$\mathbf{y}[n] = \mathbf{F}[n] \star \mathbf{s}[n] + \mathbf{w}[n] \tag{5.94}$$

where

$$\begin{aligned}\mathbf{F}[n] &= (\mathbf{f}_1[n], \mathbf{f}_2[n], ..., \mathbf{f}_K[n]) = \mathbf{H}[n] \star \mathrm{diag}\{b_1[n], b_2[n], ..., b_K[n]\} \\ &= (\mathbf{h}_1[n] \star b_1[n], \mathbf{h}_2[n] \star b_2[n], ..., \mathbf{h}_K[n] \star b_K[n])\end{aligned} \tag{5.95}$$

is the *combined MIMO system* driven by the s.i.t.i. input vector $\mathbf{s}[n]$. It follows from (5.95), Assumptions (A5-18) and (A5-19) that

$$\begin{aligned}\boldsymbol{\mathcal{F}}(z) &= (\boldsymbol{F}_1(z), \boldsymbol{F}_2(z), ..., \boldsymbol{F}_K(z)) \\ &= (\boldsymbol{H}_1(z)B_1(z), \boldsymbol{H}_2(z)B_2(z), ..., \boldsymbol{H}_K(z)B_K(z)) = \boldsymbol{\mathcal{H}}(z)\boldsymbol{\mathcal{B}}(z) \quad (5.96)\end{aligned}$$

where $\boldsymbol{\mathcal{B}}(z) = \text{diag}\{B_1(z), B_2(z), ..., B_K(z)\}$. Note that, from (5.96) and Assumption (A5-18), $B_k(z)$ is the GCD of the M components (polynomials) of $\boldsymbol{F}_k(z) = \boldsymbol{H}_k(z)B_k(z)$. Thus, one can estimate the unknown system $\boldsymbol{\mathcal{H}}(z)$ by estimating $\boldsymbol{\mathcal{F}}(z)$ (with s.i.t.i. inputs) followed by finding the GCD of the M components of the estimated $\boldsymbol{F}_k(z)$ for all k, leading to the equalization-GCD MIMO blind system estimation and equalization algorithm to be introduced next.

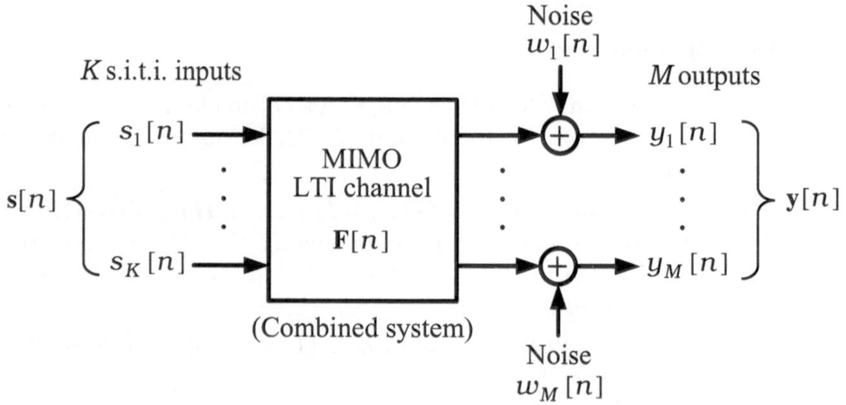

Fig. 5.5 The equivalent MIMO system model

System Estimation and Equalization

The equalization-GCD algorithm is illustrated in Fig. 5.6. From this figure, one can see that the measurements $\mathbf{y}[n]$ given by (5.94) are processed by an MIMO equalization algorithm giving rise to the equalized signal $e[n]$ as given by (5.19), which is an estimate of $s_\ell[n]$, $\ell \in \{1, 2, ..., K\}$ rather than $u_\ell[n]$. As mentioned in Section 5.4.1, the ℓth column of $\mathbf{F}[n]$ can be accordingly estimated using the IOCC system estimation method (see (5.90)). Through K stages of the MSC procedure, all the estimates of $s_k[n]$ and $\mathbf{f}_k[n]$ can be obtained. Then the unknown system $\boldsymbol{\mathcal{H}}(z)$ and $\widehat{u}_k[n]$ can be obtained from $\mathbf{F}[n]$ and $s_k[n]$ through a GCD computation. The equalization-GCD algorithm is further summarized as follows:

Equalization-GCD Algorithm: (Fig. 5.6)

(S1) Given $\mathbf{y}[n], n = 0, 1, ..., N$, obtain estimates $\widehat{\mathbf{F}}[n] = (\widehat{\mathbf{f}}_1[n], \widehat{\mathbf{f}}_2[n], ..., \widehat{\mathbf{f}}_K[n])$ and $\widehat{s}_k[n]$ through the MSC procedure using the MIMO blind equalization algorithm introduced in Section 5.4.1. Truncate each of $\widehat{\mathbf{f}}_k[n]$, $k = 1$, 2, ..., K using a sliding window of length $T_k + 1$, with maximum energy of the truncated $\widehat{\mathbf{f}}_k[n]$ (of order equal to T_k) in the window.

(S2) Obtain $\widehat{b}_k[n]$ and $\widehat{\mathbf{h}}_k[n]$ from $\widehat{\mathbf{f}}_k[n], k = 1, 2, ..., K$ using the GCD computation algorithm (proposed by Qiu *et al.* [42]) introduced above. Then the $\widehat{\mathbf{h}}_k[n]$ obtained constitute $\widehat{\mathbf{H}}[n]$ (up to a permutation matrix, a multiple-scaling matrix, and a multiple-delay matrix), and the colored system inputs $u_k[n]$'s can be estimated as

$$\widehat{u}_k[n] = \widehat{b}_k[n] \star \widehat{s}_k[n], \qquad k = 1, 2, ..., K.$$

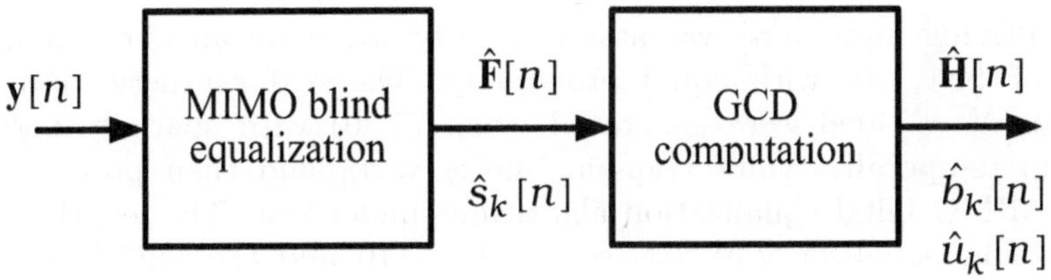

Fig. 5.6 Block diagram of the equalization-GCD algorithm

In (S1), $T_k, k = 1, 2, ..., K$ must be chosen such that $T_k \geq (L_k + M_k)$, which means that the order T_k of the truncated $\widehat{\mathbf{f}}_k[n]$ must be equal to or larger than the true order of $\mathbf{f}_k[n]$ such that the estimation error due to undermodeling can be avoided. If all the inputs $u_k[n]$ are temporally independent (i.e. $B_k(z) = 1$, $\forall k$), it can be seen from (5.95) that $\mathbf{F}[n] = \mathbf{H}[n]$. For this case, only (S1) is needed to obtain the system estimate $\widehat{\mathbf{H}}[n]$ without need of knowledge of L_k's (the order of $\mathbf{h}_k[n]$). On the other hand, it should be noted that if a GCD $C(z)$ exists among the M components of $\boldsymbol{H}_k(z)$, i.e. $\boldsymbol{H}_k(z) = \boldsymbol{H}'_k(z)C(z)$, then the $\widehat{\boldsymbol{H}}_k(z)$ and $\widehat{B}_k(z)$ obtained are estimates of $\boldsymbol{H}'_k(z)$ (rather than $\boldsymbol{H}_k(z)$) and $B_k(z)C(z)$ (rather than $B_k(z)$), respectively.

5.5 Algorithm Tests

Example 5.37 (Blind SIMO Equalization Using SOS and HOS Based Algorithms)
This example presents some simulation results of blind SIMO equalization for performance tests to the introduced noise subspace, single-stage linear prediction, MNC, SE and hybrid MNC equalization approaches. A 3×1 MA(3) system with the transfer function

$$\boldsymbol{A}(z) = \begin{pmatrix} 0.7426 + 0.7426z^{-2} \\ 0.4456z^{-1} + 0.7426z^{-2} \\ 0.8911z^{-2} + 0.5941z^{-3} \end{pmatrix}$$

was considered for the following two cases:

Case A. $\boldsymbol{H}(z) = \boldsymbol{A}(z)$ (without common subchannel zeros)
Case B. $\boldsymbol{H}(z) = (1 - 0.5z^{-1}) \cdot \boldsymbol{A}(z)$ (with a common subchannel zero)

The driving input $u[n]$ was assumed to be a zero-mean i.i.d. binary sequence of $\{+1, -1\}$ with equal probability. The real synthetic data $\mathbf{y}[n]$ of length $N = 4096$ were generated using (5.30) with spatially independent and temporally white Gaussian noise $\mathbf{w}[n]$, and then processed using the MIMO blind equalization algorithms under test. The length for the MIMO blind equalizer $\mathbf{v}[n]$ was set to $L = 10$ and the initial condition $\mathbf{v}^{[0]}[n] = (1,1,1)^T \delta[n-5]$ for the MIMO-MNC, MIMO-SE, and MIMO hybrid MNC algorithms. Additionally, other settings for each of the MIMO equalization algorithms are described below:

- *Noise subspace approach:* $N_w = 8$, $L_h = 3$ and 4 for Case A, $L_h = 4$ and 5 for Case B, and $\mathbf{v}_{\mathrm{ZF}}[n]$ obtained by (5.52) was truncated such that its length was equal to 10.
- *Single-stage linear prediction approach:* $N_w = 8$ and $\widehat{\mathbf{v}}_{\mathrm{ZF}}[n]$ in (5.60) was of length equal to $N_w + 2 = 10$ (the length of $\mathbf{V}_{\mathrm{LP}}[n]$).
- *MNC algorithm using the BFGS method:* $p = q = 2$ and convergence tolerance $\zeta = 2^{-10}$.
- *SE algorithm:* $p = q = 2$ and convergence tolerance $\zeta = 2^{-10}$.
- *Hybrid MNC algorithm:* $p = q = 2$ and convergence tolerance $\zeta = 2^{-10}$.

Thirty independent runs were performed and the averaged SINR of the equalized signal was used as the performance index. Figure 5.7a shows the simulation results (Output SINR versus SNR) for Case A associated with the noise subspace approach with the system order assumed to be 3 (indicated by 'NS (order = 3)'), the system order assumed to be 4 (indicated by 'NS (order = 4)'), the single-stage linear prediction approach (indicated by 'LP'), the MNC algorithm (indicated by 'MNC'), the SE algorithm (indicated by 'SE') and the hybrid MNC algorithm (indicated by 'hybrid MNC'). It can be seen, from this figure, that the noise subspace method with system order assumed to be 4 (i.e. order over-determined by 1) failed, verifying the fact of its high sensitivity to system order mismatch. One can also observe that the noise subspace approach (with true system order), and MNC, SE and hybrid MNC algorithms perform much better than the single-stage linear prediction approach.

On the other hand, Fig. 5.7b shows the corresponding results for Case B where the system considered has a common subchannel zero at $z = 0.5$ and thus does not satisfy the channel disparity condition (see Assumption (A5-2)). Again, one can see, from this figure, that the MNC, SE and hybrid MNC algorithms perform much better than the single-stage linear prediction approach that apparently works when the channel disparity condition is violated, whereas the noise subspace method failed in spite of the exact system order used. Finally, it can be observed, from both Fig. 5.7a, b, that the MNC, SE, and hybrid MNC equalization algorithms exhibit similar performances. □

Example 5.38 (Blind MIMO Equalization for s.i.t.i. Inputs Using HOS Based Algorithms)

This example presents some simulation results of blind MIMO equalization for performance tests to the introduced MNC, SE and hybrid MNC equalization algorithms as well as the MSC procedure. A two-input two-output system

$$\mathbf{H}(z) = \begin{pmatrix} H_{11}(z) & H_{12}(z) \\ H_{21}(z) & H_{22}(z) \end{pmatrix}$$

with

$$\begin{aligned}
H_{11}(z) &= 0.6455 - 0.3227z^{-1} + 0.6455z^{-2} - 0.3227z^{-3} \\
H_{12}(z) &= 0.6140 + 0.3684z^{-1} \\
H_{21}(z) &= 0.3873z^{-1} + 0.8391z^{-2} + 0.3227z^{-3} \\
H_{22}(z) &= -0.2579z^{-1} - 0.6140z^{-2} + 0.8842z^{-3} + 0.4421z^{-4} + 0.2579z^{-6}
\end{aligned}$$

was considered. The two system inputs $u_1[n]$ and $u_2[n]$ were assumed to be equally probable binary random sequences of $\{+1, -1\}$. The noise vector $\mathbf{w}[n]$ was assumed to be spatially independent and temporally white Gaussian. The synthetic data $\mathbf{y}[n]$ for $N = 900$ and

$$\mathrm{SNR}_k = \frac{E\{|y_k[n] - w_k[n]|^2\}}{E\{|w_k[n]|^2\}} = \mathrm{SNR} = 15 \text{ dB}, \; k = 1, 2,$$

were processed by the equalizer $\mathbf{v}[n]$ of length $L = 30$ ($L_1 = 0$ and $L_2 = 29$) associated with the MNC algorithm ($p = q = 2$) using the BFGS method, SE algorithm ($p = q = 2$), and hybrid MNC algorithm ($p = q = 2$), respectively. The same initial condition $\mathbf{v}^{[0]}[n] = (1, 1)^T \delta[n - 14]$ at the first stage and

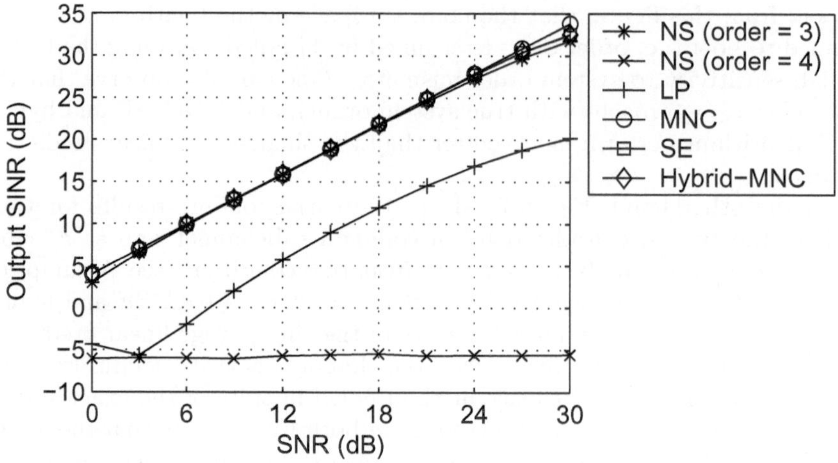

(a)

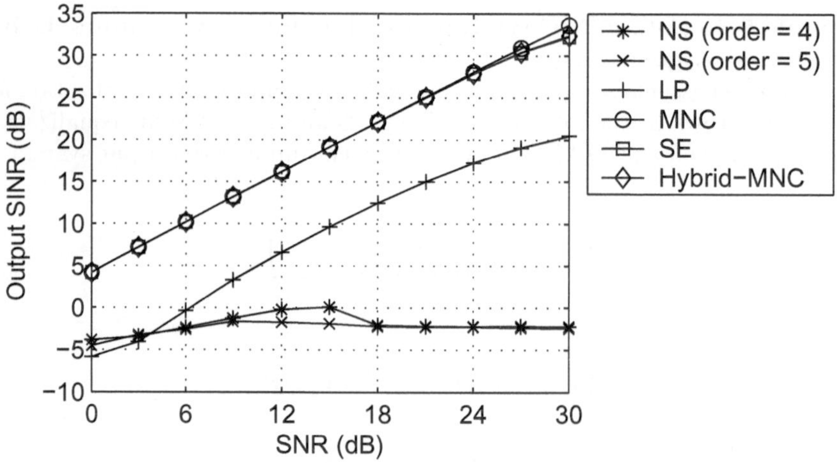

(b)

Fig. 5.7 Simulation results for SIMO equalization using the noise subspace approach, single-stage linear prediction approach, and the MNC, SE, and hybrid MNC algorithms for **(a)** Case A and **(b)** Case B in Example 5.37

$\mathbf{v}^{[0]}[n] = (1,0)^T\delta[n-14]$ at the second stage of the MSC procedure, were used in the simulation of the three algorithms.

Thirty independent realizations of the optimum $g_1[n]$ and the associated thirty ISIs versus iteration number obtained at the first stage of the MSC procedure (associated with $e[n] = \widehat{u}_1[n]$) are shown in Fig. 5.8a-f using the three algorithms, respectively. Results for $g_2[n]$ obtained at the first stage of the MSC procedure are omitted here since they are close to zero. Figures 5.8a, c, e show $g_1[n]$ associated with the MNC, SE and hybrid MNC algorithms, respectively. Figures 5.8b, d, f show ISIs associated with the MNC, SE and hybrid MNC algorithms, respectively.

The corresponding results for $g_2[n]$ and ISI obtained at the second stage of the MSC procedure (associated with $e[n] = \widehat{u}_2[n]$) are shown in Fig. 5.9a-f where the results for $g_1[n]$ are not displayed since they are close to zero. Note that the SE equalization algorithm failed to converge in one (denoted by a dashed line) of the thirty realizations (see Fig. 5.9d) and the associated $g_2[n]$ was even not a fair approximation to a Kronecker delta function at all (see Fig. 5.9c).

One can see, from Fig. 5.8 and Fig. 5.9, that the convergence speed for the SE equalization algorithm is faster than that of the MNC equalization algorithm implemented with the BFGS method. Furthermore, the MNC, SE, and hybrid MNC algorithms ended up with similar residual ISI after convergence, while the hybrid MNC algorithm converges much faster than the MNC algorithm and converges almost as fast as the SE algorithm in all thirty realizations without any divergence.

In this example, Algorithm MNC-LMMSE was also used to compute the theoretical MNC equalizer $\mathbf{v}_{\mathrm{MNC}}[n]$ with $\mathcal{N} = 64$, $\zeta = 2^{-10}$ and initial conditions $\mathbf{v}^{[0]}[n] = (1,1)^T\delta[n-14]$. Figure 5.10 shows that the MNC equalizer $\mathbf{v}_{\mathrm{MNC}}[n] = (v_1[n], v_2[n])^T$ (Fig. 5.10a, c) obtained by the MNC algorithm at the first stage of the MSC procedure, and the theoretical $\mathbf{v}_{\mathrm{MNC}}[n]$ (Fig. 5.10b, d). From Fig. 5.10a-d, one can see that all the $\mathbf{v}_{\mathrm{MNC}}[n]$ obtained by the MNC algorithm are close to the theoretical $\mathbf{v}_{\mathrm{MNC}}[n]$ obtained using Algorithm MNC-LMMSE, thereby verifying Property 5.35.

□

Example 5.39 (Blind MIMO Equalization for s.i.t.c. Inputs Using SOS and HOS Based Algorithms)

This example presents some simulation results of blind MIMO equalization for performance tests to the proposed MP method and equalization-GCD algorithm. In the example, a 3×2 MA(4) system $\mathbf{H}[n]$ whose coefficients are given in Table 5.6 was considered. The driving inputs $s_k[n]$ in (5.93) were assumed to be zero-mean i.i.d. binary sequences of $\{+1, -1\}$ with equal probability. The colored inputs $u_k[n]$ were generated by filtering $s_k[n]$ with FIR systems $B_k(z)$, $k = 1, 2$, for which the following two cases were considered:

Case A. $B_1(z) = \widetilde{B}(z) = 0.8127 - 0.2438z^{-1} + 0.5283z^{-2}$ and $B_2(z) = 0.9800 + 0.1985z^{-2}$ (i.e. two inputs with distinct power spectra).

Case B. $B_1(z) = B_2(z) = \widetilde{B}(z)$ (i.e. two inputs with the same power spectrum).

Real synthetic data $\mathbf{y}[n]$ of length $N = 10000$ were generated using (5.16) with spatially independent and temporally white Gaussian noise $\mathbf{w}[n]$, and then processed using the MP method ($l_1 = 0$, $l_2 = 1$, $N_w = 7$ and $\xi = 10^{-3}$) and the equalization-GCD algorithm ($p = q = 2$, $T_1 = T_2 = 15$,

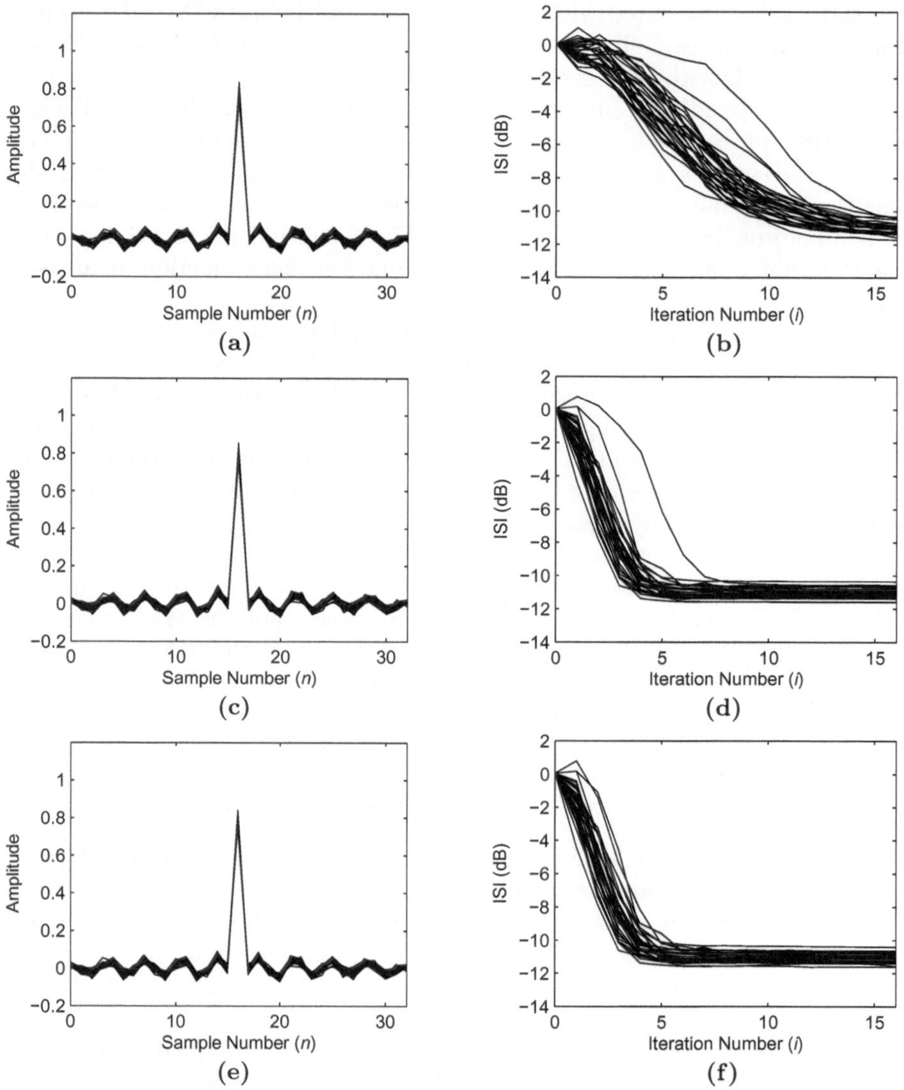

Fig. 5.8 Thirty $g_1[n]$s and ISIs versus iteration number i at the first stage of the MSC procedure of Example 5.38. **(a)** $g_1[n]$ and **(b)** ISI associated with the MNC algorithm implemented with the BFGS method; **(c)** $g_1[n]$ and **(d)** ISI associated with the SE algorithm, and **(e)** $g_1[n]$ and **(f)** ISI associated with the hybrid MNC algorithm

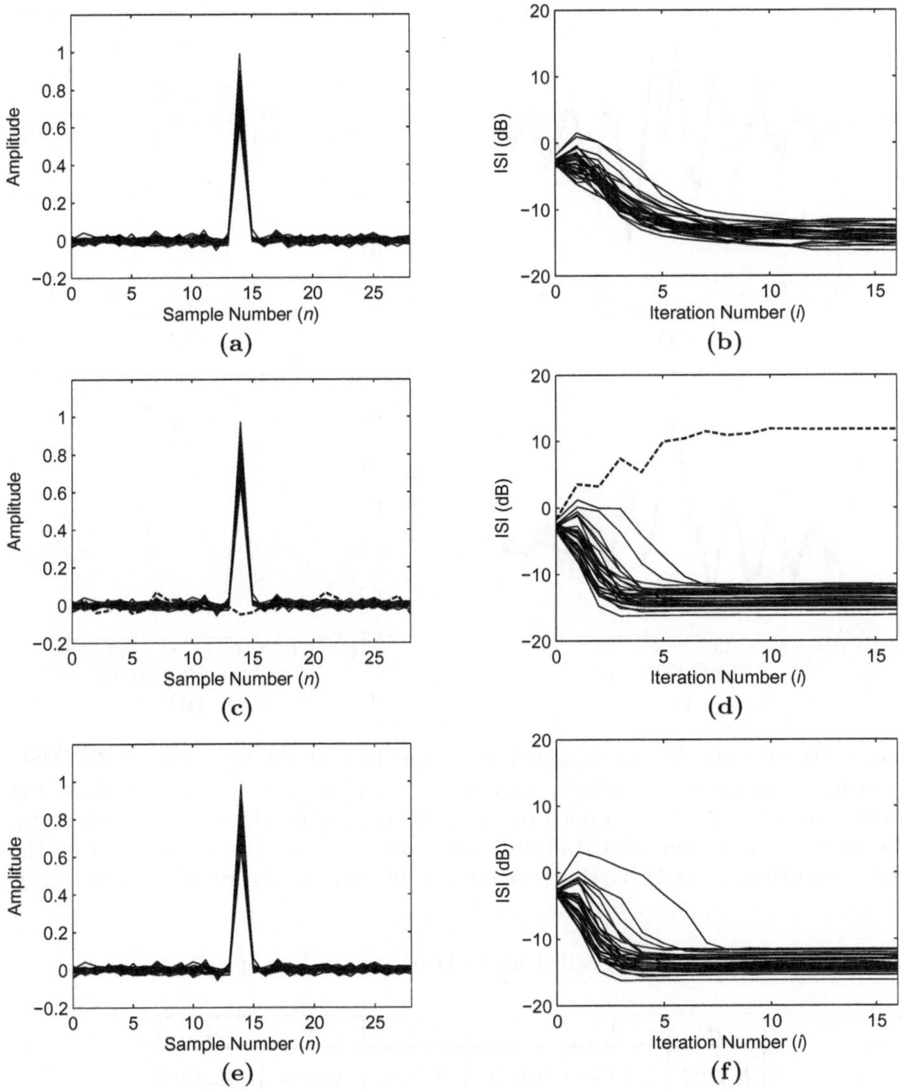

Fig. 5.9 Thirty $g_2[n]$s and ISIs versus iteration number i at the second stage of the MSC procedure of Example 5.38. **(a)** $g_2[n]$ and **(b)** ISI associated with the MNC algorithm implemented with the BFGS method; **(c)** $g_2[n]$ and **(d)** ISI associated with the SE algorithm, and **(e)** $g_2[n]$ and **(f)** ISI associated with the hybrid MNC algorithm

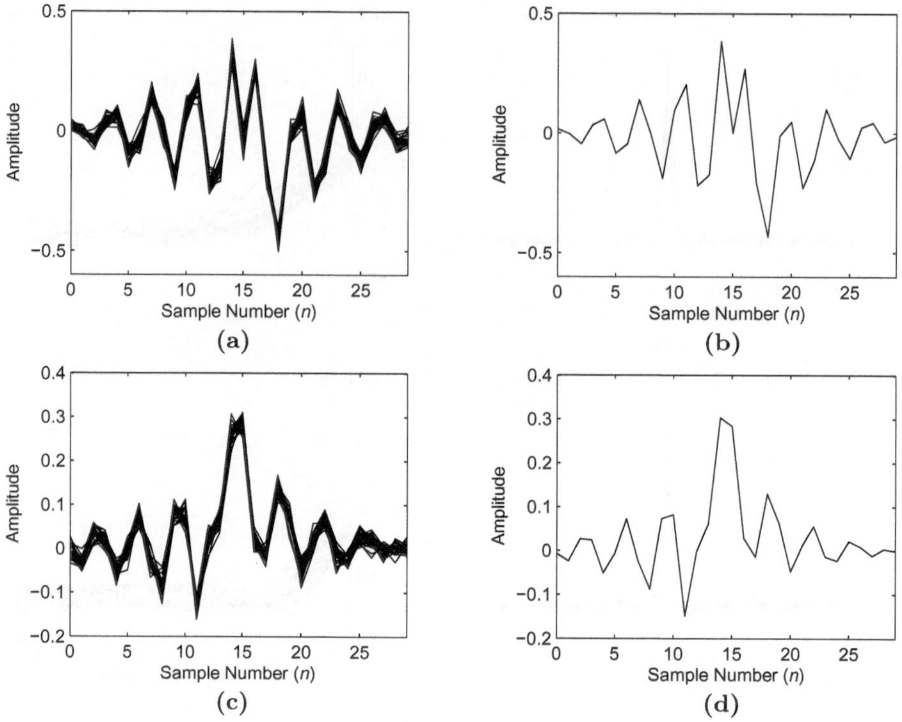

Fig. 5.10 Results for the designed MNC equalizer at the first stage of the MSC procedure and the true equalizer $\mathbf{v}_{\mathrm{MNC}}[n] = (v_1[n], v_2[n])^T$ of Example 5.38. **(a)** Thirty estimates of $v_1[n]$ obtained by the MNC algorithm, **(b)** the theoretical $v_1[n]$ obtained by Algorithm MNC-LMMSE, **(c)** thirty estimates $v_2[n]$ obtained by the MNC algorithm, and **(d)** the theoretical $v_2[n]$ obtained by Algorithm MNC-LMMSE

Table 5.6 Coefficients of $\mathbf{H}[n]$ used in Example 5.39

n	0	1	2	3	4
$h_{1,1}[n]$	0.1436	0.4620	0.0504	–0.0956	–0.2881
$h_{1,2}[n]$	0.1231	0.2294	–0.1220	–0.4818	–0.0788
$h_{2,1}[n]$	–0.0877	0.1576	0.3427	–0.1303	–0.0759
$h_{2,2}[n]$	0.1629	–0.1132	0.1333	0.2085	–0.1077
$h_{3,1}[n]$	0.2631	0.3048	0.4356	0.3576	–0.1372
$h_{3,2}[n]$	0.1717	–0.2292	–0.0327	–0.4628	0.5193

$L = 30$ and $\zeta = 2^{-10}$). Fifty independent runs were performed and then the averaged mean-square error (AMSE) of the obtained 50 estimates $\widehat{u}_1[n]$ and 50 estimates $\widehat{u}_2[n]$ after normalization by the same energy of $u_1[n]$ was calculated as the performance index for each simulated SNR.

Figure 5.11a shows the simulation results (AMSE of input estimates versus SNR) for Case A associated with the MP method and the equalization-GCD algorithm. The corresponding results for Case B are shown in Fig. 5.11b. One can observe, from Fig. 5.11a, b, that the equalization-GCD algorithm performs much better than the MP method, and that the performance of the MP method is significantly degraded in Case B because the two input power spectra are exactly the same.

□

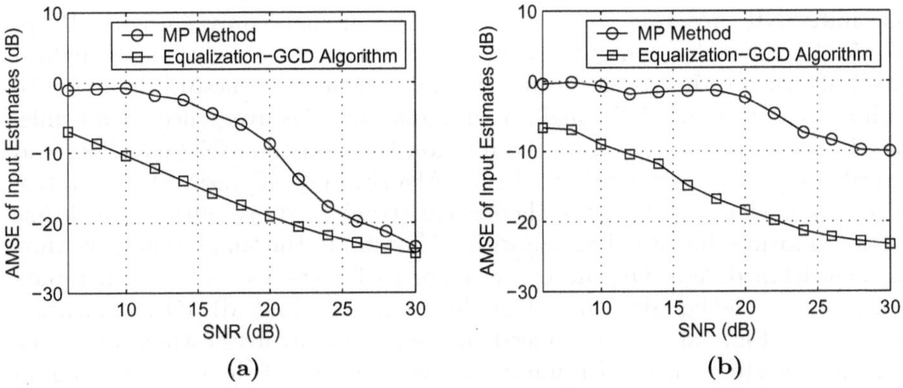

Fig. 5.11 AMSE of input estimates versus SNR using the MP method and equalization-GCD algorithm of Example 5.39 for **(a)** Case A and **(b)** Case B

5.6 Summary and Discussion

In this chapter, we began with definitions, notations and some essential properties of MIMO LTI systems. Then the MIMO equalization problem was introduced as well as ZF and MMSE criteria, and each of the associated ZF and MMSE linear equalizers is expressed by a closed-form formula in terms of the MIMO system and statistical parameters of the system inputs and the additive noises at the system outputs.

Two types of blind SIMO equalization approach using SOS were then introduced, including the noise subspace approach as summarized in Table 5.1 and the single-stage linear prediction approach as summarized in Table 5.2. Then the MP method using SOS for blind MIMO equalization was introduced as summarized in Table 5.3. Though all of these equalization algorithms are furnished with the associated closed-form formulas, their computational complexities are quite high due to calculations (e.g. eigenvalue decomposition and inversion of matrices) of high-dimension correlation matrices involved. Moreover, some conditions such as the disparity condition, the exact system order known *a priori*, and distinct input power spectra are required and these conditions may not be valid in practical applications such as wireless communications.

As for the blind MIMO equalization using HOS, the iterative MNC and SE algorithms were introduced as summarized in Tables 5.4 and 5.5, respectively, for the system inputs being s.i.t.i., followed by their properties and relations. The former is implemented through using a gradient-type optimization method with guaranteed convergence. The latter is implemented through a set of linear equations for updating the equalizer coefficients and its convergence speed is super-exponentially fast, but it may diverge for limited data length and finite SNR. According to a property regarding their equivalence (see Property 5.34), a combination of these two algorithms was introduced, the hybrid MNC algorithm, that is as computationally efficient as the SE algorithm in finding the optimum MNC equalizer with guaranteed convergence. Meanwhile, based on a property regarding the relation between the MNC equalizer and LMMSE equalizer (see Property 5.35), Algorithm MNC-LMMSE was introduced to efficiently obtain the theoretical (true) MNC equalizer (which has no closed-form solution). This algorithm is useful in the simulation stage during algorithm design. On the other hand, as for the case of system inputs being s.i.t.c., the equalization-GCD algorithm for blind MIMO equalization as shown in Fig. 5.6 was introduced that basically involves estimation of the equivalent system with s.i.t.i. inputs, i.e. convolution of the unknown system and the signal model (a diagonal MIMO LTI system), using the hybrid MNC algorithm, and GCD computation for finding the unknown system.

All the MIMO blind equalization algorithms introduced were tested by simulation for performance evaluation and comparison and verification of their analytical properties and relations. Some advanced applications of the hybrid MNC algorithm will be introduced in Chapter 6. This chapter only provides a limited introduction to MIMO blind equalization and system identification. The research on advanced MIMO blind equalization and system identification algorithms remains central to signal processing and wireless communications.

Appendix 5A
Proof of Property 5.34

In the proof of Property 5.34 below, we need $\partial\sigma_e^2/\partial\boldsymbol{v}^*$ which can easily be seen to be

$$\frac{\partial\sigma_e^2}{\partial\boldsymbol{v}^*} = E\{e[n]\boldsymbol{y}^*[n]\} = E\{\boldsymbol{y}^*[n]\boldsymbol{y}^T[n]\boldsymbol{v}\} = \mathbf{R}_{\boldsymbol{y}}^*\boldsymbol{v}. \tag{5.97}$$

Maximizing $J_{p,q}(\boldsymbol{v})$ given by (5.73) is equivalent to maximizing

$$\widetilde{J}_{p,q}(\boldsymbol{v}) = J_{p,q}^2(\boldsymbol{v}) = \frac{|C_{p,q}\{e[n]\}|^2}{|\sigma_e^2|^{p+q}} = \frac{C_{p,q}\{e[n]\}C_{q,p}\{e[n]\}}{|\sigma_e^2|^{p+q}} \tag{5.98}$$

which implies

$$\frac{\partial \widetilde{J}_{p,q}(\boldsymbol{v})}{\partial \boldsymbol{v}^*} = 2 \cdot J_{p,q}(\boldsymbol{v}) \cdot \frac{\partial J_{p,q}(\boldsymbol{v})}{\partial \boldsymbol{v}^*}. \tag{5.99}$$

Taking partial derivative of $\widetilde{J}_{p,q}(\boldsymbol{v})$ given by (5.98) with respect to $\boldsymbol{v}^*$ yields

$$\frac{\partial \widetilde{J}_{p,q}(\boldsymbol{v})}{\partial \boldsymbol{v}^*} = \widetilde{J}_{p,q}(\boldsymbol{v}) \cdot \left\{ \frac{1}{C_{q,p}\{e[n]\}} \cdot \frac{\partial C_{q,p}\{e[n]\}}{\partial \boldsymbol{v}^*} + \frac{1}{C_{p,q}\{e[n]\}} \cdot \frac{\partial C_{p,q}\{e[n]\}}{\partial \boldsymbol{v}^*} - \frac{p+q}{\sigma_e^2} \cdot \frac{\partial \sigma_e^2}{\partial \boldsymbol{v}^*} \right\} \tag{5.100}$$

which together with (5.98) and (5.99) leads to

$$\frac{\partial J_{p,q}(\boldsymbol{v})}{\partial \boldsymbol{v}^*} = \frac{J_{p,q}(\boldsymbol{v})}{2} \cdot \left\{ \frac{1}{C_{q,p}\{e[n]\}} \cdot \frac{\partial C_{q,p}\{e[n]\}}{\partial \boldsymbol{v}^*} + \frac{1}{C_{p,q}\{e[n]\}} \cdot \frac{\partial C_{p,q}\{e[n]\}}{\partial \boldsymbol{v}^*} - \frac{p+q}{\sigma_e^2} \cdot \frac{\partial \sigma_e^2}{\partial \boldsymbol{v}^*} \right\}. \tag{5.101}$$

Setting $\partial J_{p,q}/\partial \boldsymbol{v}^*$ given by (5.101) equal to zero and substituting (5.97) into the resultant equation, we obtain

$$\begin{aligned} \mathbf{R}_{\boldsymbol{y}}^* \boldsymbol{v} &= \frac{\sigma_e^2}{p+q} \cdot \frac{p}{C_{q,p}\{e[n]\}} \cdot \text{cum}\{e[n] : q, e^*[n] : p-1, \boldsymbol{y}^*[n]\} \\ &\quad + \frac{\sigma_e^2}{p+q} \cdot \frac{q}{C_{p,q}\{e[n]\}} \cdot \text{cum}\{e[n] : p, e^*[n] : q-1, \boldsymbol{y}^*[n]\} \\ &= \alpha_{q,p} \cdot \text{cum}\{e[n] : q, e^*[n] : p-1, \boldsymbol{y}^*[n]\} \\ &\quad + \alpha_{p,q} \cdot \text{cum}\{e[n] : p, e^*[n] : q-1, \boldsymbol{y}^*[n]\} \end{aligned} \tag{5.102}$$

where $\alpha_{p,q}$ is given by (5.86).

As $\mathbf{y}[n]$ is complex, letting $p = q$ in (5.102) yields

$$\mathbf{R}_{\boldsymbol{y}}^* \boldsymbol{v} = 2 \cdot \alpha_{p,p} \cdot \text{cum}\{e[n] : p, e^*[n] : p-1, \boldsymbol{y}^*[n]\}$$

which is equivalent to (5.76) with $p = q$ except for a scale factor. Similarly, as $\mathbf{y}[n]$ is real, one can also obtain from (5.102) that

$$\begin{aligned} \mathbf{R}_{\boldsymbol{y}}^* \boldsymbol{v} &= (\alpha_{p,q} + \alpha_{q,p}) \cdot \text{cum}\{e[n] : p+q-1, \boldsymbol{y}[n]\} \\ &= \frac{\sigma_e^2}{C_{p,q}\{e[n]\}} \cdot \text{cum}\{e[n] : p+q-1, \boldsymbol{y}[n]\} \end{aligned}$$

which is also equivalent to (5.76) except for a scale factor. Thus we have completed the proof.

Q.E.D.

Appendix 5B
Proof of Property 5.35

First of all, for ease of use in the proof of Property 5.35 below, let us prove that $\mathrm{cum}\{e[n]:p, e^*[n]:q-1, (y_i[n-k])^*\}$, $i = 1, 2, ..., M$, can be expressed as (5.103) as follows.

$$\begin{aligned}
&\mathrm{cum}\{e[n]:p, e^*[n]:q-1, y_i^*[n-k]\} \\
&= \mathrm{cum}\left\{\sum_{l=1}^{K}\sum_{k_1=-\infty}^{\infty} g_l[n-k_1]u_l[k_1]:p, \sum_{l=1}^{K}\sum_{k_1=-\infty}^{\infty} g_l^*[n-k_1]u_l^*[k_1]:q-1,\right. \\
&\qquad \left.\sum_{l=1}^{K}\sum_{k_1=-\infty}^{\infty} [\mathbf{H}[n-k_1-k]]_{il}^* u_l^*[k_1]\right\} \quad \text{(by (5.19))} \\
&= \sum_{l=1}^{K} C_{p,q}\{u_l[n]\} \sum_{k_1=-\infty}^{\infty} g_l^p[k_1](g_l^*[k_1])^{q-1}[\mathbf{H}[k_1-k]]_{il}^* \quad \text{(by (A5-15))} \\
&= \sum_{l=1}^{K} C_{p,q}\{u_l[n]\}\cdot\left(\widetilde{g}_{p,q}[k;l]\star[\mathbf{H}[-k]]_{il}^*\right), \quad k = L_1, L_1+1, ..., L_2 \qquad (5.103)
\end{aligned}$$

where $\widetilde{g}_{p,q}[k;l]$ is given by (5.80).

Substituting (5.103) into (5.102), one can obtain

$$\begin{aligned}
&\sum_{l=1}^{M}\sum_{k=L_1}^{L_2} r_{i,l}^*[k-n]v_l[k] = \sum_{l=1}^{M} r_{l,i}[n]\star v_l[n] \\
&= \alpha_{q,p}\cdot\sum_{l=1}^{K} C_{q,p}\{u_l[n]\}\cdot\left(\widetilde{g}_{q,p}[n;l]\star[\mathbf{H}[-n]]_{il}^*\right) \\
&\qquad + \alpha_{p,q}\cdot\sum_{l=1}^{K} C_{p,q}\{u_l[n]\}\cdot\left(\widetilde{g}_{p,q}[n;l]\star[\mathbf{H}[-n]]_{il}^*\right), \\
&\qquad\qquad i = 1, ..., M, \quad n = L_1, ..., L_2 \qquad (5.104)
\end{aligned}$$

where $r_{i,l}[k] = E[y_i[n]y_l^*[n-k]] = r_{l,i}^*[-k]$. Letting $L_1 \to -\infty$, $L_2 \to \infty$ and then taking the discrete-time Fourier transform of both sides of (5.104), one can obtain

$$\boldsymbol{\mathcal{S}}_{\mathbf{y}}^T(\omega)\boldsymbol{V}_{\mathrm{MNC}}(\omega) = \alpha_{q,p}\boldsymbol{\mathcal{H}}^*(\omega)\mathbf{D}_{q,p}\widetilde{\boldsymbol{G}}_{q,p}(\omega) + \alpha_{p,q}\boldsymbol{\mathcal{H}}^*(\omega)\mathbf{D}_{p,q}\widetilde{\boldsymbol{G}}_{p,q}(\omega) \quad (5.105)$$

where $\mathbf{D}_{p,q}$ and $\widetilde{\boldsymbol{G}}_{p,q}(\omega)$ are given by (5.79) and (5.81), respectively. Finally, from (5.105) and (5.82), we obtain

$$\begin{aligned}
\boldsymbol{V}_{\mathrm{MNC}}(\omega) &= \left[\boldsymbol{\mathcal{S}}_{\mathbf{y}}^T(\omega)\right]^{-1}\cdot\boldsymbol{\mathcal{H}}^*(\omega)\cdot\left(\alpha_{q,p}\mathbf{D}_{q,p}\widetilde{\boldsymbol{G}}_{q,p}(\omega) + \alpha_{p,q}\mathbf{D}_{p,q}\widetilde{\boldsymbol{G}}_{p,q}(\omega)\right) \\
&= \boldsymbol{\mathcal{V}}_{\mathrm{MS}}^T(\omega)\cdot\mathbf{D}_{1,1}^{-1}\cdot\left(\alpha_{q,p}\mathbf{D}_{q,p}\widetilde{\boldsymbol{G}}_{q,p}(\omega) + \alpha_{p,q}\mathbf{D}_{p,q}\widetilde{\boldsymbol{G}}_{p,q}(\omega)\right) \\
&= \boldsymbol{\mathcal{V}}_{\mathrm{MS}}^T(\omega)\cdot\mathbf{Q}(\omega)
\end{aligned}$$

where $\mathbf{D}_{p,q}$ and $\mathbf{Q}(\omega)$ are given by (5.79) and (5.85), respectively. Thus we have completed the proof.

Q.E.D.

Appendix 5C
A GCD Computation Algorithm

Assume that

$$F_i(z) = \sum_{n=0}^{T} f_i[n] z^{-n}, \quad i = 1, 2, ..., M,$$

$$H_i(z) = \sum_{n=0}^{L} h_i[n] z^{-n}, \quad i = 1, 2, ..., M.$$

Let $\mathbf{F}_i$ denote a $(T+1) \times (2T+1)$ Sylvester matrix formed by $\{f_i[0], f_i[1], ..., f_i[T]\}$ which is defined as

$$\mathbf{F}_i = \begin{pmatrix} f_i[0] & f_i[1] & \cdots & f_i[T] & 0 & 0 & \cdots & 0 \\ 0 & f_i[0] & f_i[1] & \cdots & f_i[T] & 0 & \cdots & 0 \\ & & \ddots & & & \ddots & & \\ 0 & 0 & \cdots & 0 & f_i[0] & f_i[1] & \cdots & f_i[T] \end{pmatrix},$$

and $\boldsymbol{\mathcal{F}}$ denote the $M(T+1) \times (2T+1)$ generalized Sylvester matrix associated with $\boldsymbol{F}(z)$ defined as

$$\boldsymbol{\mathcal{F}} = \left(\mathbf{F}_1^H, \mathbf{F}_2^H, ..., \mathbf{F}_M^H \right)^H . \tag{5.106}$$

Let $\mathbf{u}_m$ and $\mathbf{v}_m$, $m = 1, 2, ..., (T-L)$ denote the left and right singular vectors of $\boldsymbol{\mathcal{F}}^H$, respectively, associated with the $(T-L)$ smallest singular values. Let $u_{m,i}$ and $v_{m,i}$ denote the ith entries of $\mathbf{u}_m$ and $\mathbf{v}_m$, respectively, and let $r = (T-L+1)$ and $r' = (T+L+1)$. Define

$$\mathbf{R}_u = \sum_{m=1}^{T-L} \mathbf{U}_m^H \mathbf{U}_m \tag{5.107}$$

where

$$\mathbf{U}_m = \begin{pmatrix} u_{m,1} & u_{m,2} & \cdots & u_{m,r} \\ u_{m,2} & u_{m,3} & \cdots & u_{m,r+1} \\ \cdots & \cdots & \cdots & \cdots \\ u_{m,r'} & u_{m,r'+1} & \cdots & u_{m,2T+1} \end{pmatrix}$$

is a $r' \times r$ matrix, and define

$$\mathbf{R}_v = \sum_{m=1}^{T-L} \mathbf{V}_m \mathbf{V}_m^H \tag{5.108}$$

where

$$\mathbf{V}_m = \left(\mathbf{V}_{m,1}^H, \mathbf{V}_{m,2}^H, ..., \mathbf{V}_{m,M}^H\right)^H$$

in which $\mathbf{V}_{m,i}$ is the $(L+1) \times r'$ Sylvester matrix formed by $\{v_{m,(i-1)(T+1)+1}, v_{m,(i-1)(T+1)+2}, ..., v_{m,i(T+1)}\}$.

The GCD computation algorithm reported in [42] for obtaining $\widehat{\mathbf{h}}[n]$ and $\widehat{b}[n]$ from $\mathbf{f}[n]$ is summarized as follows:

GCD Computation Algorithm

(S1) Form $\boldsymbol{\mathcal{F}}$ by (5.106) and perform the singular value decomposition (SVD) of $\boldsymbol{\mathcal{F}}^H$.

(S2) Form $\mathbf{R}_u$ and $\mathbf{R}_v$ by (5.107) and (5.108), respectively.

(S3) Estimate $\widehat{\boldsymbol{h}} = \left(\widehat{h}_1[0], \widehat{h}_1[1], ..., \widehat{h}_1[L], \widehat{h}_2[0], ..., \widehat{h}_2[L], ..., \widehat{h}_P[0], ..., \widehat{h}_M[L]\right)^T$ and $\widehat{\boldsymbol{b}} = \left(\widehat{b}[0], \widehat{b}[1], ..., \widehat{b}[T-L]\right)^T$ as the eigenvectors associated with the minimum eigenvalues of $\mathbf{R}_v$ and $\mathbf{R}_u$, respectively. Then the estimate $\widehat{\mathbf{h}}[n] = \left(\widehat{h}_1[n], \widehat{h}_2[n], ..., \widehat{h}_M[n]\right)^T$ can be obtained from $\widehat{\boldsymbol{h}}$.

Problems

5.1. Derive the LMMSE equalizer $\boldsymbol{\mathcal{V}}_{\mathrm{MS}}(\omega)$ as given by (5.29).

5.2. Derive the linear prediction filter and the optimum linear prediction error vector given by (5.57) and (5.56), respectively.

5.3. Prove Theorem 5.31.

5.4. Prove Theorem 5.32.

5.5. Prove Property 5.36.

Computer Assignments

5.1. Write a computer program and perform the same simulation as presented in Example 5.37 for blind SIMO equalization using each of the following equalization algorithms:
(a) Noise subspace approach
(b) Single-stage linear prediction approach

(c) MIMO-MNC equalization algorithm
(d) MIMO-SE equalization algorithm
(e) MIMO hybrid MNC equalization algorithm.

5.2. Write a computer program and perform the same simulation as presented in Example 5.38 for blind MIMO equalization using each of the following equalization algorithms:
(a) MIMO-MNC equalization algorithm
(b) MIMO-SE equalization algorithm
(c) MIMO hybrid MNC equalization algorithm.
Illustrate the MNC equalizers obtained at the first stage of the MSC procedure in Part (a) and compare them with the theoretical MNC equalizer obtained by Algorithm MNC-LMMSE.

5.3. Write a computer program and perform the same simulation as presented in Example 5.39 for blind MIMO equalization using each of the following equalization algorithms:
(a) MP method
(b) Equalization-GCD equalization algorithm.

References

1. G. C. Goodwin, S. F. Graebe, and M. E. Salgado, *Control System Design.* New Jersey: Prentice-Hall, 2001.
2. J. M. Maciejowski, *Multivariable Feedback Design.* Wokingham, England: Addison-Wesley, 1989.
3. J. K. Tugnait, "Identification and deconvolution of multichannel linear non-Gaussian processes using higher-order statistics and inverse filter criteria," *IEEE Trans. Signal Processing*, vol. 45, no. 3, pp. 658–672, Mar. 1997.
4. Z. Ding and Y. Li, *Blind Equalization and Identification.* New York: Marcel Dekker, Inc., 2001.
5. J. K. Tugnait, "Blind spatio-temporal equalization and impulse response estimation for MIMO channels using a Godard cost function," *IEEE Trans. Signal Processing*, vol. 45, no. 1, pp. 268–271, Jan. 1997.
6. L. Tong, G. Xu, and T. Kailath, "A new approach to blind identification and equalization of multipath channels," in *Record of the 25th Asilomar Conference on Signals, Systems and Computers*, Pacific Grove, CA, Nov. 4–6, 1991, pp. 856–860.
7. L. Tong, G. Xu, and T. Kailath, "Blind identification and equalization based on second-order statistics: A time domain approach," *IEEE Trans. Information Theory*, vol. 40, no. 2, pp. 340–349, Mar. 1994.
8. E. Moulines, P. Duhamel, J.-F. Cardoso, and S. Mayrargue, "Subspace methods for the blind identification of multichannel FIR filters," *IEEE Trans. Signal Processing*, vol. 43, no. 2, pp. 516–525, Feb. 1995.

9. C. B. Papadias and D. T. M. Slock, "Fractionally spaced equalization of linear polyphase channels and related blind techniques based on multichannel linear prediction," *IEEE Trans. Signal Processing*, vol. 47, no. 3, pp. 641–654, Mar. 1999.
10. K. Abed-Meraim, E. Moulines, and P. Loubaton, "Prediction error methods for second-order blind identification," *IEEE Trans. Signal Processing*, vol. 45, no. 3, pp. 694–705, Mar. 1997.
11. L. Tong and S. Perreau, "Multichannel blind identification: From subspace to maximum likelihood methods," *Proceedings of the IEEE*, vol. 86, no. 10, pp. 1951–1968, Oct. 1998.
12. J. K. Tugnait, "On blind identifiability of multipath channels using fractional sampling and second-order cyclostationary statistics," *IEEE Trans. Information Theory*, vol. 41, no. 1, pp. 308–311, Jan. 1995.
13. T. Endres, B. D. O. Anderson, C. R. Johnson, and L. Tong, "On the robustness of FIR channel identification from second-order statistics," *IEEE Signal Processing Letters*, vol. 3, no. 5, pp. 153–155, May 1996.
14. H. Ali, J. H. Manton, and Y. Hua, "Modified channel subspace method for identification of SIMO FIR channels driven by a trailing zero filter band precoder," in *Proc. IEEE International Conference on Acoustics, Speech, and Signal Processing*, Salt Lake City, UT, USA, May 7–11, 2001, pp. 2053–2056.
15. T. Kailath, *Linear Systems*. New Jersey: Prentice-Hall, 1980.
16. Z. Ding, "Matrix outer-product decomposition method for blind multiple channel identification," *IEEE Trans. Signal Processing*, vol. 45, no. 12, pp. 3053–3061, Dec. 1997.
17. D. Gesbert and P. Duhamel, "Robust blind channel identification and equalization based on multi-step predictors," in *Proc. IEEE International Conference on Acoustics, Speech, and Signal Processing*, Munich, Germany, Apr. 21–24, 1997, pp. 3621–3624.
18. Y. Hua and J. K. Tugnait, "Blind identifiability of FIR-MIMO systems with colored input using second-order statistics," *IEEE Signal Processing Letters*, vol. 7, no. 12, pp. 348–350, Dec. 2000.
19. A. Gorokhov and P. Loubaton, "Subspace-based techniques for blind separation of convolutive mixtures with temporally correlated sources," *IEEE Trans. Circuits and Systems I*, vol. 44, no. 9, pp. 813–820, Sept. 1997.
20. K. Abed-Meraim and Y. Hua, "Blind identification of multi-input multi-output system using minimum noise subspace," *IEEE Trans. Signal Processing*, vol. 45, no. 1, pp. 254–258, Jan. 1997.
21. C. T. Ma, Z. Ding, and S. F. Yau, "A two-stage algorithm for MIMO blind deconvolution of nonstationary colored signals," *IEEE Trans. Signal Processing*, vol. 48, no. 4, pp. 1187–1192, Apr. 2000.
22. Y. Hua, S. An, and Y. Xiang, "Blind identification and equalization of FIR MIMO channels by BIDS," in *Proc. IEEE International Conference on Acoustics, Speech, and Signal Processing*, Salt Lake City, UT, USA, May 7–11, 2001, pp. 2157–2160.
23. S. An and Y. Hua, "Blind signal separation and blind system identification of irreducible MIMO channels," in *Proc. Sixth IEEE International Symposium on Signal Processing and its Applications*, Kuala Lumpur, Malaysia, Aug. 13–16, 2001, pp. 276–279.
24. D. T. M. Slock, "Blind fractionally-spaced equalization, perfect-reconstruction filter banks and multichannel linear prediction," in *Proc. IEEE International*

Conference on Acoustics, Speech, and Signal Processing, Adelaide, SA, Australia, Apr. 19–22 1994, pp. 585–588.
25. R. O. Duda and P. E. Hart, *Pattern Classification and Scene Analysis.* New York: John Wiley, 1973.
26. S. Choi, A. Cichocki, and A. Belouchrani, "Second order nonstationary source separation," *Journal of VLSI Signal Processing*, vol. 32, no. 1-2, pp. 93–104, Aug. 2002.
27. J. K. Tugnait, "Adaptive blind separation of convolutive mixtures of independent linear signals," *Signal Processing*, vol. 73, no. 1-2, pp. 139–152, Feb. 1999.
28. Y. Inouye and T. Sato, "Iterative algorithms based on multistage criteria for multichannel blind deconvolution," *IEEE Trans. Signal Processing*, vol. 47, no. 6, pp. 1759–1764, June 1999.
29. Y. Inouye, "Criteria for blind deconvolution of multichannel linear time-invariant systems," *IEEE Trans. Signal Processing*, vol. 46, no. 12, pp. 3432–3436, Dec. 1998.
30. K. L. Yeung and S. F. Yau, "A cumulant-based super-exponential algorithm for blind deconvolution of multi-input multi-output systems," *Signal Processing*, vol. 67, no. 2, pp. 141–162, 1998.
31. Y. Inouye and K. Tanebe, "Super-exponential algorithms for multichannel blind deconvolution," *IEEE Trans. Signal Processing*, vol. 48, no. 3, pp. 881–888, Mar. 2000.
32. Y. Li and Z. Ding, "Global convergence of fractionally spaced Godard (CMA) adaptive equalizers," *IEEE Trans. Signal Processing*, vol. 44, no. 4, pp. 818–826, Apr. 1996.
33. Z. Ding and T. Nguyen, "Stationary points of a kurtosis maximization algorithm for blind signal separation and antenna beamforming," *IEEE Trans. Signal Processing*, vol. 48, no. 6, pp. 1587–1596, June 2000.
34. Y. Li and K. J. R. Liu, "Adaptive blind source separation and equalization for multiple-input/multiple-output systems," *IEEE Trans. Information Theory*, vol. 44, no. 7, pp. 2864–2876, Nov. 1998.
35. J. Gomes and V. Barroso, "A super-exponential algorithm for blind fractionally spaced equalization," *IEEE Signal Processing Letters*, vol. 3, no. 10, pp. 283–285, Oct. 1996.
36. C. B. Papadias and A. J. Paulraj, "A constant modulus algorithm for multiuser signal separation in presence of delay spread using antenna arrays," *IEEE Signal Processing Letters*, vol. 4, no. 6, pp. 178–181, June 1997.
37. C.-Y. Chi and C.-H. Chen, "Cumulant based inverse filter criteria for MIMO blind deconvolution: Properties, algorithms, and application to DS/CDMA systems in multipath," *IEEE Trans. Signal Processing*, vol. 49, no. 7, pp. 1282–1299, July 2001.
38. O. Shalvi and E. Weinstein, "Super-exponential methods for blind deconvolution," *IEEE Trans. Information Theory*, vol. 39, no. 2, pp. 504–519, Mar. 1993.
39. C.-H. Chen, "Blind Multi-channel Equalization and Two-dimensional System Identification Using Higher-order Statistics," Ph.D. Dissertation, Department of Electrical Engineering, National Tsing Hua University, Taiwan, R.O.C., June 2001.
40. A. Touzni, I. Fijalkow, M. G. Larimore, and J. R. Treichler, "A globally convergent approach for blind MIMO adaptive deconvolution," *IEEE Trans. Signal Processing*, vol. 49, no. 6, pp. 1166–1178, June 2001.

41. C. B. Papadias, "Globally convergent blind source separation based on a multiuser kurtosis maximination criterion," *IEEE Trans. Signal Processing*, vol. 48, no. 12, pp. 3508–3519, Dec. 2000.
42. W. Qiu, Y. Hua, and K. Abed-Meraim, "A subspace method for the computation of the GCD of polynomials," *Automatica*, vol. 33, no. 4, pp. 741–743, Apr. 1997.

6

Applications of MIMO Blind Equalization Algorithms

This chapter introduces some applications of the MIMO hybrid MNC blind equalization algorithm presented in Chapter 5 in areas of signal processing and digital communications according to either the SIMO or MIMO system model involved in the application of interest.

Straightforward applications for the SIMO case include fractionally spaced equalization (FSE) and blind maximum ratio combining (BMRC), while advanced applications in blind system identification (BSI) and multiple time delay estimation (MTDE) further involve nonlinear relations between the system and the MNC equalizer. On the other hand, applications of the MIMO hybrid MNC blind equalization algorithm for the MIMO case include blind beamforming for source separation and multiuser detection in wireless communications, and meanwhile other signal processing schemes such as signal classification are also needed besides channel equalization and system identification.

6.1 Fractionally Spaced Equalization in Digital Communications

Consider a digital communication system with a received baseband signal $y(t)$ modeled by (4.182). For simplicity, let us assume that there is no residual phase offset (i.e. $\Delta\theta(t) = 0$) such that $y(t)$ can be modeled as

$$y(t) = \sum_{k=-\infty}^{\infty} u[k]h(t - kT - \varepsilon) + w(t). \tag{6.1}$$

Conventionally, as introduced in Section 4.5.3, the continuous-time signal $y(t)$ is converted into a discrete-time signal $y[n] = y(t = nT + \widehat{\varepsilon})$ through baud-rate sampling with sampling instants $t = nT + \widehat{\varepsilon}$ followed by discrete-time signal processing for channel equalization and symbol detection. When the bandwidth of the continuous-time signal $y(t)$ is wider than $1/(2T)$ (i.e.

excess bandwidth exists), baud-rate sampling will cause aliasing in the signal spectrum, and the aliasing effects depend on actual sampling instants and the channel $h(t)$. Typically, a mechanism for adjusting the timing phase (i.e. symbol timing synchronization) to maximize the SNR of the resultant $y[n]$ is used for the design of baud-rate sampling based receivers. Nevertheless, a certain level of aliasing due to excess bandwidth is inevitable even though poor timing phase can be avoided through proper symbol timing synchronization. Consequently, a sampling rate higher than the baud rate has been of importance in modern receiver design because both the aliasing effects and the timing phase issue no longer exist. Moreover, as the channel bandwidth is wider than $1/(2T)$, sampling rates higher than $1/T$ bring channel diversity, and the discrete-time signal obtained at a multiple of the baud rate can be equivalently represented by a discrete-time SIMO model. Therefore, an SIMO blind equalization algorithm can be applied.

Let $T_s = T/M$ denote the sampling interval where M is a positive integer. Then the discrete-time signal obtained by sampling the received signal $y(t)$ given by (6.1) can be expressed as

$$y[n] = y(t = nT_s + \widehat{\varepsilon}) = \sum_{k=-\infty}^{\infty} u[k]h(nT_s - kT + \Delta\varepsilon) + w(nT_s + \widehat{\varepsilon}) \qquad (6.2)$$

where $\Delta\varepsilon = \widehat{\varepsilon} - \varepsilon$ is the timing phase error. Let

$$y_i[n] = y[nM + i], \quad i = 1, 2, ..., M \qquad (6.3)$$
$$h_i[n] = h(nT + iT_s + \Delta\varepsilon), \quad i = 1, 2, ..., M \qquad (6.4)$$
$$w_i[n] = w(nT + iT_s), \quad i = 1, 2, ..., M. \qquad (6.5)$$

It is easy to see, from (6.2), that

$$y_i[n] = \sum_{k=-\infty}^{\infty} u[k]h_i[n-k] + w_i[n], \qquad i = 1, 2, ..., M,$$

or equivalently, in vector form

$$\begin{aligned}\mathbf{y}[n] &= (y_1[n], y_2[n], ..., y_M[n])^T \\ &= \mathbf{h}[n] \star u[n] + \mathbf{w}[n] = \sum_{k=-\infty}^{\infty} \mathbf{h}[n-k]u[k] + \mathbf{w}[n]\end{aligned} \qquad (6.6)$$

where $\mathbf{h}[n] = (h_1[n], h_2[n], ..., h_M[n])^T$ and $\mathbf{w}[n] = (w_1[n], w_2[n], ..., w_M[n])^T$. Obviously, in light of (6.6), samples of $y(t)$ with sampling interval equal to $T_s = T/M$ lead to a diversity of M discrete-time subchannels. Hence an SIMO blind equalization algorithm can be employed to process $\mathbf{y}[n]$ to estimate the symbol sequence $u[n]$. However, it should be noted that SOS based approaches require that all the subchannels $h_i[n]$ satisfy the channel disparity condition (no common subchannel zeros), which may be violated or nearly violated whenever the excess bandwidth is small. Therefore, SIMO blind equalization algorithms using HOS such as the MIMO hybrid MNC algorithm which allows common subchannel zeros is preferable to SOS based approaches.

Example 6.1 (Effects of Timing Phase Error)
With the received continuous-time signal $y(t)$ given by (6.1) where the noise $w(t)$ is white Gaussian with power spectral density σ_w^2, this example shows some simulation results of baud spaced equalization (BSE) (i.e. $M = 1$) and FSE (i.e. $M > 1$) using the hybrid MNC equalizers for different values of the normalized timing phase error $\Delta\varepsilon/T$. Each synthetic symbol sequence $u[n]$ (with $\sigma_u^2 = E\{|u[n]|^2\} = 1$) was assumed to be a 16-QAM signal of length equal to $N = 4096$ and the LTI channel $h(t)$ was assumed to be a truncated raised cosine pulse given by

$$h(t) = p(t) \cdot \text{rect}\left(\frac{t}{6T}\right) \tag{6.7}$$

where $p(t)$ and $\text{rect}(t)$ are defined by (4.188) and (4.186), respectively, and the roll-off factor α in $p(t)$ was chosen to be equal to 0.5 in this example. Note that the domain of support for $h(t)$ is $[-3T, 3T]$. The T-spaced discrete-time channel impulse responses $h[n] = h(nT + \Delta\varepsilon)$ and their magnitude responses $|H(\omega)|$ for $\Delta\varepsilon/T = 0, 0.1, ..., 0.5$ are shown in Figs. 6.1a, b, respectively. The synthetic $y[n]$ was generated according to (6.2) with $\sigma_w^2 = 0.01$ (i.e. SNR= 20 dB for $y[n] = y(nT)$).

The causal FIR hybrid MNC equalizer of length L was used to process the synthetic T-spaced signal $y[n]$ for both $L = 6$ and $L = 12$ and $T/2$-spaced signal $\mathbf{y}[n]$ for $L = 6$. Note that the degree of freedom of the equalizer used in the BSE for $L = 12$ is the same as used in the FSE for $L = 6$ since $M = 2$. The resultant averaged output SINRs and ISIs for $\Delta\varepsilon/T = 0, 0.1, ..., 0.5$ calculated from thirty independent runs are shown in Figs. 6.2a, b, respectively. It can be seen, from Figs. 6.2a, b, that the performance of the FSE is insensitive to the timing phase error and superior to that of the BSE for all $\Delta\varepsilon/T$ in terms of averaged output SINR, although the latter itself is also better for $L = 12$ than for $L = 6$ as $\Delta\varepsilon/T > 0.2$.

□

Example 6.2 (FSE of Real Digital Communications Data)
This example shows some results of fractionally spaced equalization using the hybrid MNC equalization algorithm ($p = q = 2$, $L = 15$, $M = 2$) to process real data taken from the on-line Signal Processing Information Base (SPIB) for the following two cases:

Modem data: A sequence of received data (about 4.75 seconds of random data) from V.29, 9600 bps modems are recorded with $T/2$ sampling at 4800 Hz. The total data length N is equal to 22800

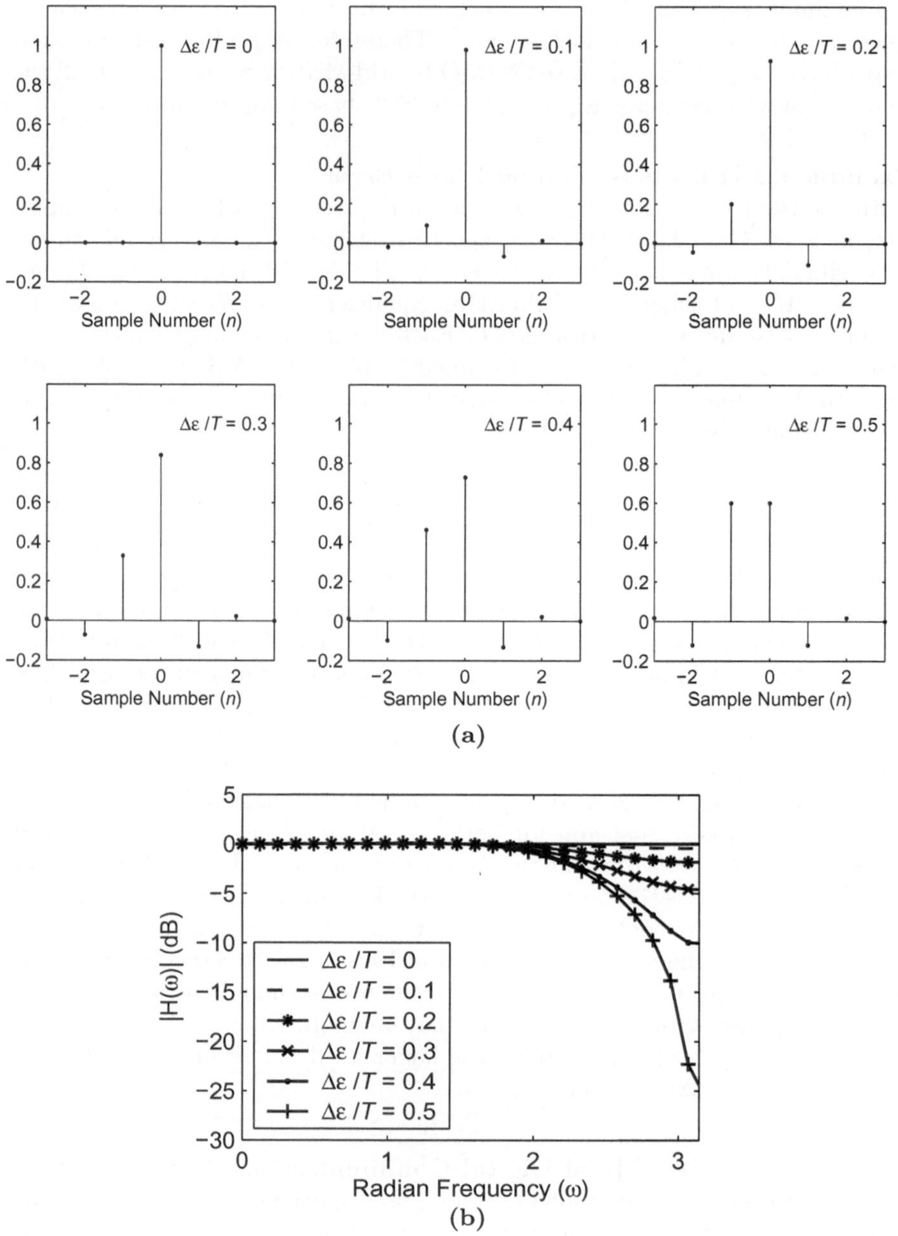

Fig. 6.1 **(a)** The T-spaced discrete-time channel impulse responses $h[n] = h(nT + \Delta\varepsilon)$ for $\Delta\varepsilon/T = 0$, 0.1, ..., 0.5 and **(b)** their magnitude responses $|H(\omega)|$

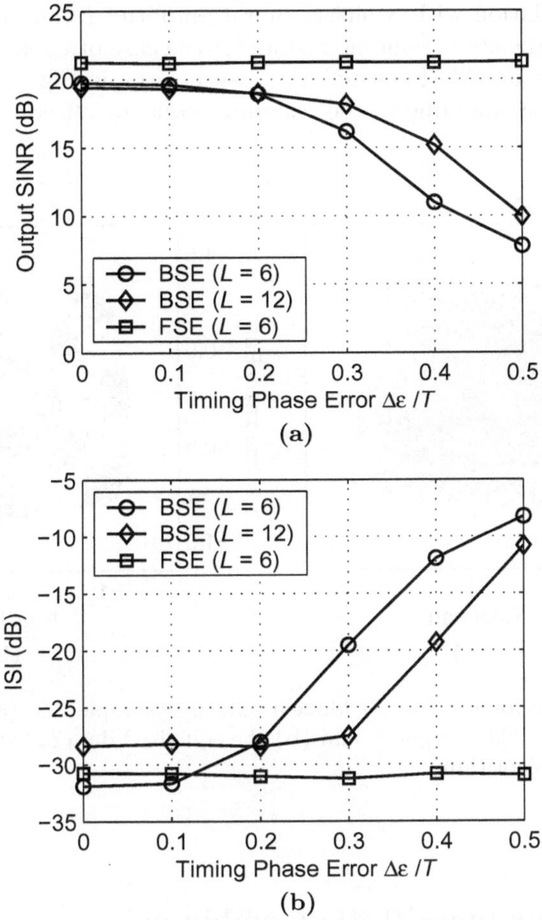

Fig. 6.2 Simulation results of Example 6.1 for BSE and FSE using the hybrid MNC algorithm. **(a)** Averaged output SINRs and **(b)** averaged ISIs versus the timing phase error $\Delta\varepsilon/T$

samples. The carrier has been removed (but some offset perhaps remains) (taken from http://spib.rice.edu/spib/modem.html).

Cable data: A sequence of received cable channel data are recorded via sampling at twice per symbol. The total data length N is equal to 65536 samples. The carrier frequency has been down converted close to 0 Hz with some residual offset (taken from http://spib.rice.edu/spib/cable.html).

For the case of Modem data, Fig. 6.3a shows 4000 samples of the received data and Fig. 6.3b shows 2000 symbols of the equalized data where a specific

16-QAM constellation with 8 phases and 4 amplitudes can be observed. On the other hand, the corresponding results for the case of Cable data are shown in Fig. 6.4a, b. A 64-QAM constellation can be seen in Fig. 6.4b as well as a rotation of each symbol due to the residual frequency offset.

□

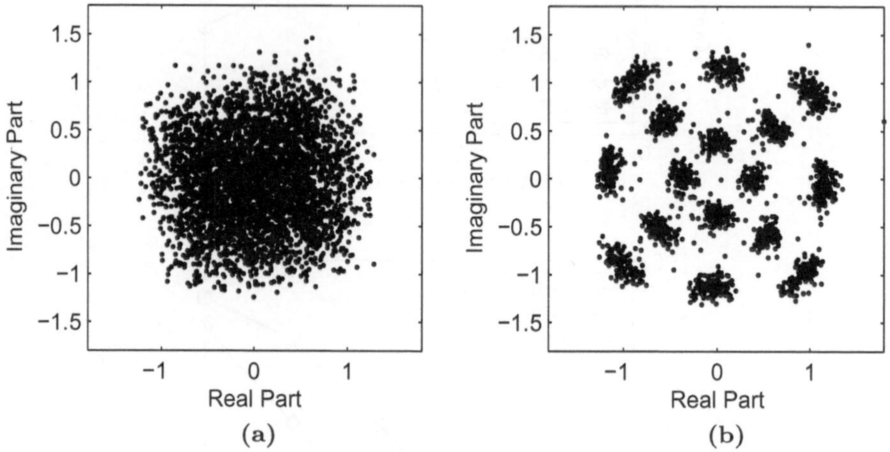

Fig. 6.3 Results for the case of Modem data in Example 6.2. **(a)** The received $T/2$ sampled data (4000 samples), and **(b)** the equalized data (2000 symbols) using the hybrid MNC algorithm

6.2 Blind Maximum Ratio Combining

Consider a zero-mean colored nonGaussian source signal $x[n]$ modeled as

$$x[n] = h[n] \star u[n] \tag{6.8}$$

where $u[n]$ is an i.i.d. zero-mean nonGaussian random sequence, and the measurement vector $\mathbf{y}[n]$ is given by

$$\begin{aligned}\mathbf{y}[n] &= (y_1[n], y_2[n], ..., y_M[n])^T \\ &= \boldsymbol{a} \cdot x[n] + \mathbf{w}[n] = \mathbf{H}[n] \star u[n] + \mathbf{w}[n]\end{aligned} \tag{6.9}$$

where $\boldsymbol{a}$ is an $M \times 1$ unknown column vector,

$$\mathbf{H}[n] = \boldsymbol{a}h[n] \tag{6.10}$$

and $\mathbf{w}[n]$ is a zero-mean Gaussian $M \times 1$ vector noise and statistically independent of the desired signal $x[n]$. The data vector $\mathbf{y}[n]$ can be thought of

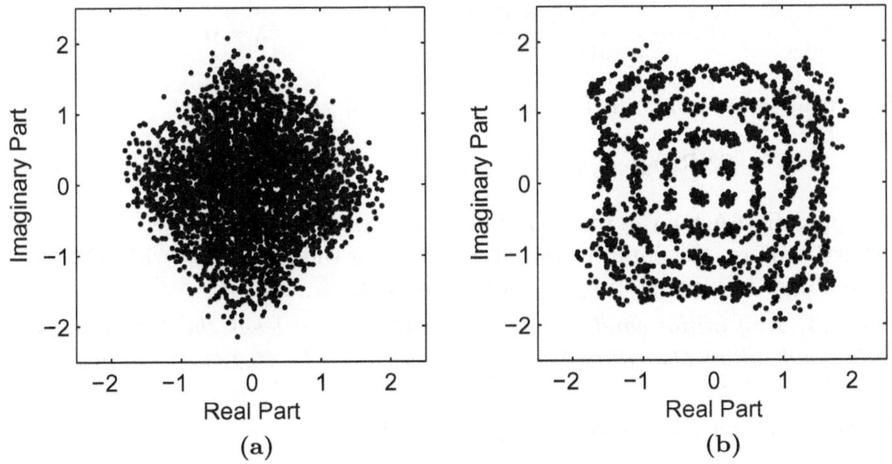

Fig. 6.4 Results for the case of Cable data in Example 6.2. **(a)** The received $T/2$ sampled data (4000 samples), and **(b)** the equalized data (2000 symbols) using the hybrid MNC algorithm

as the received signals from M diversities (such as multiple paths or multiple sensors or antennas).

It is desired to estimate $x[n]$ with maximum SNR through spatial linear processing of $\mathbf{y}[n]$ without the information of $\mathbf{H}[n]$. Let $e[n]$ be the spatial filter output, i.e.

$$e[n] = \boldsymbol{v}^T\mathbf{y}[n] = \boldsymbol{v}^T\boldsymbol{a} \cdot x[n] + w[n] \tag{6.11}$$

where $w[n] = \boldsymbol{v}^T\mathbf{w}[n]$. We would like to find the optimum $\boldsymbol{v}$ such that the SNR associated with $e[n]$ defined as

$$\mathrm{SNR}(\boldsymbol{v}) = \frac{E\{|\boldsymbol{v}^T\boldsymbol{a} \cdot x[n]|^2\}}{E\{|w[n]|^2\}} = \sigma_x^2 \cdot \frac{\boldsymbol{v}^T\boldsymbol{a}\boldsymbol{a}^H\boldsymbol{v}^*}{\boldsymbol{v}^T\mathbf{R}_{\mathbf{w}}[0]\boldsymbol{v}^*} \tag{6.12}$$

is maximum.

The MISO equalizer $\mathbf{v}[n]$ introduced in Chapter 5 consists of a bank of linear FIR filters, with $\mathbf{v}[n] \neq \mathbf{0}_M$ for $n = L_1, L_1+1, ..., L_2$ and length $L = L_2 - L_1 + 1$. As $L_1 = L_2 = 0$, the MISO equalizer $\mathbf{v}[n]$ of length $L = 1$ reduces to a linear combiner $\boldsymbol{v} = \mathbf{v}[0]$ ($M \times 1$ column vector). Therefore, the MIMO hybrid MNC equalization algorithm can be employed to design the optimum linear combiner $\boldsymbol{v}$ with no need for any information about $\boldsymbol{a}$. The resultant MIMO-MNC BMRC method is also supported by the following theorem [1,2].

Theorem 6.3 (BMRC). *Suppose that* $\mathbf{y}[n]$ *given by (6.9) satisfies Assumptions* (A5-13) *through* (A5-17). *The optimum MIMO-MNC linear combiner is given by*

$$\boldsymbol{v}_{\mathrm{MNC}} = \frac{(\mathbf{R}_{\mathbf{w}}^{*}[0])^{-1}\boldsymbol{a}^{*}}{\|(\mathbf{R}_{\mathbf{w}}^{*}[0])^{-1}\boldsymbol{a}^{*}\|} = \lambda \boldsymbol{v}_{\mathrm{MS}}, \qquad \lambda \neq 0 \tag{6.13}$$

where $\boldsymbol{v}_{\mathrm{MS}}$ *is the nonblind LMMSE estimator of* $x[n]$, *and*

$$\mathrm{SNR}(\boldsymbol{v}_{\mathrm{MNC}}) = \mathrm{SNR}_{\max} = \sigma_x^2 \cdot \boldsymbol{a}^H(\mathbf{R}_{\mathbf{w}}[0])^{-1}\boldsymbol{a}. \tag{6.14}$$

See Appendix 6A for the proof of Theorem 6.3.

Furthermore, the computational efficiency of the MIMO hybrid MNC linear combiner is supported by the following fact.

Fact 6.4. *Any initial condition* $\boldsymbol{v}^{[0]}$ *with* $\boldsymbol{a}^T\boldsymbol{v}^{[0]} \neq 0$ *leads the MIMO hybrid MNC equalization algorithm to* $\boldsymbol{v}^{[1]} = \boldsymbol{v}_{\mathrm{MNC}}$ *given by (6.13) at the first iteration.*

The proof of Fact 6.4 is given in Appendix 6B.

Example 6.5 (BMRC)

This example shows some simulation results for BMRC with the signals received by four sensors (i.e. $M = 4$) using the MIMO hybrid MNC linear combiner. The desired signal $x[n]$ was assumed to be a 16-QAM signal, the unknown column vector $\boldsymbol{a} = (1, 0.80902-j0.58779, -0.30902-j0.95106, -1)^T$ and the noise $\mathbf{w}[n]$ was assumed to be spatially independent and temporally white Gaussian. The synthetic data $\mathbf{y}[n]$ for $N = 1024$ and $SNR = 14$ dB, shown in Fig. 6.5a, were processed using the 4×1 hybrid MNC linear combiner. Figure 6.5b depicts the constellation of the combiner output. Evidently, the enhanced signal shown in Fig. 6.5b exhibits a substantial SNR improvement compared with each received signal $y_i[n]$ shown in Fig. 6.5a.

□

6.3 SIMO Blind System Identification

Blind identification of SIMO systems deals with the problem of estimating an $M \times 1$ LTI system, denoted by $\mathbf{h}[n] = (h_1[n], h_2[n], ..., h_M[n])^T$, with only the $M \times 1$ output vector measurements $\mathbf{y}[n] = (y_1[n], y_2[n], ..., y_M[n])^T$, $n = 0, 1, ..., N-1$, generated from the following convolutional model

$$\mathbf{y}[n] = \mathbf{h}[n] \star u[n] + \mathbf{w}[n] \tag{6.15}$$

where $u[n]$ is the system input and $\mathbf{w}[n]$ is the vector noise. Let us make the following assumptions for $\mathbf{y}[n]$ given by (6.15)

(A6-1) The system input $u[n]$ is a zero-mean, i.i.d., nonGaussian random process with variance σ_u^2.

(A6-2) The SIMO system $\mathbf{h}[n]$ is an FIR system with frequency response $\boldsymbol{H}(\omega) \neq \mathbf{0}$ for all $-\pi < \omega \leq \pi$.

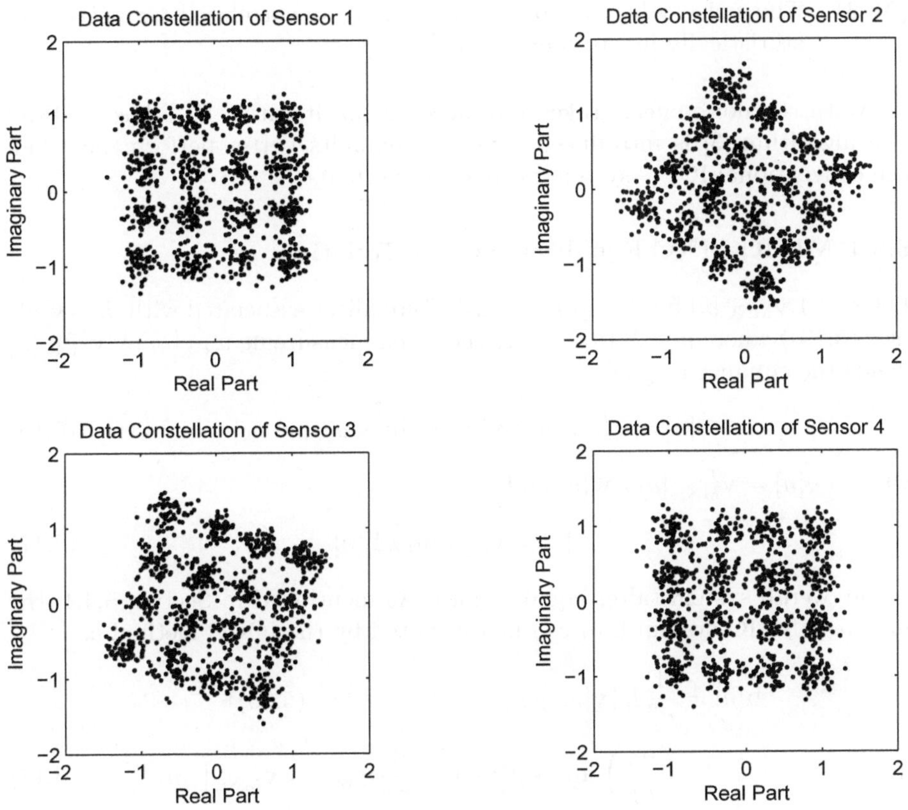

(a)

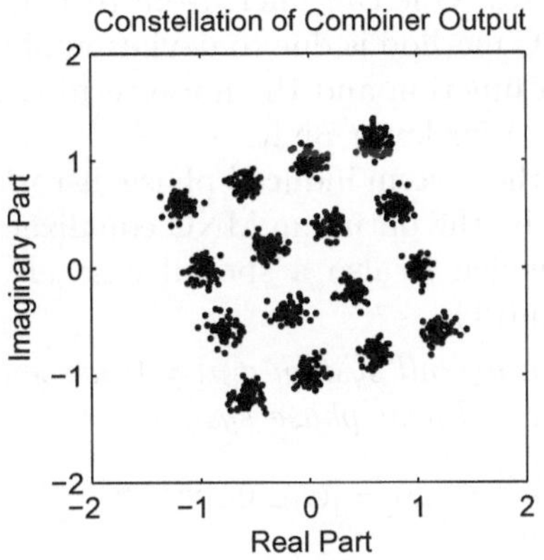

(b)

Fig. 6.5 Results of Example 6.5. The constellations of **(a)** the data received by the four sensors, and **(b)** the combiner output

(A6-3) The noise $\mathbf{w}[n]$ is zero-mean spatially uncorrelated Gaussian and statistically independent of $u[n]$.

With the noise effects taken into account, an iterative FFT-based SIMO BSI algorithm [3] is introduced next that exhibits better performance than the conventional IOCC system estimation method.

6.3.1 MIMO-MNC Equalizer–System Relation

Let $M \times 1$ $\mathbf{v}_{\text{MNC}}[n]$ be the optimum MNC equalizer associated with $J_{p,p}(\mathbf{v}[n])$ (see (5.73)) where $p \geq 2$. Processing the given measurements $\mathbf{y}[n]$ by $\mathbf{v}_{\text{MNC}}[n]$ yields the equalized signal

$$e[n] = \mathbf{v}_{\text{MNC}}^T[n] \star \mathbf{y}[n] = g[n] \star u[n] + e_{\text{N}}[n] \tag{6.16}$$

where $e_{\text{N}}[n] = \mathbf{v}_{\text{MNC}}^T[n] \star \mathbf{w}[n]$ and

$$g[n] = \mathbf{v}_{\text{MNC}}^T[n] \star \mathbf{h}[n] \tag{6.17}$$

is the overall system (after equalization). As mentioned in Section 5.4.1, the unknown SIMO system $\mathbf{h}[n]$ can be estimated by the IOCC method as

$$\begin{aligned}\widehat{\mathbf{h}}[n] &= \frac{1}{\sigma_e^2} E\{\mathbf{y}[m+n]e^*[m]\} \qquad \text{(by (5.90))} \\ &= \left(\frac{\sigma_u^2}{\sigma_e^2}\right) \mathbf{h}[n] \star g^*[-n] + \frac{1}{\sigma_e^2} \mathbf{R}_{\mathbf{w}}[n] \star \mathbf{v}_{\text{MNC}}^*[-n] \end{aligned} \tag{6.18}$$

where the second line of (6.18) was obtained by substituting (6.15) and (6.16) into the first line of (6.18). One can infer, from (6.18), that the performance degradation of the IOCC method is due to deviation of the overall system $g[n]$ from a Kronecker delta function and the noise term (the second term on the right-hand side of (6.18)) for lower SNR.

On the other hand, the system induced phase distortion for finite SNR can be completely removed by the optimum MNC equalizer $\mathbf{v}_{\text{MNC}}[n]$ as described in the following fact, which is also a special case ($K = 1$, and $p = q$) of Property 5.36 (Problem 6.1).

Fact 6.6. *The optimum overall system $g[n] \neq 0$ associated with the $\mathbf{v}_{\text{MNC}}[n]$ ($p = q$) for finite SNR is a linear phase system, i.e.*

$$G(\omega) = |G(\omega)|e^{j(\omega\tau+\varphi)} \tag{6.19}$$

where τ and φ are an integer and a real constant, respectively.

Moreover, a relation between the system $\mathbf{h}[n]$ and the $\mathbf{v}_{\text{MNC}}[n]$ ($p = q$) for any SNR can easily be established from Property 5.35 and is summarized in the following fact (Problem 6.2):

Fact 6.7. *With sufficient length for* $\mathbf{v}_{\text{MNC}}[n]$*, the system* $\boldsymbol{H}(\omega)$ *is related to* $\mathbf{v}_{\text{MNC}}[n]$ $(p = q)$ *for any SNR via*

$$G_p(\omega) \cdot \boldsymbol{H}^*(\omega) = \beta \cdot \boldsymbol{\mathcal{S}}_{\mathbf{y}}^T(\omega)\boldsymbol{V}_{\text{MNC}}(\omega) \tag{6.20}$$

where $G_p(\omega)$ *is the DTFT of*

$$g_p[n] = g^p[n](g^*[n])^{p-1}, \tag{6.21}$$

and

$$\beta = \frac{1}{\sigma_e^2}\left(\frac{C_{p,p}\{e[n]\}}{C_{p,p}\{u[n]\}}\right) = \frac{1}{\sigma_e^2}\sum_{n=-\infty}^{\infty} |g[n]|^{2p} > 0 \tag{6.22}$$

is a real positive constant.

Note that besides optimum MNC equalizer $\mathbf{v}_{\text{MNC}}[n]$, all other local optimum $\mathbf{v}[n]$ of $J_{p,p}(\mathbf{v}[n])$ and the trivial solution $\mathbf{v}[n] = \mathbf{0}$ satisfy (6.20) but they are not the solutions of interest.

6.3.2 Analysis on System Identification Based on MIMO-MNC Equalizer–System Relation

It can be observed, from the relation (6.20), that given the MNC equalizer $\mathbf{v}_{\text{MNC}}[n]$ and the power spectral matrix $\boldsymbol{\mathcal{S}}_{\mathbf{y}}(\omega)$, the system $\mathbf{h}[n]$ can be estimated by solving (6.20) if the solution for $\mathbf{h}[n]$ of the nonlinear equation (6.20) is unique over the class that the overall system is linear phase (motivated by Fact 6.6). Meanwhile, noise effects embedded in $\boldsymbol{\mathcal{S}}_{\mathbf{y}}(\omega)$ and $\mathbf{v}_{\text{MNC}}[n]$ can also be taken into account in the estimation of $\mathbf{h}[n]$ regardless of the value of SNR. Prior to presenting the algorithm for obtaining $\mathbf{h}[n]$, let us present an analysis of the solution set of $\mathbf{h}[n]$ solved from the nonlinear equation (6.20) with the following constraint:

(C-SIMO) The overall system $g[n] \neq 0$ is linear phase.

With the relation given by (6.20), the system $\mathbf{h}[n]$ can be solved up to a linear phase ambiguity as revealed in the following property.

Property 6.8. *Any SIMO system*

$$\boldsymbol{H}'(\omega) = \boldsymbol{H}(\omega) \cdot e^{j(\omega\tau+\varphi)} \tag{6.23}$$

for any integer τ *and real* φ *satisfies the relation given by (6.20).*

The proof of Property 6.8 is left as an exercise (Problem 6.3). Property 6.8 and the constraint (C-SIMO) imply that $g[n]$ can be zero phase. Furthermore, $g_p[n]$ given by (6.21) can also be shown to be zero phase without zeros on the unit circle as $g[n]$ is zero phase, as stated in the following property.

Property 6.9. *Assume that* $g[n] \neq 0$ *is zero phase and that*

(A6-4) *the number of zeros of* $G(z)$ *on the unit circle is finite.*

Then the associated system $g_p[n] \neq 0$ *given by (6.21) is a positive definite sequence, i.e.* $G_p(\omega) > 0$*, for all* $-\pi < \omega \leq \pi$.

The proof of Property 6.9 is left as an exercise (Problem 6.4). Next, let us present the following property regarding the solution set of $\mathbf{h}[n]$ from (6.20) under the constraint (C-SIMO).

Property 6.10. *The system* $\boldsymbol{H}(\omega)$ *can be identified up to a linear phase ambiguity by solving (6.20) provided that* $g[n] \neq 0$ *is zero phase and satisfies Assumption* (A6-4).

See Appendix 6C for the proof of Property 6.10 in which Property 6.9 and the Lemma below are needed.

Lemma 6.11. *Assume that* $a > 0$, $b > 0$, $c > 0$ *and* m *is a positive integer. If* $a(a^m + c^m) = b(b^m + c^m)$, *then* $a = b$.

The proof of Lemma 6.11 is straightforward and thus omitted here.

It can be inferred, by Property 6.10, that the solution of (6.20) under the constraint (C-SIMO) and Assumption (A6-4) is exactly the true system $\boldsymbol{H}(\omega)$ except for a time delay (due to a linear phase ambiguity) as long as the true $\mathbf{v}_{\mathrm{MNC}}[n]$ and $\boldsymbol{\mathcal{S}}_{\mathbf{y}}(\omega)$ are given. Assumption (A6-4) generally holds true in practical applications and is never an issue in the design of BSI algorithms. In general, the system estimate $\widehat{\mathbf{h}}[n]$ solved from (6.20) is consistent provided that an estimate $\widehat{\boldsymbol{\mathcal{S}}}_{\mathbf{y}}(\omega)$ can be obtained using a consistent multichannel power spectral estimator.

On the other hand, one can see, from (6.21) and (6.17), that the left-hand side of (6.20) is a highly nonlinear function of $\mathbf{h}[n]$, implying that determining a closed-form solution of (6.20) for $\mathbf{h}[n]$ is formidable under the constraint (C-SIMO). Next, let us introduce an iterative FFT-based nonparametric algorithm for estimating $\mathbf{h}[n]$ under the constraint (C-SIMO).

6.3.3 SIMO Blind System Identification Algorithm

Let $G_p[k]$, $\boldsymbol{H}[k]$, $\boldsymbol{\mathcal{S}}_{\mathbf{y}}[k]$, and $\boldsymbol{V}_{\mathrm{MNC}}[k]$ denote the $\mathcal{N}$-point DFTs of $g_p[n]$, $\mathbf{h}[n]$, $\mathbf{R}_{\mathbf{y}}[n]$, and $\mathbf{v}_{\mathrm{MNC}}[n]$, respectively. Let $\boldsymbol{a}_k$ and $\boldsymbol{b}_k$ be $M \times 1$ vectors defined as

$$\boldsymbol{a}_k = G_p[k]\boldsymbol{H}^*[k], \qquad k = 0, 1, ..., \mathcal{N} - 1, \tag{6.24}$$

$$\boldsymbol{b}_k = \boldsymbol{\mathcal{S}}_{\mathbf{y}}^T[k]\boldsymbol{V}_{\mathrm{MNC}}[k], \qquad k = 0, 1, ..., \mathcal{N} - 1. \tag{6.25}$$

Then, according to the relation given by (6.20), one can obtain

$$\mathbf{a} = \beta \mathbf{b} \tag{6.26}$$

where $\mathbf{a} = \left(\boldsymbol{a}_0^T, \boldsymbol{a}_1^T, ..., \boldsymbol{a}_{\mathcal{N}-1}^T\right)^T$, $\mathbf{b} = \left(\boldsymbol{b}_0^T, \boldsymbol{b}_1^T, ..., \boldsymbol{b}_{\mathcal{N}-1}^T\right)^T$, and $\beta > 0$ is given by (6.22). Let us emphasize that the positive constant β in (6.26) is unknown, and this implies that either β must be estimated together with the estimation of $\mathbf{h}[n]$ or its role must be virtual during the estimation of $\mathbf{h}[n]$. Consequently, the equations (6.24), (6.25) and (6.26) reveal that the true system $\boldsymbol{H}(\omega)$ is the one such that the angle between the vector $\mathbf{a}$ and the vector $\mathbf{b}$ is zero (in phase). Thus, $\boldsymbol{H}[k]$ at $k = 0, 1, ..., \mathcal{N} - 1$ can be estimated by maximizing

$$\mathcal{C}(\boldsymbol{H}[k]) \triangleq \frac{\mathrm{Re}\{\mathbf{a}^H \mathbf{b}\}}{||\mathbf{a}|| \cdot ||\mathbf{b}||}. \tag{6.27}$$

Note that $-1 \leq \mathcal{C}(\boldsymbol{H}[k]) = \mathcal{C}(\alpha \boldsymbol{H}[k]) \leq 1$ for any $\alpha \neq 0$ and that $\mathcal{C}(\boldsymbol{H}[k]) = 1$ if and only if (6.26) holds.

However, it is quite involved to obtain the gradient $\partial \mathcal{C}(\boldsymbol{H}[k])/\partial \boldsymbol{H}[k]$ because $\mathcal{C}(\boldsymbol{H}[k])$ is a highly nonlinear function of $\boldsymbol{H}[k]$, so gradient based optimization methods are not considered for finding the maximum of $\mathcal{C}(\boldsymbol{H}[k])$. Instead, an iterative FFT-based BSI algorithm [3], which can also be thought of as a numerical optimization approach, is introduced below for the estimation of $\mathbf{h}[n]$. This algorithm, as shown in Fig. 6.6, includes two steps, one to obtain the MNC equalizer and estimate the power spectral matrix of $\mathbf{y}[n]$ and the other to obtain the estimate of $\boldsymbol{H}[k]$ by maximizing $\mathcal{C}(\boldsymbol{H}[k])$ given by (6.27), as described as follows.

SIMO BSI Algorithm (Fig. 6.6)

(T1) Blind Deconvolution and Power Spectral Matrix Estimation.
With finite data $\mathbf{y}[n]$, obtain the MNC equalizer $\mathbf{v}_{\mathrm{MNC}}[n]$ with $p = q$, and compute its $\mathcal{N}$-point FFT $\boldsymbol{V}_{\mathrm{MNC}}[k]$. Obtain the power spectral matrix estimate $\boldsymbol{\mathcal{S}}_{\mathbf{y}}[k]$, using a multichannel power spectral estimator. Form the vector $\boldsymbol{b}_k$ via $\boldsymbol{\mathcal{S}}_{\mathbf{y}}[k]$ and $\boldsymbol{V}_{\mathrm{MNC}}[k]$ according to (6.25).

(T2) System Identification.

(S1) Set the iteration number $i = 0$ and the initial condition $\boldsymbol{H}^{[0]}[k]$. Obtain $G^{[0]}[k] = \boldsymbol{V}_{\mathrm{MNC}}^T[k]\boldsymbol{H}^{[0]}[k]$ followed by its $\mathcal{N}$-point inverse FFT $g^{[0]}[n]$.

(S2) Update i to $i + 1$. Compute

$$g[n] = \frac{1}{2}\left(g^{[i-1]}[n] + \left(g^{[i-1]}[-n]\right)^*\right) \quad \text{(zero-phase)} \tag{6.28}$$

and update $g^{[i-1]}[n]$ to $g^{[i-1]}[n] = g[n]$. Then obtain

$$\boldsymbol{H}^{[i]}[k] = \frac{\boldsymbol{b}_k^*}{G_p^{[i-1]}[k]} \quad \text{(by (6.24) and (6.26))}$$

where $G_p^{[i-1]}[k]$ is the $\mathcal{N}$-point FFT of $g_p^{[i-1]}[n]$ obtained by (6.21). Normalize $\boldsymbol{H}^{[i]}[k]$ such that $\sum_k \|\boldsymbol{H}^{[i]}[k]\|^2 = 1$.

(S3) Obtain $G^{[i]}[k] = \boldsymbol{V}_{\mathrm{MNC}}^T[k]\boldsymbol{H}^{[i]}[k]$ followed by its $\mathcal{N}$-point inverse FFT $g^{[i]}[n]$. If $\mathcal{C}(\boldsymbol{H}^{[i]}[k]) > \mathcal{C}(\boldsymbol{H}^{[i-1]}[k])$, go to (S4). Otherwise, compute $\Delta\boldsymbol{H}[k] = \boldsymbol{H}^{[i]}[k] - \boldsymbol{H}^{[i-1]}[k]$, and update $\boldsymbol{H}^{[i]}[k]$ via

$$\boldsymbol{H}^{[i]}[k] = \boldsymbol{H}^{[i-1]}[k] + \mu \cdot \Delta\boldsymbol{H}[k]$$

where the step size μ is chosen such that $\mathcal{C}(\boldsymbol{H}^{[i]}[k]) > \mathcal{C}(\boldsymbol{H}^{[i-1]}[k])$. Normalize $\boldsymbol{H}^{[i]}[k]$ such that $\sum_k \|\boldsymbol{H}^{[i]}[k]\|^2 = 1$.

(S4) If

$$\frac{\mathcal{C}(\boldsymbol{H}^{[i]}[k]) - \mathcal{C}(\boldsymbol{H}^{[i-1]}[k])}{\left|\mathcal{C}(\boldsymbol{H}^{[i-1]}[k])\right|} > \zeta$$

where ζ is a pre-assigned convergence tolerance, then go to (S2); otherwise, the frequency response estimate $\widehat{\boldsymbol{H}}(\omega) = \boldsymbol{H}^{[i]}(\omega)$ at $\omega = 2\pi k/\mathcal{N}$ for $k = 0, 1, ..., \mathcal{N} - 1$, and its $\mathcal{N}$-point inverse FFT $\widehat{\mathbf{h}}[n]$ are obtained.

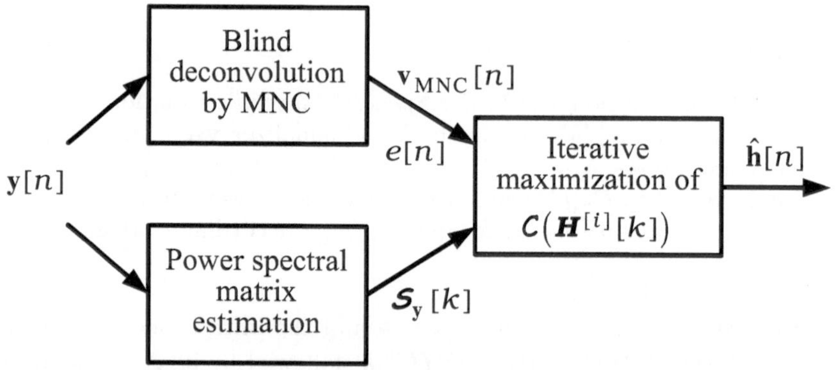

Fig. 6.6 Block diagram of the SIMO BSI algorithm

In (T1) of the above SIMO BSI algorithm, the MNC equalizer $\mathbf{v}_{\mathrm{MNC}}[n]$ can be obtained efficiently using the MIMO hybrid MNC equalization algorithm, and the AR spectral estimators such as the multichannel Levinson recursion algorithm [4] summarized in Appendix 6D can be employed to estimate $\boldsymbol{\mathcal{S}}_{\mathbf{y}}[k]$.

In (T2) of the above SIMO BSI algorithm, local convergence is guaranteed because $\mathcal{C}(\boldsymbol{H}^{[i]}[k])$ is upper bounded by unity and its value is increased at each iteration before convergence. Therefore, the closer to unity the $\mathcal{C}(\widehat{\boldsymbol{H}}[k])$, the more accurate the system estimate obtained.

With the estimate of $\widehat{\mathbf{h}}[n]$ obtained by (6.18) as the initial condition for $\boldsymbol{H}^{[0]}[k]$ in (S1) of (T2), $\mathcal{C}(\widehat{\boldsymbol{H}}[k]) = 1$ can usually be obtained after convergence of (T2). Meanwhile, the associated overall system $\widehat{g}[n]$ is guaranteed zero phase due to (6.28) in (S2) of (T2). Moreover, in (S3) of (T2), the step size $\mu = \pm 1/2^l$ can be used with a suitable $l \in \{0, 1, ..., \mathbb{K}\}$ where $\mathbb{K}$ is a preassigned positive integer, such that $\mathcal{C}(\boldsymbol{H}^{[i]}[k]) > \mathcal{C}(\boldsymbol{H}^{[i-1]}[k])$ at each iteration.

It should be noted that an upper bound of the system order is needed though the exact system order is not required. The FFT size $\mathcal{N}$ should be chosen to be larger than the upper bound of the order of $\mathbf{h}[n]$ such that aliasing effects on the resultant $\widehat{\mathbf{h}}[n]$ are negligible. Surely the larger the FFT size, the larger the computational load of the SIMO BSI algorithm, while the estimation error of the resultant $\widehat{\mathbf{h}}[n]$ is almost the same. If the true system is an IIR system, a finite-length approximation of $\mathbf{h}[n]$ will be obtained by the above SIMO BSI algorithm.

Example 6.12 (SIMO Blind System Identification)
This example continues Example 5.37, where a 3×1 system is considered for Case A (without common subchannel zeros) and for Case B (with a common subchannel zero), by further showing some simulation results using the noise subspace approach, the IOCC method and the SIMO BSI algorithm introduced above.

Let $\widehat{h}_i[n; r]$ denote the estimate of the ith subchannel $h_i[n]$ (the ith entry of $\mathbf{h}[n]$) obtained at the rth run with the time delay and the scale factor between them artificially removed. The normalized mean square error (NMSE) associated with $h_i[n]$ defined as

$$\text{NMSE}_i = \frac{1}{30} \cdot \frac{\displaystyle\sum_{r=1}^{30} \sum_{n=0}^{15} \left| \widehat{h}_i[n; r] - h_i[n] \right|^2}{\displaystyle\sum_{n=0}^{15} |h_i[n]|^2} \tag{6.29}$$

was computed for all the subchannels and then averaged to obtain the overall NMSE (ONMSE)

$$\text{ONMSE} = \frac{1}{M} \sum_{i=1}^{M} \text{NMSE}_i \tag{6.30}$$

which was used as the performance index.

Figure 6.7a shows the simulation results (ONMSEs versus SNR) for Case A associated with the noise subspace approach (indicated by 'NS'), the IOCC method (indicated by 'IOCC') and the SIMO BSI algorithm (indicated by 'SIMO BSI'), and Fig. 6.7b shows the corresponding results for Case B. One can observe, from Fig. 6.7a, b, that the performance of the SIMO BSI algorithm is uniformly superior to that of the IOCC method, and that the values of ONMSE reach a minimum (a floor) as SNR increases for each N, which is smaller for larger data length N. On the other hand, the noise subspace approach (with the true order) performs much better than both the SIMO BSI algorithm and the IOCC method for high SNR in Case A, whereas it failed in Case B due to violation of the channel disparity condition.

□

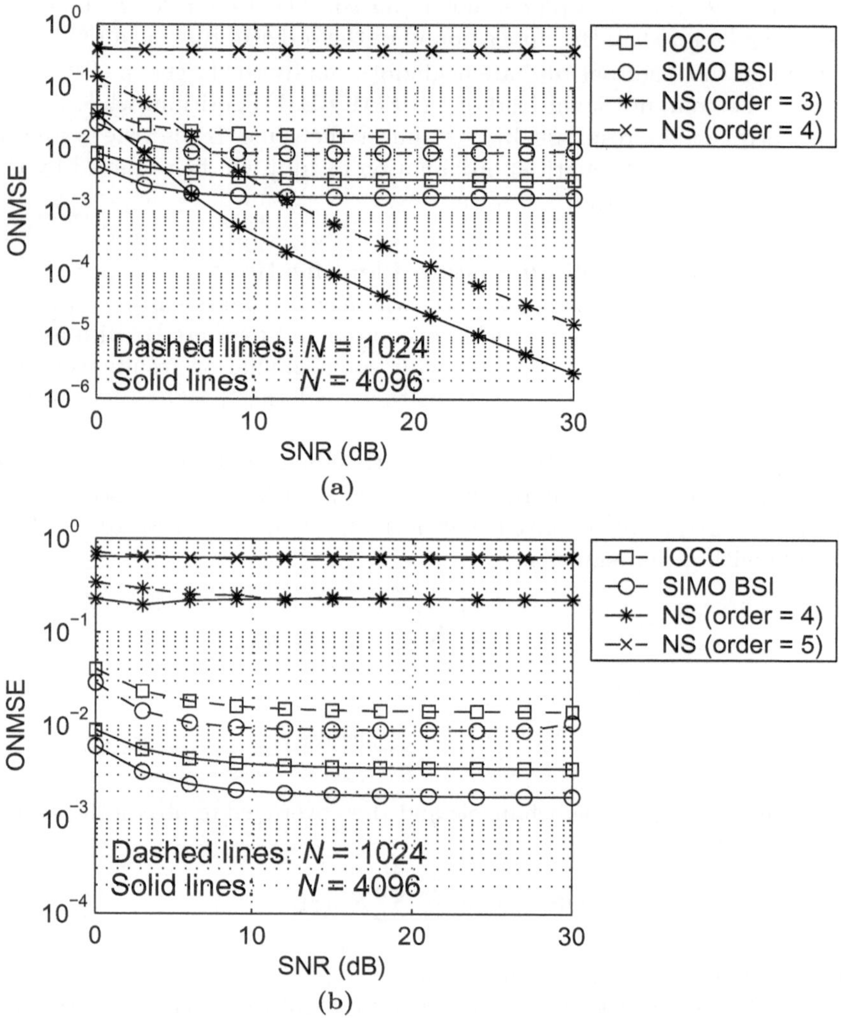

Fig. 6.7 Simulation results of blind SIMO system identification for **(a)** Case A and **(b)** Case B in Example 6.12

6.4 Multiple Time Delay Estimation

Estimation of time delay(s) between the measurements received by two (or more) sensors is crucial in many signal processing areas such as direction of arrival and range estimation in sonar, radar, and geophysics, etc. The conventional methods estimate a single time delay with the associated two sensor measurements at a time (instead of simultaneous estimation of multiple time delays) assuming that the two channel gains are in phase. By the fact that the signal model for the time delay estimation (TDE) problem can be formulated as an SIMO model and the fact that complete information of all the time delays is contained in the phase of the SIMO system, an FFT-based TDE algorithm is introduced next for simultaneous estimation of all time delays (with respect to a reference sensor) using all the available sensor measurements, not only leading to space diversity gain but also allowing the unknown channel gains to be independent.

6.4.1 Model Assumptions

Assume that the measurements $\mathbf{y}[n] = (y_1[n], y_2[n], ..., y_M[n])^T$ are measured from M (≥ 2) spatially separated sensors that satisfy

$$\begin{cases} y_1[n] = x[n] + w_1[n], \\ y_i[n] = x[n] \star \mathtt{d}_i[n] + w_i[n], \quad i = 2, 3, ..., M \end{cases} \tag{6.31}$$

where $x[n]$ is the source signal (could be temporally colored), and $\mathtt{d}_i[n]$, $i \in \{2, 3, ..., M\}$ is an LTI system of frequency response

$$\mathtt{D}_i(\omega) = A_i e^{-j\omega d_i} \tag{6.32}$$

where A_i and d_i, $i = 2, 3, ..., M$, are (real or complex) gains and real numbers, respectively, and $\mathbf{w}[n] = (w_1[n], w_2[n], ..., w_M[n])^T$ is an $M \times 1$ vector noise whose components can be spatially correlated and temporally colored. Note that $\mathtt{D}_i(\omega)$ given by (6.32) is called the (scaled) ideal delay system of d_i samples delay, even if d_i is not an integer. Therefore, for notational convenience, let us rewrite the signal model given by (6.31) as

$$\mathbf{y}[n] = (x[n], A_2 x[n - d_2], ..., A_M x[n - d_M])^T + \mathbf{w}[n], \tag{6.33}$$

where $A_i x[n - d_i]$ stands for the signal $x[n] \star \mathtt{d}_i[n]$. The goal is to estimate all the $(M - 1)$ time delays $\{d_2, d_3, ..., d_M\}$ simultaneously from the measurements $\mathbf{y}[n]$.

For ease of later use, let us define the SNR associated with $y_i[n]$ as follows

$$SNR_i = \frac{E\{|y_i[n] - w_i[n]|^2\}}{E\{|w_i[n]|^2\}}. \tag{6.34}$$

It should be noted that SNR_i denotes the signal quality of the ith subchannel (or sensor) while SNR defined by (5.18) denotes the total signal quality associated with $\mathbf{y}[n]$, and that there exist multiple distributions of $\{SNR_i, i = 1, 2, ..., M\}$ corresponding to the same SNR.

Let us further assume that the source signal $x[n]$ is stationary colored nonGaussian and can be modeled as

$$x[n] = \sum_{k=-\infty}^{\infty} \widetilde{h}[k]u[n-k] \tag{6.35}$$

where $\widetilde{h}[n]$ is a stable LTI system and $u[n]$ is a stationary zero-mean, temporally i.i.d. nonGaussian random sequence. Substituting (6.35) into (6.33) gives rise to an SIMO system model as given by (6.15) with

$$\mathbf{h}[n] = (\widetilde{h}[n], A_2\widetilde{h}[n-d_2], ..., A_M\widetilde{h}[n-d_M])^T, \tag{6.36}$$

where $A_i\widetilde{h}[n-d_i]$ stands for $\widetilde{h}[n]\star\mathrm{d}_i[n]$. Note that all of the M subchannels of the SIMO system $\mathbf{h}[n]$ given by (6.36) have the same zeros as $\widetilde{h}[n]$. The time delays $\{d_2, d_3, ..., d_M\}$ can be extracted from the SIMO system $\mathbf{h}[n]$ which can be estimated from the measurements $\mathbf{y}[n]$ ahead of time using a preferred blind SIMO system identification method. In light of the specific form of $\mathbf{h}[n]$ given by (6.36) and the fact that the phase information of $\mathbf{h}[n]$ is sufficient for retrieving multiple time delays, the estimation of $\{d_2, d_3, ..., d_M\}$ can be quite efficient as illustrated below.

6.4.2 MTDE with Space Diversity Gain

Let $\mathbf{v}_{\mathrm{MNC}}[n]$ denote the $M \times 1$ optimum MIMO-MNC equalizer with the measurements $\mathbf{y}[n]$ given by (6.15) where $\mathbf{h}[n]$ is given by (6.36). The specific form of $\mathbf{h}[n]$ given by (6.36) leads to the phase of $\boldsymbol{H}(\omega)$ given by

$$\begin{aligned}\measuredangle\{\boldsymbol{H}(\omega)\} = (&\measuredangle\{\widetilde{H}(\omega)\}, \measuredangle\{\widetilde{H}(\omega)\} + \theta_2 - \omega d_2, \\ &..., \measuredangle\{\widetilde{H}(\omega)\} + \theta_M - \omega d_M)^T\end{aligned} \tag{6.37}$$

where θ_i is the phase of A_i. On the other hand, it can easily be seen, from the relation given by (6.20) ($\beta > 0$), that

$$\begin{aligned}\measuredangle\{\boldsymbol{H}(\omega)\} &= \measuredangle\{\boldsymbol{S}_{\mathbf{y}}^H(\omega)\boldsymbol{V}_{\mathrm{MNC}}^*(\omega)\} - \measuredangle\{G_p^*(\omega)\} \\ &= \left(\Phi_1(\omega) - \measuredangle\{G_p^*(\omega)\}, \Phi_2(\omega) - \measuredangle\{G_p^*(\omega)\}, ..., \Phi_M(\omega) - \measuredangle\{G_p^*(\omega)\}\right)^T\end{aligned} \tag{6.38}$$

where

$$(\Phi_1(\omega), \Phi_2(\omega), ..., \Phi_M(\omega))^T \triangleq \measuredangle\{\boldsymbol{S}_{\mathbf{y}}^H(\omega)\boldsymbol{V}_{\mathrm{MNC}}^*(\omega)\}. \tag{6.39}$$

By subtracting $\measuredangle\{\widetilde{H}(\omega)\}$ and $(\Phi_1(\omega) - \measuredangle\{G_p^*(\omega)\})$ (i.e. the first element of $\measuredangle\{\boldsymbol{H}(\omega)\}$) from each element of $\measuredangle\{\boldsymbol{H}(\omega)\}$ given by (6.37) and (6.38), respectively, one can obtain

$$(0, \Phi_2(\omega) - \Phi_1(\omega), ..., \Phi_M(\omega) - \Phi_1(\omega))^T = (0, \theta_2 - \omega d_2, ..., \theta_M - \omega d_M)^T. \quad (6.40)$$

Then,

$$\begin{aligned} \boldsymbol{F}(\omega) &\triangleq \left(1, e^{j[\Phi_2(\omega)-\Phi_1(\omega)]}, ..., e^{j[\Phi_M(\omega)-\Phi_1(\omega)]}\right)^T \\ &= \left(1, e^{j(\theta_2-\omega d_2)}, ..., e^{j(\theta_M-\omega d_M)}\right)^T \quad \text{(by (6.40))} \end{aligned} \quad (6.41)$$

and the inverse DTFT of $F_i(\omega) = e^{j(\theta_i - \omega d_i)}$, $i \in \{2, 3, ..., M\}$ is given by

$$f_i[n] = e^{j\theta_i} \cdot \frac{\sin(\pi(n - d_i))}{\pi(n - d_i)},$$

which can be thought of as a sequence obtained by sampling a continuous-time signal

$$f_i(t) = e^{j\theta_i} \cdot \frac{\sin(\pi(t - d_i T)/T)}{\pi(t - d_i T)/T}$$

with sampling period T. Note that $|f_i(t)| \le |f_i(d_i T)| = 1, \ \forall t$ and thus d_i can be estimated by

$$\widehat{d}_i = \frac{1}{T} \arg\{\max_t |f_i(t)|\}.$$

Instead, d_i can be estimated within a preferred resolution by sampling $f_i(t)$ at a higher sampling rate as described below.

Consider another sequence $b_i[n]$ which is obtained by sampling $f_i(t)$ at sampling interval $T/\mathcal{P}$, i.e. $b_i[n] = f_i(t = nT/\mathcal{P})$, where $\mathcal{P}$ is a positive integer such that $\mathcal{P}d_i$ is an integer. Then, it can be shown that $|b_i[n]| \le |b_i[\mathcal{P}d_i]| = |f_i(d_i T)| = 1, \forall n$, which implies that d_i can be estimated by finding the time index associated with the maximum value of $|b_i[n]|$. Because $b_i[n]$ (with sampling period $T/\mathcal{P}$) is actually an interpolated version of $f_i[n]$ (with sampling period T), the above idea can be implemented efficiently using zero-padded FFT computation.

Assume that $F_i[k] = F_i(\omega = 2\pi k/\mathcal{N})$, $k = 0, 1, ..., \mathcal{N} - 1$, have been obtained by (6.39) and (6.41) for a chosen $\mathcal{N}$. Define

$$B_i[k] \triangleq \begin{cases} F_i[k], & 0 \le k \le (\mathcal{N}/2) - 1, \\ F_i[k + \mathcal{N} - \mathcal{L}], & \mathcal{L} - (\mathcal{N}/2) \le k \le \mathcal{L} - 1, \\ 0, & \text{otherwise} \end{cases} \quad (6.42)$$

where $\mathcal{L} = \mathcal{P}\mathcal{N}$. Then, the $\mathcal{L}$-point inverse DFT of $B_i[k]$, denoted by $b_i[n], n = 0, 1, ..., \mathcal{L} - 1$, is exactly the interpolated version of $f_i[n], n = 0, 1, ..., \mathcal{N} - 1$. Then the time delays can be estimated as

$$\widehat{d}_i = \begin{cases} \frac{n_i}{\mathcal{P}}, & 0 \le n_i \le \mathcal{L}/2 - 1, \\ \frac{n_i - \mathcal{L}}{\mathcal{P}}, & \mathcal{L}/2 \le n_i \le \mathcal{L} - 1 \end{cases} \quad (6.43)$$

where

$$n_i = \arg\{\max\{|b_i[n]|,\ 0 \le n \le \mathcal{L}-1\}\}. \tag{6.44}$$

Note that (6.43) means that the resolution of the estimated time delays is $T/\mathcal{P}$ rather than T. The larger the $\mathcal{P}$, the better the estimation accuracy of $\widehat{d}_i$, especially for the case of noninteger time delays. The above procedure for the estimation of multiple d_is constitutes the FFT-based MTDE algorithm which is summarized as follows.

MTDE Algorithm

(S1) With finite data $\mathbf{y}[n]$, obtain the optimum MNC equalizer $\mathbf{v}_{\mathrm{MNC}}[n]$ ($p = q$), and compute its $\mathcal{N}$-point FFT $\boldsymbol{V}_{\mathrm{MNC}}[k]$. Obtain the power spectral matrix $\boldsymbol{\mathcal{S}}_{\mathbf{y}}[k]$ using a multichannel power spectral estimator.

(S2) Compute $\boldsymbol{F}[k]$ according to (6.39) and the first line of (6.41), and $B_i[k]$ given by (6.42). Then, obtain $b_i[n]$ ($\mathcal{L}$-point inverse DFT of $B_i[k]$) for $i = 2, 3, \ldots, M$.

(S3) Estimate $\{d_2, d_3, \ldots, d_M\}$ by (6.43) and (6.44).

Unlike conventional TDE methods (which estimate a single time delay with the two associated sensor measurements at a time), the MTDE algorithm simultaneously processes all the measurements received at all the sensors. In other words, the space diversity of multiple sensors has been exploited by the MTDE algorithm, implying that its performance is robust to the distribution of $\{SNR_i, i = 1, 2, \ldots, M\}$ due to channel fading as long as the overall SNR stays the same. On the other hand, the MTDE algorithm can also be employed to estimate each time delay using the associated two sensor measurements as with conventional TDE methods. In this case, the MTDE algorithm reduces to a single time delay estimation algorithm, referred to as the "MTDE-1 algorithm" to distinguish it from the MTDE algorithm using all the sensor measurements. However, the performance of the MTDE-1 algorithm is inferior to that of the MTDE algorithm because of the reduced space diversity exploited by the former.

Example 6.13 (MTDE)
This example considers the problem of simultaneously estimating two time delays (d_2 and d_3) of a nonGaussian source signal impinging upon three sensors (i.e. $M = 3$) using the MTDE algorithm. For comparison, d_2 and d_3 are also separately estimated by the MTDE-1 algorithm as well as Ye and Tugnait's integrated bispectrum based time delay estimator (IBBTDE) [5] (summarized in Appendix 6E) with the associated two sensor measurements used. That is, for the MTDE-1 algorithm and the IBBTDE, d_2 and d_3 are also separately estimated with the data sets $\{y_1[n], y_2[n]\}$ and $\{y_1[n], y_3[n]\}$, respectively. Let us emphasize that the performance of the IBBTDE has been shown to as-

ymptotically approach the Cramér–Rao bound of a single time delay estimate using the associated two sensor measurements [5].

The source signal $x[n]$ was assumed to be a zero-mean, i.i.d. one-sided exponentially distributed random sequence. The noise sequence $w_1[n]$ (in the first sensor) was assumed to be a zero-mean colored Gaussian sequence generated as the output of the MA(1) system $H_w(z) = 1 + 0.8z^{-1}$ driven by a real white Gaussian sequence. The noise sequence $w_2[n] = w_1[n]$ was perfectly correlated with $w_1[n]$, and the noise sequence $w_3[n] = \gamma(w_1[n] - w_1[n-6])$, where γ was chosen such that $E\{|w_3[n]|^2\} = E\{|w_1[n]|^2\}$, was also correlated with $w_1[n]$, and thus all three sensor noises have the same power. The data $\mathbf{y}[n]$ were synthesized according to (6.33) with the two time delays $\{d_2 = 2.5, d_3 = 11.4\}$ and two gains $\{A_2, A_3\}$ given as follows:

Case A: $A_2 = A_3 = 1$ (i.e. $SNR_1 = SNR_2 = SNR_3$)
Case B: $A_2 = 3$ and $A_3 = -2$ (i.e. $SNR_1 < SNR_3 = 4SNR_1 < SNR_2 = 9SNR_1$).

The synthetic data $\mathbf{y}[n]$ of length $N = 4096$ were then processed by the MTDE and MTDE-1 algorithms with the following settings: a 3×1 causal FIR filter (or a 2×1 causal FIR filter) of length $L = 10$ for the MNC equalizer $\mathbf{v}_{\mathrm{MNC}}[n]$ with $p = q = 2$, an AR multichannel power spectral estimator of order equal to 12 for power spectral matrix estimation, the FFT size $\mathcal{N} = 128$, and $\mathcal{P} = 100$. The same values of $\mathcal{N}$ and $\mathcal{P}$ were also used for the IBBTDE.

The simulation results for Case A and Case B are displayed in Fig. 6.8 and Fig. 6.9, respectively. These figures show the root-mean-square errors (RMSE) of the time delay estimates $\widehat{d_2}$ and $\widehat{d_3}$ associated with the MTDE algorithm, the MTDE-1 algorithm and the IBBTDE.

From Fig. 6.8, one can see that the MTDE algorithm performs better than the MTDE-1 algorithm due to the exploitation of more space diversity by the former. The MTDE algorithm performs much better than the IBBTDE for lower SNR, whereas the latter performs better than the former for higher SNR. Furthermore, comparing Fig. 6.8 and Fig. 6.9, one can observe that the performances of the MTDE algorithm, the MTDE-1 algorithm and the IBBTDE for Case B are worse than those for Case A. Nevertheless, the MTDE algorithm still works well for a wide range of overall SNR for Case B, justifying that the performance of the MTDE algorithm is less sensitive to the nonuniform SNR_i among sensors because of space diversity gain. On the other hand, the IBBTDE failed to estimate d_3 for Case B because the phase shift of π (since $A_1 > 0$ and $A_3 < 0$) between the received measurements $y_1[n]$ and $y_3[n]$ is never considered in the model in [5], whereas the model given by (6.33) does.

□

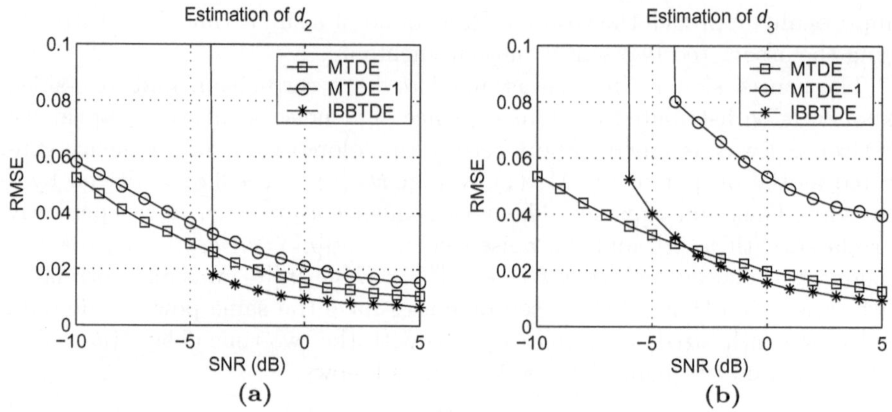

Fig. 6.8 Simulation results of Case A in Example 6.13 for estimating two time delays of $d_2 = 2.5$ and $d_3 = 11.4$. **(a)** and **(b)** Plots of the RMSEs of $\widehat{d}_2$ and $\widehat{d}_3$, respectively, associated with the MTDE algorithm, the MTDE-1 algorithm and the IBBTDE

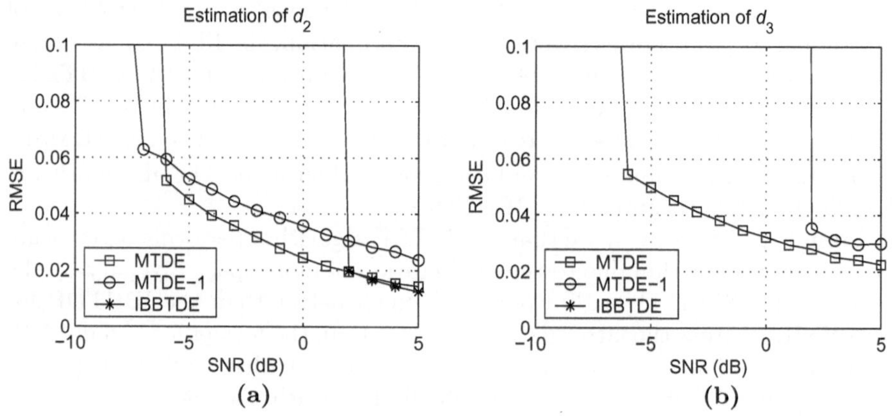

Fig. 6.9 Simulation results of Case B in Example 6.13 for estimating two time delays of $d_2 = 2.5$ and $d_3 = 11.4$. **(a)** and **(b)** Plots of the RMSEs of $\widehat{d}_2$ and $\widehat{d}_3$, respectively, associated with the MTDE algorithm, the MTDE-1 algorithm and the IBBTDE. Note that IBBTDE failed in the estimation of d_3, so its RMSEs are not displayed in **(b)**

6.5 Blind Beamforming for Source Separation

Blind source separation under multipath environments is essential in the areas of array signal processing, wireless communications and biomedical signal processing. In this section, applications of the MIMO-MNC equalization algorithm to blind beamforming and source separation are introduced.

6.5.1 Model Assumptions

Consider the case of K sources arriving at M sensors in the presence of multiple paths where the received $M \times 1$ signal vector is modeled as

$$\mathbf{y}[n] = \sum_{k=1}^{K} \sum_{l=1}^{L_k} \alpha_{k,l} \mathbf{a}(\theta_{k,l}) u_k[n - \tau_{k,l}] + \mathbf{w}[n] \tag{6.45}$$

in which

$$\begin{aligned} u_k[n] &: \text{signal of source } k \\ \mathbf{a}(\theta_{k,l}) &: M \times 1 \text{ steering vector of the } l\text{th path of source } k \\ \alpha_{k,l} &: \text{fading factor} \\ \tau_{k,l} &: \text{time delay} \\ \mathbf{w}[n] &: M \times 1 \text{ additive noise vector} \\ L_k &: \text{number of paths from source } k. \end{aligned}$$

Let us make the following assumptions:

(A6-5) $u_k[n]$ are stationary zero-mean, s.i.t.i. nonGaussian random processes with $C_{p,q}\{u_k[n]\} \neq 0$, for all k.

(A6-6) $\mathbf{w}[n]$ is zero-mean Gaussian and statistically independent of $u_k[n]$ for all k.

(A6-7) $\tau_{k,l} \neq \tau_{k,m}$ for $l \neq m$.

The whiteness assumption (A6-5) for the K source signals is generally valid in wireless multiuser communications. Assumption (A6-7) includes the case that as $\tau_{k,l} = \tau_{k,m}$ for some l and m, $u_k[n - \tau_{k,l}]$ and $u_k[n - \tau_{k,m}]$ will merge with the combined steering vector $\mathbf{a}(\theta_{k,l}) + \mathbf{a}(\theta_{k,m})$ and thus the resultant number of paths associated with $u_k[n]$ becomes $L_k - 1$.

Let $\mathcal{L} = L_1 + L_2 + ... + L_K$ and

$$\begin{aligned} u_{k,l}[n] &= u_k[n - \tau_{k,l}], \ l = 1, ..., L_k, \ k = 1, ..., K \\ \widetilde{\mathbf{u}}[n] &= (\widetilde{u}_1[n], \widetilde{u}_2[n], ..., \widetilde{u}_{\mathcal{L}}[n])^T \\ &= (u_{1,1}[n], u_{1,2}[n], ..., u_{1,L_1}[n], u_{2,1}[n], u_{2,2}[n], ..., u_{2,L_2}[n], ..., u_{K,L_K}[n])^T. \end{aligned}$$

The model for $\mathbf{y}[n]$ given by (6.45) can also be expressed as

$$\mathbf{y}[n] = \mathbf{A}\widetilde{\mathbf{u}}[n] + \mathbf{w}[n] \tag{6.46}$$

where $\mathbf{A}$ is an $M \times \mathcal{L}$ matrix as follows:

$$\begin{aligned}\mathbf{A} = (&\alpha_{1,1}\mathbf{a}(\theta_{1,1}), \alpha_{1,2}\mathbf{a}(\theta_{1,2}), ..., \alpha_{1,L_1}\mathbf{a}(\theta_{1,L_1}), \alpha_{2,1}\mathbf{a}(\theta_{2,1}), \alpha_{2,2}\mathbf{a}(\theta_{2,2}), ..., \\ &\alpha_{2,L_2}\mathbf{a}(\theta_{2,L_2}), ..., \alpha_{K,L_K}\mathbf{a}(\theta_{K,L_K})).\end{aligned} \tag{6.47}$$

Note that the $M \times \mathcal{L}$ system given by (6.46) is nothing but a special case of the MIMO model given by (5.16) with $\mathbf{H}[n] = \mathbf{A}\delta[n]$ and $K = \mathcal{L}$.

6.5.2 Blind Beamforming

Given $\mathbf{y}[n]$ for $n = 1, 2, ..., N$, the task of beamforming is to extract one of the $\mathcal{L}$ signals $u_{k,l}[n]$ (up to a scale factor) through linear spatial processing of $\mathbf{y}[n]$. The beamformer output is given by

$$e[n] = \boldsymbol{v}^T\mathbf{y}[n] = \mathbf{g}^T\widetilde{\mathbf{u}}[n] + \widetilde{w}[n] \tag{6.48}$$

where $\mathbf{g} = \mathbf{A}^T\boldsymbol{v}$ and $\widetilde{w}[n] = \boldsymbol{v}^T\mathbf{w}[n]$. The MIMO hybrid MNC equalization algorithm can be employed to design the optimum beamformer $\boldsymbol{v}$ with no need for any information about $\mathbf{A}$ (i.e. the direction-of-arrival information of the source signals) which is supported by the following theorem [1, 2].

Theorem 6.14 (Beamforming). *Suppose that* $\mathbf{y}[n]$ *given by (6.45) satisfies Assumptions* (A6-5), (A6-7), *the assumption that the* $M \times \mathcal{L}$ *matrix* $\mathbf{A}$ *given by (6.47) is of full column rank with* $M \geq \mathcal{L}$, *and the noise-free assumption. The optimum MIMO-MNC linear beamformer* $\boldsymbol{v}_{\mathrm{MNC}}$ *by maximizing* $J_{p,q}(e[n])$ *given by (5.73) results in*

$$e[n] = \alpha u_{k,l}[n] \tag{6.49}$$

where $\alpha \neq 0$ *is an unknown scalar,* l *is an arbitrary integer belonging to* $\{1, 2, ..., L_k\}$, *and*

$$k = \arg\max_i \{J_{p,q}(u_i[n])\}\,. \tag{6.50}$$

The proof of Theorem 6.14 is left as an exercise (Problem 6.5).

Note that Theorem 6.14 refers to the received signal $\mathbf{y}[n]$ that is actually a superposition of K colored source signals from K multipath channels to which K s.i.t.i. source signals $u_k[n]$'s are input, while all the $\mathcal{L}$ source components $u_{k,l}[n]$ for $l = 1, 2, ..., L_k$ and $k = 1, 2, ..., K$ arrive at the array from different directions such that $\mathbf{A}$ is of full column rank. Specifically, as $\mathbf{a}(\theta_{\kappa,l})$ for some κ are identical for all l, $\mathrm{rank}\{\mathbf{A}\}$ reduces to $\mathcal{L} - L_\kappa + 1$ implying that only the colored signal $\widetilde{u}_\kappa[n] = \sum_l \alpha_{\kappa,l}u_{\kappa,l}[n]$ (resultant from multipath effects) can be extracted. Therefore, Theorem 6.14 can be extended as the following fact (Problem 6.6):

Fact 6.15. *Theorem 6.14 is also true if Assumption* (A6-5) *is replaced with that among all the mutually independent nonGaussian sources* $u_k[n]$, *some are* *temporally colored without multipath (i.e. $L_k = 1$) and the other sources are s.i.t.i.*

Example 6.16 (Blind Beamforming)
Consider the case of three independent nonGaussian source signals with zero mean and unity variance arriving at ten sensors ($M = 10$) as planewaves. Source 1 ($u_1[n]$) is a 16-QAM signal and the other two sources ($u_2[n]$ and $u_3[n]$) are 4-QAM signals. All the 10 sensors are uniformly separated by half wavelength under the following multipath channel parameters:

$$\begin{aligned}
(L_1, L_2, L_3) &= (2, 3, 1) \\
(\alpha_{1,1}, \tau_{1,1}, \theta_{1,1}) &= (1.2, 0, -75^\circ) \\
(\alpha_{1,2}, \tau_{1,2}, \theta_{1,2}) &= (0.8, 2, 0^\circ) \\
(\alpha_{2,1}, \tau_{2,1}, \theta_{2,1}) &= (1.0, 0, -45^\circ) \\
(\alpha_{2,2}, \tau_{2,2}, \theta_{2,2}) &= (0.3, 1, 50^\circ) \\
(\alpha_{2,3}, \tau_{2,3}, \theta_{2,3}) &= (0.6, 2, 5^\circ) \\
(\alpha_{3,1}, \tau_{3,1}, \theta_{3,1}) &= (1.1, 0, 75^\circ).
\end{aligned}$$

With the noise sequence $\mathbf{w}[n]$ assumed to be spatially independent and temporally white Gaussian, the synthetic data $\mathbf{y}[n]$ for $N = 1024$ (number of snapshots) and SNR = 20 dB were processed using the MIMO-MNC equalization algorithm and all the $\mathcal{L} = 6$ sources were obtained through the MSC procedure.

Figure 6.10a displays the constellation of the synthetic $y_1[n]$ received by the first sensor. Fig. 6.10b displays the constellations of the extracted six source signals, each corresponding to one source signal component from one path. Obviously, all of the six signals $u_{k,l}[n]$ are correctly captured.

In the next subsection, we will introduce a source separation algorithm including classification of these captured $\mathcal{L}$ signals, followed by a proper combination of all the extracted L_k source components associated with the same source signal $u_k[n]$ such that the SNR of the resultant estimate $\widehat{u}_k[n]$ is maximum.

□

6.5.3 Multistage Source Separation

With $\mathbf{y}[n]$ given by (6.45) for $n = 1, 2, \ldots, N-1$, the task of source separation is to further restore all the source signals $u_k[n], k = 1, 2, \ldots, K$ (up to a scale factor) with little interference from the $\mathcal{L}$ components $u_{k,l}[n]$ extracted by the MNC beamformer. The multistage source separation (MSS) algorithm reported in [6] includes the following four signal processing steps at each stage:

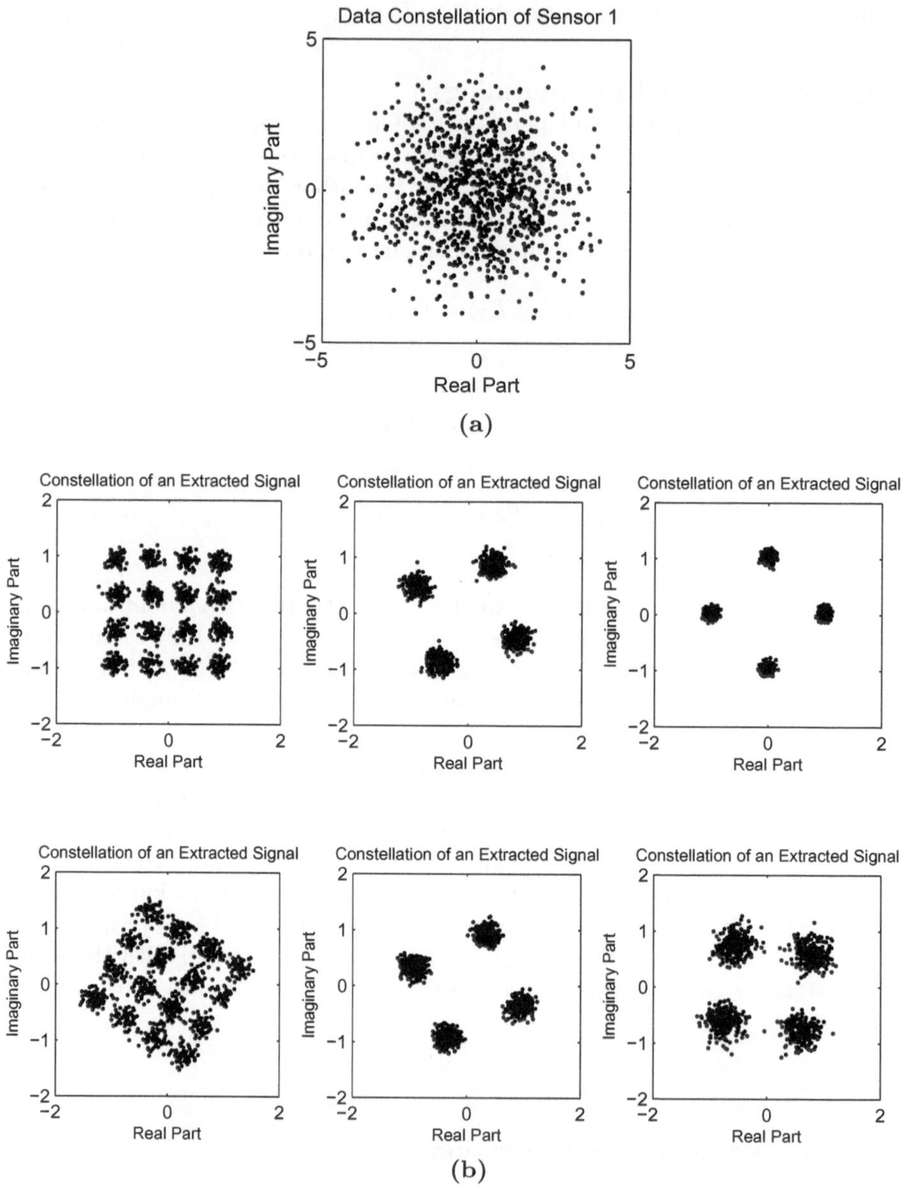

Fig. 6.10 The constellations of **(a)** the synthetic $y_1[n]$ received by the first sensor, and **(b)** the extracted six signals each corresponding to one source signal component from one path

MSS Algorithm

(S1) Source extraction.
Obtain a source signal $e[n]$ using the MIMO-MNC beamformer $\boldsymbol{v}_{\mathrm{MNC}}$.

(S2) Classification.
Compute the normalized cross-correlation $\rho_k[\tau]$ between $e[n]$ and the detected source signal $\widehat{u}_k[n-\tau]$ at the previous stage, defined as

$$\rho_k[\tau] = \frac{E\{e[n]\widehat{u}_k^*[n-\tau]\}}{\sigma_e \sigma_{\widehat{u}_k}}, \quad k = 1, 2, ..., \mathcal{K},$$

where $\mathcal{K}$ denotes the number of sources detected before the present stage, and find

$$|\rho_\kappa[\widehat{\tau}]| = \max_{k,\tau}\{|\rho_k[\tau]|\}.$$

If $|\rho_\kappa[\widehat{\tau}]| > \eta$ (a threshold), then classify $e[n]$ as a signal from one path of the source signal $u_\kappa[n]$, i.e. $e[n]$ can be approximated as

$$e[n] = au_\kappa[n-\widehat{\tau}] + w[n],$$

otherwise, $\widehat{u}_{\mathcal{K}+1}[n] = e[n]$ is a newly detected source signal, update $\mathcal{K}$ by $\mathcal{K}+1$ and then go to (S4).

(S3) BMRC.
Process $x_1[n] = e[n+\widehat{\tau}]$ and $x_2[n] = \widehat{u}_\kappa[n]$ to obtain the optimum estimate $\widetilde{u}_\kappa[n]$ using the BMRC method introduced in Section 6.2, and then update $\widehat{u}_\kappa[n]$ by $\widehat{u}_\kappa[n] = \widetilde{u}_\kappa[n]$.

(S4) Deflation (source cancellation).
Update $\mathbf{y}[n]$ by $\mathbf{y}[n] - \mathbf{y}'[n]$ where

$$\mathbf{y}'[n] = \frac{E\{\mathbf{y}[n]e^*[n]\}}{E\{|e[n]|^2\}} \cdot e[n]$$

is the estimated contribution of the extracted source signal $e[n]$ (the beamformer output) to the array measurement vector $\mathbf{y}[n]$.

The signal processing procedure of the MSS Algorithm at each stage is illustrated in Fig. 6.11. Because L_k, $k = 1, 2, ..., K$, are unknown, the MSS Algorithm ends as the value of $J_{p,q}(e[n])$ in (S1) is below a threshold (i.e. all the source signals have been extracted), and the resultant $\mathcal{K}$ is an estimated number of independent sources as K is unknown. The signal classification together with the time delay estimation in (S2) is based on the cross-correlation magnitudes between the extracted source component and all the restored source estimates $\widehat{u}_k[n]$ before the present stage. Notice that (S2) and (S3) are not needed for the first stage. As a final remark, the computational load is proportional to the total path number $\mathcal{L}$ indicating that the MSS Algorithm is practical for $\mathcal{L}$ not too large such as wireless communication channels with finite (but not too many) multiple paths (due to remote reflections).

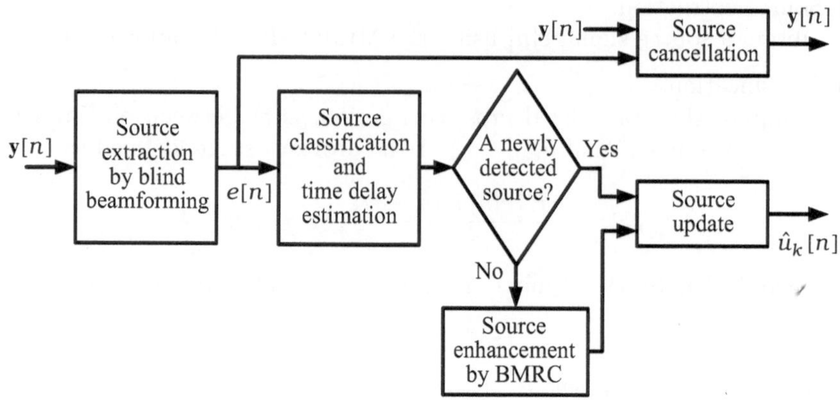

Fig. 6.11 Signal processing procedure of the MSS Algorithm at each stage

Example 6.17 (Blind Source Separation in Multipath)
The same synthetic data $\mathbf{y}[n]$ used in Example 6.16 were further processed by the MSS Algorithm where threshold $\eta = 0.1$ in (S2), and the restored signals are exhibited in Fig. 6.12.

Figure 6.12a displays the constellation of the synthetic $y_1[n]$ received by the first sensor. Figure 6.12b-d depict the constellations of the estimates $\widehat{u}_1[n]$, $\widehat{u}_2[n]$ and $\widehat{u}_3[n]$, respectively. Notice from these figures that $u_1[n]$ and $u_2[n]$ (with multipath diversity) are estimated more accurately than $u_3[n]$ (without multipath diversity). At each stage of the MSS Algorithm, the iterative MIMO hybrid MNC algorithm spent no more than 7 and 3 iterations in (S1) and (S3), respectively, in obtaining the results shown in these figures.

□

6.6 Multiuser Detection in Wireless Communications

Blind equalization algorithms [7–15] of MIMO systems have been widely used for the suppression of MAI and ISI, crucial to the receiver design of multiuser communication systems. Recently, for code-division multiple-access (CDMA) systems in the presence of multiple paths, many algorithms for simultaneously suppressing MAI and removing ISI have been reported [11–13, 16]. Tsatsanis and Giannakis [11] proposed an MMSE decorrelating receiver for asynchronous direct sequence/CDMA (DS/CDMA) systems. Tsatsanis [12] also proposed a near–far resistant minimum-output-energy (MOE) receiver for asynchronous DS/CDMA systems assuming that a path of the desired user is

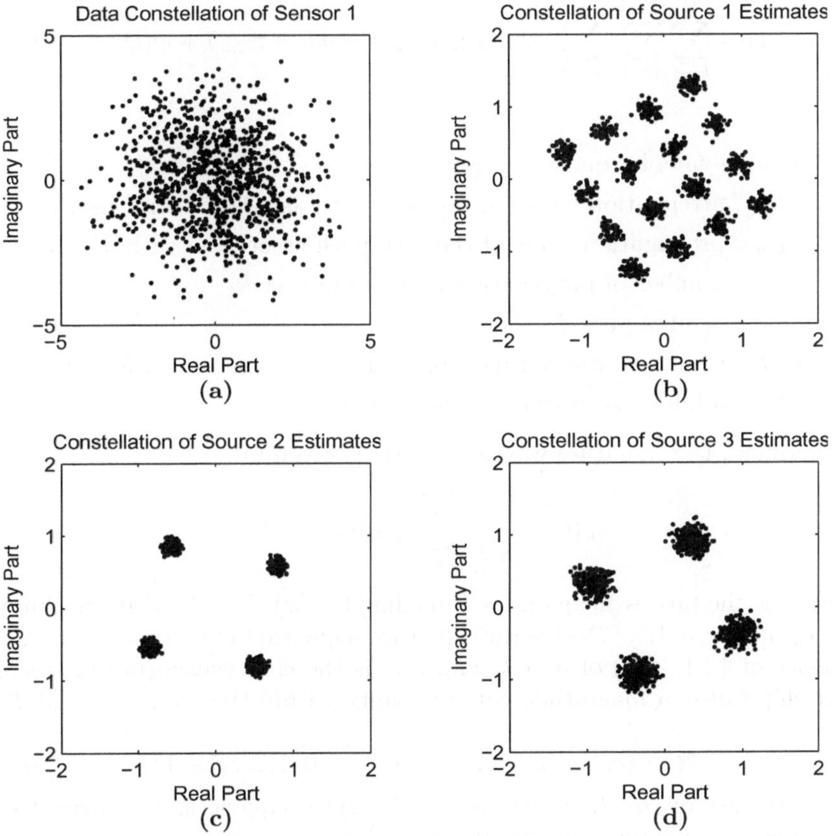

Fig. 6.12 Simulation results of Example 6.17. **(a)** The constellation of the synthetic data $y_1[n]$ received by the first sensor of the 10-element sensor array; **(b)**, **(c)** and **(d)** are constellations of source estimates $\widehat{u}_1[n]$, $\widehat{u}_2[n]$ and $\widehat{u}_3[n]$, respectively, obtained by the MSS Algorithm

known ahead of time. Then Tsatsanis and Xu [13] further proposed a blind minimum variance (MV) receiver that is near–far resistant with performance close to the MMSE decorrelating receiver for high SNR, and estimation of the multipath channel of the desired user is also included. In this section, let us address how the MIMO hybrid MNC algorithm can be applied to the multiuser detection with suppression of both MAI and ISI for asynchronous DS/CDMA systems.

6.6.1 Model Assumptions and Problem Statement

For a K-user asynchronous DS/CDMA communication system in the presence of multiple paths, the received baseband continuous-time signal can be represented as

$$y(t) = \sum_{k=1}^{K} \sum_{m=1}^{M_k} \sum_{n=-\infty}^{\infty} A_{k,m} u_k[n] s_k(t - nT - \tau_{k,m}) + w(t) \tag{6.51}$$

where

$u_k[n]$: symbol sequence of user k

$A_{k,m}$: attenuation factor of the mth path associated with user k

$\tau_{k,m}$: propagation delay of the mth path associated with user k

M_k : number of propagation paths from user k

T : symbol period

$s_k(t)$: signature waveform of unity energy associated with user k

$w(t)$: additive zero-mean Gaussian noise.

Furthermore, the signature waveform $s_k(t)$ is given by

$$s_k(t) = \frac{1}{\sqrt{T}} \sum_{n=0}^{\mathcal{P}-1} c_k[n] p(t - nT_c)$$

where $\mathcal{P}$ is the processing gain (or spreading factor), $T_c = T/\mathcal{P}$ is the chip period, $c_k[n], n = 0, 1, ..., \mathcal{P}-1$ is the signature sequence (a binary pseudorandom sequence of $\{+1, -1\}$) of user k, and $p(t)$ is the chip waveform (e.g. rectangular chip pulse of magnitude equal to unity within the interval $t \in [0, T_c)$). Let

$$\mathcal{R} = \{c_k[n],\ k = 1, 2, ..., K,\ n = 0, 1, ..., \mathcal{P} - 1\} \tag{6.52}$$

denote the set of the K active users' signature sequences. Moreover, let us make a general assumption that $0 \le \tau_{k,1} \le \tau_{k,2} \le \cdots \le \tau_{k,M_k} \le T + \tau_{k,1}, \forall k$, i.e. the delay spread of all the channels $\tau_{k,M_k} - \tau_{k,1} \le T, \forall k$ and $0 \le \tau_{1,1} \le \tau_{2,1} \le \cdots \le \tau_{K,1} \le T$.

The objective of the blind multiuser detection is either to estimate the symbol sequence of the desired user (e.g. $u_1[n]$) or to estimate all the symbol sequences $\{u_1[n], u_2[n], ..., u_K[n]\}$ with only the received signal $\mathbf{y}(t)$. The continuous-time signal $y(t)$, however, needs to be transformed into an equivalent discrete-time MIMO model first before the use of the MIMO hybrid MNC algorithm. As will be illustrated below, two discrete-time MIMO models, denoted as $\mathbf{y}^{(1)}[n]$ and $\mathbf{y}^{(2)}[n]$, can be formed through signature waveform matched filtering of $y(t)$, and one, denoted as $\mathbf{y}^{(3)}[n]$, through chip waveform matched filtering of $y(t)$.

6.6.2 Signature Waveform Matched Filtering Based Multiuser Detection

MIMO Models

Let $y_{k,m}[n]$ be the signature waveform matched filter output associated with the mth path of the kth user assuming perfect synchronization, and $w_{k,m}[n]$

be the noise term in $y_{k,m}[n]$ due to $w(t)$, i.e.

$$y_{k,m}[n] = \int_{nT+\tau_{k,m}}^{(n+1)T+\tau_{k,m}} y(t)s_k^*(t - nT - \tau_{k,m})dt,$$

$$w_{k,m}[n] = \int_{nT+\tau_{k,m}}^{(n+1)T+\tau_{k,m}} w(t)s_k^*(t - nT - \tau_{k,m})dt.$$

It can be easily shown that, for $1 \leq \mathcal{M}_k \leq M_k$,

$$\begin{aligned} \mathbf{y}_k[n] &\triangleq (y_{k,1}[n], y_{k,2}[n], ..., y_{k,\mathcal{M}_k}[n])^T \\ &= \mathbf{H}_k[n] \star \mathbf{u}[n] + \mathbf{w}_k[n] \\ &= \sum_{i=1}^{K} \mathbf{h}_{k;i}[n] \star u_i[n] + \mathbf{w}_k[n] \end{aligned} \tag{6.53}$$

where $\mathbf{u}[n] = (u_1[n], ..., u_K[n])^T$, $\mathbf{w}_k[n] = (w_{k,1}[n], ..., w_{k,\mathcal{M}_k}[n])^T$ is spatially correlated and temporally colored Gaussian noise, $\mathbf{h}_{k;i}[n]$ is the ith column of $\mathbf{H}_k[n]$ and $\mathbf{H}_k[n]$ is an $\mathcal{M}_k \times K$ FIR system of length five with the (l, i)th element

$$h_{k;l,i}[n] \triangleq [\mathbf{H}_k[n]]_{li} = \begin{cases} \sum_{m=1}^{M_k} \rho_{k,i;l,m}^{(n)} \cdot A_{i,m}, & n = -2, -1, 0, 1, 2, \\ 0, & \text{otherwise} \end{cases} \tag{6.54}$$

in which

$$\rho_{k,i;l,m}^{(n)} = \int_0^T s_k^*(t)s_i(t + nT + \tau_{k,l} - \tau_{i,m})dt. \tag{6.55}$$

In general, $\rho_{k,i;l,m}^{(n)} \approx 0$ for $i \neq k$ due to low cross-correlation between waveforms $s_i(t)$ and $s_k(t)$.

Model I: Concatenation of Matched Filter Output Vectors

By concatenation of $\mathbf{y}_k[n], k = 1, 2, ..., K$, each comprising $\mathcal{M}_k$ matched filter outputs as shown in Fig. 6.13, a discrete-time MIMO model (Model I) can be established as

$$\mathbf{y}^{(1)}[n] = \left(\mathbf{y}_1^T[n], \mathbf{y}_2^T[n], ..., \mathbf{y}_K^T[n]\right)^T = \mathbf{H}^{(1)}[n] \star \mathbf{u}[n] + \mathbf{w}^{(1)}[n] \tag{6.56}$$

where $\mathbf{w}^{(1)}[n] = \left(\mathbf{w}_1^T[n], \mathbf{w}_2^T[n], ..., \mathbf{w}_K^T[n]\right)^T$ and

$$\mathbf{H}^{(1)}[n] \triangleq \left(\mathbf{h}_1^{(1)}[n], \mathbf{h}_2^{(1)}[n], ..., \mathbf{h}_K^{(1)}[n]\right) = \left(\mathbf{H}_1^T[n] \ \mathbf{H}_2^T[n] \ \cdots \ \mathbf{H}_K^T[n]\right)^T \tag{6.57}$$

is an $\mathcal{M} \times K$ FIR system where

$$\mathcal{M} = \sum_{k=1}^{K} \mathcal{M}_k$$

and

$$\mathbf{h}_k^{(1)}[n] = \left((\mathbf{h}_{1,k}^{(1)}[n])^T, (\mathbf{h}_{2,k}^{(1)}[n])^T, ..., (\mathbf{h}_{K,k}^{(1)}[n])^T\right)^T$$
$$= \left(\mathbf{h}_{1;k}^T[n], \mathbf{h}_{2;k}^T[n], ..., \mathbf{h}_{K;k}^T[n]\right)^T \tag{6.58}$$

in which

$$\mathbf{h}_{i,k}^{(1)}[n] = \mathbf{h}_{i;k}[n] \tag{6.59}$$

is the ith subvector of $\mathbf{h}_k^{(1)}[n]$ with dimension $\mathcal{M}_i \times 1$.

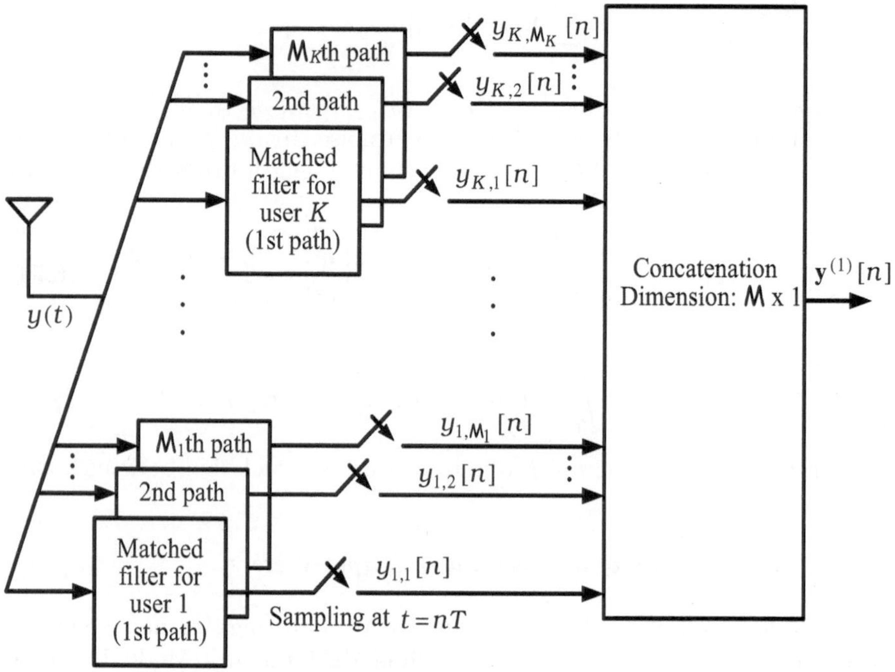

Fig. 6.13 Model I: Establishment of $\mathbf{y}^{(1)}[n]$ from $y(t)$

Model II: Concatenation of Combiner's Output Signals

Alternatively, $\mathbf{y}_k[n]$ given by (6.53) can be expressed as

$$\mathbf{y}_k[n] = \mathbf{a}_k \cdot u_k[n] + \boldsymbol{\mathcal{I}}_k[n] + \mathbf{w}_k[n] \tag{6.60}$$

where $\boldsymbol{\mathcal{I}}_k[n]$ is the co-channel interference plus intersymbol interference with respect to $u_k[n]$ and $\mathbf{a}_k = (A_{k,1}, A_{k,2}, ..., A_{k,\mathcal{M}_k})^T$. By treating the $\mathbf{y}_k[n]$ given by (6.60) as an approximation to the $\mathbf{y}[n]$ given by (6.9) with $M = \mathcal{M}_k$, $\boldsymbol{a} = \mathbf{a}_k$, $h[n] = \delta[n]$ and $\mathbf{w}[n] = \boldsymbol{\mathcal{I}}_k[n] + \mathbf{w}_k[n]$, the BMRC method introduced in Section 6.2 can be used to obtain the optimum

$$e_k[n] = \boldsymbol{v}_k^T \mathbf{y}_k[n], \ k = 1, 2, ..., K \tag{6.61}$$

such that its output SNR is "maximum" by Theorem 6.3.

By concatenation of $e_k[n], k = 1, 2, ..., K$ as shown in Fig. 6.14, the second discrete-time MIMO model (Model II) can be established as

$$\mathbf{y}^{(2)}[n] = (e_1[n], e_2[n], ..., e_K[n])^T = \mathbf{H}^{(2)}[n] \star \mathbf{u}[n] + \mathbf{w}^{(2)}[n] \tag{6.62}$$

where $\mathbf{H}^{(2)}[n] = \left(\mathbf{h}_1^{(2)}[n], \mathbf{h}_2^{(2)}[n], ..., \mathbf{h}_K^{(2)}[n]\right)$ is a $K \times K$ matrix and $\mathbf{w}^{(2)}[n]$ is spatially correlated and temporally colored Gaussian. Note that Model II can be thought of as a dimension reduced version of Model I, and that $\mathbf{y}^{(2)}[n] = \mathbf{y}^{(1)}[n]$ when $\mathcal{M}_k = 1, \ \forall k$.

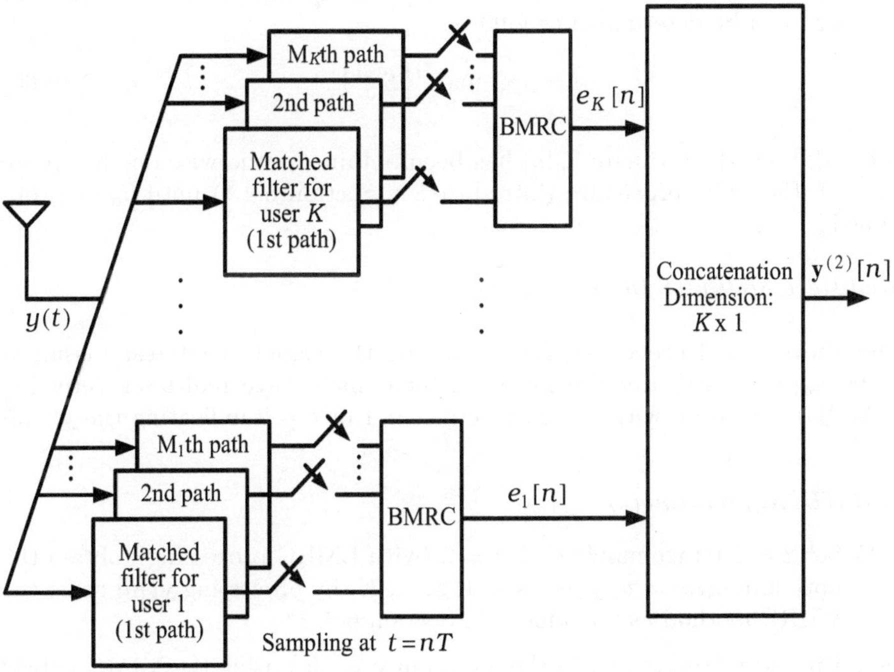

Fig. 6.14 Model II: Establishment of $\mathbf{y}^{(2)}[n]$ from $y(t)$

Multiuser Detection

User Identification

Assume that the user of interest is user d. One can employ the MIMO hybrid MNC algorithm to process $\mathbf{y}^{(\ell)}[n]$, $\ell \in \{1, 2\}$, but the optimum $e[n]$ obtained is an estimate of $u_l[n]$ where l is unknown by Theorems 5.32 and 5.33. Therefore, a user identification mechanism is needed to ascertain the estimated data sequence $u_l[n]$.

Let $h_{i,l}^{(\ell)}[n]$, $\ell = 1, 2$, denote the (i, l)th component of $\mathbf{H}^{(\ell)}[n]$. It can be shown, from (6.59), (6.54) and (6.55), that

$$\mathcal{E}_{l,l}^{(1)} = \sum_n \left\| \mathbf{h}_{l,l}^{(1)}[n] \right\|^2 \gg \mathcal{E}_{i,l}^{(1)} = \sum_n \left\| \mathbf{h}_{i,l}^{(1)}[n] \right\|^2, \ \forall\, i \neq l, \tag{6.63}$$

$$\mathcal{E}_{l,l}^{(2)} = \sum_n \left| h_{l,l}^{(2)}[n] \right|^2 \gg \mathcal{E}_{i,l}^{(2)} = \sum_n \left| h_{i,l}^{(2)}[n] \right|^2, \ \forall\, i \neq l, \tag{6.64}$$

which further imply that, with an estimate of $\mathbf{h}_l^{(\ell)}[n]$ (see (5.90)), the user number l can be determined as follows:

$$\widehat{l} = \arg \max_{1 \leq i \leq K} \left\{ \mathcal{E}_{i,l}^{(\ell)} \right\}. \tag{6.65}$$

If $\widehat{l} = d$, then the estimate $\widehat{u}_d[n]$ has been obtained; otherwise one has to go through the MSC procedure (introduced in Section 5.4.1) until $\widehat{u}_d[n]$ is obtained.

Multistage Multiuser Detection Algorithm

The above signal processing for estimating the signal of interest (assumed to be $u_d[n]$) constitutes the following blind multistage multiuser detection (BMMD) algorithm with the parameter $\ell = 1$ or $\ell = 2$ indicating the $\mathbf{y}^{(\ell)}[n]$ used:

BMMD Algorithm-(ℓ)

(S1) Set $\mathbb{k} = 1$ (stage number). If $\ell = 2$ (with BMRC processing), obtain the optimum $e_k[n] = \boldsymbol{v}_k^T \mathbf{y}_k[n]$, $k = 1, 2, ..., K$, by processing $\mathbf{y}_k[n]$ using the BMRC method as introduced in Section 6.2.

(S2) Process $\mathbf{y}^{(\ell)}[n]$ to obtain the optimum $\mathbf{v}_{\mathrm{MNC}}[n]$ using the MIMO hybrid MNC algorithm, and the associated $e[n]$ and $\widehat{\mathbf{h}}_l^{(\ell)}[n]$ (where l is unknown) (see (5.90)).

(S3) Update $\mathbf{y}^{(\ell)}[n]$ by $\mathbf{y}^{(\ell)}[n] - \widehat{\mathbf{h}}_l^{(\ell)}[n] \star e[n]$ (see (5.91)).

(S4) Determine $\widehat{l}$ using (6.65). If $\widehat{l} \neq d$, update $\Bbbk$ by $\Bbbk + 1$ and go to (S2), otherwise $\widehat{u}_d[n] = e[n]$ has been obtained.

As $\mathcal{M}_i > 1$ for some i, the computational complexity of the BMMD Algorithm-(ℓ) for $\ell = 1$ is higher than that for $\ell = 2$ due to higher model dimension ($\mathcal{M} \times K$) associated with $\mathbf{y}^{(1)}[n]$. However, the performance of the BMMD Algorithm-(ℓ) is similar for both $\ell = 1$ and $\ell = 2$ as demonstrated by a simulation example below.

In the simulation examples that follow, the power of the received signal from user k is defined as

$$\mathcal{E}_k^{(\ell)} = E\left\{ \left\| \mathbf{h}_k^{(\ell)}[n] \star u_k[n] \right\|^2 \right\}. \tag{6.66}$$

Assume that user k is the desired user, and $\mathcal{E}_i^{(\ell)} = \mathcal{E}$ for all $i \neq k$. The near–far ratio (NFR) is defined as

$$\text{NFR} = \frac{\mathcal{E}}{\mathcal{E}_k^{(\ell)}}. \tag{6.67}$$

Example 6.18 (BMMD over Signature Waveform Matched Filtering)

Let us show some simulation results using the BMMD Algorithm-(ℓ) with $\mathbf{y}^{(1)}[n]$ (Model I) and $\mathbf{y}^{(2)}[n]$ (Model II) for a five-user ($K = 5$) asynchronous DS/CDMA system with three paths for each user ($M_k = 3\ \forall k$). Thirty independent runs for data length $N = 2000$ were performed with Gold codes of length $\mathcal{P} = 31$ for users' spreading codes $c_k[n]$. The synthetic symbol sequences $u_k[n]$, $k = 1, 2, ..., 5$ were generated as equally probable binary random sequences whose amplitudes were adjusted such that $\mathcal{E}_k^{(1)} = \mathcal{E}$, $k = 2, ..., 5$.

The averaged output SINRs of user 1 (the weak user) associated with Models I and II are shown in Fig. 6.15. One can observe, from Fig. 6.15, that the performance of the BMMD Algorithm-(ℓ) is quite close to that of the nonblind LMMSE equalizer, and better for larger $\mathcal{M}_k$ (i.e. more multipath diversity gain). The similar performance for NFR$= \mathcal{E}/\mathcal{E}_1^{(1)}$ equal to 0 dB and 9 dB also implies that the BMMD Algorithm-(ℓ) is near–far resistant. □

6.6.3 Chip Waveform Matched Filtering Based Multiuser Detection

MIMO Model

Through chip waveform matched filtering and chip rate sampling of the continuous-time signal $y(t)$ given by (6.51), the discrete-time signal can be obtained as

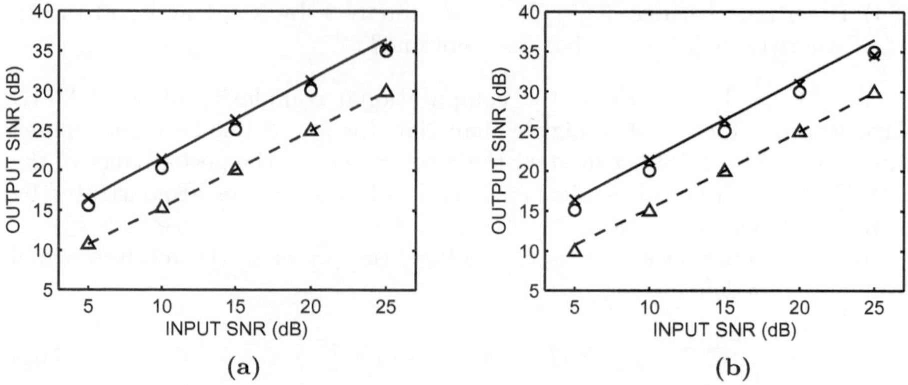

Fig. 6.15 Averaged output SINR of user 1 for **(a)** NFR = 0 dB and **(b)** NFR = 9 dB, respectively, associated with the nonblind LMMSE equalizer for $\mathcal{M}_k = 1\ \forall k$ (dashed lines) and $\mathcal{M}_k = 3\ \forall k$ (solid lines), the BMMD Algorithm-(1) for $\mathcal{M}_k = 1$ $\forall k$ ('$\triangle$') and $\mathcal{M}_k = 3\ \forall k$ ('$\times$'), and the BMMD Algorithm-(2) for $\mathcal{M}_k = 3\ \forall k$ ('$\bigcirc$')

$$\begin{aligned} \mathtt{y}[n] &= y(t) \star p^*(-t)|_{t=nT_c} \\ &= \sum_{k=1}^{K} \sum_{l=-\infty}^{\infty} u_k[l] h_k[n - l\mathcal{P}] + \mathtt{w}[n] = \sum_{k=1}^{K} \mathtt{y}_k[n] + \mathtt{w}[n] \end{aligned} \tag{6.68}$$

where

$$\begin{aligned} \mathtt{y}_k[n] &= \sum_{l=-\infty}^{\infty} u_k[l] h_k[n - l\mathcal{P}], \\ \mathtt{w}[n] &= w(t) \star p^*(-t)|_{t=nT_c}, \end{aligned}$$

and

$$h_k[n] = c_k[n] \star f_k[n] = \sum_{i=0}^{\mathcal{P}-1} c_k[i] f_k[n - i] \tag{6.69}$$

is the "signature waveform" of user k in which $f_k[n]$ denotes the discrete-time multipath channel from user k to the receiver.

Model III: Polyphase Decomposition of Chip Waveform Matched Filter Outputs

Through polyphase decomposition of $\mathtt{y}[n]$ given by (6.68) with dimension $\mathcal{P}$, one can obtain the following MIMO signal model shown in Fig. 6.16:

$$\mathbf{y}^{(3)}[n] = \mathbf{H}^{(3)}[n] \star \mathbf{u}[n] + \mathbf{w}^{(3)}[n] \tag{6.70}$$

where

$$\mathbf{y}^{(3)}[n] = (\mathtt{y}[n\mathcal{P}], \mathtt{y}[n\mathcal{P}+1], ..., \mathtt{y}[n\mathcal{P}+\mathcal{P}-1])^T,$$
$$\mathbf{u}[n] = (u_1[n], u_2[n], ..., u_K[n])^T,$$
$$\mathbf{w}^{(3)}[n] = (\mathtt{w}[n\mathcal{P}], \mathtt{w}[n\mathcal{P}+1], ..., \mathtt{w}[n\mathcal{P}+\mathcal{P}-1])^T,$$

and $\mathbf{H}^{(3)}[n] = \left(\mathbf{h}_1^{(3)}[n], \mathbf{h}_2^{(3)}[n], ..., \mathbf{h}_K^{(3)}[n]\right)$ is a $\mathcal{P} \times K$ matrix with the (i,k)th entry equal to

$$h_{i,k}^{(3)}[n] \triangleq \left[\mathbf{H}^{(3)}[n]\right]_{ik} = h_k[n\mathcal{P}+i-1] \tag{6.71}$$

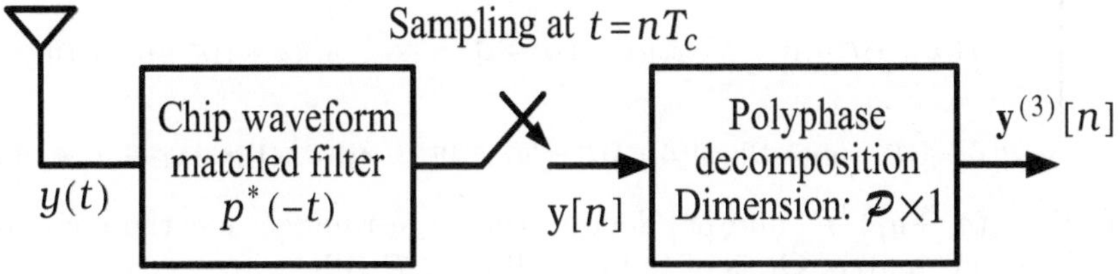

Fig. 6.16 Model III: Establishment of $\mathbf{y}^{(3)}[n]$ from $y(t)$

Multiuser Detection

The MIMO hybrid MNC algorithm introduced in Section 5.4.1 can be employed to estimate $\mathbf{u}[n]$ with $\mathbf{y}^{(3)}[n]$ given by (6.70), but a good initial condition for the equalizer $\boldsymbol{v}^{[0]}$ is needed because despreading has not been performed, leading to low SNR in $\mathbf{y}^{(3)}[n]$. Let us first present an initial condition of $\boldsymbol{v}^{[0]}$ by making use of the channel structure involving users' spreading sequences (see (6.69)).

Initial Condition for MIMO-MNC Criterion

Assume that each multipath channel $f_k[n]$ (see (6.69)) is an FIR channel of order equal to $q_f < \mathcal{P}$ which occurs in most asynchronous DS/CDMA channels [13] and user 1 (i.e. $u_1[n]$) is the user of interest.

Let $\mathbf{v}[n] = (v_1[n], v_2[n], ..., v_{\mathcal{P}}[n])^T$ ($\mathcal{P} \times 1$ vector) be an FIR equalizer with $\mathbf{v}[n] \neq 0$ for $n = L_1, L_1+1, ..., L_2$, and define the following notation:

$$\boldsymbol{v}_i = (v_i[L_1], v_i[L_1+1], ..., v_i[L_2])^T$$
$$\boldsymbol{v} = (\boldsymbol{v}_1^T, \boldsymbol{v}_2^T, ..., \boldsymbol{v}_{\mathcal{P}}^T)^T$$
$$q_v = (L_2 - L_1 + 1) \cdot \mathcal{P} - 1$$
$$\widetilde{\boldsymbol{v}} = (v_{\mathcal{P}}[L_1], v_{\mathcal{P}-1}[L_1], ..., v_1[L_1], ..., v_{\mathcal{P}}[L_2], v_{\mathcal{P}-1}[L_2], ..., v_1[L_2])^T.$$

Note that either of $\widetilde{\boldsymbol{v}}$ and $\boldsymbol{v}$ includes all the coefficients of the equalizer $\mathbf{v}[n]$. Let

$$\mathbf{C}_k = \begin{pmatrix} c_k[\mathcal{P}-1] & c_k[\mathcal{P}-2] & \cdots & c_k[\mathcal{P}-(q_v+1)] \\ c_k[\mathcal{P}-2] & c_k[\mathcal{P}-3] & \cdots & c_k[\mathcal{P}-(q_v+2)] \\ \vdots & \vdots & & \vdots \\ c_k[\mathcal{P}-(q_f+1)] & c_k[\mathcal{P}-(q_f+2)] & \cdots & c_k[\mathcal{P}-(q_f+q_v+1)] \\ c_k[2\mathcal{P}-1] & c_k[2\mathcal{P}-2] & \cdots & c_k[2\mathcal{P}-(q_v+1)] \\ \vdots & \vdots & & \vdots \\ c_k[2\mathcal{P}-(q_f+1)] & c_k[2\mathcal{P}-(q_f+2)] & \cdots & c_k[2\mathcal{P}-(q_f+q_v+1)] \\ \vdots & \vdots & & \vdots \\ c_k[(\mathcal{L}+1)\mathcal{P}-1] & c_k[(\mathcal{L}+1)\mathcal{P}-2] & \cdots & c_k[(\mathcal{L}+1)\mathcal{P}-(q_v+1)] \\ \vdots & \vdots & & \vdots \\ c_k[(\mathcal{L}+1)\mathcal{P}-(q_f+1)] & c_k[(\mathcal{L}+1)\mathcal{P}-(q_f+2)] & \cdots & c_k[(\mathcal{L}+1)\mathcal{P}-(q_f+q_v+1)] \end{pmatrix}$$

where $\mathcal{L} = \lfloor (q_v+q_f)/\mathcal{P} \rfloor$ and $\lfloor x \rfloor$ denotes the largest integer less than or equal to x. Note that matrix $\mathbf{C}_k$ is of full rank if $q_f < \mathcal{P}$ [12].

Chi *et al.* [14] reported a set of linear constraints on $\mathbf{v}[n]$ as given below

$$\mathbf{g}[n] = \left(\mathbf{H}^{(3)}[n]\right)^T \mathbf{v}[n] = \boldsymbol{\eta}_1 \cdot f_1[k_0] \cdot \delta[n-n_0] \tag{6.72}$$

which, by (6.71) and (6.69), or equivalently

$$\mathcal{C}\widetilde{\boldsymbol{v}} = \boldsymbol{\eta}_{n_0\cdot(q_f+1)+k_0+1} \tag{6.73}$$

where

$$\mathcal{C} = (\mathbf{C}_1^T \mathbf{C}_2^T \ \cdots \ \mathbf{C}_K^T)^T. \tag{6.74}$$

Then the LS solution to (6.73) is given by

$$\widetilde{\boldsymbol{v}}_{\text{LS}}(n_0, k_0) = \mathcal{C}^+ \cdot \boldsymbol{\eta}_{n_0\cdot(q_f+1)+k_0+1} \tag{6.75}$$

where $\mathcal{C}^+$ is the pseudoinverse of $\mathcal{C}$. The initial condition $\widetilde{\boldsymbol{v}}^{[0]}$ for MIMO-MNC criterion $J_{p,q}(\boldsymbol{v})$ is suggested as follows:

$$\widetilde{\boldsymbol{v}}^{[0]} = \widetilde{\boldsymbol{v}}_{\text{LS}}(\widetilde{n}_0, \widetilde{k}_0) \tag{6.76}$$

where

$$(\widetilde{n}_0, \widetilde{k}_0) = \arg\max_{(n_0,k_0)} J_{p,q}(\widetilde{\boldsymbol{v}}_{\text{LS}}(n_0, k_0)). \tag{6.77}$$

Remarkably, the $\widetilde{\boldsymbol{v}}^{[0]}$ given by (6.76) not only minimizes the error squares of the decorrelating constraint (6.73) (indexed by (n_0, k_0)) associated with the desired user (user 1) but also maximizes the associated $J_{p,q}$ with respect to (n_0, k_0) such that a transmission path with "large" path gain associated with the desired user is detected without the need for any prior multipath information concerning the desired user. Consequently, the estimate of interest

($e[n] = \widehat{u}_1[n]$) can usually be obtained using the MIMO hybrid MNC algorithm with the use of $\widetilde{\boldsymbol{v}}^{[0]}$ given by (6.76).

User Identification

First of all, let us present two facts on which the UIA to be introduced below is based. The first fact is regarding the relation between the phase and higher-order moments of a stable sequence. Let $a[n]$ (i.e. $\sum_n |a[n]| < \infty$) be a stable sequence with a certain amplitude spectrum $|A(\omega)|$. Define

$$\lambda(a[n]) = \int_{-\pi}^{\pi} |A(\omega)| \cdot [\phi_a(\omega)]^2 d\omega, \tag{6.78}$$

$$\Lambda(a[n]) = \sum_{n=-\infty}^{\infty} |a[n]|^{2m}, \tag{6.79}$$

where $\phi_a(\omega) = \arg\{A(\omega)\}$ with linear phase term removed[1] and $m \geq 2$. Note that

$$\Lambda(\alpha a[n-\tau]) = |\alpha|^{2m} \Lambda(a[n]) \tag{6.80}$$

implying that $\Lambda(a[n])$ is invariant for any linear phase change in $\arg\{A(\omega)\}$ as long as $|\alpha| = 1$. Chien *et al.* [17] have shown the following fact for real $a[n]$.

Fact 6.19. *The smaller $\lambda(a[n])$, the larger $\Lambda(a[n])$. In other words, $\Lambda(a[n])$ is maximum as $\phi_a(\omega) = 0$ for all ω.*

Following the same procedure for proving Fact 6.19 as presented in [17], one can easily show that Fact 6.19 is also true if $a[n]$ is complex (Problem 6.7). Some properties of users' spreading sequences [7, 8], which are needed to prove the identification criterion given by (6.83)) below, are given in the following fact (Problem 6.8):

Fact 6.20. *Each spreading sequence $c_k[n] \in \mathcal{R}$ (see (6.52)) is basically a pseudorandom (approximately allpass) sequence with approximately random phase and autocorrelation function $c_k[n] \star c_k[-n] \simeq \mathcal{P}\delta[n]$ (or $|C_k(\omega)|^2 \simeq \mathcal{P}$), and uncorrelated with $c_i[n] \in \mathcal{R}$ for $i \neq k$. Moreover, $c_i[n] = c_k[n]$ if $\phi_{c_i}(\omega) = \phi_{c_k}(\omega)$ (with linear phase term removed) where $\phi_{c_i}(\omega) = \arg\{\mathscr{F}\{c_i[n]\}\}$.*

Assume that $e[n]$ and $\widehat{\mathbf{h}}_k^{(3)}[n]$ are the estimates of $\alpha_k u_k[n-\tau_k]$ and $\mathbf{h}_k^{(3)}[n+\tau_k]/\alpha_k$, respectively, obtained using the MIMO hybrid MNC algorithm, where the user number k is unknown. Let $\widehat{h}_k[n]$ be the (chip rate) "signature waveform" estimate associated with $\widehat{\mathbf{h}}_k^{(3)}[n]$, i.e.

$$\widehat{h}_k[n\mathcal{P} + i - 1] = \widehat{h}_{i,k}^{(3)}[n] \quad \text{(by (6.71))}.$$

Therefore,

$$\widehat{h}_k[n] \simeq \frac{1}{\alpha_k} h_k[n + \tau_k \mathcal{P}] \tag{6.81}$$

[1] That is, the linear term in the Taylor series expansion of $\phi_a(\omega)$ is equal to zero.

where $h_k[n]$ is the true (chip rate) signature waveform for user k (see (6.69)). Let

$$a_{k,i}[n] = \frac{\widehat{h}_k[n]}{\sqrt{\sum_n \left|\widehat{h}_k[n]\right|^2}} \star c_i[-n], \quad c_i[n] \in \mathcal{R}. \tag{6.82}$$

Then $\lambda(a_{k,i}[n])$ (see (6.78)) can be expressed as

$$\begin{aligned}
\lambda(a_{k,i}[n]) &\simeq \frac{1}{\alpha_k \sqrt{\sum_n \left|\widehat{h}_k[n]\right|^2}} \cdot \int_{-\pi}^{\pi} |F_k(\omega)| \cdot [\phi_{f_k}(\omega) + \phi_{c_k}(\omega) - \phi_{c_i}(\omega)]^2 d\omega \\
&\simeq \frac{1}{\sqrt{\sum_n |h_k[n]|^2}} \cdot (\lambda_1 + \lambda_2 + \lambda_3) \quad \text{(by (6.69) and (6.81))}
\end{aligned}$$

where $F_k(\omega) = \mathcal{F}\{f_k[n]\}$, $\phi_{f_k}(\omega) = \arg\{F_k(\omega)\}$ and

$$\begin{aligned}
\lambda_1 &= \int_{-\pi}^{\pi} |F_k(\omega)| \cdot [\phi_{c_k}(\omega) - \phi_{c_i}(\omega)]^2 d\omega, \\
\lambda_2 &= \int_{-\pi}^{\pi} |F_k(\omega)| \cdot [\phi_{f_k}(\omega)]^2 d\omega, \\
\lambda_3 &= 2\int_{-\pi}^{\pi} |F_k(\omega)| \cdot \phi_{f_k}(\omega) \cdot [\phi_{c_k}(\omega) - \phi_{c_i}(\omega)] d\omega.
\end{aligned}$$

Note that λ_2 is a constant (not a function of $\phi_{c_i}(\omega)$) and that $|\lambda_3| \ll \lambda_1$ when $\phi_{c_k}(\omega) \neq \phi_{c_i}(\omega)$ due to approximately random phases $\phi_{c_k}(\omega)$ and $\phi_{c_i}(\omega)$ by Fact 6.20. Therefore, $\lambda(a_{k,i}[n])$ is minimum and $\Lambda(a_{k,i}[n])$ is maximum when $c_i[n] = c_k[n]$ by Fact 6.19 that leads to the following UIA:

UIA

(S1) Calculate $\Lambda(a_{k,i}[n])$ $\forall$ $c_i[n] \in \mathcal{R}$ using (6.79) and (6.82).
(S2) Identify $e[n]$ with $\widehat{u}_k[n]$ where the user number associated with the signature sequence $c_k[n]$ is determined by

$$\widehat{k} = \arg\max_i \{\Lambda(a_{k,i}[n]), \ \forall \ c_i[n] \in \mathcal{R}\}. \tag{6.83}$$

Multistage Multiuser Detection Algorithm

The above signal processing for estimating the symbol sequence of interest (assumed to be $u_1[n]$) constitutes the following BMMD algorithm with the $\mathbf{y}^{(3)}[n]$ used:

BMMD Algorithm-(3)

(S1) Set $\Bbbk = 1$ (stage number).

(S2) Find the optimum $\mathbf{v}_{\rm MNC}[n]$ using the MIMO hybrid MNC algorithm from $\mathbf{y}^{(3)}[n]$, $n = 0, 1, ..., N-1$, with the initial condition $\widetilde{\boldsymbol{v}}^{[0]}$ given by (6.76), and the associated $e[n]$, and obtain the associated $\widehat{\mathbf{h}}_k^{(3)}[n]$ using the IOCC method.

(S3) Update $\mathbf{y}^{(3)}[n]$ by $\mathbf{y}^{(3)}[n] - \widehat{\mathbf{h}}_k^{(3)}[n] \star e[n]$.

(S4) Determine the user number k using the UIA. If $\widehat{k} \neq 1$, update $\Bbbk$ by $\Bbbk+1$ and $\boldsymbol{\mathcal{C}}$ (given by (6.74)) as the one with $\mathbf{C}_{\widehat{k}}$ removed and then go to (S2); otherwise $\widehat{u}_1[n] = e[n]$ has been obtained at stage $\Bbbk$.

The $\widehat{u}_1[n]$ obtained is free from error propagation as $\Bbbk = 1$. As the power of user 1 is sufficient, $\widehat{u}_1[n]$ can always be obtained at stage $\Bbbk = 1$ due to the initial condition $\widetilde{\boldsymbol{v}}^{[0]}$ given by (6.76) used in (S2). However, it may happen that $\Bbbk > 1$ as user 1 is a weak user and NFR is high. The smaller the $\Bbbk$, the more accurate the estimate $\widehat{u}_1[n]$. In other words, $\Bbbk$ also provides some information for power control, i.e. demand for raising the power of the desired user for larger $\Bbbk$.

Example 6.21 (BMMD over Chip Waveform Matched Filtering)
Let us present some simulation results using the BMMD Algorithm-(3) (associated with Model III) for a six-user ($K = 6$) asynchronous DS/CDMA system with three paths for each user ($M_k = 3, \forall k$). Thirty independent runs for data length $N = 2500$ were performed with Gold codes of length $\mathcal{P} = 31$ for users' spreading codes. The symbol sequences $u_k[n]$, $k = 1, 2, ..., 6$ were assumed to be equally probable binary random sequences of $\{+1, -1\}$, and the synthetic signal $\mathbf{y}^{(3)}[n]$ was generated with $\mathcal{E}_k^{(3)} = \mathcal{E}$, $k = 2, 3, ..., 6$.

The averaged output SINRs of user 1 are shown in Fig. 6.17. One can observe, from Fig. 6.17, that the performance of the BMMD Algorithm-(3) is close to that of the nonblind LMMSE equalizer for NFR$= \mathcal{E}/\mathcal{E}_1^{(3)} = 0$ dB (i.e. the near–far problem does not exist), but the latter is superior to the former for NFR $= 10$ dB (high NFR). The performance degradation of the BMMD Algorithm-(3) for high NFR (see Fig. 6.17b) results from the error propagation in the MSC procedure because of more stages in the MSC procedure involved.

□

6.6.4 Multiple Antennas Based Multiuser Detection

Consider a K-user asynchronous DS/CDMA communication system in the presence of multiple paths where the source signals sent from the K active users arrive as planewaves at a J-element antenna array. The received baseband continuous-time $\mathrm{J} \times 1$ signal vector $\mathbf{y}(t) = (y_1(t), y_2(t), ..., y_{\mathrm{J}}(t))^T$ can be expressed as

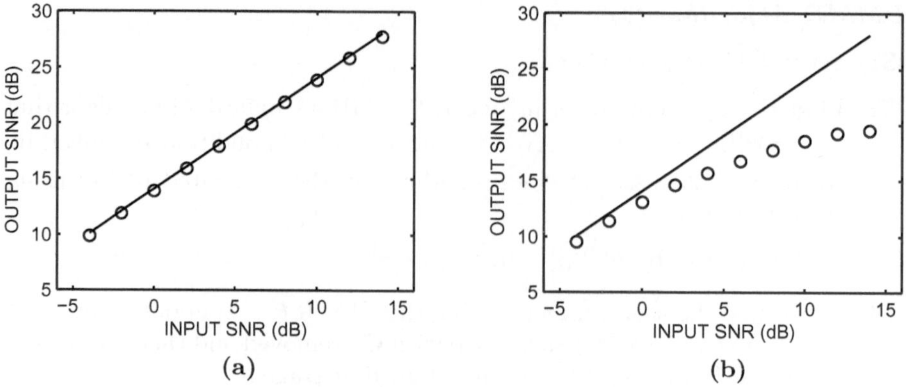

Fig. 6.17 Averaged output SINR for user 1 associated with the nonblind LMMSE equalizer (solid line) and the BMMD Algorithm-(3) ('○') for **(a)** NFR = 0 dB and **(b)** NFR = 10 dB, respectively

$$\mathbf{y}(t) = \sum_{k=1}^{K} \sum_{m=1}^{M_k} \sum_{n=-\infty}^{\infty} \mathbf{a}(\theta_{k,m}) A_{k,m} u_k[n] s_k(t - nT - \tau_{k,m}) + \mathbf{w}(t) \quad (6.84)$$

where $\theta_{k,m}$ and $\mathbf{a}(\theta_{k,m}) = (\mathrm{a}_1(\theta_{k,m}), \mathrm{a}_2(\theta_{k,m}), ..., \mathrm{a}_\mathrm{J}(\theta_{k,m}))^T$ are the direction of arrival and the $\mathrm{J} \times 1$ steering vector of the mth path associated with user k, respectively, $\mathbf{w}(t) = (w_1(t), w_2(t), ..., w_\mathrm{J}(t))^T$ is a zero-mean white Gaussian noise vector, and all the other parameters have been defined in (6.51). A discrete-time model can be obtained from $\mathbf{y}(t)$ as

$$\begin{aligned}\mathbf{y}^{(\ell)}[n] &= \left(\mathbf{y}_1^{(\ell)}[n], \mathbf{y}_2^{(\ell)}[n], ..., \mathbf{y}_\mathrm{J}^{(\ell)}[n]\right)^T \\ &= \mathbf{H}^{(\ell)}[n] \star \mathbf{u}[n] + \mathbf{w}^{(\ell)}[n], \quad \ell = 1, 2, 3,\end{aligned} \quad (6.85)$$

where

$$\mathbf{y}_j^{(\ell)}[n] = \mathbf{H}_j^{(\ell)}[n] \star \mathbf{u}[n] + \mathbf{w}_j^{(\ell)}[n] \quad (6.86)$$

is obtained through the same procedure as obtained for the single antenna (see (6.56), (6.62) and (6.70)), and

$$\begin{aligned}\mathbf{H}^{(\ell)}[n] &\triangleq \left(\mathbf{h}_1^{(\ell)}[n], \mathbf{h}_2^{(\ell)}[n], ..., \mathbf{h}_K^{(\ell)}[n]\right) \\ &= \left((\mathbf{H}_1^{(\ell)}[n])^T, (\mathbf{H}_2^{(\ell)}[n])^T, ..., (\mathbf{H}_\mathrm{J}^{(\ell)}[n])^T\right)^T.\end{aligned} \quad (6.87)$$

When j ($\leq \mathrm{J}$) antennas are used for multiuser detection, a straightforward algorithm, called the BMMD-BMRC(j) Algorithm, is introduced as depicted in Fig. 6.18, which comprises j parallel signal processing channels using the BMMD Algorithm-(ℓ) (with only temporal processing involved) followed by appropriate compensation of relative time delays (as presented in Section 6.5.3) and a BMRC procedure (as introduced in Section 6.2) over j antennas (with only spatial processing involved) to obtain the maximum SNR estimate of the desired user's symbol sequence. Note that the BMMD-BMRC(1) Algorithm is exactly the same BMMD Algorithm-(ℓ).

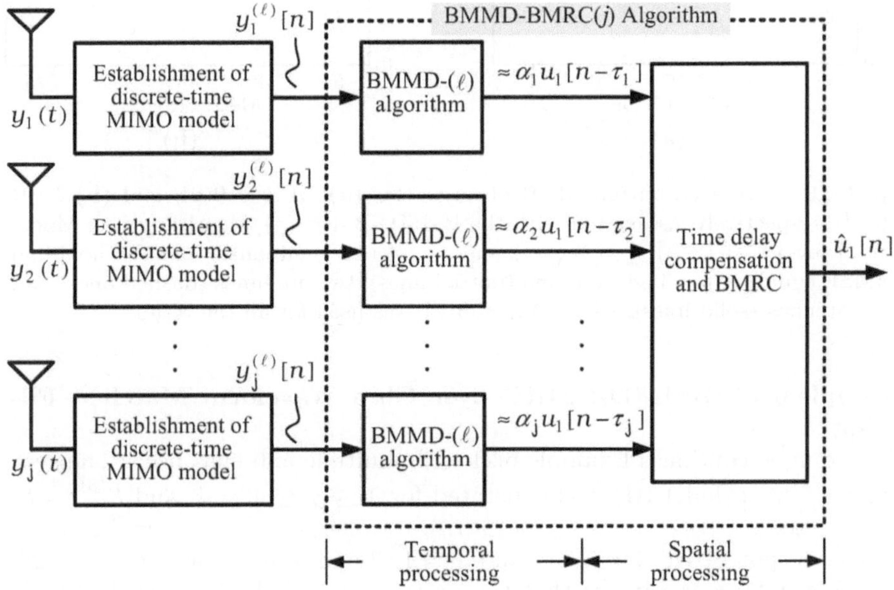

Fig. 6.18 The BMMD-BMRC(j) Algorithm using j antennas

Example 6.22 (BMMD-BMRC over Signature Waveform Matched Filtering)

This example continues Example 6.18 with multiple antennas used. The synthetic $\mathbf{y}_j^{(1)}[n]$ (Model I) and $\mathbf{y}_j^{(2)}[n]$ (Model II) were generated for $1 \leq j \leq \mathtt{J} = 4$, and $\mathcal{E}_k^{(1)} = \mathcal{E}$, $k = 2, ..., 5$.

The averaged output SINRs of user 1 (the weak user) are shown in Fig. 6.19 associated with Model II for J = 4 and j = 1, 2 and 4. One can observe, from Fig. 6.19, that the performance of the BMMD-BMRC(j) algorithm is close to that of the nonblind LMMSE equalizer, and that the output SINR of user 1 is higher for larger j (i.e. more antennas or more space diversity). □

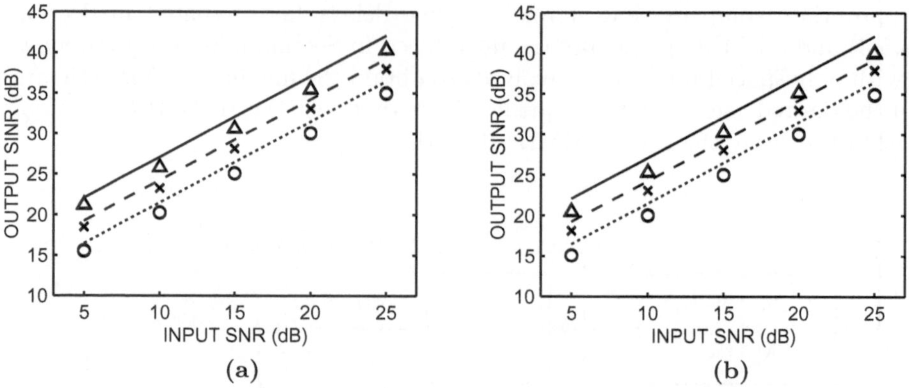

Fig. 6.19 Averaged output SINR of user 1 for **(a)** NFR = 0 dB and **(b)** NFR = 9 dB, respectively, associated with the BMMD-BMRC(j) Algorithm with Model II used for j = 1 ('○'), j = 2 ('×') and j = 4 ('△') antennas, and the nonblind LMMSE equalizer for single antenna (dotted lines), two antennas (dashed lines) and four antennas (solid lines), where $\mathcal{M}_k = 3\ \forall k$ was used for all the results

Example 6.23 (BMMD-BMRC over Chip Waveform Matched Filtering)
This example continues Example 6.21 with multiple antennas used. The synthetic $\mathbf{y}_j^{(3)}[n]$ (Model III) was generated for $1 \leq j \leq \mathtt{J} = 4$, and $\mathcal{E}_k^{(3)} = \mathcal{E}$, $k = 2, ..., 6$.

The output SINRs for J = 4 and j = 1, 2 and 4 are shown in Fig. 6.20. From Fig. 6.20, one can see that the performance of the BMMD-BMRC(j) Algorithm and the nonblind LMMSE equalizer can be significantly improved by using more antennas, even though the latter is superior to the former for high NFR. The performance degradation of the BMMD-BMRC(j) Algorithm for high NFR (see Fig. 6.20b) results from the error propagation in the MSC procedure because of more stages involved in the MSC procedure associated with the BMMD Algorithm-(3). Specifically, the antenna 2 performs significantly worse than the other three antennas for high NFR, leading to less contribution to performance improvement, as observed in Fig. 6.20b.

□

6.7 Summary and Discussion

This chapter introduced some applications of the MIMO hybrid MNC equalization algorithm in signal processing and wireless communications. For the SIMO case, we began with the FSE for digital communications where a set of real cable data and a set of real modem data were used to validate the efficacy of this algorithm. The applications of this equalization algorithm to BMRC,

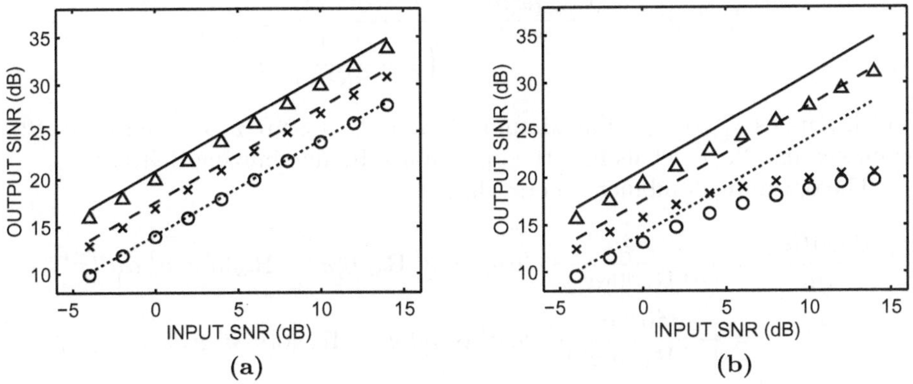

Fig. 6.20 Averaged output SINR of user 1 for **(a)** NFR = 0 dB and **(b)** NFR = 10 dB, respectively, associated with the BMMD-BMRC(j) Algorithm with Model III used for j = 1 ('○'), j = 2 ('×') and j = 4 ('△') antennas, and the nonblind LMMSE equalizer for single antenna (dotted lines), two antennas (dashed lines) and four antennas (solid lines)

BSI and MTDE were then introduced. On the other hand, for the MIMO case, the applications of this equalization algorithm include blind beamforming for source separation and blind multiuser detection in DS/CDMA wireless communications. Some simulation examples were provided to justify the good performance of this equalization algorithm.

In each of the applications presented, a discrete-time SIMO or an MIMO model must be established before use of the MIMO hybrid MNC equalization algorithm, in addition to certain constraints, structures, and considerations on the channel or the equalizer. This implies that proper MIMO model establishment is paramount and crucial to whether the application of this equalization algorithm is successful, and analytic relations if any are always a boost to the results obtained using this blind equalization algorithm. Moreover, other signal processing techniques depending on the specific application may also be needed in conjunction with the MIMO hybrid MNC equalization algorithm. The applications of the MIMO hybrid MNC equalization algorithm described here are not comprehensive, and many other applications in science and engineering involving MIMO blind equalization are continually reported in the open literature. Certainly, the MIMO hybrid MNC equalization algorithm is never the unique choice, although it can be an excellent candidate.

Appendix 6A Proof of Theorem 6.3

Substituting (6.11) into (5.73) and simplifying the resultant equation, one can obtain

$$J_{p,q}(e[n]) = \frac{|C_{p,q}\{x[n]\}|}{\sigma_x^{(p+q)}} \cdot \frac{1}{\left(1 + \dfrac{1}{\mathrm{SNR}(\boldsymbol{v})}\right)^{(p+q)/2}}$$

which implies $J_{p,q}(e[n])$ is maximum if and only if $\mathrm{SNR}(\boldsymbol{v})$ defined by (6.12) is maximum. Next, let us find the optimum $\boldsymbol{v}$ by maximizing $\mathrm{SNR}(\boldsymbol{v})$.

The optimum combiner occurs when

$$\begin{aligned}\frac{\partial \mathrm{SNR}(\boldsymbol{v})}{\partial \boldsymbol{v}} &= \frac{\sigma_x^2}{(\boldsymbol{v}^T \mathbf{R_w}[0]\boldsymbol{v}^*)^2} \cdot [\boldsymbol{a}\boldsymbol{a}^H \boldsymbol{v}^* \boldsymbol{v}^T \mathbf{R_w}[0]\boldsymbol{v}^* - \mathbf{R_w}[0]\boldsymbol{v}^*\boldsymbol{v}^T \boldsymbol{a}\boldsymbol{a}^H \boldsymbol{v}^*] \\ &= \frac{\sigma_x^2 \boldsymbol{a}^H \boldsymbol{v}^*}{(\boldsymbol{v}^T \mathbf{R_w}[0]\boldsymbol{v}^*)^2} \cdot [\boldsymbol{a}\boldsymbol{v}^T \mathbf{R_w}[0]\boldsymbol{v}^* - \mathbf{R_w}[0]\boldsymbol{v}^*\boldsymbol{v}^T \boldsymbol{a}] = 0\end{aligned}$$

or

$$(\boldsymbol{v}^T \boldsymbol{a}) \cdot \mathbf{R_w}[0]\boldsymbol{v}^* = (\boldsymbol{v}^T \mathbf{R_w}[0]\boldsymbol{v}^*) \cdot \boldsymbol{a}$$

which leads to the optimum $\boldsymbol{v}_{\mathrm{MNC}}$ given by (6.13) with $\|\boldsymbol{v}_{\mathrm{MNC}}\| = 1$. Then, substituting $\boldsymbol{v}_{\mathrm{MNC}}$ given by (6.13) into (6.12) gives rise to the maximum $\mathrm{SNR}(\boldsymbol{v})$ given by (6.14).

It can easily be shown using the orthogonality principle that the LMMSE estimator which minimizes $E\{|e[n] - x[n]|^2\}$ is given by

$$\boldsymbol{v}_{\mathrm{MS}} = \left\{\mathbf{R_y}^{-1} E\{\mathbf{y}[n]x^*[n]\}\right\}^* = \sigma_x^2 \cdot \left(\sigma_x^2 \cdot \boldsymbol{a}^* \boldsymbol{a}^T + \mathbf{R}_\mathbf{w}^*[0]\right)^{-1} \cdot \boldsymbol{a}^*. \tag{6.88}$$

With the use of Woodbury's identity (see Corollary 2.6), $\boldsymbol{v}_{\mathrm{MS}}$ given by (6.88) can easily be shown to be

$$\begin{aligned}\boldsymbol{v}_{\mathrm{MS}} &= \frac{\sigma_x^2}{1 + \sigma_x^2 \cdot \boldsymbol{a}^T (\mathbf{R}_\mathbf{w}^*[0])^{-1} \boldsymbol{a}^*} \cdot (\mathbf{R}_\mathbf{w}^*[0])^{-1} \boldsymbol{a}^* \\ &= \frac{1}{\lambda} \cdot \boldsymbol{v}_{\mathrm{MNC}} \text{ (by (6.13))},\end{aligned}$$

where

$$\lambda = \frac{1 + \sigma_x^2 \cdot \boldsymbol{a}^T (\mathbf{R}_\mathbf{w}^*[0])^{-1} \boldsymbol{a}^*}{\sigma_x^2 \cdot \|(\mathbf{R}_\mathbf{w}^*[0])^{-1} \boldsymbol{a}^*\|}. \tag{6.89}$$

Thus we have completed the proof.

Q.E.D.

Appendix 6B Proof of Fact 6.4

Let $\beta = \boldsymbol{a}^T \boldsymbol{v}^{[0]} \neq 0$. Then

$$e^{[0]}[n] = (\boldsymbol{v}^{[0]})^T \mathbf{y}[n] = \beta x[n] + (\boldsymbol{v}^{[0]})^T \mathbf{w}[n]. \tag{6.90}$$

At the first iteration of the MIMO hybrid MNC equalization algorithm with $p = q = 2$, $\boldsymbol{v}^{[1]}$ is updated by

$$\boldsymbol{v}^{[1]} = \frac{(\mathbf{R}_{\mathbf{y}}^*)^{-1}\mathbf{d}_{\mathbf{ey}}^{[0]}}{\left\|(\mathbf{R}_{\mathbf{y}}^*)^{-1}\mathbf{d}_{\mathbf{ey}}^{[0]}\right\|} \quad \text{(by (5.76))} \tag{6.91}$$

where

$$\begin{aligned}\mathbf{d}_{\mathbf{ey}}^{[0]} &= \mathrm{cum}\{e^{[0]}[n], e^{[0]}[n], (e^{[0]}[n])^*, \mathbf{y}^*[n]\} \\ &= |\beta|^2\beta C_{2,2}\{x[n]\} \cdot \boldsymbol{a}^*. \quad \text{(by (6.90))}\end{aligned} \tag{6.92}$$

Moreover, it can easily be seen from (6.88), (6.91) and (6.92) that

$$\boldsymbol{v}^{[1]} = \lambda \boldsymbol{v}_{\mathrm{MS}} = \boldsymbol{v}_{\mathrm{MNC}} \quad \text{(by Theorem 6.3)}$$

where λ is given by (6.89). Therefore, we have completed the proof.

Q.E.D.

Appendix 6C
Proof of Property 6.10

Let $\widetilde{\mathbf{h}}[n]$ be an arbitrary system satisfying (6.20) and thus we have

$$G_p(\omega)\boldsymbol{H}^*(\omega) = \widetilde{G}_p(\omega)\widetilde{\boldsymbol{H}}^*(\omega) = \beta \boldsymbol{\mathcal{S}}_{\mathbf{y}}^T(\omega)\boldsymbol{V}(\omega). \tag{6.93}$$

Without loss of generality, let us assume that both $G(\omega)$ and $\widetilde{G}(\omega)$ are zero phase with positive $g[0]$ and $\widetilde{g}[0]$, i.e.

$$g[n] = g^*[-n], \quad \text{with } g[0] > 0, \tag{6.94}$$
$$\widetilde{g}[n] = \widetilde{g}^*[-n], \quad \text{with } \widetilde{g}[0] > 0. \tag{6.95}$$

It can be obtained, from (6.93), that

$$\widetilde{\boldsymbol{H}}(\omega) = \Gamma(\omega) \cdot \boldsymbol{H}(\omega) \tag{6.96}$$

where

$$\Gamma(\omega) = \frac{G_p^*(\omega)}{\widetilde{G}_p^*(\omega)} = \frac{G_p(\omega)}{\widetilde{G}_p(\omega)} > 0 \quad \text{(by Property 6.9)} \tag{6.97}$$

and that

$$\begin{aligned}G_p(\omega)\boldsymbol{V}^H(\omega)\boldsymbol{H}^*(\omega) &= G_p(\omega)G^*(\omega) = \widetilde{G}_p(\omega)\boldsymbol{V}^H(\omega)\widetilde{\boldsymbol{H}}^*(\omega) \\ &= \widetilde{G}_p(\omega)\widetilde{G}^*(\omega) \geq 0 \quad \text{(by (C-SIMO) and Property 6.9)}\end{aligned} \tag{6.98}$$

Let $s[n]$ be the inverse Fourier transform of $G_p(\omega)G^*(\omega)$, i.e.

$$s[n] = g_p[n] \star g^*[-n] = \sum_{l=-\infty}^{\infty} |g[l]|^m g[l] g^*[l-n] \quad \text{(by (6.21))} \tag{6.99}$$

where $m = 2(p-1)$ and $p \geq 2$. One can easily infer from (6.98) that

$$s[n] = \widetilde{s}[n] \tag{6.100}$$

where $s[n] = g_p[n] \star g^*[-n]$ and $\widetilde{s}[n] = \widetilde{g}_p[n] \star \widetilde{g}^*[-n]$ as given by (6.99).

Let us further assume that $g[n] \neq 0$ for all $n \in [-L, L]$ and thus $g_p[n] \neq 0$ only for $n \in [-L, L]$ by (6.21), and $s[n] \neq 0$ only for $n \in [-2L, 2L]$. Then $s[n]$ given by (6.99) can be expressed as

$$s[n] = \sum_{l=-L}^{L} |g[l]|^m g[l] g^*[l-n], \quad n = -2L, -2L+1, ..., 2L-1, 2L, \tag{6.101}$$

and the equality $s[n] = \widetilde{s}[n]$ implies that $\widetilde{g}[n] \neq 0$ and $\widetilde{g}_p[n] \neq 0$ only for $n \in [-L, L]$, and $\widetilde{s}[n] \neq 0$ only for $n \in [-2L, 2L]$. Furthermore, it can be seen from (6.101) and (6.100) that

$$s[2L] = |g[L]|^m g^2[L] = \widetilde{s}[2L] = |\widetilde{g}[L]|^m \widetilde{g}^2[L],$$

which implies

$$g[L] = \widetilde{g}[L] \text{ or } -\widetilde{g}[L]. \tag{6.102}$$

Again, by (6.101), simplifying $s[2L-1] = \widetilde{s}[2L-1]$ (by (6.100)) results in

$$g[L]g[L-1]\{|g[L]|^m + |g[L-1]|^m\} = \widetilde{g}[L]\widetilde{g}[L-1]\{|\widetilde{g}[L]|^m + |\widetilde{g}[L-1]|^m\}. \tag{6.103}$$

It can be shown from (6.102), (6.103) and Lemma 6.11 (with $a = |g[L-1]|$, $b = |\widetilde{g}[L-1]|$, and $c = |g[L]| = |\widetilde{g}[L]|$), that

$$|g[L-1]| = |\widetilde{g}[L-1]|. \tag{6.104}$$

Moreover, it can be inferred from (6.102), (6.103) and (6.104) that

$$\frac{\widetilde{g}[L-1]}{g[L-1]} = \frac{g[L]}{\widetilde{g}[L]}.$$

By the same fashion, simplifying $s[n] = \widetilde{s}[n]$ (by (6.100)) for $n = 2L-2, 2L-3, ..., L$, one can also prove, by (6.101) and Lemma 6.11, that

$$|g[n]| = |\widetilde{g}[n]|, \quad n = L, L-1, ..., 1, 0$$

and

$$\frac{\widetilde{g}[n]}{g[n]} = \frac{g[L]}{\widetilde{g}[L]}, \quad n = L-1, L-2, ..., 1, 0,$$

which together with (6.102) leads to

$$g[n] = \widetilde{g}[n], \quad \forall n \in [0, L] \quad (\text{since } g[0] > 0 \text{ and } \widetilde{g}[0] > 0). \tag{6.105}$$

Moreover, one can infer from (6.94), (6.95) and (6.105) that

$$g[n] = \widetilde{g}[n], \quad \forall n \in [-L, L]. \tag{6.106}$$

It can easily be seen from (6.106) and (6.21) that $G_p(\omega) = \widetilde{G}_p(\omega)$, which gives rise to $\Gamma(\omega) = 1, \forall\omega$ (by (6.97)). Therefore, one can obtain from (6.96) that

$$\widetilde{\boldsymbol{H}}(\omega) = \boldsymbol{H}(\omega) \tag{6.107}$$

under the zero-phase assumption for both $g[n]$ and $\widetilde{g}[n]$. Furthermore, by Property 6.8 and (6.107), $\boldsymbol{H}'(\omega) = \widetilde{\boldsymbol{H}}(\omega) \cdot e^{j(\omega\tau+\varphi)} = \boldsymbol{H}(\omega) \cdot e^{j(\omega\tau+\varphi)}$ is also a solution of (6.20) under the constraint (C-SIMO). The assumption that $g[n] \neq 0$ only for $n \in [-L, L]$ can be relaxed by allowing $L \to \infty$. However, a general proof without the assumption that $g[n] \neq 0$ for all $n \in [-L, L]$ is still unknown.

Q.E.D.

Appendix 6D Multichannel Levinson Recursion Algorithm

Assume that a multivariable process $\mathbf{y}[n]$ can be described by a multichannel AR(L_p) process as

$$\mathbf{y}[n] = -\sum_{i=1}^{L_p} \mathbf{A}[i]\mathbf{y}[n-i] + \mathbf{s}[n] \tag{6.108}$$

where $\mathbf{s}[n]$ is a zero-mean white input process with

$$\mathbf{R_s}[l] = \mathbf{R_s}[0]\delta[l]. \tag{6.109}$$

Hence the power spectral matrix of $\mathbf{y}[n]$ is given by

$$\boldsymbol{\mathcal{S}}_{\mathbf{y}}(\omega) = \boldsymbol{\mathcal{A}}^{-1}(\omega)\mathbf{R_s}[0]\left(\boldsymbol{\mathcal{A}}^H(\omega)\right)^{-1} \tag{6.110}$$

where

$$\boldsymbol{\mathcal{A}}(\omega) = \mathbf{I} + \sum_{n=1}^{L_p} \mathbf{A}[n]e^{-j\omega n}. \tag{6.111}$$

To estimate $\boldsymbol{\mathcal{S}}_{\mathbf{y}}(\omega)$ using (6.110) requires estimating $\mathbf{A}[n]$ and $\mathbf{R_s}[0]$. Let us define the following notation:

$\mathbf{A}_k^f[n]$: kth order multichannel forward linear predictor

$\mathbf{A}_k^b[n]$: kth order multichannel backward linear predictor

$\mathbf{K}_k^f$: multichannel reflection coefficient matrix associated with $\mathbf{A}_k^f[n]$

$\mathbf{K}_k^b$: multichannel reflection coefficient matrix associated with $\mathbf{A}_k^b[n]$

$\boldsymbol{\Sigma}_k^f$: multichannel prediction error power matrix associated with $\mathbf{A}_k^f[n]$

$\boldsymbol{\Sigma}_k^b$: multichannel prediction error power matrix associated with $\mathbf{A}_k^b[n]$.

The multichannel Levinson recursion algorithm [18], an efficient algorithm to obtain $\mathbf{A}[n]$ and $\mathbf{R}_{\mathbf{s}}[0]$ from the correlation matrix $\mathbf{R}_{\mathbf{y}}[l]$, is summarized as follows:

Multichannel Levinson Recursion Algorithm:

(T1) Initialization.
Set

$$\begin{aligned}\mathbf{A}_1^f[0] &= \mathbf{A}_1^b[0] = \mathbf{I}\\ \mathbf{A}_1^f[1] &= \mathbf{K}_1^f = -\mathbf{R}_{\mathbf{y}}[1]\mathbf{R}_{\mathbf{y}}^{-1}[0]\\ \mathbf{A}_1^b[1] &= \mathbf{K}_1^b = -\mathbf{R}_{\mathbf{y}}[-1]\mathbf{R}_{\mathbf{y}}^{-1}[0]\\ \boldsymbol{\Sigma}_1^f &= \left(\mathbf{I} - \mathbf{K}_1^f\mathbf{K}_1^b\right)\mathbf{R}_{\mathbf{y}}[0]\\ \boldsymbol{\Sigma}_1^b &= \left(\mathbf{I} - \mathbf{K}_1^b\mathbf{K}_1^f\right)\mathbf{R}_{\mathbf{y}}[0].\end{aligned}$$

(T2) Recursion.
For $k = 2, 3, ..., L_p$, compute

(S1) Reflection coefficient matrices:

$$\begin{aligned}\mathbf{K}_k^f &= -\boldsymbol{\Delta}_k(\boldsymbol{\Sigma}_{k-1}^b)^{-1}\\ \mathbf{K}_k^b &= -\boldsymbol{\Delta}_k^H(\boldsymbol{\Sigma}_{k-1}^f)^{-1}\end{aligned}$$

where

$$\boldsymbol{\Delta}_k = \sum_{i=0}^{k-1}\mathbf{A}_{k-1}^f[i]\mathbf{R}_{\mathbf{y}}[k-i].$$

(S2) Predictor coefficient matrices:

$$\mathbf{A}_k^f[i] = \begin{cases}\mathbf{A}_{k-1}^f[i] + \mathbf{K}_k^f\mathbf{A}_{k-1}^b[k-i], & i = 1, 2, ..., k-1,\\ \mathbf{K}_k^f, & i = k.\end{cases}$$

$$\mathbf{A}_k^b[i] = \begin{cases}\mathbf{A}_{k-1}^b[i] + \mathbf{K}_k^b\mathbf{A}_{k-1}^f[k-i], & i = 1, 2, ..., k-1,\\ \mathbf{K}_k^b, & i = k.\end{cases}$$

(S3) Prediction error power matrices:

$$\mathbf{\Sigma}_k^f = \left(\mathbf{I} - \mathbf{K}_k^f \mathbf{K}_k^b\right) \mathbf{\Sigma}_{k-1}^f.$$

$$\mathbf{\Sigma}_k^b = \left(\mathbf{I} - \mathbf{K}_k^b \mathbf{K}_k^f\right) \mathbf{\Sigma}_{k-1}^b.$$

Then, $\mathbf{A}[n] = \mathbf{A}_{L_p}^f[n]$ and $\mathbf{R}_\mathbf{s}[0] = \mathbf{\Sigma}_{L_p}^f$ are obtained.

Appendix 6E Integrated Bispectrum Based Time Delay Estimation

Consider the case of the model given by (6.33) for $M = 2$ as follows

$$\mathbf{y}[n] = (y_1[n], y_2[n])^T = (x[n], x[n-d])^T + \mathbf{w}[n]. \tag{6.112}$$

For ease of later use, let $S_{y_1 y_2}(\omega)$ denote the cross-spectrum of $y_1[n]$ and $y_2[n]$ and $S_{y_1 y_2}[k] = S_{y_1 y_2}(\omega_k)$ where $\omega_k = 2\pi k/\mathcal{N}$, and $\alpha_k = S^*_{y_1^2 y_2}[k]/S^*_{y_1^2 y_1}[k]$.

The IBBTDE proposed by Ye and Tugnait [5] estimates the time delay d in (6.112) by minimizing the following cost:

$$J_{\rm IBBTDE}(d) = \sum_{k=1}^{\frac{\mathcal{N}}{2}-1} \frac{1}{\sigma_k^2} \cdot \left| \widehat{H}[k] - e^{-j\omega_k d} \right|^2 \tag{6.113}$$

where

$$\widehat{H}[k] = \left|\widehat{H}[k]\right| \cdot e^{j\widehat{\phi}[k]} = \frac{S_{y_1^2 y_2}[k]}{S_{y_1^2 y_1}[k]}, \tag{6.114}$$

and

$$\sigma_k^2 = \frac{S_{y_1^2 y_1^2}[k] S_{y_2 y_2}[k]}{\left|S_{y_1^2 y_1}[k]\right|^2} \cdot \left(1 + |\alpha_k|^2 \cdot \frac{S_{y_1^2 y_1^2}[k]}{S_{y_2 y_2}[k]} - 2\mathrm{Re}\left\{\alpha_k \cdot \frac{S_{y_1^2 y_2}[k]}{S_{y_2 y_2}[k]}\right\}\right). \tag{6.115}$$

Specifically, instead of minimizing $J_{\rm IBBTDE}(d)$ given by (6.113) via nonlinear iterative optimization for a continuous range of values of d, a closed-form solution of d can be obtained through the following steps [5]:

(S1) Estimate $S_{y_1^2 y_2}[k]$, $S_{y_1^2 y_1}[k]$ and σ_k^2 from the finite data $\mathbf{y}[n]$, $n = 0, 1, ..., N-1$. Compute $\widehat{\phi}[k] = \measuredangle\{S_{y_1^2 y_2}[k]/S_{y_1^2 y_1}[k]\}$.

(S2) Compute

$$F[k] = \begin{cases} -\widetilde{F}[k], & k \text{ is odd}, \\ \widetilde{F}[k], & k \text{ is even} \end{cases} \tag{6.116}$$

where

$$\widetilde{F}[k] = \begin{cases} \frac{1}{\sigma_k^2} \cdot \frac{S_{y_1^2 y_2}[k]}{S_{y_1^2 y_1}[k]}, & 1 \le k \le \frac{\mathcal{N}}{2} - 1, \\ 0, & \frac{\mathcal{N}}{2} \le k \le \mathcal{L} \end{cases} \tag{6.117}$$

in which $\mathcal{L} = \mathcal{P}\mathcal{N}$ and $\mathcal{P}$ is a positive integer.

(S3) Find $n_{\min}$ such that

$$J(n) = -\text{Re}\{f[n]\}, \ 0 \le n \le \mathcal{L} - 1, \tag{6.118}$$

is minimum for $n = n_{\min}$, where $f[n]$ is the inverse $\mathcal{L}$-point DFT of $F[k]$. Then, compute

$$\overline{d} = \frac{n_{\min} - \mathcal{L}/2}{\mathcal{L}}. \tag{6.119}$$

(S4) Obtain the time delay estimate $\widehat{d}$ as

$$\widehat{d} = \frac{\sum_{k=1}^{\frac{\mathcal{N}}{2}-1} \omega_k \widehat{\phi}'[k]/\sigma_k^2}{\sum_{k=1}^{\frac{\mathcal{N}}{2}-1} \omega_k^2/\sigma_k^2} + \overline{d}, \tag{6.120}$$

where

$$\widehat{\phi}'[k] = \widehat{\phi}[k] - \omega_k \overline{d}. \tag{6.121}$$

The cross-periodogram can be used for the estimation of cross-spectra, $S_{y_1^2 y_2}[k]$ and $S_{y_1^2 y_1}[k]$ in (S1), from the given data $\mathbf{y}[n]$, $n = 0, 1, ..., N-1$. Divide the given sample sequence of length N into $\mathbb{B}$ nonoverlapping segments each of length $\mathcal{N}$ so that $N = \mathbb{B}\mathcal{N}$. Let $A^{(i)}[k]$ denote the DFT of the ith segment $\{y_1^2[n + (i-1)\mathcal{N}], 0 \le n \le \mathcal{N} - 1\}$, $i \in \{1, 2, ..., \mathbb{B}\}$ given by

$$A^{(i)}[k] = \sum_{n=0}^{\mathcal{N}-1} y_1^2[n + (i-1)\mathcal{N}] e^{-j2\pi kn/\mathcal{N}}, \quad k = 0, 1, ..., \mathcal{N} - 1. \tag{6.122}$$

Similarly, let $B^{(i)}[k]$ denote the DFT of the ith segment $\{y_2[n + (i-1)\mathcal{N}], 0 \le n \le \mathcal{N} - 1\}$, $i \in \{1, 2, ..., \mathbb{B}\}$. Then, the cross-periodogram for the ith segment of data is given by

$$\widehat{S}_{y_1^2 y_2}^{(i)}[k] = \frac{1}{\mathcal{N}} A^{(i)}[k] \left(B^{(i)}[k]\right)^*. \tag{6.123}$$

Finally, the estimate $\widehat{S}_{y_1^2 y_2}[k]$ is given by averaging over $\mathbb{B}$ segments as

$$\widehat{S}_{y_1^2 y_2}[k] = \frac{1}{\mathbb{B}} \sum_{i=1}^{\mathbb{B}} \widehat{S}^{(i)}_{y_1^2 y_2}[k]. \tag{6.124}$$

On the other hand, the cross-periodogram associated with $S_{y_1^2 y_1}[k]$ can be obtained in the same fashion with $y_2[n]$ replaced by $y_1[n]$ in computing $B^{(i)}[k]$ above.

Problems

6.1. Prove Fact 6.6.

6.2. Prove Fact 6.7.

6.3. Prove Property 6.8.

6.4. Prove Property 6.9.

6.5. Prove Theorem 6.14.

6.6. Prove Fact 6.15.

6.7. Prove Fact 6.19 for complex $a[n]$.

6.8. Prove Fact 6.20.

Computer Assignments

6.1. Perform fractionally spaced equalization using the hybrid MIMO-MNC equalization algorithm with the real data posted on the following on-line Signal Processing Information Base (SPIB) web sites:
(a) Modem data (http://spib.rice.edu/spib/modem.html)
(b) Cable data (http://spib.rice.edu/spib/cable.html)

6.2. Write a computer program to perform the same simulation as presented in Example 6.5 for BMRC using the hybrid MIMO-MNC equalization algorithm.

6.3. Write a computer program to perform the same simulation as presented in Example 6.12 for blind SIMO system identification using the SIMO BSI algorithm introduced in Section 6.3.3.

6.4. Write a computer program to perform the same simulation as presented in Example 6.13 for multiple time delay estimation using the MTDE algorithm introduced in Section 6.4.2.

6.5. Write a computer program to perform the same simulation as presented in Example 6.16 for blind beamforming using the MIMO hybrid MNC equalization algorithm.

6.6. Write a computer program to perform the same simulation as presented in Example 6.17 for blind source separation using the MSS Algorithm introduced in Section 6.5.3.

6.7. Write a computer program to perform the same simulation as presented in Example 6.18 for multiuser detection in DS/CDMA systems using the BMMD Algorithm-(ℓ) with $\ell = 1$ and 2 introduced in Section 6.6.2.

6.8. Write a computer program to perform the same simulation as presented in Example 6.21 for multiuser detection in DS/CDMA systems using the BMMD Algorithm-(ℓ) with $\ell = 3$ introduced in Section 6.6.3.

6.9. Write a computer program to perform the same simulation as presented in Examples 6.22 and 6.23 for blind multiuser detection in DS/CDMA systems using the BMMD-BMRC Algorithm introduced in Section 6.6.4.

References

1. C.-H. Chen, "Blind Multi-channel Equalization and Two-dimensional System Identification Using Higher-order Statistics," Ph.D. Dissertation, Department of Electrical Engineering, National Tsing Hua University, Taiwan, R.O.C., June 2001.
2. Z. Ding and T. Nguyen, "Stationary points of a kurtosis maximization algorithm for blind signal separation and antenna beamforming," *IEEE Trans. Signal Processing*, vol. 48, no. 6, pp. 1587–1596, June 2000.
3. C.-Y. Chi, C.-Y. Chen, C.-H. Chen, C.-C. Feng, and C.-H. Peng, "Blind identification of SIMO systems and simultaneous estimation of multiple time delays from HOS-based inverse filter criteria," *IEEE Trans. Signal Processing*, vol. 52, no. 10, pp. 2749–2761, Oct. 2004.
4. S. M. Kay, *Modern Spectral Estimation.* New Jersey: Prentice-Hall, 1998.
5. Y. Ye and J. K. Tugnait, "Performance analysis of integrated polyspectrum based time delay estimators," in *Proc. IEEE International Conference on Acoustics, Speech, and Signal Processing*, Detroit, MI, USA, May 9–12, 1995, pp. 3147–3150.

6. C.-Y. Chi and C.-Y. Chen, "Blind beamforming and maximum ratio combining by kurtosis maximization for source separation in multipath," in *Proc. Third IEEE Workshop on Signal Processing Advances in Wireless Communications*, Taoyuan, Taiwan, March 20–23, 2001, pp. 243–246.
7. S. Verdu, *Multiuser Detection.* Cambridge: Cambridge University Press, 1998.
8. H. V. Poor and G. W. Wornell, *Wireless Communications: Signal Processing Perspective.* New Jersey: Prentice-Hall, 1998.
9. H. Liu, *Signal Processing Applications in CDMA Communications.* Boston: Artech House, Inc., 2000.
10. A. J. Paulraj and C. B. Padadias, "Space-time processing for wireless communications," *IEEE Signal Processing Magazine*, vol. 14, no. 6, pp. 49–83, Nov. 1997.
11. M. K. Tsatsanis and G. B. Giannakis, "Optimal decorrelating receivers for DS-CDMA systems: A signal processing framework," *IEEE Trans. Signal Processing*, vol. 44, no. 12, pp. 3044–3055, Dec. 1996.
12. M. K. Tsatsanis, "Inverse filtering criteria for CDMA systems," *IEEE Trans. Signal Processing*, vol. 45, no. 1, pp. 102–112, Jan. 1997.
13. M. K. Tsatsanis and Z. Xu, "Performance analysis of minimum variance CDMA receivers," *IEEE Trans. Signal Processing*, vol. 46, no. 11, pp. 3014–3022, Nov. 1998.
14. C.-Y. Chi, C.-H. Chen, and C.-Y. Chen, "Blind MAI and ISI suppression for DS/CDMA systems using HOS-based inverse filter criteria," *IEEE Trans. Signal Processing*, vol. 50, no. 6, pp. 1368–1381, June 2002.
15. G. Leus, "Signal Processing Algorithms for CDMA-Based Wireless Communications," Ph.D. Dissertation, Faculty of Applied Sciences, K. U. Leuven, Leuven, Belgium, 2000.
16. M. K. Tsatsanis and G. B. Giannakis, "Blind estimation of direct sequence spread spectrum signals in multipath," *IEEE Trans. Signal Processing*, vol. 45, no. 5, pp. 1241–1251, May 1997.
17. H.-M. Chien, H.-L. Yang, and C.-Y. Chi, "Parametric cumulant based phase estimation of 1-D and 2-D nonminimum phase systems by allpass filtering," *IEEE Trans. Signal Processing*, vol. 45, no. 7, pp. 1742–1762, July 1997.
18. R. A. Wiggins and E. A. Robinson, "Recursive solution to the multichannel filtering problem," *Journal of Geophysical Research*, vol. 70, pp. 1885–1891, Apr. 1965.

7

Two-Dimensional Blind Deconvolution Algorithms

Two-dimensional (2-D) blind deconvolution processing is essential in 2-D statistical signal processing areas such as *image restoration*, *image model identification*, *texture synthesis*, *texture image classification*, and so forth. This chapter provides an introduction to some 2-D deconvolution algorithms that we believe are effective in these applications. This chapter begins with a review of 2-D deterministic signals, systems and linear random processes. Two well-known 2-D deconvolution criteria, PD criterion and MMSE criterion are presented followed by the widely used 2-D LPE filtering approach using SOS. Then 2-D deconvolution algorithms using HOS are introduced, including MNC and SE algorithms, and their properties and relations to the MMSE algorithm. Finally, a 2-D hybrid blind deconvolution algorithm based on these properties and relations is introduced. Applications of these algorithms are left to the next chapter.

7.1 Two-Dimensional Discrete-Space Signals, Systems and Random Processes

7.1.1 2-D Deterministic Signals

A 2-D discrete-space signal is discrete in space and will be denoted by a function whose two arguments are integers. For instance, $x[n_1, n_2]$ represents a 2-D signal defined for all integer values of n_1 and n_2, while $x[n_1, n_2]$ is not defined for nonintegers n_1 and n_2. The notation $x[n_1, n_2]$ refers to either the discrete-space function x or the value of the function x at a specific $[n_1, n_2]$.

Some specific 2-D signals, which are indispensable in 2-D signal processing, including the 2-D *Kronecker delta function*, *separable signals*, *periodic signals* and *stable signals* are defined as follows.

2-D Kronecker Delta Function

The 2-D Kronecker delta function (or unit sample signal), denoted by $\delta[n_1, n_2]$, is defined by

$$\delta[n_1, n_2] = \begin{cases} 1, & n_1 = n_2 = 0 \\ 0, & \text{otherwise.} \end{cases} \tag{7.1}$$

Any sequence $x[n_1, n_2]$ can be represented by a linear combination of shifted-versions of $\delta[n_1, n_2]$ as follows:

$$x[n_1, n_2] = \sum_{k_1=-\infty}^{\infty} \sum_{k_2=-\infty}^{\infty} x[k_1, k_2]\delta[n_1 - k_1, n_2 - k_2]. \tag{7.2}$$

2-D Separable Signals

A 2-D signal $x[n_1, n_2]$ is said to be a separable signal if it can be expressed as

$$x[n_1, n_2] = a[n_1]b[n_2] \tag{7.3}$$

where $a[n_1]$ and $b[n_2]$ are functions of only n_1 and n_2, respectively. For instance, the 2-D Kronecker delta function $\delta[n_1, n_2]$ is a separable signal because

$$\delta[n_1, n_2] = \delta[n_1]\delta[n_2] \tag{7.4}$$

where $\delta[n]$ is the 1-D Kronecker delta function.

2-D Periodic Signals

A 2-D signal $x[n_1, n_2]$ is said to be periodic with a period of $N_1 \times N_2$ (where N_1 and N_2 are nonnegative integers) if $x[n_1, n_2]$ satisfies

$$x[n_1, n_2] = x[n_1 + k_1 N_1, n_2 + k_2 N_2] \quad \text{for all } [n_1, n_2] \tag{7.5}$$

where k_1 and k_2 are both integers. For example, $x[n_1, n_2] = \cos[(\pi/2)n_1 + (\pi/3)n_2] = x[n_1 + 4k_1, n_2 + 6k_2]$ for any integers k_1 and k_2 is a periodic signal with period 4×6.

2-D Stable Signals

A 2-D signal $x[n_1, n_2]$ is said to be stable if it is absolutely summable, i.e.

$$\sum_{k_1=-\infty}^{\infty} \sum_{k_2=-\infty}^{\infty} |x[k_1, k_2]| < \infty. \tag{7.6}$$

For instance, any finite-length 2-D signal is stable because it satisfies (7.6).

7.1.2 2-D Transforms

2-D z-Transform

The 2-D *z-transform* of a discrete-space signal $x[n_1, n_2]$ is defined as

$$X(z_1, z_2) = \mathcal{Z}\{x[n_1, n_2]\} = \sum_{n_1=-\infty}^{\infty} \sum_{n_2=-\infty}^{\infty} x[n_1, n_2] z_1^{-n_1} z_2^{-n_2} \tag{7.7}$$

where z_1 and z_2 are complex variables. From (7.7), it follows that

$$|X(z_1, z_2)| \le \sum_{n_1=-\infty}^{\infty} \sum_{n_2=-\infty}^{\infty} |x[n_1, n_2]| \cdot |z_1|^{-n_1} |z_2|^{-n_2}, \tag{7.8}$$

which implies that the convergence of the series given by (7.7) depends on $|z_1|$ and $|z_2|$. For instance, if $x[n_1, n_2]$ is a stable signal, the ROC of $X(z_1, z_2)$ includes the *unit surface* (i.e. $z_1 = e^{j\omega_1}, z_2 = e^{j\omega_2}$). The *inverse z-transform* $x[n_1, n_2] = \mathcal{Z}^{-1}\{X(z_1, z_2)\}$ can be uniquely specified by both $X(z_1, z_2)$ and its ROC.

2-D Discrete-Space Fourier Transform

The *2-D discrete-space Fourier transform (DSFT)* of a 2-D discrete-space signal $x[n_1, n_2]$ is defined as

$$X(\omega_1, \omega_2) = \mathcal{F}\{x[n_1, n_2]\} = \sum_{n_1=-\infty}^{\infty} \sum_{n_2=-\infty}^{\infty} x[n_1, n_2] e^{-j\omega_1 n_1} e^{-j\omega_2 n_2}. \tag{7.9}$$

It is obvious that $X(\omega_1, \omega_2)$ is periodic with period $2\pi \times 2\pi$, i.e. $X(\omega_1, \omega_2) = X(\omega_1 + 2k_1\pi, \omega_2 + 2k_2\pi)$ for all integers k_1 and k_2. The 2-D DSFT is also called the 2-D Fourier Transform (FT) for short. The 2-D *inverse DSFT (IDSFT)* of $X(\omega_1, \omega_2)$ is nothing but the Fourier series expansion of $X(\omega_1, \omega_2)$ given by

$$\begin{aligned} x[n_1, n_2] &= \mathcal{F}^{-1}\{X(\omega_1, \omega_2)\} \\ &= \frac{1}{(2\pi)^2} \int_{-\pi}^{\pi} \int_{-\pi}^{\pi} X(\omega_1, \omega_2) e^{j\omega_1 n_1} e^{j\omega_2 n_2} d\omega_1 d\omega_2. \end{aligned} \tag{7.10}$$

As $x[n_1, n_2]$ is stable,

$$\sum_{n_1=-N_1}^{N_1} \sum_{n_2=-N_2}^{N_2} x[n_1, n_2] e^{-j\omega_1 n_1} e^{-j\omega_2 n_2}$$

converges uniformly and absolutely to $X(\omega_1, \omega_2)$ as $N_1 \to \infty$ and $N_2 \to \infty$, and meanwhile $X(\omega_1, \omega_2)$ is continuous. As $x[n_1, n_2]$ is absolutely square summable but not stable,

$$\int_{-\pi}^{\pi}\int_{-\pi}^{\pi}\left|\sum_{n_1=-N_1}^{N_1}\sum_{n_2=-N_2}^{N_2} x[n_1,n_2]e^{-j\omega_1 n_1}e^{-j\omega_2 n_2} - X(\omega_1,\omega_2)\right|^2 d\omega_1 d\omega_2 \to 0 \tag{7.11}$$

as $N_1 \to \infty$ and $N_2 \to \infty$, (i.e. mean-square convergence), and then $X(\omega_1, \omega_2)$ must have discontinuities. On the other hand, it can be inferred that if $X(\omega_1, \omega_2)$ is continuous, then $x[n_1, n_2]$ is stable, or equivalently absolutely summable. Moreover, comparing (7.9) with (7.7), one can see that if the ROC of $X(z_1, z_2)$ includes the unit surface, then $X(\omega_1, \omega_2) = X(z_1 = e^{j\omega_1}, z_2 = e^{j\omega_2})$ exists.

2-D Discrete Fourier Transform

The DSFT and the IDSFT are very useful in theoretical analyses and algorithm development, but calculations according to (7.9) and (7.10) need substantial computation power and thus are not very practical since ω_1 and ω_2 are continuous variables. Assume that $x[n_1, n_2]$, $n_1 = 0, 1, ..., N_1 - 1$, $n_2 = 0, 1, ..., N_2 - 1$ is a finite-duration sequence of length $N_1 \times N_2$, its DSFT $X(\omega_1, \omega_2)$ can be computed at the discrete values of radian frequency $(\omega_1 = 2\pi k_1/N_1, \omega_2 = 2\pi k_2/N_2)$ via the $N_1 \times N_2$-point DFT defined as

$$X[k_1,k_2] = \text{DFT}\{x[n_1,n_2]\} = \sum_{n_1=0}^{N_1-1}\sum_{n_2=0}^{N_2-1} x[n_1,n_2]W_{N_1}^{k_1 n_1}W_{N_2}^{k_2 n_2},$$
$$0 \le k_1 \le N_1 - 1, \;\; 0 \le k_2 \le N_2 - 1, \tag{7.12}$$

where $W_N = \exp\{-j2\pi/N\}$ is the twiddle factor. The 2-D IDFT of $X[k_1, k_2]$, $k_1 = 0, 1, ..., N_1 - 1$, $k_2 = 0, 1, ..., N_2 - 1$ is given by

$$\begin{aligned} x[n_1,n_2] &= \text{IDFT}\{X[k_1,k_2]\} \\ &= \frac{1}{N_1 N_2}\sum_{k_1=0}^{N_1-1}\sum_{k_2=0}^{N_2-1} X[k_1,k_2]W_{N_1}^{-k_1 n_1}W_{N_2}^{-k_2 n_2}, \end{aligned}$$
$$0 \le n_1 \le N_1 - 1, \;\; 0 \le n_2 \le N_2 - 1. \tag{7.13}$$

As $x[n_1, n_2]$ is a finite-extent sequence with length $N_1' \times N_2'$ smaller than or equal to $N_1 \times N_2$ ($N_1' \le N_1$ and $N_2' \le N_2$), $X[k_1, k_2]$ given by (7.12) can be used to completely specify $x[n_1, n_2]$ via the $N_1 \times N_2$-point IDFT.

One useful property of the 2-D DFT is the *circular convolution* of two finite-extent sequences $x_1[n_1, n_2]$ with length $N_1' \times N_2'$ and $x_2[n_1, n_2]$ with length $N_1'' \times N_2''$. Assuming that $N_1 \ge N_1'$, $N_1 \ge N_1''$ and $N_2 \ge N_2'$, $N_2 \ge N_2''$, the $N_1 \times N_2$-point circular convolution of $x_1[n_1, n_2]$ and $x_2[n_1, n_2]$ is defined by

$$x_1[n_1, n_2] \circledast x_2[n_1, n_2] = \sum_{k_1=0}^{N_1-1} \sum_{k_2=0}^{N_2-1} x_1[k_1, k_2] x_2[((n_1 - k_1))_{N_1}, ((n_2 - k_2))_{N_2}]. \quad (7.14)$$

Let $X_1[k_1, k_2]$ and $X_2[k_1, k_2]$ denote the $N_1 \times N_2$-point DFT of $x_1[n_1, n_2]$ and $x_2[n_1, n_2]$, respectively. Then

$$\mathrm{DFT}\{x_1[n_1, n_2] \circledast x_2[n_1, n_2]\} = X_1[k_1, k_2] X_2[k_1, k_2].$$

7.1.3 2-D Linear Shift-Invariant Systems

For a 2-D LSI system $h[n_1, n_2]$, the relation between the input $x[n_1, n_2]$ and output $y[n_1, n_2]$ can be simply expressed as the following 2-D linear convolution

$$y[n_1, n_2] = x[n_1, n_2] \star h[n_1, n_2] = \sum_{k_1=-\infty}^{\infty} \sum_{k_2=-\infty}^{\infty} x[k_1, k_2] h[n_1 - k_1, n_2 - k_2]. \quad (7.15)$$

As in the 1-D case, 2-D convolution also possesses several properties such as *commutativity, associativity* and *distributivity.* For simplicity, the 2-D system also refers to the 2-D LSI system hereafter.

Let $X[k_1, k_2]$ and $H[k_1, k_2]$ denote the $N_1 \times N_2$-point DFTs of $x[n_1, n_2]$ with length $N_1' \times N_2'$ and $h[n_1, n_2]$ with length $N_1'' \times N_2''$, respectively. Then, $y[n_1, n_2] = x[n_1, n_2] \star h[n_1, n_2] = x[n_1, n_2] \circledast h[n_1, n_2]$ for $0 \le n_1 \le N_1 - 1$, $0 \le n_2 \le N_2 - 1$ (i.e. the linear convolution and the circular convolution are equivalent) as long as $N_1 \ge N_1' + N_1'' - 1$ and $N_2 \ge N_2' + N_2'' - 1$.

2-D Stable Systems

For a 2-D system with impulse response $h[n_1, n_2]$, it is BIBO stable if and only if $h[n_1, n_2]$ is absolutely summable, i.e.

$$\sum_{k_1=-\infty}^{\infty} \sum_{k_2=-\infty}^{\infty} |h[k_1, k_2]| < \infty. \quad (7.16)$$

Although the above condition (7.16) for system stability is a straightforward extension of the 1-D case, the stability test for 2-D systems is generally much more complicated than for 1-D systems and is beyond the scope of this book.

2-D Causal Systems and Special Support Systems

A 1-D causal LTI system with impulse response $h[n]$ zero for $n < 0$ is preferred in applications where real-time processing is needed. In contrast to the 1-D case, the causality for 2-D systems is based on how "*past*", "*present*" and "*future*" are defined in terms of *region of support*, whereas real-time processing may not be a major concern in most 2-D signal processing applications.

For a specific space index $[k_1, k_2]$, the past of $[k_1, k_2]$ is defined as the set [1]

$$\Omega_{\mathrm{P}}[k_1, k_2] = \{[n_1, n_2] | n_1 = k_1, n_2 < k_2; n_1 < k_1, -\infty < n_2 < \infty\} \tag{7.17}$$

and the future of $[k_1, k_2]$ is defined as the set

$$\Omega_{\mathrm{F}}[k_1, k_2] = \{[n_1, n_2] | n_1 = k_1, n_2 > k_2; n_1 > k_1, -\infty < n_2 < \infty\} \tag{7.18}$$

as shown in Fig. 7.1. A 2-D system $h[n_1, n_2]$ is said to be causal (or *one-sided*) when $h[n_1, n_2] = 0$ for all $[n_1, n_2] \in \Omega_{\mathrm{P}}[0, 0]$, or its region of support is $\Omega_{\mathrm{F}}[0, -1]$.

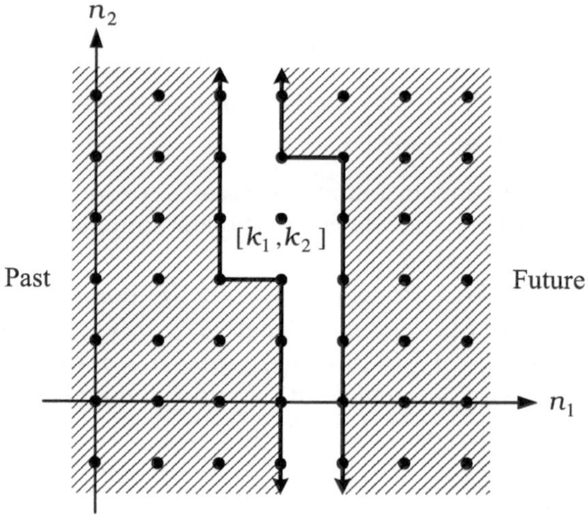

Fig. 7.1 The definition of "present", "future" and "past" of a specific space index $[k_1, k_2]$

For ease of later use, define the *nonsymmetric half plane (NSHP)* as

$$\Omega_{\mathrm{NSHP}} = \Omega_{\mathrm{F}}[0, -1], \tag{7.19}$$

and the *truncated NSHP (TNSHP)* as [1–4]

$$\Omega_{\mathrm{TNSHP}}[p_1, p_2] = \Omega_{\mathrm{NSHP}} \cap \{[n_1, n_2] : n_1 \leq p_1, |n_2| \leq p_2\}. \tag{7.20}$$

Note that Ω_{NSHP} is also the region of support of a 2-D causal system $h[n_1, n_2]$. Define the *quarter plane (QP)* as

$$\Omega_{\mathrm{QP}} = \{[n_1, n_2] : n_1 \geq 0, n_2 \geq 0\}, \tag{7.21}$$

and the *truncated QP (TQP)* as [1–4]

$$\Omega_{\mathrm{TQP}}[p_1, p_2] = \Omega_{\mathrm{QP}} \cap \{[n_1, n_2] : n_1 \le p_1, n_2 \le p_2\}. \tag{7.22}$$

Figure 7.2 shows an example of the region of support of $\Omega_{\mathrm{TNSHP}}[3, 2]$ and $\Omega_{\mathrm{TQP}}[3, 2]$. It is easy to see that $\Omega_{\mathrm{TQP}}[3, 2]$ is a subset of $\Omega_{\mathrm{TNSHP}}[3, 2]$.

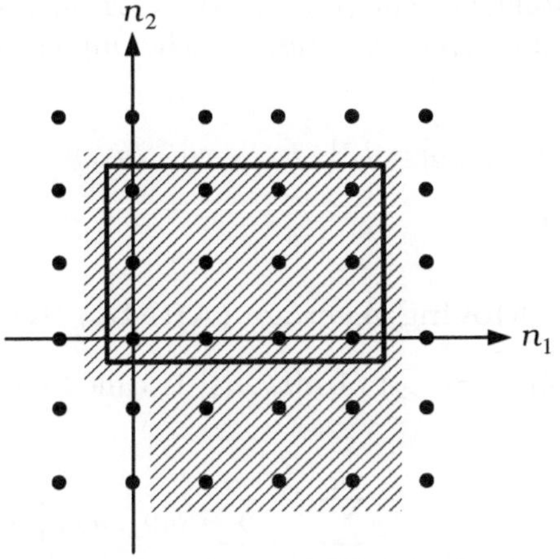

Fig. 7.2 Example of regions of support for $\Omega_{\mathrm{TNSHP}}[3, 2]$ (shaded region) and $\Omega_{\mathrm{TQP}}[3, 2]$ (inside the rectangle)

2-D Minimum-Phase and Nonminimum-Phase Systems

A 2-D system $h[n_1, n_2]$ is said to be *minimum phase* if $h[n_1, n_2]$ is causal (with region of support Ω_{NSHP}) and stable and its inverse system $h_{\mathrm{INV}}[n_1, n_2]$ is also casual and stable.

Assume that $h_{\mathrm{MP}}[n_1, n_2]$ is a minimum-phase system and $h_{\mathrm{NMP}}[n_1, n_2]$ is a nonminimum-phase system. As in the 1-D case, $h_{\mathrm{NMP}}[n_1, n_2]$ can be decomposed as

$$h_{\mathrm{NMP}}[n_1, n_2] = h_{\mathrm{MP}}[n_1, n_2] \star h_{\mathrm{AP}}[n_1, n_2] \tag{7.23}$$

where $h_{\mathrm{AP}}[n_1, n_2]$ is an *allpass system* with magnitude response $|H_{\mathrm{AP}}(\omega_1, \omega_2)|$ equal to a constant. The *minimum-energy delay property* of a 1-D minimum-phase system also applies to $h_{\mathrm{MP}}[n_1, n_2]$, that is, [1]

$$\sum\sum_{[i_1,i_2]\in\Omega[p_1,p_2]} |h_{\mathrm{MP}}[i_1, i_2]|^2 \ge \sum\sum_{[i_1,i_2]\in\Omega[p_1,p_2]} |h_{\mathrm{NMP}}[i_1, i_2]|^2 \tag{7.24}$$

where $\Omega[p_1, p_2] = \Omega_{\mathrm{NSHP}} \cap \Omega_{\mathrm{P}}[p_1, p_2 + 1]$.

2-D Linear Phase Systems

A 2-D LSI system $h[n_1, n_2]$ is said to be *linear phase* if its frequency response $H(\omega_1, \omega_2)$ can be expressed as

$$H(\omega_1, \omega_2) = |H(\omega_1, \omega_2)| e^{j(\omega_1 \tau_1 + \omega_2 \tau_2 + \varsigma)} \tag{7.25}$$

where τ_1, τ_2 and ς are real constants. Furthermore, as $(\tau_1, \tau_2, \varsigma) = (0, 0, 0)$ or $(0, 0, \pi)$, the resultant system ($H(\omega_1, \omega_2) = |H(\omega_1, \omega_2)|$ or $H(\omega_1, \omega_2) = -|H(\omega_1, \omega_2)|$) is called a *zero-phase system*. The impulse response of a zero-phase system satisfies

$$h[n_1, n_2] = h^*[-n_1, -n_2], \ \forall n_1, n_2 \tag{7.26}$$

since $H(\omega_1, \omega_2)$ is real.

2-D Autoregressive Moving-Average (ARMA) Systems

A rational z-transform $H(z_1, z_2)$ of $h[n_1, n_2]$ is called an *ARMA model* if it can be expressed as

$$H(z_1, z_2) = \frac{B(z_1, z_2)}{A(z_1, z_2)} = \frac{\displaystyle\sum_{n_1=-p_1}^{p_1} \sum_{n_2=-p_2}^{p_2} b[n_1, n_2] z_1^{-n_1} z_2^{-n_2}}{\displaystyle\sum_{n_1=-p_1}^{p_1} \sum_{n_2=-p_2}^{p_2} a[n_1, n_2] z_1^{-n_1} z_2^{-n_2}} \tag{7.27}$$

where p_1 and p_2 are nonnegative integers and $A(z_1, z_2)$ and $B(z_1, z_2)$ are z-transforms of $a[n_1, n_2]$ and $b[n_1, n_2]$, respectively. As $B(z_1, z_2)$ is a nonzero constant, $H(z_1, z_2)$ reduces to an AR model, while it reduces to an MA model for $A(z_1, z_2)$ equal to a nonzero constant. The MA model is guaranteed stable. Stability testing of AR and ARMA models can be conducted by one-dimensional root-finding of the 2-D polynomial $A(z_1, z_2)$ [2], which is usually tedious. An alternative model with stability guaranteed is a frequency-domain 2-D FSBM [5–7], to be introduced next.

2-D Fourier Series Based Model

The 2-D parametric nonminimum-phase FSBM for 2-D stable LSI systems is nothing but an extension of the 1-D FSBM introduced in Chapter 3. As in the 1-D FSBM, the 2-D FSBM is guaranteed stable and its complex cepstrum can be obtained from its amplitude and phase parameters by a closed-form formula without involving complicated 2-D phase unwrapping and polynomial rooting.

Assume that $h[n_1, n_2]$ is a real stable 2-D LSI system with frequency response $H(\omega_1, \omega_2)$. The 2-D FSBM for $H(\omega_1, \omega_2)$ can be expressed by the following MG-PS decomposition:

$$H(\omega_1,\omega_2) = H^*(-\omega_1,-\omega_2) = H_{\mathrm{MG}}(\omega_1,\omega_2)\cdot H_{\mathrm{PS}}(\omega_1,\omega_2) \tag{7.28}$$

where $H_{\mathrm{MG}}(\omega_1,\omega_2)$ is a 2-D zero-phase FSBM given by

$$H_{\mathrm{MG}}(\omega_1,\omega_2) = \exp\left\{ \sum\sum_{[i_1,i_2]\in\Omega'_{\mathrm{TNSHP}}[p_1,p_2]} \alpha_{i_1,i_2}\cos(i_1\omega_1 + i_2\omega_2) \right\} \tag{7.29}$$

and $H_{\mathrm{PS}}(\omega_1,\omega_2)$ is a 2-D allpass FSBM given by

$$H_{\mathrm{PS}}(\omega_1,\omega_2) = \exp\left\{ j \sum\sum_{[i_1,i_2]\in\Omega'_{\mathrm{TNSHP}}[p_1,p_2]} \beta_{i_1,i_2}\sin(i_1\omega_1 + i_2\omega_2) \right\} \tag{7.30}$$

in which

$$\begin{aligned} \Omega'_{\mathrm{TNSHP}}[p_1,p_2] &= \{[n_1,n_2]\in\Omega_{\mathrm{TNSHP}}[p_1,p_2], [n_1,n_2]\neq[0,0]\} \\ &= \Omega_{\mathrm{TNSHP}}[p_1,p_2]\cap\Omega_{\mathrm{F}}[0,0] \end{aligned} \tag{7.31}$$

is the region of support associated with both the real amplitude parameters α_{i_1,i_2} and real phase parameters β_{i_1,i_2}. Similar to its 1-D counterpart, the 2-D FSBM given by (7.28) can also be expressed by MP-AP decomposition as follows:

$$H(\omega_1,\omega_2) = H^*(-\omega_1,-\omega_2) = H_{\mathrm{MP}}(\omega_1,\omega_2)\cdot H_{\mathrm{AP}}(\omega_1,\omega_2) \tag{7.32}$$

where $H_{\mathrm{MP}}(\omega_1,\omega_2)$ is a 2-D minimum-phase FSBM given by

$$H_{\mathrm{MP}}(\omega_1,\omega_2) = \exp\left\{ \sum\sum_{[i_1,i_2]\in\Omega'_{\mathrm{TNSHP}}[p_1,p_2]} \alpha_{i_1,i_2} e^{-j(i_1\omega_1+i_2\omega_2)} \right\} \tag{7.33}$$

and $H_{\mathrm{AP}}(\omega_1,\omega_2)$ is a 2-D allpass FSBM given by

$$H_{\mathrm{AP}}(\omega_1,\omega_2) = \exp\left\{ j \sum\sum_{[i_1,i_2]\in\Omega'_{\mathrm{TNSHP}}[p_1,p_2]} (\alpha_{i_1,i_2}+\beta_{i_1,i_2})\sin(i_1\omega_1 + i_2\omega_2) \right\}. \tag{7.34}$$

It can easily be shown that the region of support of the minimum-phase system $h_{\mathrm{MP}}[n_1,n_2]$ given by (7.33) is the right half plane (i.e. Ω_{NSHP}) and $h_{\mathrm{MP}}[0,0] = 1$. This is left as an exercise (Problem 7.1).

The 2-D FSBM given by (7.28) and (7.32) is potentially a better choice for modeling arbitrary 2-D LSI systems than the 2-D ARMA model in statistical signal processing applications thanks to its stability and closed-form formulas for computing its complex cepstrum . The complex cepstrum [8–10] of $h[n_1,n_2]$, denoted $\mathtt{h}[n_1,n_2]$, can easily be shown from the MP-AP decomposition of the 2-D FSBM (see (7.32)) to be

$$\mathtt{h}[n_1,n_2] = \mathscr{F}^{-1}\{\ln\{H(\omega_1,\omega_2)\}\} = \mathtt{h}_{\mathrm{MP}}[n_1,n_2] + \mathtt{h}_{\mathrm{AP}}[n_1,n_2] \tag{7.35}$$

where

$$\mathtt{h}_{\mathrm{MP}}[n_1,n_2] = \begin{cases} \alpha_{n_1,n_2}, & [n_1,n_2] \in \Omega'_{\mathrm{TNSHP}}[p_1,p_2] \\ 0, & \text{otherwise} \end{cases} \tag{7.36}$$

and

$$\mathtt{h}_{\mathrm{AP}}[n_1,n_2] = \begin{cases} -\frac{1}{2}(\alpha_{n_1,n_2} + \beta_{n_1,n_2}), & [n_1,n_2] \in \Omega'_{\mathrm{TNSHP}}[p_1,p_2] \\ \frac{1}{2}(\alpha_{-n_1,-n_2} + \beta_{-n_1,-n_2}), & [-n_1,-n_2] \in \Omega'_{\mathrm{TNSHP}}[p_1,p_2] \\ 0, & \text{otherwise.} \end{cases} \tag{7.37}$$

Similarly, the complex cepstrum $\mathtt{h}[n_1,n_2]$ using the MG-PS decomposition of the 2-D FSBM given by (7.28) can also be expressed as a simple closed-form formula of α_{n_1,n_2} and β_{n_1,n_2} (Problem 7.2). As the 2-D LSI system $h[n_1,n_2]$ (with frequency response $\mathcal{H}(\omega_1,\omega_2)$), however, is not a 2-D FSBM (e.g. a 2-D ARMA model), the larger the chosen values for p_1 and p_2 of the 2-D FSBM $H(\omega_1,\omega_2)$, the better the approximation $H(\omega_1,\omega_2)$ to the true system $\mathcal{H}(\omega_1,\omega_2)$.

As a remark, complex cepstra of speech signals with the vocal tract-filter modeled as a minimum-phase AR model have been widely used in *speech recognition* and *speaker identification* [8–10]. Similarly, the 2-D FSBM model can also be used for modeling texture images [11], and meanwhile its complex cepstrum obtained by (7.35) can be used as features for classification of texture images that will be introduced in Chapter 8.

7.1.4 2-D Stationary Random Processes

Definitions and Notations

Let $x[n_1,n_2]$ be a 2-D stationary complex random process (or random field). The mean and variance of $x[n_1,n_2]$ are respectively defined as

$$m_x = E\{x[n_1,n_2]\},$$
$$\sigma_x^2 = \mathrm{Var}(x[n_1,n_2]) = E\{|x[n_1,n_2]-m_x|^2\} = E\{|x[n_1,n_2]|^2\} - |m_x|^2.$$

The $(p+q)$th-order cumulant of $x[n_1,n_2]$ introduced in Chapter 3 is repeated here for convenience

$$C_{p,q}\{x[n_1,n_2]\} = \mathrm{cum}\{x[n_1,n_2] : p, x^*[n_1,n_2] : q\} \tag{7.38}$$

where p and q are nonnegative integers. Note that $m_x = C_{1,0}\{x[n_1,n_2]\}$ and $\sigma_x^2 = C_{1,1}\{x[n_1,n_2]\}$. For real $x[n_1,n_2]$, $C_{p,q}\{x[n_1,n_2]\}$ is invariant for all (p,q) as long as $(p+q)$ is constant, and can be simplified as

$$\begin{aligned} C_{p+q}\{x[n_1,n_2]\} &= \text{cum}\{x[n_1,n_2]:p, x^*[n_1,n_2]:q\} \\ &= \text{cum}\{x[n_1,n_2]:p+q\}. \end{aligned} \tag{7.39}$$

The *correlation function* of $x[n_1, n_2]$ is defined as

$$r_x[l_1,l_2] = E\{x[n_1,n_2]x^*[n_1-l_1,n_2-l_2]\} \tag{7.40}$$

and its *power spectrum* $S_x(\omega_1,\omega_2) = \mathscr{F}\{r_x[l_1,l_2]\}$.

For ease of later use, we further define

$$\mathbf{n} = [n_1,n_2],\ \mathbf{k} = [k_1,k_2],\ \boldsymbol{l}_i = [l_{i1},l_{i2}],\ \boldsymbol{\omega} = [\omega_1,\omega_2],\ \boldsymbol{\omega}_i = [\omega_{i1},\omega_{i2}]. \tag{7.41}$$

Let $C^x_{p,q}(\boldsymbol{l}_1,...,\boldsymbol{l}_{p+q-1})$ denote the $(p+q)$th-order cumulant function of a 2-D random process $x[\mathbf{n}]$ defined as

$$\begin{aligned} &C^x_{p,q}(\boldsymbol{l}_1,...,\boldsymbol{l}_{p+q-1}) \\ &= \text{cum}\{x[\mathbf{n}],x[\mathbf{n}-\boldsymbol{l}_1],...,x[\mathbf{n}-\boldsymbol{l}_{p-1}],x^*[\mathbf{n}-\boldsymbol{l}_p],...,x^*[\mathbf{n}-\boldsymbol{l}_{p+q-1}]\}. \end{aligned} \tag{7.42}$$

Its $2(p+q-1)$-dimensional FT, denoted $S^x_{p,q}(\boldsymbol{\omega}_1,...,\boldsymbol{\omega}_{p+q-1})$, is called the $(p+q)$th-order *polyspectrum* of $x[\mathbf{n}]$ [12].

As $x[\mathbf{n}]$ is i.i.d. with zero-mean, variance σ_x^2 and nonzero $C_{p,q}\{x[n_1,n_2]\}$, both its power spectrum and $(p+q)$th-order polyspectrum are flat and given by $S_x(\boldsymbol{\omega}) = \sigma_x^2$ and $S^x_{p,q}(\boldsymbol{\omega}_1,...,\boldsymbol{\omega}_{p+q-1}) = C_{p,q}\{x[n_1,n_2]\}$, respectively.

2-D Linear Processes

Consider a 2-D stable LSI system $h[\mathbf{n}]$ driven by a stationary input $x[\mathbf{n}]$ with power spectrum $S_x(\boldsymbol{\omega})$ as depicted by (7.15). The output $y[\mathbf{n}]$ is also a stationary random process. Then the power spectrum $S_y(\boldsymbol{\omega})$ and the $(p+q)$th-order polyspectrum of $y[\mathbf{n}]$ [13] are continuous and given by

$$S_y(\boldsymbol{\omega}) = S_x(\boldsymbol{\omega})|H(\boldsymbol{\omega})|^2 \tag{7.43}$$

and

$$\begin{aligned} &S^y_{p,q}(\boldsymbol{\omega}_1,...,\boldsymbol{\omega}_{p+q-1}) = S^x_{p,q}(\boldsymbol{\omega}_1,...,\boldsymbol{\omega}_{p+q-1}) \\ &\qquad \cdot H\left(\sum_{i=1}^{p+q-1}\boldsymbol{\omega}_i\right)\cdot\prod_{i=1}^{p-1}H(-\boldsymbol{\omega}_i)\cdot\prod_{i=p}^{p+q-1}H^*(\boldsymbol{\omega}_i), \end{aligned} \tag{7.44}$$

respectively.

2-D Sample Correlations and Cumulants

In practice, we are usually given a set of data $y[n_1,n_2]$, $n_1 = 0, 1, ..., N_1-1$, and $n_2 = 0, 1, ..., N_2-1$ for the estimation of correlations and cumulants of $y[n_1, n_2]$. Biased sample correlations and cumulants are widely used as estimates of correlations and cumulants. For instance, assume that $y[n_1, n_2]$ is a 2-D zero-mean stationary random process and then the sample correlation $\widehat{r}_y[l_1, l_2]$, the sample cumulants $\widehat{C}_{2,1}\{y[n_1, n_2]\}$ and $\widehat{C}_{2,2}\{y[n_1, n_2]\}$ can be obtained as follows

$$\widehat{r}_y[l_1, l_2] = \frac{1}{N_1 N_2} \sum_{k_1=0}^{N_1-1} \sum_{k_2=0}^{N_2-1} y[k_1, k_2] y^*[k_1 - l_1, k_2 - l_2], \text{ (by (7.40))} \tag{7.45}$$

$$\widehat{C}_{2,1}\{y[n_1, n_2]\} = \frac{1}{N_1 N_2} \sum_{k_1=0}^{N_1-1} \sum_{k_2=0}^{N_2-1} y^2[k_1, k_2] y^*[k_1, k_2], \text{ (by (3.207))} \tag{7.46}$$

$$\begin{aligned} \widehat{C}_{2,2}\{y[n_1, n_2]\} &= \frac{1}{N_1 N_2} \sum_{k_1=0}^{N_1-1} \sum_{k_2=0}^{N_2-1} |y[k_1, k_2]|^4 - 2 \cdot \widehat{r}_y^2[0, 0] \\ &\quad - \left| \frac{1}{N_1 N_2} \sum_{k_1=0}^{N_1-1} \sum_{k_2=0}^{N_2-1} y^2[k_1, k_2] \right|^2 . \text{ (by (3.208))} \end{aligned} \tag{7.47}$$

As mentioned in Chapter 3, the biased sample correlations and cumulants are consistent and asymptotically unbiased under certain conditions [12]. In other words, the consistency property of the biased sample correlations and cumulants is shared by both 1-D and 2-D stationary linear processes.

7.2 2-D Deconvolution

This section begins with an introduction to the 2-D blind deconvolution problem followed by two widely used deconvolution criteria, i.e. PD and MMSE deconvolution criteria.

7.2.1 Blind Deconvolution Problem

Assume that $y[n_1, n_2]$ is a 2-D discrete-space signal of length $N \times N$ given by

$$y[n_1, n_2] = x[n_1, n_2] + w[n_1, n_2], n_1 = 0, ..., N-1, n_2 = 0, ..., N-1 \tag{7.48}$$

where $x[n_1, n_2]$ is the noise-free output signal of an LSI system $h[n_1, n_2]$ driven by an unknown input signal $u[n_1, n_2]$, i.e.

$$\begin{aligned} x[n_1, n_2] &= u[n_1, n_2] \star h[n_1, n_2] \\ &= \sum_{k_1=-\infty}^{\infty} \sum_{k_2=-\infty}^{\infty} h[k_1, k_2] u[n_1 - k_1, n_2 - k_2] \end{aligned} \tag{7.49}$$

and $w[n_1, n_2]$ is additive noise. Let us make the following general assumptions about $u[n_1, n_2]$, $h[n_1, n_2]$ and $w[n_1, n_2]$, respectively.

(A7-1) The input signal $u[n_1,n_2]$ is stationary complex, zero-mean, i.i.d., nonGaussian with variance σ_u^2 and nonzero $(p+q)$th-order cumulant $C_{p,q}\{u[n_1,n_2]\}$ for nonnegative integers p and q and $(p+q)\geq 3$.

(A7-2) Both $h[n_1,n_2]$ and its inverse system $h_{\rm INV}[n_1,n_2]$ are stable LSI systems.

(A7-3) Noise $w[n_1,n_2]$ is stationary complex, zero-mean, colored Gaussian with variance σ_w^2 and power spectrum $S_w(\omega_1,\omega_2)$. Moreover, $w[n_1,n_2]$ is statistically independent of $u[n_1,n_2]$.

The signal model under the above three assumptions is depicted in Fig. 7.3.

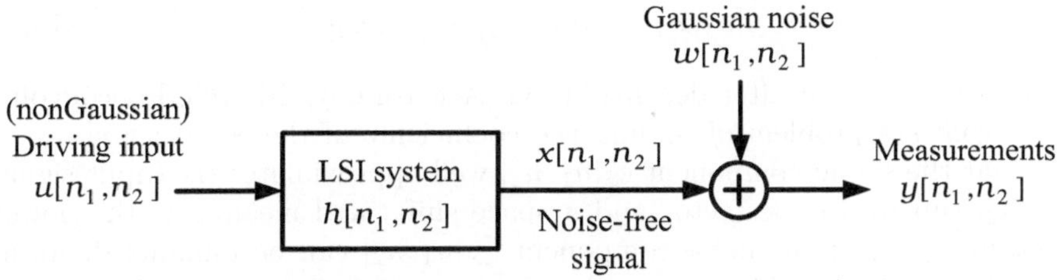

Fig. 7.3 Block diagram for 2-D signal model

Under the Assumptions (A7-1) – (A7-3), the 2-D signal $y[n_1,n_2]$ can be shown to be a stationary nonGaussian random process (or random field) with the following correlation function

$$r_y[l_1,l_2]=\sigma_u^2 h[l_1,l_2]\star h^*[-l_1,-l_2]+r_w[l_1,l_2] \tag{7.50}$$

and power spectrum (2-D DSFT of $r_y[l_1,l_2]$)

$$S_y(\omega_1,\omega_2)=S_x(\omega_1,\omega_2)+S_w(\omega_1,\omega_2)=\sigma_u^2|H(\omega_1,\omega_2)|^2+S_w(\omega_1,\omega_2). \tag{7.51}$$

The signal quality of the 2-D signal given by (7.48) can be quantified by SNR defined as

$$\text{SNR}=\frac{E\{|x[n_1,n_2]|^2\}}{E\{|w[n_1,n_2]|^2\}}=\frac{r_x[0,0]}{r_w[0,0]}. \tag{7.52}$$

2-D Linear Blind Deconvolution

Two-dimensional blind deconvolution is a signal processing procedure for estimating the input signal $u[n_1,n_2]$ up to a scale factor and a space shift only with the given measurements $y[n_1,n_2]$ (without training signals). The estimation of $u[n_1,n_2]$ can be achieved by processing $y[n_1,n_2]$ with a designed linear *deconvolution filter* $v[n_1, n_2]$. Then the deconvolution filter output $e[n_1, n_2]$ can be expressed as

$$e[n_1, n_2] = v[n_1, n_2] \star y[n_1, n_2] = e_S[n_1, n_2] + e_N[n_1, n_2] \tag{7.53}$$

where

$$e_N[n_1, n_2] = v[n_1, n_2] \star w[n_1, n_2] \quad \text{(by (7.48))} \tag{7.54}$$

is the noise component and

$$e_S[n_1, n_2] = v[n_1, n_2] \star x[n_1, n_2] = g[n_1, n_2] \star u[n_1, n_2] \quad \text{(by (7.49))} \tag{7.55}$$

is the corresponding signal component in which

$$g[n_1, n_2] = h[n_1, n_2] \star v[n_1, n_2] \tag{7.56}$$

is the overall system after deconvolution. Accordingly, 2-D blind deconvolution becomes a problem of finding the coefficients of the equalizer $v[n_1, n_2]$ such that the signal component $e_S[n_1, n_2]$ well approximates the input signal $u[n_1, n_2]$ (up to a scale factor and a space shift) and meanwhile the power $E\{|e_N[n_1, n_2]|^2\}$ of the noise component $e_N[n_1, n_2]$ can be minimized. As in the SISO case (see (4.13)), an index for quantifying the approximation of $e_S[n_1, n_2]$ to the input signal $u[n_1, n_2]$ is the amount of ISI defined as [14]

$$\mathrm{ISI}\{g[n_1, n_2]\} = \frac{\displaystyle\sum_{n_1}\sum_{n_2} |g[n_1, n_2]|^2 - \max_{[n_1, n_2]} \{|g[n_1, n_2]|^2\}}{\displaystyle\max_{[n_1, n_2]} \{|g[n_1, n_2]|^2\}}. \tag{7.57}$$

Note that $\mathrm{ISI}\{\alpha\delta[n_1 - \tau_1, n_2 - \tau_2]\} = 0$ and that the smaller the value of $\mathrm{ISI}\{g[n_1, n_2]\}$, the better the overall system $g[n_1, n_2]$ approximates $\alpha\delta[n_1 - \tau_1, n_2 - \tau_2]$.

7.2.2 Peak Distortion and Minimum Mean-Square-Error Deconvolution Criteria

2-D Peak Distortion Deconvolution

The goal of PD deconvolution criterion is to design a filter $v[n_1, n_2]$ such that

$$J_{\mathrm{PD}}(v[n_1, n_2]) = \mathrm{ISI}\{g[n_1, n_2]\} \tag{7.58}$$

is minimum. The optimum deconvolution filter, also referred to as 2-D ZF deconvolution filter, is known as

$$V_{\mathrm{ZF}}(\omega_1, \omega_2) = \alpha \frac{e^{-j(\omega_1\tau_1 + \omega_2\tau_2)}}{H(\omega_1, \omega_2)} = \alpha e^{-j(\omega_1\tau_1 + \omega_2\tau_2)} H_{\mathrm{INV}}(\omega_1, \omega_2) \tag{7.59}$$

where α is a nonzero constant, and τ_1 and τ_2 are integers. Note that $V_{\rm ZF}(\omega_1, \omega_2)$ is usually an IIR filter (with infinite region of support) which is also a 2-D stable LSI filter by Assumption (A7-2), and that $J_{\rm PD}(v_{\rm ZF}[n_1, n_2]) = {\rm ISI}\{\alpha\delta[n_1 - \tau_1, n_2 - \tau_2]\} = 0$.

In practical applications where SNR is finite, the ZF deconvolution filter may result in significant noise enhancement that causes performance degradation. From (7.53), (7.54) and (7.59), the output of the ZF deconvolution filter is given by

$$\begin{aligned} e_{\rm ZF}[n_1, n_2] &= e_{\rm S}[n_1, n_2] + e_{\rm N}[n_1, n_2] \\ &= \alpha u[n_1 - \tau_1, n_2 - \tau_2] + v_{\rm ZF}[n_1, n_2] \star w[n_1, n_2] \end{aligned} \qquad (7.60)$$

and the signal and noise power spectra are given by $S_{e_S}(\omega_1, \omega_2) = |\alpha|^2 \sigma_u^2$ and

$$S_{e_N}(\omega_1, \omega_2) = |V_{\rm ZF}(\omega_1, \omega_2)|^2 S_w(\omega_1, \omega_2) = \frac{|\alpha|^2 S_w(\omega_1, \omega_2)}{|H(\omega_1, \omega_2)|^2}, \qquad (7.61)$$

respectively. It can be seen, from (7.61), that $S_{e_N}(\omega_1, \omega_2)$ can be very large as $|H(\omega_1, \omega_2)|$ is small for some frequencies (ω_1, ω_2), consequently leading to the noise power $E\{|e_{\rm N}[n_1, n_2]|^2\}$ significantly enhanced. In other words, the ZF deconvolution filter never takes the noise reduction into account. The MMSE deconvolution filter, which performs both the ISI suppression and noise reduction simultaneously, is introduced next.

2-D MMSE Deconvolution

The MMSE deconvolution filter $v_{\rm MS}[n_1, n_2]$, known as a Wiener filter, is designed by minimizing the following mean-square-error

$$J_{\rm MSE}(v[n_1, n_2]) = E\{|\alpha u[n_1 - \tau_1, n_2 - \tau_2] - e[n_1, n_2]|^2\} \qquad (7.62)$$

where $e[n_1, n_2]$ is the deconvolution filter output given by (7.53). As $v_{\rm MS}[n_1, n_2]$ is noncausal IIR, by the orthogonality principle, the MMSE deconvolution filter can be easily shown to be

$$\begin{aligned} V_{\rm MS}(\omega_1, \omega_2) &= \frac{\alpha \sigma_u^2 \cdot H^*(\omega_1, \omega_2) e^{-j(\omega_1 \tau_1 + \omega_2 \tau_2)}}{S_y(\omega_1, \omega_2)} \\ &= \frac{\alpha \sigma_u^2 \cdot H^*(\omega_1, \omega_2) e^{-j(\omega_1 \tau_1 + \omega_2 \tau_2)}}{S_x(\omega_1, \omega_2) + S_w(\omega_1, \omega_2)} \end{aligned} \qquad (7.63)$$

which is independent of the choices of $[\tau_1, \tau_2]$ if the linear phase term is ignored. On the other hand, as $v_{\rm MS}[n_1, n_2]$ is confined to an FIR filter, different MMSE deconvolution filters can be obtained for different choices of $[\tau_1, \tau_2]$.

Without loss of generality, let us assume $\tau_1 = \tau_2 = 0$ and $\alpha = 1$. Then $V_{\rm MS}(\omega_1, \omega_2)$ in (7.63) can be further expressed as

$$V_{\rm MS}(\omega_1,\omega_2) = \frac{\sigma_u^2 \cdot |H(\omega_1,\omega_2)|^2}{S_y(\omega_1,\omega_2)} \cdot \frac{1}{H(\omega_1,\omega_2)}$$
$$= V_{\rm NR}(\omega_1,\omega_2) V_{\rm ZF}(\omega_1,\omega_2) \tag{7.64}$$

where

$$V_{\rm NR}(\omega_1,\omega_2) = \frac{\sigma_u^2 \cdot |H(\omega_1,\omega_2)|^2}{S_y(\omega_1,\omega_2)} = \frac{S_x(\omega_1,\omega_2)}{S_y(\omega_1,\omega_2)} \tag{7.65}$$

is actually the optimum *nosie-reduction filter* by minimizing $E\{|x[n_1,n_2] - y[n_1,n_2] \star v[n_1,n_2]|^2\}$. From (7.64), one can observe that the processing of the MMSE deconvolution consists of a signal enhancement processing using the optimum Wiener filter $V_{\rm NR}(\omega_1,\omega_2)$ and an ISI suppression processing using the ZF deconvolution filter in cascade as shown in Fig. 7.4. As a remark, as SNR $= \infty$, $V_{\rm NR}(\omega_1,\omega_2) = 1$, it is easy to see that $V_{\rm MS}(\omega_1,\omega_2)$ reduces to $V_{\rm ZF}(\omega_1,\omega_2)$.

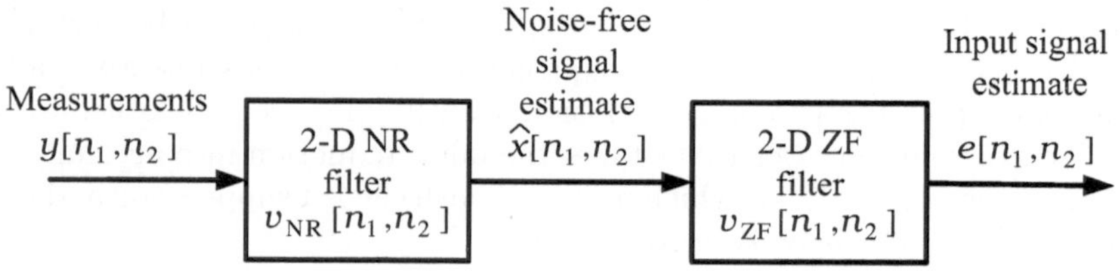

Fig. 7.4 Signal processing of the 2-D MMSE deconvolution

7.3 SOS Based Blind Deconvolution Approach: Linear Prediction

Two-dimensional *linear prediction* concerns the prediction of a stationary random process $y[n_1,n_2]$ through a linear combination of its "*past*" samples. Assume that $v[n_1,n_2]$ is an LSI filter with region of support $\Omega_{\rm TNSHP}[p_1,p_2]$ given by (7.20) and the leading coefficient $v[0,0] = 1$. Let $\mathbf{v}$ be a vector containing all the LPE filter coefficients except $v[0,0]$ (i.e. the coefficients $v[n_1,n_2], \forall [n_1,n_2] \in \Omega'_{\rm TNSHP}[p_1,p_2]$ given by (7.31)) as

$$\mathbf{v} = (\mathrm{v}^T[0] \ \mathrm{v}^T[1] \ \cdots \ \mathrm{v}^T[p_1])^T \tag{7.66}$$

in which

$$\mathrm{v}[0] = (v[0,1], ..., v[0,p_2])^T \tag{7.67}$$
$$\mathrm{v}[i] = (v[i,-p_2], v[i,-p_2+1], ..., v[i,p_2])^T, \ i = 1, \ ..., \ p_1. \tag{7.68}$$

Let $\widehat{y}[n_1, n_2]$ denote the predicted $y[n_1, n_2]$ as follows

$$\begin{aligned}\widehat{y}[n_1, n_2] &= - \sum\sum_{[i_1,i_2]\in\Omega'_{\mathrm{TNSHP}}[p_1,p_2]} v[i_1, i_2] y[n_1 - i_1, n_2 - i_2] \\ &= -\mathbf{v}^T \mathbf{y}[n_1, n_2] \end{aligned} \tag{7.69}$$

where

$$\mathbf{y}[n_1, n_2] = (\mathbf{y}_1^T[n_1, n_2], \mathbf{y}_2^T[n_1 - 1, n_2], ..., \mathbf{y}_2^T[n_1 - p_1, n_2])^T \tag{7.70}$$

in which

$$\mathbf{y}_1[n_1, n_2] = (y[n_1, n_2 - 1], y[n_1, n_2 - 2], ..., y[n_1, n_2 - p_2])^T \tag{7.71}$$

and

$$\mathbf{y}_2[k, n_2] = (y[k, n_2 + p_2], y[k, n_2 + p_2 - 1], ..., y[k, n_2 - p_2])^T, \quad n_1 - p_1 \le k \le n_1 - 1. \tag{7.72}$$

Then the prediction error, denoted by $\varepsilon[n_1, n_2]$, is thus the output of the 2-D filter $v[n_1, n_2]$ driven by $y[n_1, n_2]$, i.e.

$$\begin{aligned}\varepsilon[n_1, n_2] &= y[n_1, n_2] - \widehat{y}[n_1, n_2] = y[n_1, n_2] + \mathbf{v}^T \mathbf{y}[n_1, n_2] \\ &= \sum\sum_{[k_1,k_2]\in\Omega_{\mathrm{TNSHP}}[p_1,p_2]} v[k_1, k_2] y[n_1 - k_1, n_2 - k_2]. \end{aligned} \tag{7.73}$$

The LPE filter $v_{\mathrm{LPE}}[n_1, n_2]$ is also designed by minimizing the following MSE criterion

$$J_{\mathrm{MSE}}(v[n_1, n_2]) = E\{|\varepsilon[n_1, n_2]|^2\}. \tag{7.74}$$

Again, by applying the *orthogonality principle*, the LPE filter $v_{\mathrm{LPE}}[n_1, n_2]$ can be shown to satisfy the following linear 2-D *normal equations* [1, 15]

$$E\{\varepsilon[n_1, n_2]\mathbf{y}^*[n_1, n_2]\} = \mathbf{r} + \mathcal{R}\mathbf{v} = \mathbf{0} \tag{7.75}$$

where

$$\mathcal{R} = E\{\mathbf{y}^*[n_1, n_2]\mathbf{y}^T[n_1, n_2]\} \tag{7.76}$$

is a $[p_2 + (2p_2 + 1)p_1] \times [p_2 + (2p_2 + 1)p_1]$ correlation matrix and

$$\mathbf{r} = E\{y[n_1, n_2]\mathbf{y}^*[n_1, n_2]\} \tag{7.77}$$

is a $[p_2+(2p_2+1)p_1]\times 1$ correlation vector. By (7.75), the LPE filter coefficients $\mathbf{v}$ can then be obtained by

$$\mathbf{v} = -\mathcal{R}^{-1}\mathbf{r} \tag{7.78}$$

where $\mathcal{R}^{-1}$ can be obtained by SVD to avoid numerical problems as $\mathcal{R}$ is rank deficient. On the other hand, for sufficiently large p_1 and p_2, the optimum LPE filter $v_{\rm LPE}[n_1, n_2]$ has been shown to be minimum phase and meanwhile performs as a whitening filter (which corresponds to an "amplitude equalizer"), i.e.

$$|V_{\rm LPE}(\omega_1, \omega_2)|^2 \propto \frac{1}{S_y(\omega_1, \omega_2)} = \frac{1}{S_x(\omega_1, \omega_2) + S_w(\omega_1, \omega_2)} \tag{7.79}$$

as long as the autocorrelation function $r_y[k_1, k_2]$ is *analytic positive definite* [1] for which $S_y(z_1, z_2)$ is analytic over a set of (z_1, z_2) with the unit surface included and meanwhile its value is not equal to zero on the unit surface.

If the region of support of the LPE filter $v_{\rm LPE}[n_1, n_2]$ is $\Omega_{\rm TQP}[p_1, p_2]$ (see (7.73)) rather than $\Omega_{\rm TNSHP}[p_1, p_2]$, one can also obtain a set of linear equations similar to (7.75). The LPE filter $v_{\rm LPE}[n_1, n_2]$ with a sufficiently large region of support $\Omega_{\rm TQP}[p_1, p_2]$ also performs as a whitening filter which tries to make the resultant power spectral density $S_\varepsilon(\omega_1, \omega_2)$ of $\varepsilon[n_1, n_2]$ flat.

The design of the LPE filter with region of support $\Omega_{\rm TNSHP}[p_1, p_2]$ is summarized in Table 7.1. Note that for a given set of finite data $y[n_1, n_2]$, all the correlations $r_y[l_1, l_2]$ needed by (7.78) can be replaced by the associated sample correlations (consistent estimates of $r_y[l_1, l_2]$) given by (7.45). Therefore, the designed LPE filter is also a consistent estimate of the true LPE filter by Slutsky's Theorem (Theorem 3.58).

Table 7.1 2-D LPE filter design

Parameter setting	Set the order (p_1, p_2) of the LPE filter for $\Omega'_{\rm TNSHP}[p_1, p_2]$ given by (7.31).
Steps	(S1) Form the vector $\mathbf{y}[n_1, n_2]$ given by (7.70) and estimate the correlation matrix $\mathcal{R}$ and correlation vector $\mathbf{r}$ $$\mathcal{R} = E\{\mathbf{y}^*[n_1, n_2]\mathbf{y}^T[n_1, n_2]\}$$ $$\mathbf{r} = E\{y[n_1, n_2]\mathbf{y}^*[n_1, n_2]\}$$ using sample correlations $$\widehat{r}_y[l_1, l_2] = \frac{1}{N_1 N_2} \sum_{k_1=0}^{N_1-1} \sum_{k_2=0}^{N_2-1} y[k_1, k_2] y^*[k_1 - l_1, k_2 - l_2].$$
	(S2) Calculate $\mathbf{v}$ by $$\mathbf{v} = -\mathcal{R}^{-1}\mathbf{r}$$ and obtain the associated $v_{\rm LPE}[n_1, n_2]$.

7.4 HOS Based Blind Deconvolution Approaches

Higher-order statistics have been used for the estimation of 2-D AR or ARMA models that can be nonminimum phase, asymmetric and noncausal [11,16–20]. As will be introduced in this section, HOS can also be used for 2-D blind deconvolution.

Let $v[n_1, n_2]$ be a deconvolution filter with region of support $\Omega_{\mathrm{TQP}}[p_1, p_2]$ to be designed, and $\boldsymbol{v}$ be a vector consisting of the filter coefficients $v[n_1, n_2]$ for all $[n_1, n_2] \in \Omega_{\mathrm{TQP}}[p_1, p_2]$ as follows

$$\boldsymbol{v} = (\mathbf{v}^T[0], \mathbf{v}^T[1], ..., \mathbf{v}^T[p_1])^T \tag{7.80}$$

in which

$$\mathbf{v}[i] = (v[i, 0], v[i, 1], ..., v[i, p_2])^T, \ i = 0, \ 1, \ ..., \ p_1. \tag{7.81}$$

By processing the given noisy data $y[n_1, n_2]$ modeled by (7.48) and (7.49), the deconvolution filter output $e[n_1, n_2]$ can be expressed as

$$e[n_1, n_2] = v[n_1, n_2] \star y[n_1, n_2] = \boldsymbol{v}^T \boldsymbol{y}[n_1, n_2] \tag{7.82}$$

where

$$\boldsymbol{y}[n_1, n_2] = (\mathrm{y}^T[n_1, n_2], \mathrm{y}^T[n_1 - 1, n_2], ..., \mathrm{y}^T[n_1 - p_1, n_2])^T \tag{7.83}$$

in which

$$\mathrm{y}[k, n_2] = (y[k, n_2], y[k, n_2 - 1], ..., y[k, n_2 - p_2])^T, \ n_1 - p_1 \le k \le n_1. \tag{7.84}$$

In general, a blind deconvolution algorithm tries to find an optimum $\boldsymbol{v}$ based on an optimization criterion such that the deconvolved signal $e[n_1, n_2]$ approximates $\alpha u[n_1 - \tau_1, n_2 - \tau_2]$. Two deconvolution algorithms using cumulants of $e[n_1, n_2]$ and cross-cumulants of $e[n_1, n_2]$ and $\boldsymbol{y}[n_1, n_2]$ are introduced next.

7.4.1 2-D Maximum Normalized Cumulant Deconvolution Algorithm

The MNC criterion presented in Chapter 4 for 1-D blind deconvolution of SISO systems is also applicable to the 2-D SISO case. For convenience, the MNC criterion is repeated as follows

$$J_{p,q}(\boldsymbol{v}) = J_{p,q}(e[n_1, n_2]) = |\gamma_{p,q}\{e[n_1, n_2]\}| = \frac{|C_{p,q}\{e[n_1, n_2]\}|}{\sigma_e^{p+q}} \tag{7.85}$$

where p and q are nonnegative integers and $(p + q) \ge 3$. When $y[n_1, n_2]$ is real, $J_{p,q}(\boldsymbol{v})$ are the same for all (p, q) as long as $(p + q)$ is fixed. Therefore, for simplicity, $J_{p+q}(\boldsymbol{v})$ rather than $J_{p,q}(\boldsymbol{v})$ is used for the real case.

Similar to the 1-D SISO case, the *2-D MNC blind deconvolution filter* obtained by maximizing $J_{p,q}(\boldsymbol{v})$ is a *perfect deconvolution filter* in the absence of noise as stated in the following theorem.

Theorem 7.1 (Deconvolution Capability). *Assume that the given* $y[n_1, n_2]$ *can be modeled by (7.48) and (7.49) under the Assumptions* (A7-1), (A7-2) *and the assumption of* $SNR = \infty$. *Let* $v[n_1, n_2]$ *be a 2-D LSI filter with region of support* $\Omega_{\text{TQP}}[\infty, \infty]$. *Then* $J_{p,q}(\boldsymbol{v})$ *is maximum if and only if* $v[n_1, n_2] = v_{\text{MNC}}[n_1, n_2]$ *and*

$$e[n_1, n_2] = v_{\text{MNC}}[n_1, n_2] \star y[n_1, n_2] = \alpha u[n_1 - \tau_1, n_2 - \tau_2] \tag{7.86}$$

where $\alpha \neq 0$ *is a complex scale factor and* τ_1 *and* τ_2 *are unknown integers, and*

$$\begin{aligned} \max\{J_{p,q}(\boldsymbol{v})\} &= J_{p,q}(u[n_1, n_2]) \\ &= |\gamma_{p,q}\{u[n_1, n_2]\}| = \frac{|C_{p,q}\{u[n_1, n_2]\}|}{\sigma_u^{p+q}}. \end{aligned} \tag{7.87}$$

The proof of Theorem 7.1 (through using Theorem 2.33) is similar to the proof of Theorem 4.14 for the 1-D SISO case and is left as an exercise (Problem 7.3).

Similar to the 1-D SISO MNC equalization algorithm introduced in Chapter 4, the iterative approximate BFGS algorithm (introduced in Chapter 2) can be employed to find a local maximum of $J_{p,q}(\boldsymbol{v})$. At the ith iteration, $\boldsymbol{v}$ is updated by

$$\boldsymbol{v}^{[i+1]} = \boldsymbol{v}^{[i]} + \mu^{[i]} \mathbf{Q}^{[i]} \left. \frac{\partial J_{p,q}(\boldsymbol{v})}{\partial \boldsymbol{v}^*} \right|_{\boldsymbol{v}=\boldsymbol{v}^{[i]}} \tag{7.88}$$

where $\mu^{[i]}$ is a step size and the matrix $\mathbf{Q}^{[i+1]}$ (see (2.160) and (2.173)) is updated by

$$\begin{aligned} \mathbf{Q}^{[i+1]} = \mathbf{Q}^{[i]} + \frac{1}{\text{Re}\{\mathbf{r}_{i+1}^H \mathbf{s}_{i+1}\}} \Bigg\{ & (1+\beta_i)\, \mathbf{r}_{i+1} \mathbf{r}_{i+1}^H - \mathbf{r}_{i+1} \mathbf{s}_{i+1}^H \mathbf{Q}^{[i]} - \mathbf{Q}^{[i]} \mathbf{s}_{i+1} \mathbf{r}_{i+1}^H \\ & - \frac{1-\alpha}{(\alpha+\beta_i)(1+\beta_i)} \mathbf{Q}^{[i]} \mathbf{s}_{i+1} \mathbf{s}_{i+1}^H \mathbf{Q}^{[i]} \Bigg\} \end{aligned} \tag{7.89}$$

in which $\alpha = 1$ and $1/2$ for real and complex $y[n_1, n_2]$, β_i (see (2.161) and (2.174)) is given by

$$\beta_i = \frac{\text{Re}\left\{\mathbf{s}_{i+1}^H \mathbf{Q}^{[i]} \mathbf{s}_{i+1}\right\}}{\text{Re}\left\{\mathbf{s}_{i+1}^H \mathbf{r}_{i+1}\right\}}, \tag{7.90}$$

vectors $\mathbf{r}_{i+1}$ and $\mathbf{s}_{i+1}$ are defined by

$$\mathbf{r}_{i+1} = \boldsymbol{v}^{[i+1]} - \boldsymbol{v}^{[i]} \tag{7.91}$$

and

$$\mathbf{s}_{i+1} = \left. \frac{\partial J_{p,q}(\boldsymbol{v})}{\partial \boldsymbol{v}^*} \right|_{\boldsymbol{v}=\boldsymbol{v}^{[i+1]}} - \left. \frac{\partial J_{p,q}(\boldsymbol{v})}{\partial \boldsymbol{v}^*} \right|_{\boldsymbol{v}=\boldsymbol{v}^{[i]}}, \tag{7.92}$$

respectively. The step size $\mu^{[i]}$ can be chosen as

$$\mu^{[i]} = \frac{\mu_0}{2^l} \tag{7.93}$$

where $l \in [0, K]$ is determined such that $\mu^{[i]}$ is the maximum step size leading to $J_{p,q}(\boldsymbol{v}^{[i+1]}) > J_{p,q}(\boldsymbol{v}^{[i]})$, in which μ_0 and K are a constant and a positive integer assigned ahead of time, respectively.

The gradient $\partial J_{p,q}(\boldsymbol{v})/\partial \boldsymbol{v}^*$ in (7.88), which is the same for both 1-D and 2-D SISO cases, has been shown in Chapter 4 to be

$$\begin{aligned}\frac{\partial J_{p,q}(\boldsymbol{v})}{\partial \boldsymbol{v}^*} = \frac{J_{p,q}(\boldsymbol{v})}{2} \cdot \Bigg\{ & \frac{1}{C_{q,p}\{e[n_1,n_2]\}} \cdot \frac{\partial C_{q,p}\{e[n_1,n_2]\}}{\partial \boldsymbol{v}^*} \\ & + \frac{1}{C_{p,q}\{e[n_1,n_2]\}} \cdot \frac{\partial C_{p,q}\{e[n_1,n_2]\}}{\partial \boldsymbol{v}^*} - \frac{p+q}{\sigma_e^2} \cdot \frac{\partial \sigma_e^2}{\partial \boldsymbol{v}^*} \Bigg\}\end{aligned} \tag{7.94}$$

where

$$\frac{\partial C_{p,q}\{e[n_1,n_2]\}}{\partial \boldsymbol{v}^*} = q \cdot \mathrm{cum}\{e[n_1,n_2] : p, e^*[n_1,n_2] : q-1, \boldsymbol{y}^*[n_1,n_2]\}. \tag{7.95}$$

For instance,

$$J_{2,1}(\boldsymbol{v}) = \frac{|C_{2,1}\{e[n_1,n_2]\}|}{\sigma_e^3} = \frac{|E\{|e[n_1,n_2]|^2 e[n_1,n_2]\}|}{\sigma_e^3} \tag{7.96}$$

and

$$\begin{aligned}\frac{\partial J_{2,1}(\boldsymbol{v})}{\partial \boldsymbol{v}^*} = \frac{J_{2,1}(\boldsymbol{v})}{2} \cdot \Bigg\{ & \frac{2E\{|e[n_1,n_2]|^2 \boldsymbol{y}^*[n_1,n_2]\}}{E\{|e[n_1,n_2]|^2 e^*[n_1,n_2]\}} \\ & + \frac{E\{e^2[n_1,n_2]\boldsymbol{y}^*[n_1,n_2]\}}{E\{|e[n_1,n_2]|^2 e[n_1,n_2]\}} - \frac{3E\{e[n_1,n_2]\boldsymbol{y}^*[n_1,n_2]\}}{\sigma_e^2} \Bigg\}.\end{aligned} \tag{7.97}$$

The resultant 2-D MNC deconvolution algorithm is summarized in Table 7.2. However, for a given set of data, the $(p+q)$th-order sample cumulant $\widehat{C}_{p,q}\{e[n_1,n_2]\}$ and sample cross-cumulants of $e[n_1,n_2]$ and $y[n_1,n_2]$ must be used to compute $J_{p,q}(\boldsymbol{v})$ and $\partial J_{p,q}(\boldsymbol{v})/\partial \boldsymbol{v}^*$ needed by the 2-D MNC deconvolution algorithm. Because $\widehat{C}_{p,q}\{e[n_1,n_2]\}$ is a consistent estimate of $C_{p,q}\{e[n_1,n_2]\}$ as mentioned in Section 7.1.4, $\widehat{J}_{p,q}(\boldsymbol{v})$ is therefore also a consistent estimate of $J_{p,q}(\boldsymbol{v})$ by Slutsky's Theorem (see Theorem 3.58), implying that the optimum MNC deconvolution filter $\widehat{v}_{\mathrm{MNC}}[n_1,n_2]$ is a consistent estimate of the true optimum $v_{\mathrm{MNC}}[n_1,n_2]$ as well.

Properties of the 2-D MNC Deconvolution Filter

As mentioned in Section 7.2, the MMSE deconvolution filter $V_{\mathrm{MS}}(\omega_1,\omega_2)$ given by (7.64), which requires the system $h[n_1,n_2]$ given ahead of time, is exactly

Table 7.2 2-D MNC deconvolution algorithm

Parameter setting	Choose filter order (p_1, p_2), cumulant order (p, q) for the region of support $\Omega_{\mathrm{TQP}}[p_1, p_2]$, initial conditions $\boldsymbol{v}^{[0]}$ and $\mathbf{Q}^{[0]}$, convergence tolerance $\zeta > 0$, initial step size μ_0 and a positive integer K for the step size search range $[\mu_0/2^0, \mu_0/2^K]$.
Steps	(S1) Set the iteration number $i = 0$. (S2) Compute the deconvolved signal $e^{[0]}[n_1, n_2] = (\boldsymbol{v}^{[0]})^T \boldsymbol{y}[n_1, n_2]$ and obtain the corresponding gradient function $\left.\frac{\partial J_{p,q}(\boldsymbol{v})}{\partial \boldsymbol{v}^*}\right\vert_{\boldsymbol{v}=\boldsymbol{v}^{[0]}}$ using (7.94). (S3) Update $\boldsymbol{v}^{[i+1]}$ by $$\boldsymbol{v}^{[i+1]} = \boldsymbol{v}^{[i]} + \mu^{[i]} \mathbf{Q}^{[i]} \left.\frac{\partial J_{p,q}(\boldsymbol{v})}{\partial \boldsymbol{v}^*}\right\vert_{\boldsymbol{v}=\boldsymbol{v}^{[i]}}$$ where an integer $l \in [0, K]$ is determined such that $\mu^{[i]} = \mu_0/2^l$ is the maximum step size leading to $J_{p,q}(\boldsymbol{v}^{[i+1]}) > J_{p,q}(\boldsymbol{v}^{[i]})$. (S4) If $$\frac{\left\vert J_{p,q}(\boldsymbol{v}^{[i+1]}) - J_{p,q}(\boldsymbol{v}^{[i]})\right\vert}{J_{p,q}(\boldsymbol{v}^{[i]})} \geq \zeta,$$ then go to Step (S5); otherwise, obtain a (local) maximum solution $\boldsymbol{v} = \boldsymbol{v}^{[i+1]}$ for the 2-D MNC deconvolution filter $v_{\mathrm{MNC}}[n_1, n_2]$. (S5) Update $\mathbf{r}_{i+1}$, $\mathbf{s}_{i+1}$ and $\mathbf{Q}^{[i+1]}$ by (7.91), (7.92) and (7.89), respectively. (S6) Compute the deconvolved signal $$e^{[i+1]}[n_1, n_2] = (\boldsymbol{v}^{[i+1]})^T \cdot \boldsymbol{y}[n_1, n_2]$$ and the gradient function $\left.\frac{\partial J_{p,q}(\boldsymbol{v})}{\partial \boldsymbol{v}^*}\right\vert_{\boldsymbol{v}=\boldsymbol{v}^{[i+1]}}$ using (7.94). (S7) Update the iteration number i by $(i + 1)$ and go to Step (S3).

an optimum Wiener filter performing ISI suppression and noise reduction simultaneously. The optimum blind deconvolution filter $v_{\mathrm{MNC}}[n_1, n_2]$ obtained by the 2-D MNC deconvolution algorithm relates to $V_{\mathrm{MS}}(\omega_1, \omega_2)$ in a nonlinear manner as stated in the following property.

Property 7.2 (Relation of the 2-D MNC and LMMSE Deconvolution Filters). *The optimum MNC deconvolution filter $V_{\mathrm{MNC}}(\omega_1, \omega_2)$ with region of support $\Omega_{\mathrm{TQP}}[\infty, \infty]$ is related to the 2-D MMSE deconvolution fil-*

ter $V_{\rm MS}(\omega_1, \omega_2)$ *given by (7.64) via*

$$V_{\rm MNC}(\omega_1, \omega_2) = \left[\alpha_{p,q}\widetilde{G}_{p,q}(\omega_1, \omega_2) + \alpha_{q,p}\widetilde{G}_{q,p}(\omega_1, \omega_2)\right] \cdot V_{\rm MS}(\omega_1, \omega_2) \quad (7.98)$$

where $\widetilde{G}_{p,q}(\omega_1, \omega_2)$ *is the 2-D DSFT of the 2-D sequence* $\widetilde{g}_{p,q}[n_1, n_2]$ *given by*

$$\widetilde{g}_{p,q}[n_1, n_2] = g^q[n_1, n_2](g^*[n_1, n_2])^{p-1} \quad (7.99)$$

in which $g[n_1, n_2] = v_{\rm MNC}[n_1, n_2] \star h[n_1, n_2]$ *is the overall system after deconvolution (see (7.56)), and* $\alpha_{p,q}$ *is a nonzero constant given by*

$$\alpha_{p,q} = \frac{q}{p+q} \cdot \frac{\sigma_e^2}{\sigma_u^2} \cdot \frac{C_{p,q}\{u[n_1, n_2]\}}{C_{p,q}\{e[n_1, n_2]\}} . \quad (7.100)$$

The proof of Property 7.2 for the 2-D SISO case is also similar to that of the corresponding Theorem 4.21 for the 1-D SISO case, and is left as an exercise (Problem 7.4).

By Theorem 7.1, the 2-D MNC deconvolution filter is a perfect deconvolution filter (i.e. a 2-D ZF deconvolution filter which is a perfect amplitude and phase deconvolution filter) as SNR is infinite. However, for finite SNR, the 2-D MNC deconvolution filter is still a perfect phase deconvolution filter as stated in the following property.

Property 7.3 (Linear Phase Property of the 2-D MNC Deconvolution Filter). *The phase response* $\arg[V_{\rm MNC}(\omega_1, \omega_2)]$ *of the 2-D MNC deconvolution filter* $v_{\rm MNC}[n_1, n_2]$ *with region of support* $\Omega_{\rm TQP}[\infty, \infty]$ *is related to the system phase* $\arg[H(\omega_1, \omega_2)]$ *by*

$$\arg[V_{\rm MNC}(\omega_1, \omega_2)] = -\arg[H(\omega_1, \omega_2)] + \omega_1\tau_1 + \omega_2\tau_2 + \varsigma \quad (7.101)$$

where τ_1 *and* τ_2 *are unknown integers and* ς *is an unknown real constant. As* $y[n_1, n_2]$ *is real,* $\varsigma = 0$.

The proof of Property 7.3 for the 2-D SISO case is similar to that of the corresponding Property 4.17 for the 1-D SISO case, and is left as an exercise (Problem 7.5).

7.4.2 2-D Super-Exponential Deconvolution Algorithm

Two-dimensional SE deconvolution algorithm [21] for SISO systems, that is a direct 2-D extension of the 1-D SE equalization algorithm for 1-D SISO systems introduced in Chapter 4, iteratively finds the 2-D deconvolution filter $v[n_1, n_2]$ with region of support $\Omega_{\rm TQP}[p_1, p_2]$ by solving the following linear equations

$$\frac{1}{\sigma_u^2}\sum_{k=0}^{p_1}\sum_{l=0}^{p_2} v[k,l] r_y[n_1-k,n_2-l]$$
$$= \frac{1}{C_{p,q}\{u[n_1,n_2]\}} \operatorname{cum}\{e[m,n]:p, e^*[m,n]:q-1, y^*[m-n_1,n-n_2]\},$$
$$\forall [n_1,n_2] \in \Omega_{\mathrm{TQP}}[p_1,p_2] \tag{7.102}$$

where p and $q-1$ are nonnegative integers and $p+q \geq 3$.

At the ith iteration, by expressing (7.102) alternatively in a matrix form, the 2-D SE deconvolution algorithm updates the unknown parameter vector $\boldsymbol{v}$ (with $\|\boldsymbol{v}\| = 1$) via

$$\boldsymbol{v}^{[i+1]} = \frac{\mathbb{R}^{-1}\cdot \mathbf{d}_{ey}^{[i]}}{\left\|\mathbb{R}^{-1}\cdot \mathbf{d}_{ey}^{[i]}\right\|} \tag{7.103}$$

where $\mathbb{R}$ is a $[(p_1+1)(p_2+1)] \times [(p_1+1)(p_2+1)]$ autocorrelation matrix given by

$$\mathbb{R} = \left(E\{\boldsymbol{y}[n_1,n_2]\boldsymbol{y}^H[n_1,n_2]\}\right)^* \tag{7.104}$$

and $\mathbf{d}_{ey}^{[i]}$ is a $(p_1+1)(p_2+1)\times 1$ vector given by

$$\mathbf{d}_{ey}^{[i]} = \operatorname{cum}\{e^{[i]}[n_1,n_2]:p, (e^{[i]}[n_1,n_2])^*:q-1, \boldsymbol{y}^*[n_1,n_2]\} \tag{7.105}$$

where $e^{[i]}[n_1,n_2]$ is the deconvolved signal obtained at the ith iteration, i.e.

$$e^{[i]}[n_1,n_2] = y[n_1,n_2]\star v^{[i]}[n_1,n_2] = (\boldsymbol{v}^{[i]})^T\boldsymbol{y}[n_1,n_2]. \tag{7.106}$$

In updating $\boldsymbol{v}^{[i]}$ by (7.103), the SVD is also suggested for computing the inverse matrix $\mathbb{R}^{-1}$ to avoid potential numerical problems caused by rank deficiency of $\mathbb{R}$. The 2-D SE deconvolution filter obtained after convergence is denoted as $v_{\mathrm{SE}}[n_1,n_2]$. Note that as $y[n_1,n_2]$ is real, the $v_{\mathrm{SE}}[n_1,n_2]$ will be the same for the same value of $(p+q)$.

Similar to the 1-D SISO case, for SNR $=\infty$ and p_1 and p_2 sufficiently large for the region of support $\Omega_{\mathrm{TQP}}[p_1,p_2]$, the 2-D SE algorithm, which is basically developed by forcing the amount of ISI to decrease at a super-exponential speed, will converge at the super-exponential rate and end up with the deconvolved signal

$$e[n_1,n_2] = v_{\mathrm{SE}}[n_1,n_2]\star y[n_1,n_2] = \alpha u[n_1-\tau_1,n_2-\tau_2] \tag{7.107}$$

where $\alpha \neq 0$ is a complex scale factor and $[\tau_1,\tau_2]$ is an unknown space shift. In other words, the 2-D SE deconvolution filter $v_{\mathrm{SE}}[n_1,n_2]$ is the same as the 2-D ZF deconvolution filter (with zero ISI) for SNR $=\infty$.

In spite of the normalization $\|\boldsymbol{v}\| = 1$ at each iteration, a complex scale factor $e^{j\phi}$ may still exist between $\boldsymbol{v}^{[i+1]}$ and $\boldsymbol{v}^{[i]}$. Therefore, similar to the 1-D SE equalization algorithm, the convergence rule for the 2-D SE deconvolution algorithm can be chosen as

$$\left|(\boldsymbol{v}^{[i+1]})^H \boldsymbol{v}^{[i]}\right| > 1 - \zeta/2 \tag{7.108}$$

where $\zeta > 0$ is a preassigned convergence tolerance.

Again, for a given set of data $y[n_1, n_2]$, all the cumulants and correlations needed by the 2-D SE deconvolution algorithm must be replaced by the associated sample cumulants and correlations which are consistent as presented in Section 7.1.4. Accordingly, the $v_{\rm SE}[n_1, n_2]$ obtained is also a consistent estimate of the one satisfying (7.102). The 2-D SE deconvolution algorithm is summarized in Table 7.3. Because no explicit objective function (function of $y[n_1, n_2]$ or $e[n_1, n_2]$ or both) is minimized or maximized by the 2-D SE deconvolution algorithm, the algorithm may diverge for finite SNR and data length.

Table 7.3 2-S SE deconvolution algorithm

Parameter setting		Choose filter order (p_1, p_2) for the region of support $\Omega_{\rm TQP}[p_1, p_2]$, cumulant order (p, q), initial condition $\boldsymbol{v}^{[0]}$ and convergence tolerance $\zeta > 0$.
Steps	(S1)	Set the iteration number $i = 0$.
	(S2)	Estimate the correlation function of $y[n_1, n_2]$ by (7.45) and form $\mathbb{R}$ by (7.104) and obtain $\mathbb{R}^{-1}$ by SVD.
	(S3)	Compute the deconvolved signal $e^{[0]}[n_1, n_2] = (\boldsymbol{v}^{[0]})^T \boldsymbol{y}[n_1, n_2]$ and estimate the vector $\mathbf{d}_{ey}^{[0]}$ (see (7.105)) from $e^{[0]}[n_1, n_2]$ and $y[n_1, n_2]$,
	(S4)	Update the parameter vector $\boldsymbol{v}$ at the ith iteration via $$\boldsymbol{v}^{[i+1]} = \frac{\mathbb{R}^{-1} \cdot \mathbf{d}_{ey}^{[i]}}{\left\|\mathbb{R}^{-1} \cdot \mathbf{d}_{ey}^{[i]}\right\|}.$$
	(S5)	If $$\left|\left(\boldsymbol{v}^{[i+1]}\right)^H \boldsymbol{v}^{[i]}\right| \leq 1 - \frac{\zeta}{2},$$ then go to Step (S6); otherwise, obtain the parameter vector $\boldsymbol{v} = \boldsymbol{v}^{[i+1]}$ for the 2-D SE deconvolution filter $v_{\rm SE}[n_1, n_2]$.
	(S6)	Compute the deconvolved signal $$e^{[i+1]}[n_1, n_2] = \left(\boldsymbol{v}^{[i+1]}\right)^T \boldsymbol{y}[n_1, n_2]$$ and estimate $\mathbf{d}_{ey}^{[i+1]}$ from $e^{[i+1]}[n_1, n_2]$ and $y[n_1, n_2]$.
	(S7)	Update the iteration number i by $(i + 1)$ and go to Step (S4).

Properties of the 2-D SE Deconvolution Filters

The optimum deconvolution filter $V_{\rm SE}(\omega_1,\omega_2)$ obtained by the 2-D SE deconvolution algorithm is also nonlinearly related to $V_{\rm MS}(\omega_1,\omega_2)$ as stated in the following property.

Property 7.4 (Relation of the 2-D SE and LMMSE Deconvolution Filters). *The optimum deconvolution filter* $V_{\rm SE}(\omega_1,\omega_2)$ *with region of support* $\Omega_{\rm TQP}[\infty,\infty]$ *is related to* $V_{\rm MS}(\omega_1,\omega_2)$ *given by (7.64) via*

$$\begin{aligned} V_{\rm SE}(\omega_1,\omega_2) &= \widetilde{G}_{p,q}(\omega_1,\omega_2)V_{\rm MS}(\omega_1,\omega_2) \\ &= \widetilde{G}_{p,q}(\omega_1,\omega_2)\frac{\sigma_u^2 H^*(\omega_1,\omega_2)}{S_y(\omega_1,\omega_2)} \end{aligned} \tag{7.109}$$

where $\widetilde{G}_{p,q}(\omega_1,\omega_2)$ *is the 2-D DSFT of the 2-D sequence* $\widetilde{g}_{p,q}[n_1,n_2]$ *given by (7.99) in which* $g[n_1,n_2] = v_{\rm SE}[n_1,n_2] \star h[n_1,n_2]$.

The proof of Property 7.4 for the 2-D SISO case is also similar to that of the corresponding Property 4.22 for the 1-D SISO case, and is left as an exercise (Problem 7.6).

As mentioned above, the 2-D SE deconvolution filter $v_{\rm SE}[n_1,n_2]$ is a perfect deconvolution filter as SNR is infinite (see (7.107)). Besides the relation between the 2-D SE deconvolution filter and MMSE deconvolution filter for finite SNR presented in Property 7.4, the linear phase property of the 2-D MNC deconvolution filter (Property 7.3) is also shared by the 2-D SE deconvolution filter thanks to their equivalence under certain conditions, to be introduced next.

7.4.3 Improvements on 2-D MNC Deconvolution Algorithm

For SNR $=\infty$ and sufficiently large data length $N \times N$, the preceding 2-D SE deconvolution algorithm is computationally efficient, and meanwhile it is a fast algorithm thanks to the super-exponential convergence rate. However, due to lack of an explicit objective function iteratively directing it towards the optimum $v_{\rm SE}[n_1,n_2]$, it may diverge for finite SNR and $N \times N$ as presented in Chapter 4. On the other hand, the convergence of the gradient-type iterative 2-D MNC deconvolution algorithm can be guaranteed thanks to the objective function $J_{p,q}(\boldsymbol{v})$ used, though its convergence speed is slower and its computational load (for each iteration) larger than the 2-D SE deconvolution algorithm. Therefore, a combination of the two algorithms to share their advantages is introduced below, based on a relation between them as stated in the following property.

Property 7.5 (Equivalence of the 2-D MNC and SE Deconvolution Filters). *The optimum $V_{\mathrm{MNC}}(\omega_1, \omega_2)$ with region of support $\Omega_{\mathrm{TQP}}[\infty, \infty]$ and the optimum $V_{\mathrm{SE}}(\omega_1, \omega_2)$ with the same region of support are the same (up to a scale factor and space shift) for (i) $y[n_1, n_2]$ is real and (ii) $y[n_1, n_2]$ is complex and $p = q$.*

The proof of Property 7.5 for the 2-D SISO case is similar to that of Corollary 4.24 for the 1-D SISO case and is omitted here.

Hybrid MNC Deconvolution Algorithm

The hybrid algorithm introduced in Chapter 4 for 1-D SISO case is applicable to the 2-D SISO case due to Property 7.5. For convenience, it is summarized in the following table.

Table 7.4 2-D hybrid MNC deconvolution algorithm

Procedure for Obtaining the 2-D MNC Deconvolution Filter at Iteration i
(S1) Update the parameter vector $\boldsymbol{v}^{[i+1]}$ via the update equations of the 2-D SE deconvolution algorithm given by (7.103), and obtain the associated deconvolved signal $e^{[i+1]}[n_1, n_2]$.
(S2) If $J_{p,q}(\boldsymbol{v}^{[i+1]}) > J_{p,q}(\boldsymbol{v}^{[i]})$, then go to the next iteration; otherwise, update $\boldsymbol{v}^{[i+1]}$ through a gradient-type optimization method such that $J_{p,q}(\boldsymbol{v}^{[i+1]}) > J_{p,q}(\boldsymbol{v}^{[i]})$, and obtain the associated $e^{[i+1]}[n_1, n_2]$.

Specifically, as $y[n_1, n_2]$ is real and $y[n_1, n_2]$ is complex for $p = q$, it can be shown that

$$\left.\frac{\partial J_{p,q}(\boldsymbol{v})}{\partial \boldsymbol{v}^*}\right|_{\boldsymbol{v}=\boldsymbol{v}^{[i]}} = \frac{p+q}{2} \cdot J_{p,q}(\boldsymbol{v}^{[i]}) \cdot \left(\frac{\mathbf{d}_{ey}^{[i]}}{C_{p,q}\{e^{[i]}[n_1, n_2]\}} - \frac{\mathbb{R}\boldsymbol{v}^{[i]}}{E\{|e^{[i]}[n_1, n_2]|^2\}} \right) \tag{7.110}$$

where $\mathbf{d}_{ey}^{[i]}$ (see (7.105)) has been obtained in (S1) and $\mathbb{R}$ (see (7.104)) is the same at each iteration, indicating simple and straightforward computation for obtaining $\partial J_{p,q}(\boldsymbol{v})/\partial \boldsymbol{v}^*$ in (S2).

According to Property 7.5, the 2-D hybrid MNC deconvolution algorithm is always applicable to the case of real signals and only applicable to the case of complex signals if $p = q \geq 2$. The 2-D hybrid MNC deconvolution algorithm uses the 2-D SE deconvolution algorithm in (S1) for fast convergence (basically with super-exponential rate) which usually happens in most iterations before convergence, and a gradient-type algorithm in (S2) for the guaranteed convergence regardless of data length and SNR. It is worth mentioning that the deconvolution filter obtained by the 2-D hybrid MNC deconvolution algorithm is the optimum $v_{\mathrm{MNC}}[n_1, n_2]$ because the objective function $J_{p,q}(\boldsymbol{v})$ is used, while the 2-D SE deconvolution algorithm in (S1) plays the role of a fast algorithm for finding the optimum $v_{\mathrm{MNC}}[n_1, n_2]$.

7.5 Simulation

This section presents some simulation results for performance tests of the 2-D MNC deconvolution algorithm, 2-D SE deconvolution algorithm and 2-D hybrid MNC deconvolution algorithm. In the simulation, $h[n_1, n_2]$ used was a 2-D MA model taken from [5] as follows

$$\begin{aligned} x[n_1, n_2] &= u[n_1, n_2] - 0.8u[n_1 - 1, n_2] + 0.2u[n_1 - 2, n_2] \\ &\quad + 1.8u[n_1, n_2 - 1] - 1.44u[n_1 - 1, n_2 - 1] + 0.36u[n_1 - 2, n_2 - 1] \\ &\quad - 0.5u[n_1, n_2 - 2] + 0.4u[n_1 - 1, n_2 - 2] - 0.1u[n_1 - 2, n_2 - 2] \\ &\quad + 0.5u[n_1, n_2 - 3] - 0.4u[n_1 - 1, n_2 - 3] + 0.1u[n_1 - 2, n_2 - 3]. \end{aligned} \tag{7.111}$$

The driving input $u[n_1, n_2]$ was a real zero-mean, exponentially distributed, i.i.d., random field with variance $\sigma_u^2 = 1$ and $J_4(u[n_1, n_2]) = 6$, and $w[n_1, n_2]$ was real zero-mean white Gaussian. Synthetic 128×128 data $y[n_1, n_2]$ were generated for both SNR = 5 dB and SNR = 20 dB, and then processed by the three deconvolution algorithms with region of support $\Omega_{\mathrm{TQP}}[5, 5]$ for the real 2-D deconvolution filter $v[n_1, n_2]$ and the initial condition $v^{[0]}[n_1, n_2] = \delta[n_1 - 2, n_2 - 2]$. Thirty independent runs were performed for performance evaluation in terms of the amount of ISI and the value of $J_4(e[n_1, n_2])$ (magnitude of the normalized kurtosis of the deconvolved signal $e[n_1, n_2]$).

Results Obtained by the MNC Deconvolution Algorithm

The 2-D MNC deconvolution algorithm with $p + q = 4$, $\mathbf{Q}^{[0]} = \mathbf{I}$, $\mu_0 = 1$ and $K = 10$ was employed to process $y[n_1, n_2]$ to find the optimum $v_{\mathrm{MNC}}[n_1, n_2]$. Figures 7.5a, b show the objective function $J_4(\boldsymbol{v})$ versus iteration number for SNR = 5 dB and the associated ISI for the first 20 iterations. Figures 7.5c, d show the corresponding results for SNR = 20 dB. From Fig. 7.5a, c, one can observe that all the values of $J_4(\boldsymbol{v})$, which are larger for SNR = 20 dB than for SNR = 5 dB after convergence, increase with iteration number. Note that the values of $J_4(\boldsymbol{v}) = J_4(e[n_1, n_2])$ are much smaller than $J_4(u[n_1, n_2]) = 6$ for SNR = 5 dB due to low SNR, and that some values of $J_4(\boldsymbol{v}) = J_4(e[n_1, n_2])$ are even larger than $J_4(u[n_1, n_2]) = 6$ due to insufficient data length. Figures

7.5b, d show that the values of ISI basically decrease, but not monotonically, with iteration number, and that the values of ISI for SNR = 20 dB are much smaller than those for SNR = 5 dB after convergence.

Instead of displaying the 30 independent estimates of the overall system $g[n_1, n_2]$ obtained at the 20th iteration in an overlay fashion, Figures 7.5e, f show the averaged phase response $\arg[G(\omega_1, \omega_2)]$ (principal values) of the 30 estimates of $g[n_1, n_2]$ for SNR = 5 dB and 20 dB, respectively. One can see from these figures that $\arg[G(\omega_1, \omega_2)]$ approximates a linear function of (ω_1, ω_2), thus verifying the linear phase property of $G(\omega_1, \omega_2)$ introduced in Property 7.3, and that the linear phase approximation for SNR = 20 dB (Fig. 7.5f) is better than that (Fig. 7.5e) for SNR = 5 dB.

Results Obtained by the SE Deconvolution Algorithm

The 2-D SE deconvolution algorithm with $p + q = 4$ was employed to process $y[n_1, n_2]$ to find the optimum $v_{\mathrm{SE}}[n_1, n_2]$. Figures 7.6a, b show the ISI versus iteration number for SNR = 5 dB and the associated $J_4(\boldsymbol{v})$ for the first 20 iterations. Figures 7.6c, d show the corresponding results for SNR = 20 dB. From Fig. 7.6a, c, one can see that all the values of ISI, which are much smaller for SNR = 20 dB than for SNR = 5 dB, decrease rapidly with iteration number and reach minimum values within five iterations. These results support the fast convergence of the 2-D SE deconvolution algorithm. Let us emphasize that the SE deconvolution algorithm for either of 1-D and 2-D cases is theoretically developed such that the ISI decreases super-exponentially in the absence of noise. Moreover, it can be seen from Fig. 7.6b, d, that the values of $J_4(\boldsymbol{v})$, which are much larger for SNR = 20 dB than for SNR = 5 dB after convergence, also converge rapidly within five iterations.

Figures 7.6e, f show the averaged phase response of $\arg[G(\omega_1, \omega_2)]$ (principal values) of the 30 estimates of $g[n_1, n_2]$ obtained at the 20th iteration for SNR = 5 dB and 20 dB, respectively. One can see from these figures that $\arg[G(\omega_1, \omega_2)]$ approximates a linear function of (ω_1, ω_2), thus verifying the linear phase property of $G(\omega_1, \omega_2)$ (by Properties 7.3 and 7.5), and that the linear phase approximation for SNR = 20 dB (Fig. 7.6f) is better than that (Fig. 7.6e) for SNR = 5 dB.

Results Obtained by the Hybrid MNC Deconvolution Algorithm

The 2-D hybrid MNC deconvolution algorithm with $p + q = 4$ was employed to process $y[n_1, n_2]$ to find the optimum $v_{\mathrm{MNC}}[n_1, n_2]$. Figures 7.7a, b show the objective function $J_4(\boldsymbol{v})$ versus iteration number for SNR = 5 dB and the associated ISI for the first 20 iterations. Figures 7.7c, d show the corresponding results for SNR = 20 dB. All the observations associated with Fig. 7.6a–d also apply to those associated with Fig. 7.7a–d. These results also demonstrate that $v_{\mathrm{NNC}}[n_1, n_2]$ and $v_{\mathrm{SE}}[n_1, n_2]$ are similar in the simulation. Comparing Fig. 7.7a–d and Fig. 7.5a–d, one can see that the 2-D hybrid MNC deconvolution

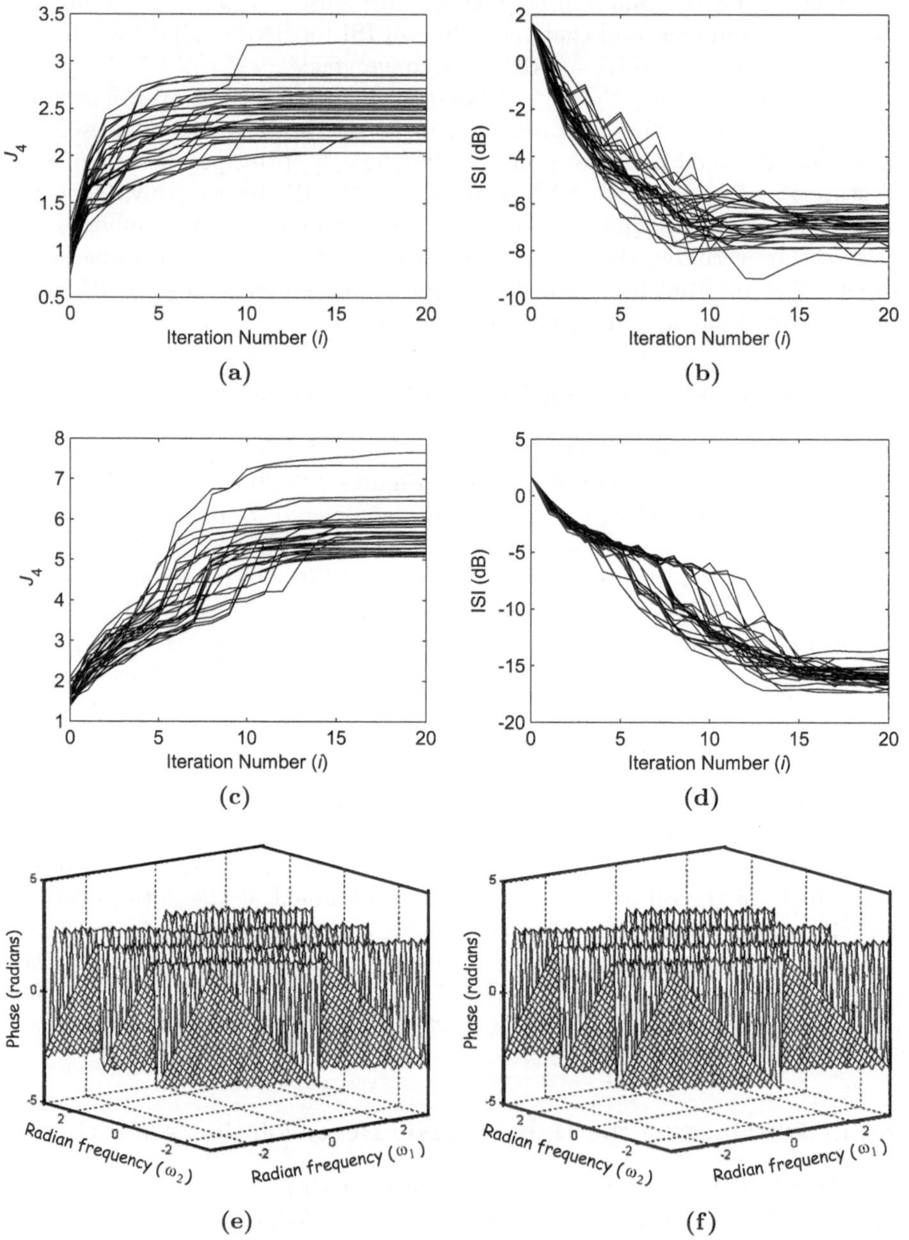

Fig. 7.5 Simulation results obtained by the 2-D MNC deconvolution algorithm. (**a**) and (**c**) show $J_4(\boldsymbol{v})$ versus iteration number and (**b**) and (**d**) show the associated ISI versus iteration number for SNR = 5 dB and SNR = 20 dB, respectively. (**e**) and (**f**) show the averaged $\arg[G(\omega_1, \omega_2)]$ for SNR = 5 dB and 20 dB, respectively

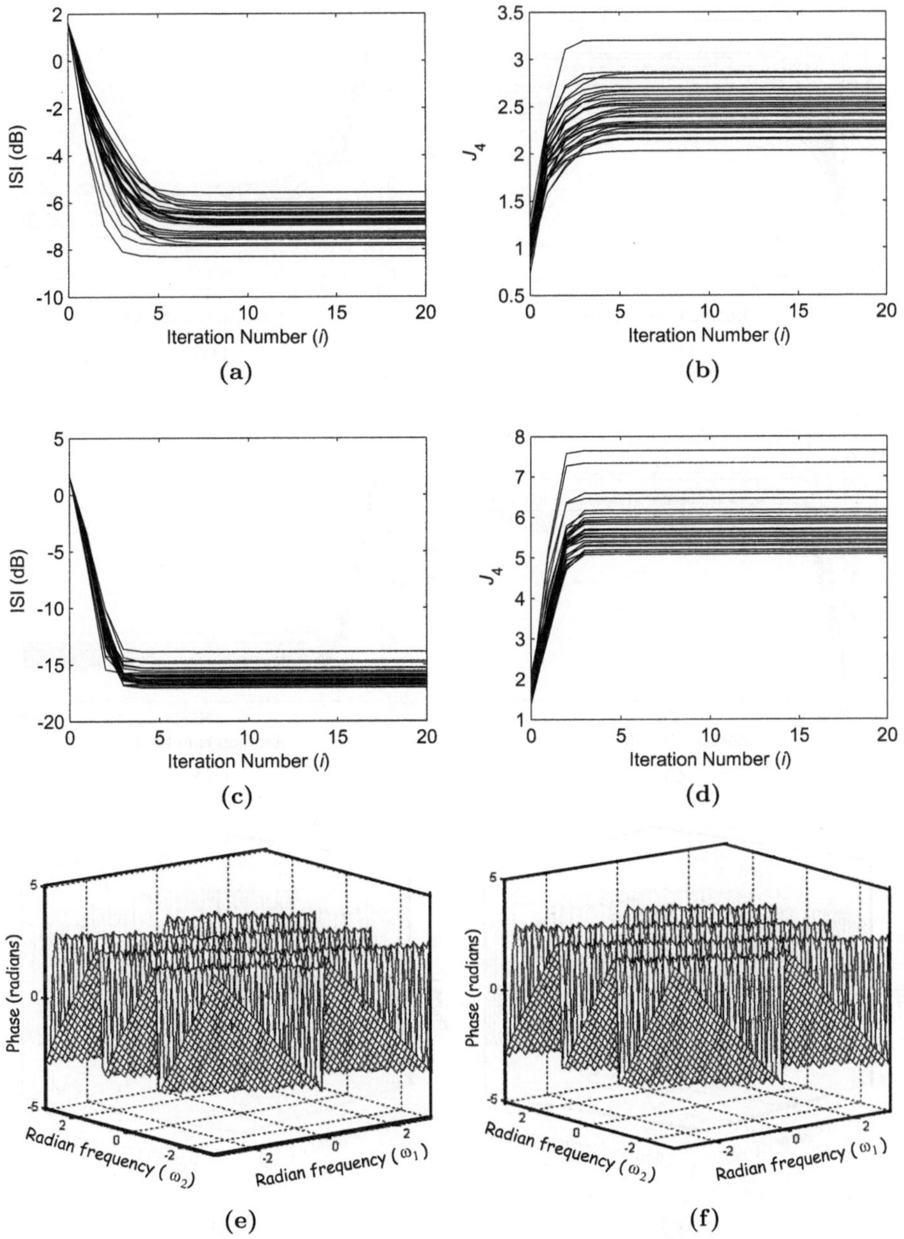

Fig. 7.6 Simulation results obtained by the 2-D SE deconvolution algorithm. (**a**) and (**c**) show the values of ISI versus iteration number and (**b**) and (**d**) show the associated $J_4(\boldsymbol{v})$ versus iteration number for SNR = 5 dB and SNR = 20 dB, respectively. (**e**) and (**f**) show the averaged $\arg[G(\omega_1, \omega_2)]$ for SNR = 5 dB and 20 dB, respectively

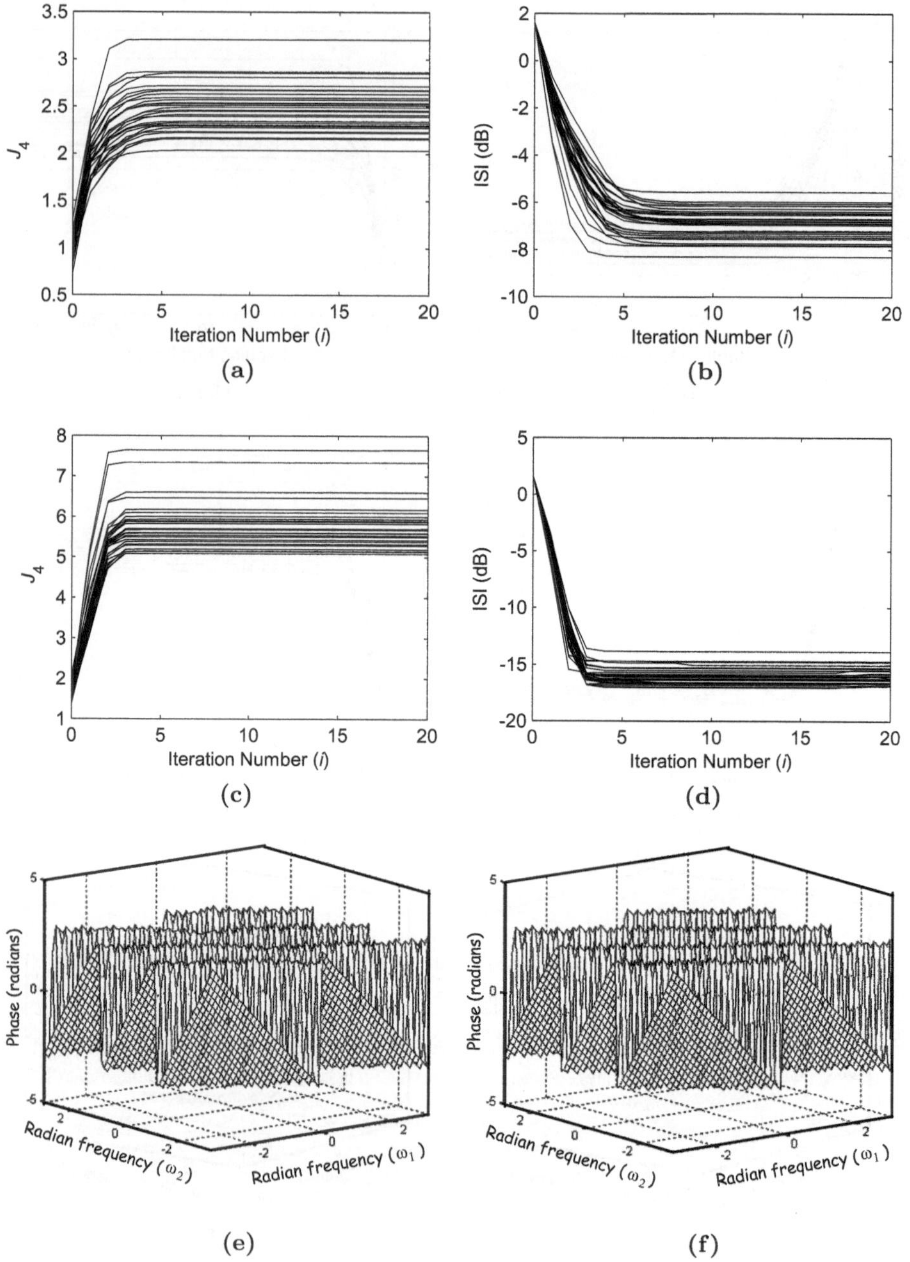

Fig. 7.7 Simulation results obtained by the 2-D hybrid-MNC deconvolution algorithm. (**a**) and (**c**) show $J_4(\boldsymbol{v})$ versus iteration number and (**b**) and (**d**) show the associated ISI versus iteration number for SNR = 5 dB and SNR = 20 dB, respectively. (**e**) and (**f**) show the averaged $\arg[G(\omega_1, \omega_2)]$ for SNR = 5 dB and 20 dB, respectively

algorithm is computationally much more efficient and faster than the 2-D MNC deconvolution algorithm for the design of $v_{\rm NNC}[n_1, n_2]$.

Figures 7.7e, f exhibit the averaged phase response of $\arg[G(\omega_1, \omega_2)]$ (principal values) of the 30 estimates of $g[n_1, n_2]$ obtained at the 20th iteration for SNR = 5 dB and 20 dB, respectively. Again, one can see from these figures that $\arg[G(\omega_1, \omega_2)]$ approximates a linear function of (ω_1, ω_2), thus verifying the linear phase property of $G(\omega_1, \omega_2)$ introduced in Property 7.3, and that the linear phase approximation for SNR = 20 dB (Fig. 7.7f) is better than that (Fig. 7.7e) for SNR = 5 dB.

7.6 Summary and Discussion

In this chapter, after a review of basic 2-D signals and systems, and 2-D random processes, the 2-D nonblind deconvolution algorithm design was introduced including the ZF deconvolution filter and the MMSE deconvolution filter in order to extract the input signal distorted by an unknown 2-D LSI system (channel). Then we introduced the 2-D LPE filter, a blind deconvolution filter using SOS, that performs as a minimum-phase whitening filter like an amplitude equalizer used in communications. Therefore, the phase distortion cannot be suppressed by the LPE filter as the unknown system (channel) is nonminimum phase.

Two 2-D blind deconvolution algorithms using HOS, the 2-D MNC and SE deconvolution algorithms, were then introduced. The former is a gradient-type algorithm with guaranteed convergence that, however, is not very computationally efficient even though the efficient BFGS optimization algorithm introduced in Chapter 2 has been utilized to find the optimum deconvolution filter $v_{\rm NNC}[n_1, n_2]$ associated with the objective function $J_{p,q}(\boldsymbol{v})$ given by (7.85). The latter is a fast and computationally efficient deconvolution algorithm but it may diverge for finite SNR and limited data. These two 2-D blind deconvolution algorithms have some interesting properties as presented in Section 7.4. Among these properties, Property 7.5, regarding the equivalence of the MNC deconvolution filter $v_{\rm NNC}[n_1, n_2]$ and the SE deconvolution filter $v_{\rm SE}[n_1, n_2]$ under some conditions, leads to the 2-D hybrid MNC deconvolution algorithm that shares all the advantages of the two deconvolution algorithms, including fast convergence, computational efficiency, and convergence guarantee. Some simulation results were also presented to demonstrate their performance. Some advanced applications of the 2-D hybrid MNC deconvolution algorithm will be discussed in Chapter 8.

Problems

7.1. Prove that the region of support for the minimum-phase system $h_{\mathrm{MP}}[n_1, n_2]$ given by (7.33) is the half plane Ω_{NSHP} and $h_{\mathrm{MP}}[0, 0] = 1$.

7.2. Derive the complex cepstrum $\mathtt{h}[n_1, n_2]$ of $H_{\mathrm{MG}}(\omega_1, \omega_2)$ given by (7.29) and $H_{\mathrm{PS}}(\omega_1, \omega_2)$ given by (7.30) in the MG-PS decomposition (7.28).

7.3. Prove Theorem 7.1.

7.4. Prove Property 7.2.

7.5. Prove Property 7.3.

7.6. Prove Property 7.4.

Computer Assignments

7.1. 2-D Deconvolution Filter Design
Consider that $y[n_1, n_2] = h[n_1, n_2] \star u[n_1, n_2] + w[n_1, n_2]$ where $h[n_1, n_2]$ is a separable system whose transfer function is given by

$$H(z_1, z_2) = (1 - 0.8z_1^{-1}) \cdot (1 - 0.8z_2^{-1}),$$

$u[n_1, n_2]$ is a zero-mean, i.i.d., exponentially distributed stationary process with variance $\sigma_u^2 = 1$, normalized skewness $\gamma_3\{u[n_1, n_2]\} = 2$ and normalized kurtosis $\gamma_4\{u[n_1, n_2]\} = 6$ and $w[n_1, n_2]$ is white Gaussian noise. Generate thirty independent sets of synthetic data $\{y[n_1, n_2], n_1 = 0, ..., 127, n_2 = 0, ..., 127\}$. With the generated synthetic data, we desire to design the 2-D deconvolution filter $v[n_1, n_2]$ with the region of support $\Omega_{\mathrm{TQP}}[5, 5]$ using the presented 2-D deconvolution filter algorithms with $v^{[0]}[n_1, n_2] = \delta[n_1 - 2, n_2 - 2]$.

(a) With cumulant order $p+q = 4$, initial step size $\mu_0 = 1$, $\mathbf{Q}^{[0]} = \mathbf{I}$ and $K = 10$ for step size search range, write a computer program to implement the 2-D MNC deconvolution algorithm. By performing thirty independent runs, obtain thirty MNC deconvolution filters $v_{\mathrm{MNC}}[n_1, n_2]$ for SNR = 20 dB and SNR = 5 dB and exhibit the associated $J_4(\boldsymbol{v})$ and ISI for the first 20 iterations. Explain what you observe from these results.

(b) With cumulant order $p + q = 4$, write a computer program to implement the 2-D SE deconvolution algorithm. By performing thirty independent runs, obtain thirty SE deconvolution filters $v_{\mathrm{SE}}[n_1, n_2]$ for SNR = 20 dB and SNR = 5 dB and exhibit the associated ISI and $J_4(\boldsymbol{v})$ for the first 20 iterations. Explain what you observe from these results.

(c) With cumulant order $p + q = 4$, initial step size $\mu_0 = 1$ and $K = 10$ for step size search range, write a computer program to implement the 2-D hybrid MNC deconvolution algorithm. By performing thirty independent runs, obtain thirty hybrid MNC deconvolution filters $v_{\mathrm{MNC}}[n_1, n_2]$ for SNR = 20 dB and SNR = 5 dB and exhibit the associated $J_4(\boldsymbol{v})$ and ISI for the first 20 iterations. Explain what you observe from these results.

References

1. T. L. Marzetta, "Two-dimensional linear prediction: Autocorrelation arrays, minimum-phase predictor error filters, and reflection coefficient arrays," *IEEE Trans. Acoustics, Speech and Signal Processing*, vol. 28, no. 6, pp. 725–733, June 1980.
2. J. A. Lim, *Two-dimensional Signal and Image Processing.* New Jersey: Prentice-Hall, 1990.
3. R. L. Kashyap and R. Chellappa, "Estimation and choice of neighbors in spatial-interaction models of images," *IEEE Trans. Information Theory*, vol. 29, no. 1, pp. 58–72, Jan. 1983.
4. S. R. Parker and A. H. Kayran, "Lattice parameter autoregressive modeling of two-dimensional fields-Part I: The quarter-plane case," *IEEE Trans. Acoustics, Speech and Signal Processing*, vol. 32, no. 4, pp. 872–885, Aug. 1984.
5. H.-M. Chien, H.-L. Yang, and C.-Y. Chi, "Parametric cumulant based phase estimation of 1-D and 2-D nonminimum phase systems by allpass filtering," *IEEE Trans. Signal Processing*, vol. 45, no. 7, pp. 1742–1762, July 1997.
6. C.-H. Chen, "Blind Multi-channel Equalization and Two-dimensional System Identification Using Higher-order Statistics," Ph.D. Dissertation, Department of Electrical Engineering, National Tsing Hua University, Taiwan, R.O.C., June 2001.
7. C.-H. Chen, C.-Y. Chi, and C.-Y. Chen, "2-D Fourier series based model for nonminimum-phase linear shift-invariant systems and texture image classification," *IEEE Trans. Signal Processing*, vol. 50, no. 4, pp. 945–955, April 2002.
8. L. R. Rabiner and R. W. Schafer, *Digital Processing of Speech Signals.* New Jersey: Prentice-Hall, 1978.
9. L. R. Rabiner, K. C. Pan, and F. K. Soong, "On the performance of isolated word speech recognizers using vector quantization and temporal energy contours," *AT&T Bell Laboratories Technical Journal*, vol. 63, no. 7, pp. 1245–1260, 1984.
10. A. V. Oppenheim and R. W. Schafer, *Discrete-Time Signal Processing.* New Jersey: Prentice-Hall, 1989.
11. T. E. Hall and G. B. Giannakis, "Bispectral analysis and model validation of texture images," *IEEE Trans. Image Processing*, vol. 4, no. 7, pp. 996–1009, July 1995.
12. C. L. Nikias and A. P. Petropulu, *Higher-Order Spectra Analysis: A Nonlinear Signal Processing Framework.* New Jersey: Prentice-Hall, 1993.
13. P. O. Amblard, M. Gaeta and J. L. Lacoume, "Statistics for complex variables and signals-Part II: Signals," *Signal Processing*, vol. 53, pp. 15–25, 1996.
14. O. Shalvi and E. Weinstein, Universal methods for blind deconvolution, A Chapter in *Blind Deconvolution*, S. Haykin, ed. New Jersey: Prentice-Hall, 1994.

15. D. E. Dudgeon and R. M. Mersereau, *Multidimensional Digital Signal Processing*. New Jersey: Prentice-Hall, 1984.
16. T. E. Hall and G. B. Giannakis, "Image modeling using inverse filtering criteria with application to textures," *IEEE Trans. Image Processing*, vol. 5, no. 6, pp. 938–949, June 1996.
17. M. K. Tsatsanis and G. B. Giannakis, "Object and texture classification using higher order statistics," *IEEE Trans. Pattern Analysis and Machine Intelligence*, vol. 14, no. 7, pp. 733–750, July 1992.
18. A. Swami, G. B. Giannakis, and J. M. Mendel, "Linear modeling of multidimensional non-Gaussian processes using cumulants," *Multidimensional Systems and Signal Processing*, vol. 1, pp. 11–37, 1990.
19. J. K. Tugnait, "Estimation of linear parametric models of nonGaussian discrete random fields with application to texture synthesis," *IEEE Trans. Image Processing*, vol. 3, no. 2, pp. 109–127, Feb. 1994.
20. S. Bhattacharya, N. C. Ray, and S. Sinha, "2-D signal modelling and reconstruction using third-order cumulants," *Signal Processing*, vol. 62, pp. 61–72, 1997.
21. C.-Y. Chi and C.-H. Chen, "Two-dimensional frequency-domain blind system identification using higher order statistics with application to texture synthesis," *IEEE Trans. Signal Processing*, vol. 49, no. 4, pp. 864–877, April 2001.

8

Applications of Two-Dimensional Blind Deconvolution Algorithms

Two-dimensional BSI is needed in 2-D random signal modeling, texture synthesis and classification. In the chapter, a BSI algorithm, which is an iterative FFT based nonparametric 2-D BSI algorithm using the 2-D hybrid MNC deconvolution algorithm introduced in Chapter 7, will be introduced with application to texture synthesis. With the 2-D FSBM model for 2-D LSI systems in Chapter 7, the chapter also introduces a 2-D FSBM based parametric BSI algorithm, that includes an amplitude estimator using SOS and two phase estimators using HOS, with application to texture image classification.

8.1 Nonparametric Blind System Identification and Texture Synthesis

Consider that $y[n_1, n_2]$ is a 2-D real discrete-space signal modeled by

$$\begin{aligned} y[n_1, n_2] &= x[n_1, n_2] + w[n_1, n_2] \\ &= u[n_1, n_2] \star h[n_1, n_2] + w[n_1, n_2], \end{aligned} \tag{8.1}$$

where $u[n_1, n_2]$, $h[n_1, n_2]$ and $w[n_1, n_2]$ satisfy the following assumptions:

(A8-1) $u[n_1, n_2]$ is a real, zero-mean, stationary, i.i.d., nonGaussian 2-D random field with variance σ_u^2 and pth-order ($p \geq 3$) cumulant $C_p\{u[n_1, n_2]\} \neq 0$.

(A8-2) Both $h[n_1, n_2]$ and its inverse system $h_{\mathrm{INV}}[n_1, n_2]$ are real stable 2-D LSI systems.

(A8-3) $w[n_1, n_2]$ is a real, zero-mean, stationary (white or colored) Gaussian 2-D random field with variance σ_w^2. Moreover, $w[n_1, n_2]$ is statistically independent of $u[n_1, n_2]$.

Note that the above three assumptions are nothing but the Assumptions (A7-1), (A7-2) and (A7-3) for the case of real $y[n_1, n_2]$.

Two-dimensional BSI is a problem of extracting information about the unknown 2-D LSI system $h[n_1, n_2]$ using only measurements $y[n_1, n_2]$. As introduced in Chapter 7, the 2-D MNC deconvolution filter $v_{\rm MNC}[n_1, n_2]$ can be obtained efficiently using the 2-D hybrid MNC algorithm, and the deconvolution filter $v_{\rm MNC}[n_1, n_2]$ is related to the unknown system $h[n_1, n_2]$ in a nonlinear manner as presented in Property 7.2. This relation is valid for finite SNR which implies that the unknown system $h[n_1, n_2]$ can be estimated accurately from $v_{\rm MNC}[n_1, n_2]$ and power spectral density $S_y(\omega_1, \omega_2)$ for finite SNR. Next, let us introduce the resultant nonparametric 2-D BSI algorithm based on this relation.

8.1.1 Nonparametric 2-D BSI

By Property 7.2, one can easily show that for real $y[n_1, n_2]$, the 2-D MNC deconvolution filter $V_{\rm MNC}(\omega_1, \omega_2)$ by maximizing $J_{2p}(\boldsymbol{v})$ (see (7.85)) for $p \geq 2$, is related to the unknown system $H(\omega_1, \omega_2)$ by

$$G_1(\omega_1, \omega_2)H^*(\omega_1, \omega_2) = \beta \cdot V_{\rm MNC}(\omega_1, \omega_2)S_y(\omega_1, \omega_2) \tag{8.2}$$

where $G_1(\omega_1, \omega_2)$ is the 2-D DSFT of

$$g_1[n_1, n_2] = \widetilde{g}_{p,p}[n_1, n_2] = g^{2p-1}[n_1, n_2] \tag{8.3}$$

in which $\widetilde{g}_{p,p}[n_1, n_2]$ is given by (7.99),

$$g[n_1, n_2] = v_{\rm MNC}[n_1, n_2] \star h[n_1, n_2], \tag{8.4}$$

and

$$\beta = \frac{1}{\sigma_e^2} \sum_{n_1=-\infty}^{\infty} \sum_{n_2=-\infty}^{\infty} g^{2p}[n_1, n_2] > 0. \tag{8.5}$$

Let us emphasize that the relation given by (8.2) is true for finite SNR. By Property 7.3, the overall system $g[n_1, n_2]$ is linear phase, which implies that $g_1[n_1, n_2]$ is also linear phase. The solution set for the system $h[n_1, n_2]$ satisfying the relation given by (8.2) is stated in the following property.

Property 8.1. *Any 2-D real system*

$$H'(\omega_1, \omega_2) = \alpha H(\omega_1, \omega_2) \cdot e^{j(\omega_1\tau_1 + \omega_2\tau_2)} \tag{8.6}$$

for any real $\alpha \neq 0$ and integers τ_1 and τ_2 satisfies the relationship (8.2).

The proof of Property 8.1 is left as an exercise (Problem 8.1). By Properties 8.1 and 7.3, $g[n_1, n_2]$ can be zero-phase if τ_1 and τ_2 are properly removed, leading to the following property for $g_1[n_1, n_2]$.

Property 8.2. *Assume that $g[n_1, n_2] \neq 0$ is zero-phase and that*

(A8-4) *the number of zeros of $\mathcal{G}_1(z) = G(z_1 = z, z_2)$ where $|z_2| = 1$, and that of $\mathcal{G}_2(z) = G(z_1, z_2 = z)$ where $|z_1| = 1$ are both finite on the unit circle.*

Then the associated system $g_1[n_1, n_2]$ is a positive definite sequence, i.e. $G_1(\omega_1, \omega_2) > 0, \forall -\pi \leq \omega_1 < \pi, -\pi \leq \omega_2 < \pi$.

For the proof of Property 8.2, see Appendix 8A. Next, let us present the solution set to (8.2) by the following property.

Property 8.3. *The system $H(\omega_1, \omega_2)$ can be identified up to a linear phase ambiguity by solving (8.2) provided that $g[n_1, n_2] \neq 0$ is zero-phase and satisfies Assumption* (A8-4).

See Appendix 8B for the proof of Property 8.3, which needs Lemma 6.10.

Properties 8.1 and 8.3 indicate that, under the constraint $g[n_1, n_2] \neq 0$ being zero-phase and the Assumption (A8-4), all the $H'(\omega_1, \omega_2)$ satisfying (8.6) form a solution set of (8.2). Therefore, a consistent channel estimate $\widehat{h}[n_1, n_2]$ for finite SNR can be obtained up to a scale factor and a space shift by solving (8.2) in which $V(\omega_1, \omega_2)$ and $S_y(\omega_1, \omega_2)$ are substituted by their consistent estimates. However, from (8.3) and (8.4), one can see that (8.2) is a highly nonlinear equation in $h[n_1, n_2]$ for which closed-form solution of (8.2) is formidable. Next, let us introduce an iterative FFT based algorithm for estimating $h[n_1, n_2]$ under the zero-phase constraint on $g[n_1, n_2]$.

2-D System-MNC Filter Based BSI Algorithm

A consistent estimate of $S_y(\omega_1, \omega_2)$ with measurements $y[n_1, n_2]$ can be obtained from the LPE filter $v_{\rm LPE}[n_1, n_2]$ with region of support $\Omega_{\rm TNSHP}[p_1, p_2]$ introduced in Chapter 7. For sufficiently large p_1 and p_2, $v_{\rm LPE}[n_1, n_2]$ performs as a whitening filter or an amplitude equalizer as given by (7.79), which is repeated as follows:

$$|V_{\rm LPE}(\omega_1, \omega_2)|^2 \propto \frac{1}{S_y(\omega_1, \omega_2)}. \tag{8.7}$$

From (8.2) and (8.7), we obtain

$$G_1(\omega_1, \omega_2) H^*(\omega_1, \omega_2) = \gamma \cdot \frac{V_{\rm MNC}(\omega_1, \omega_2)}{|V_{\rm LPE}(\omega_1, \omega_2)|^2} \tag{8.8}$$

where $\gamma > 0$ is a constant.

Let $G_1(k_1, k_2)$, $H[k_1, k_2]$, $V_{\rm MNC}[k_1, k_2]$ and $V_{\rm LPE}[k_1, k_2]$ denote the $L \times L$-point DFT of $g_1[n_1, n_2]$, $h[n_1, n_2]$, $v_{\rm MNC}[n_1, n_2]$ and $v_{\rm LPE}[n_1, n_2]$, respectively.

Let

$$A[k_1, k_2] = G_1[k_1, k_2] \cdot H^*[k_1, k_2] \tag{8.9}$$

and

$$B[k_1, k_2] = \frac{V_{\mathrm{MNC}}[k_1, k_2]}{|V_{\mathrm{LPE}}[k_1, k_2]|^2}. \tag{8.10}$$

Then, according to the relation between $v_{\mathrm{MNC}}[n_1, n_2]$ and the unknown system $h[n_1, n_2]$ given by (8.8), one can obtain

$$\mathsf{a} = \gamma \mathsf{b} \tag{8.11}$$

where

$$\mathsf{a} = (\mathsf{a}_0^T, ..., \mathsf{a}_{L-1}^T)^T \tag{8.12}$$

in which $\mathsf{a}_i = (A[i, 0], ..., A[i, L-1])^T$, and

$$\mathsf{b} = (\mathsf{b}_0^T, ..., \mathsf{b}_{L-1}^T)^T \tag{8.13}$$

in which $\mathsf{b}_i = (B[i, 0], ..., B[i, L-1])^T$. Note that the constant $\gamma > 0$ in (8.11) is unknown, and therefore, either it can be estimated together with the estimation of $H(\omega_1, \omega_2)$ or its role ought to be virtual during the estimation of $H(\omega_1, \omega_2)$. Consequently, (8.9), (8.10) and (8.11) imply that the true system $H(\omega_1, \omega_2)$ is the one such that the angle between the two column vectors a and b is zero. Thus, $H[k_1, k_2]$ can be estimated by maximizing

$$\mathcal{C}(H) = \frac{\mathrm{Re}\{\mathsf{a}^H \mathsf{b}\}}{\|\mathsf{a}\| \cdot \|\mathsf{b}\|}. \tag{8.14}$$

Note that $-1 \leq \mathcal{C}(H) = \mathcal{C}(\alpha H) \leq 1$ for any $\alpha > 0$ and that $\mathcal{C}(H) = 1$ if and only if (8.11) holds true.

However, it is quite involved to obtain the gradient $\partial \mathcal{C}(H)/\partial H[k_1, k_2]$ because $\mathcal{C}(H)$ is a highly nonlinear function of $H[k_1, k_2]$. Therefore, gradient based optimization methods are not considered suitable for finding the maximum of $\mathcal{C}(H)$. Instead, an iterative nonparametric BSI algorithm using FFT, called the *2-D System-MNC Filter Based BSI Algorithm*, is introduced to obtain an estimate of $H[k_1, k_2]$ as follows.

2-D System-MNC Filter Based BSI Algorithm

(T1) Blind Deconvolution.

- (S1) Obtain the 2-D LPE filter $v_{\mathrm{LPE}}[n_1, n_2]$ with region of support $\Omega_{\mathrm{TNSHP}}[p_1, p_2]$ for a chosen (p_1, p_2) by (7.78).
- (S2) Obtain the $v_{\mathrm{MNC}}[n_1, n_2]$ with region of support $\Omega_{\mathrm{TQP}}[p_1, p_2]$ for a chosen (p_1, p_2) by the 2-D hybrid MNC algorithm with cumulant order set to $2p$.
- (S3) Obtain $L \times L$-point DFTs $V_{\mathrm{MNC}}[k_1, k_2]$ and $V_{\mathrm{LPE}}[k_1, k_2]$ for a chosen L, and then compute $B[k_1, k_2]$ given by (8.10).

(T2) System Estimation.

(S1) Set the iteration number $i = 0$ and the initial values $H^{[0]}[k_1, k_2]$ for $H[k_1, k_2]$ and the convergence tolerance $\zeta > 0$.

(S2) Update i by $i + 1$. Compute

$$G^{[i-1]}[k_1, k_2] = H^{[i-1]}[k_1, k_2] \cdot V[k_1, k_2]$$

and its $L \times L$-point 2-D inverse FFT $g^{[i-1]}[n_1, n_2]$. Then update $g^{[i-1]}[n_1, n_2]$ by $g[n_1, n_2]$, where

$$g[n_1, n_2] = \frac{1}{2}\left(g^{[i-1]}[n_1, n_2] + g^{[i-1]}[-n_1, -n_2]\right). \text{ (by Property 8.3)}$$

(S3) Compute $g_1[n_1, n_2]$ using (8.4) with $g[n_1, n_2] = g^{[i-1]}[n_1, n_2]$ and its $L \times L$-point 2-D FFT $G_1[k_1, k_2]$.

(S4) Compute

$$H^{[i]}[k_1, k_2] = \frac{B^*[k_1, k_2]}{G_1[k_1, k_2]} \tag{8.15}$$

by (8.8), (8.9) and (8.10) and then normalize $H^{[i]}[k_1, k_2]$ by $\sum_{k_1=0}^{L-1}\sum_{k_2=0}^{L-1} |H^{[i]}[k_1, k_2]|^2 = 1$.

(S5) Obtain $G^{[i]}[k_1, k_2] = H^{[i]}[k_1, k_2] \cdot V_{\text{MNC}}[k_1, k_2]$ followed by its $L \times L$-point 2-D inverse FFT $g^{[i]}[n_1, n_2]$. If $\mathcal{C}(H^{[i]}) > \mathcal{C}(H^{[i-1]})$, go to (S6). Otherwise, compute $\Delta H[k_1, k_2] = H^{[i]}[k_1, k_2] - H^{[i-1]}[k_1, k_2]$ and update $H^{[i]}[k_1, k_2]$ via

$$H^{[i]}[k_1, k_2] = H^{[i-1]}[k_1, k_2] + \mu^{[i-1]} \Delta H[k_1, k_2] \tag{8.16}$$

where the step size $\mu^{[i-1]}$ is chosen such that $\mathcal{C}(H^{[i]}) > \mathcal{C}(H^{[i-1]})$. Then normalize $H^{[i]}[k_1, k_2]$ by $\sum_{k_1=0}^{L-1}\sum_{k_2=0}^{L-1} |H^{[i]}[k_1, k_2]|^2 = 1$.

(S6) If

$$\frac{\mathcal{C}(H^{[i]}) - \mathcal{C}(H^{[i-1]})}{|\mathcal{C}(H^{[i-1]})|} > \zeta$$

then go to (S2); otherwise, the frequency response estimate

$$\widehat{H}[k_1, k_2] = H^{[i]}[k_1, k_2]$$

and its $L \times L$-point 2-D inverse FFT $\widehat{h}[n_1, n_2]$ are obtained.

The above 2-D system-MNC filter based BSI algorithm is a frequency-domain nonparametric system estimation algorithm that obtains an estimate $\widehat{H}[k_1, k_2]$ by processing the given nonGaussian measurements $y[n_1, n_2]$. The

$\widehat{h}[n_1, n_2]$ obtained is an estimate of $h[n_1, n_2]$ except for a scale factor and a space shift by Property 8.1. In (T2), the local convergence is guaranteed because $\mathcal{C}(H^{[i]})$ is upper bounded by unity and its value is increased at each iteration before convergence. Therefore, the closer to unity the $\mathcal{C}(\widehat{H})$, the more reliable the system estimate obtained.

Note that in (S5) of (T2), the step size $\mu^{[i]} = \pm\mu_0/2^l$ can be used with a suitable $l \in \{0, 1, ..., K\}$ where μ_0 and K are preassigned constant and positive integer, respectively, such that $\mathcal{C}(H^{[i+1]}) > \mathcal{C}(H^{[i]})$ at each iteration. It is remarkable that an upper bound of the system order is needed though the exact system order is not required. The FFT length $L \times L$ should be chosen sufficiently large that aliasing effects on the resultant $\widehat{h}[n_1, n_2]$ are negligible. Surely the larger L, the larger the computational load of the 2-D system-MNC filter based BSI algorithm, while the estimation error of the resultant $\widehat{h}[n_1, n_2]$ is almost the same. If the true system is an IIR system, a finite-length approximation of $h[n_1, n_2]$ will be obtained by the 2-D system-MNC filter based BSI algorithm.

Estimation of the MMSE Deconvolution Filter and Noise-Reduction Filter

With the 2-D LSI system estimate $\widehat{H}(\omega_1, \omega_2)$ and the LPE filter $V_{\rm LPE}(\omega_1, \omega_2)$ obtained by the preceding 2-D system-MNC filter based BSI algorithm, the MMSE deconvolution filter $V_{\rm MS}(\omega_1, \omega_2)$ and the MMSE noise-reduction filter $V_{\rm NR}(\omega_1, \omega_2)$, can be estimated as

$$\widehat{V}_{\rm MS}(\omega_1, \omega_2) = \frac{\widehat{H}^*(\omega_1, \omega_2)}{\widehat{S}_y(\omega_1, \omega_2)} \qquad \text{(by (7.63))} \tag{8.17a}$$

$$= \widehat{H}^*(\omega_1, \omega_2)|V_{\rm LPE}(\omega_1, \omega_2)|^2 \qquad \text{(by (8.7))} \tag{8.17b}$$

and

$$\widehat{V}_{\rm NR}(\omega_1, \omega_2) = \widehat{V}_{\rm MS}(\omega_1, \omega_2) \cdot \widehat{H}(\omega_1, \omega_2), \qquad \text{(by (7.64))} \tag{8.18}$$

respectively, up to a scale factor and a space shift.

Example 8.4 (Estimation of a 2-D ARMA System)
The driving input signal $u[n_1, n_2]$ was assumed to be a real zero-mean, exponentially distributed, i.i.d., random field with variance $\sigma_u^2 = 1$. The unknown 2-D LSI system $h[n_1, n_2]$ used was a 2-D ARMA model with a 3×3 nonsymmetric region of support $\{[n_1, n_2]| -1 \le n_1 \le 1, -1 \le n_2 \le 1\}$ taken from [1] as follows

$$\begin{aligned}x[n_1, n_2] &= 0.004x[n_1 + 1, n_2 + 1] - 0.0407x[n_1 + 1, n_2] \\ &+ 0.027x[n_1 + 1, n_2 - 1] + 0.2497x[n_1, n_2 + 1] + 0.568x[n_1, n_2 - 1] \\ &- 0.1037x[n_1 - 1, n_2 + 1] + 0.3328x[n_1 - 1, n_2] - 0.1483x[n_1 - 1, n_2 - 1] \\ &+ u[n_1, n_2] - 0.5u[n_1 + 1, n_2] - 0.5u[n_1, n_2 + 1] - u[n_1, n_2 - 1] \\ &- u[n_1 - 1, n_2].\end{aligned} \tag{8.19}$$

In each of the thirty independent runs performed, a 256 × 256 synthetic $y[n_1, n_2]$ was generated for SNR = 5 dB and $w[n_1, n_2]$ being real lowpass colored Gaussian noise generated from the output of a 2-D separable filter $(1+0.8z_1^{-1}) \cdot (1+0.8z_2^{-1})$ driven by 2-D white Gaussian noise. Then the synthetic $y[n_1, n_2]$ was processed by the proposed 2-D system-MNC filter based BSI algorithm. In (T1), the parameters $p_1 = p_2 = 5$ were used in (S1) and (S2). The initial condition $v^{[0]}[n_1, n_2] = \delta[n_1 - 2, n_2 - 2]$ was used by the hybrid MNC algorithm with the cumulant order equal to $2p = 4$ in (S2), and FFT size $L \times L = 32 \times 32$ was used in (S3). In (T2), $H^{[0]}[k_1, k_2] = 1$ for all k_1 and k_2, $\mu_0 = 1$, $K = 10$ and $\zeta = 10^{-5}$ were used.

Only $\widehat{H}(\omega_1, \omega_2)$ and $\widehat{V}_{\mathrm{MS}}(\omega_1, \omega_2)$ given by (8.17) will be presented because the MMSE noise-reduction filter estimate $\widehat{V}_{\mathrm{NR}}(\omega_1, \omega_2)$ given by (8.18) is actually redundant. The true system $h[n_1, n_2]$, the true $v_{\mathrm{MS}}[n_1, n_2]$, the estimate $\widehat{h}[n_1, n_2]$, and the estimate $\widehat{v}_{\mathrm{MS}}[n_1, n_2]$ were normalized by unity energy, and the 2-D space shift between $\widehat{h}[n_1, n_2]$ and the true $h[n_1, n_2]$ and that between $\widehat{v}_{\mathrm{MS}}[n_1, n_2]$ and the true $v_{\mathrm{MS}}[n_1, n_2]$ were artificially removed before calculation of the averages and root mean square (RMS) errors of $\widehat{h}[n_1, n_2]$ and $\widehat{v}_{\mathrm{MS}}[n_1, n_2]$.

Let $\widehat{h}_i[n_1, n_2]$ and $\widehat{v}_{\mathrm{MS},i}[n_1, n_2]$ denote the $\widehat{h}[n_1, n_2]$ and $\widehat{v}_{\mathrm{MS}}[n_1, n_2]$ obtained at the ith run. As performance indices, the averages of $\widehat{h}[n_1, n_2]$ and $\widehat{v}_{\mathrm{MS}}[n_1, n_2]$, denoted $\bar{h}[n_1, n_2]$ and $\bar{v}_{\mathrm{MS}}[n_1, n_2]$,

$$\bar{h}[n_1, n_2] = \frac{1}{30} \sum_{i=1}^{30} \widehat{h}_i[n_1, n_2], \tag{8.20}$$

$$\bar{v}_{\mathrm{MS}}[n_1, n_2] = \frac{1}{30} \sum_{i=1}^{30} \widehat{v}_{\mathrm{MS},i}[n_1, n_2], \tag{8.21}$$

and RMS errors, denoted $\sigma(\widehat{h}[n_1, n_2])$ and $\sigma(\widehat{v}_{\mathrm{MS}}[n_1, n_2])$,

$$\sigma(\widehat{h}[n_1, n_2]) = \left(\frac{1}{30} \sum_{i=1}^{30} \left(h[n_1, n_2] - \widehat{h}_i[n_1, n_2] \right)^2 \right)^{1/2}, \tag{8.22}$$

$$\sigma(\widehat{v}_{\mathrm{MS}}[n_1, n_2]) = \left(\frac{1}{30} \sum_{i=1}^{30} \left(v_{\mathrm{MS}}[n_1, n_2] - \widehat{v}_{\mathrm{MS},i}[n_1, n_2] \right)^2 \right)^{1/2} \tag{8.23}$$

were calculated, respectively, from the thirty independent runs.

The 2-D system-MNC filter based BSI algorithm required only three iterations in (T2) for the estimation of $h[n_1,n_2]$ at each run in obtaining all the results in this example. Figures 8.1a, d show the true $h[n_1,n_2]$ and $v_{\rm MS}[n_1,n_2]$, respectively. Figures 8.1b, c, e, f show $\bar{h}[n_1,n_2]$, $\sigma(\widehat{h}[n_1,n_2])$, $\bar{v}_{\rm MS}[n_1,n_2]$ and $\sigma(\widehat{v}_{\rm MS}[n_1,n_2])$, respectively. One can see from these figures that $\bar{h}[n_1,n_2]$ and $\bar{v}_{\rm MS}[n_1,n_2]$ shown in Fig. 8.1b, e are quite close to the true $h[n_1,n_2]$ and $v_{\rm MS}[n_1,n_2]$ shown in Fig. 8.1a, d, respectively, and the associated RMS errors $\sigma(\widehat{h}[n_1,n_2])$ and $\sigma(\widehat{v}_{\rm MS}[n_1,n_2])$ shown in Fig. 8.1c, f are also small. These simulation results demonstrate that the 2-D system-MNC filter based BSI algorithm provides estimates of both the unknown 2-D system and MMSE deconvolution filter that are good approximations to the true ARMA system and the true MMSE deconvolution filter for low SNR (5 dB), respectively. □

8.1.2 Texture Synthesis

It has been validated [1–9] that an $N \times N$ texture image can be modeled as a 2-D stationary nonGaussian linear process as follows

$$x[n_1,n_2] = u[n_1,n_2] \star h[n_1,n_2] + m_x \tag{8.24}$$

where $x[n_1,n_2] \geq 0$ due to finite gray levels, $u[n_1,n_2]$ is zero-mean nonGaussian, $h[n_1,n_2]$ is a stable 2-D LSI system, and m_x is the mean of $x[n_1,n_2]$. The 2-D system-MNC filter based BSI algorithm introduced above can be applied to estimate both $h[n_1,n_2]$ and $u[n_1,n_2]$ from which synthetic texture images, denoted as $\widehat{x}[n_1,n_2]$, with the same statistical characteristics as $x[n_1,n_2]$ can be obtained. The resultant *texture synthesis method (TSM)* includes five steps as follows.

TSM

($\mathcal{S}1$) Obtain

$$\widehat{m}_x = \frac{1}{N^2} \sum_{n_1=0}^{N-1} \sum_{n_2=0}^{N-1} x[n_1,n_2] \tag{8.25}$$

and

$$\begin{aligned} y[n_1,n_2] &= x[n_1,n_2] - \widehat{m}_x \\ &\simeq u[n_1,n_2] \star h[n_1,n_2] \end{aligned} \tag{8.26}$$

which is an approximation of the 2-D convolutional model given by (8.1) for SNR$= \infty$.

($\mathcal{S}2$) Obtain the texture image model $\widehat{h}[n_1,n_2]$ and the 2-D LPE filter $V_{\rm LPE}(\omega_1,\omega_2)$ with region of support $\Omega_{\rm TNSHP}[p_1,p_2]$ by processing $y[n_1,n_2]$ using the 2-D system-MNC filter based BSI algorithm with a chosen set

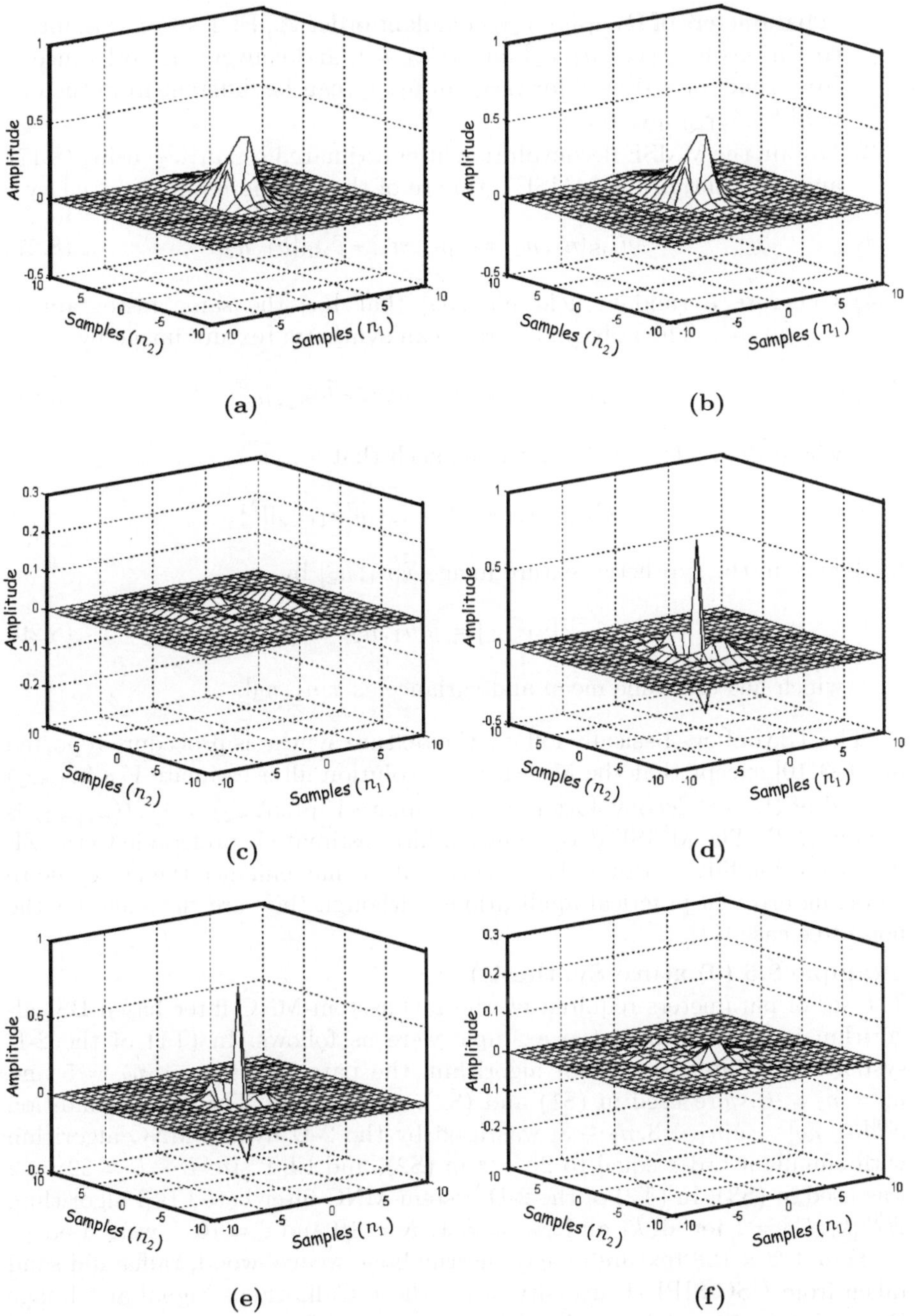

Fig. 8.1 (**a**) and (**d**) show the true $h[n_1, n_2]$ and $v_{\rm MS}[n_1, n_2]$, respectively. (**b**), (**e**), (**c**) and (**f**) show the averages $\bar{h}[n_1, n_2]$ and $\bar{v}_{\rm MS}[n_1, n_2]$ and RMS errors $\sigma(\widehat{h}[n_1, n_2])$ and $\sigma(\widehat{v}_{\rm MS}[n_1, n_2])$, respectively

of parameters of $\Omega_{\text{TQP}}[p_1, p_2]$, cumulant order $2p$, FFT size $L \times L$, initial conditions for $v_{\text{MNC}}[n_1, n_2]$ and $H[k_1, k_2]$, and convergence tolerance ζ. Note that (p_1, p_2) used for $\Omega_{\text{TNSHP}}[p_1, p_2]$ can be different from the one used for $\Omega_{\text{TQP}}[p_1, p_2]$.

($\mathcal{S}3$) Obtain the MMSE deconvolution filter estimate $\widehat{V}_{\text{MS}}(\omega_1, \omega_2)$ using (8.17b) and then obtain the MMSE estimate of the driving input $u[n_1, n_2]$ by

$$\widehat{u}_{\text{MS}}[n_1, n_2] = y[n_1, n_2] \star \widehat{v}_{\text{MS}}[n_1, n_2]. \tag{8.27}$$

($\mathcal{S}4$) Generate a random field $\widetilde{u}[n_1, n_2]$ that has the same histogram as $\widehat{u}_{\text{MS}}[n_1, n_2]$. Then obtain a zero-mean synthetic texture image by

$$\widetilde{y}[n_1, n_2] = \widetilde{u}[n_1, n_2] \star \gamma\widehat{h}[n_1, n_2] \tag{8.28}$$

where the scale factor γ is chosen such that

$$E\{|\widetilde{y}[n_1, n_2]|^2\} = E\{|y[n_1, n_2]|^2\}.$$

($\mathcal{S}5$) Obtain the synthetic texture image $\widehat{x}[n_1, n_2]$ by

$$\widehat{x}[n_1, n_2] = \widetilde{y}[n_1, n_2] + \widehat{m}_x \tag{8.29}$$

which has the same mean and variance as $x[n_1, n_2]$.

The TSM above basically follows the texture synthesis procedure reported in [1–3, 10] except that the MMSE deconvolution filter estimate $\widehat{V}_{\text{MS}}(\omega_1, \omega_2)$ instead of the ZF deconvolution filter estimate $\widehat{V}_{\text{ZF}}(\omega_1, \omega_2) = 1/\widehat{H}(\omega_1, \omega_2)$ is used in ($\mathcal{S}3$). The MMSE deconvolution filter estimate is preferable to the ZF deconvolution filter estimate because the latter may enhance the noise due to modeling error in practical applications, although they are the same for the noise-free case [11].

Example 8.5 (Texture Synthesis)

The set of parameters required by the 2-D system-MNC filter based BSI algorithm in ($\mathcal{S}2$) used in the example were as follows. In (T1) of the 2-D system-MNC filter based BSI algorithm, the parameters $p_1 = p_2 = 5$ and $p_1 = p_2 = 6$ were used in (S1) and (S2), respectively. The initial condition $v^{[0]}[n_1, n_2] = \delta[n_1 - 3, n_2 - 3]$ was used by the 2-D hybrid MNC algorithm with cumulant order equal to $2p = 4$ in (S2), and FFT size $L \times L = 32 \times 32$ was used in (S3). In (T2) of the 2-D system-MNC filter based BSI algorithm, $H^{[0]}[k_1, k_2] = 1$ for all k_1 and k_2, $\mu_0 = 1$, $K = 10$ and $\zeta = 10^{-5}$ were used.

Four 128×128 texture images, herringbone weave, wood, raffia and sand taken from USC-SIPI (University of Southern California - Signal and Image Processing Institute) Image Data Base were used for texture synthesis using the proposed TSM. The 2-D system-MNC filter based BSI algorithm spent 11, 4, 14 and 3 iterations for system estimation (performed in (T2)) in obtaining $\widehat{H}(\omega_1, \omega_2)$ associated with herringbone weave, wood, raffia and sand images,

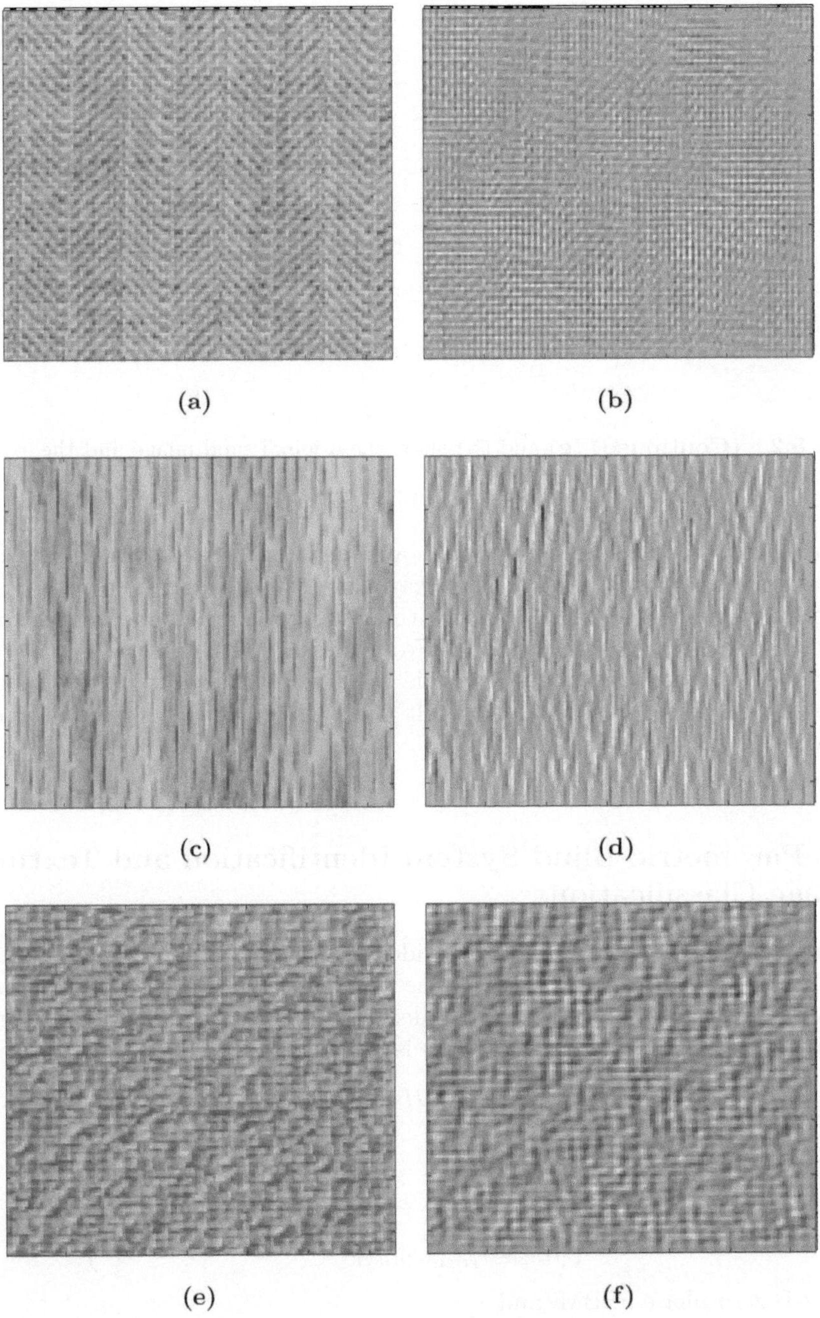

Fig. 8.2 Experimental results using TSM. (**a**), (**c**) and (**e**) show the original herringbone weave, wood and raffia images, respectively. (**b**), (**d**) and (**f**) show the synthetic herringbone weave, wood and raffia images, respectively

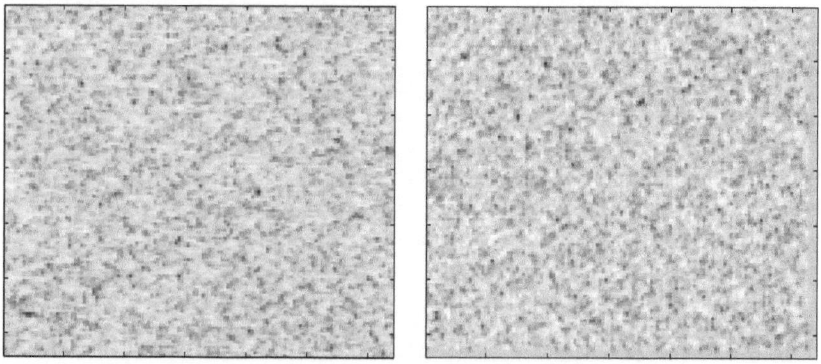

(g) (h)

Fig. 8.2 **(Continued)** (**g**) and (**h**) show the original sand image and the synthetic sand image, respectively

respectively. The experimental results are shown in Fig. 8.2. Figures 8.2a, c, e, g show the original herringbone weave, wood, raffia and sand images, respectively, and Figs. 8.2b, d, f, h show the synthetic herringbone weave, wood, raffia and sand images, respectively. From these figures, one can see that the four synthetic texture images resemble well the four respective original texture images. These experimental results support the efficacy of the proposed TSM.

□

8.2 Parametric Blind System Identification and Texture Image Classification

As reported in [12], this section considers the identification of 2-D real stable LSI systems through the use of the parametric 2-D FSBM introduced in Chapter 7 (Section 7.1). The MG-PS decomposition of the FSBM with region of support $\Omega'_{\mathrm{TNSHP}}[p_1, p_2]$ (see (7.31)) is given by

$$H(\omega_1, \omega_2) = H^*(-\omega_1, -\omega_2) = H_{\mathrm{MG}}(\omega_1, \omega_2) \cdot H_{\mathrm{PS}}(\omega_1, \omega_2) \tag{8.30}$$

where

$$H_{\mathrm{MG}}(\omega_1, \omega_2) = \exp\left\{ \sum\sum_{[i_1,i_2]\in\Omega'_{\mathrm{TNSHP}}[p_1,p_2]} \alpha_{i_1,i_2} \cos(i_1\omega_1 + i_2\omega_2) \right\} \tag{8.31}$$

is a 2-D zero-phase FSBM, and

$$H_{\mathrm{PS}}(\omega_1, \omega_2) = \exp\left\{ j \sum\sum_{[i_1,i_2]\in\Omega'_{\mathrm{TNSHP}}[p_1,p_2]} \beta_{i_1,i_2} \sin(i_1\omega_1 + i_2\omega_2) \right\} \tag{8.32}$$

is a 2-D allpass FSBM. On the other hand, the MP-AP decomposition with region of support $\Omega'_{\rm TNSHP}[p_1, p_2]$ is given by

$$H(\omega_1, \omega_2) = H^*(-\omega_1, -\omega_2) = H_{\rm MP}(\omega_1, \omega_2) \cdot H_{\rm AP}(\omega_1, \omega_2) \tag{8.33}$$

where

$$H_{\rm MP}(\omega_1, \omega_2) = \exp\left\{ \sum\sum_{[i_1,i_2]\in\Omega'_{\rm TNSHP}[p_1,p_2]} \alpha_{i_1,i_2} e^{-j(i_1\omega_1 + i_2\omega_2)} \right\} \tag{8.34}$$

is a 2-D minimum-phase FSBM and

$$H_{\rm AP}(\omega_1, \omega_2) = \exp\left\{ j \sum\sum_{[i_1,i_2]\in\Omega'_{\rm TNSHP}[p_1,p_2]} (\alpha_{i_1,i_2} + \beta_{i_1,i_2}) \sin(i_1\omega_1 + i_2\omega_2) \right\} \tag{8.35}$$

is also a 2-D allpass FSBM.

Note that $h_{\rm MP}[0, 0] = 1$ and the region of support of the 2-D system $h_{\rm MP}[n_1, n_2]$ is $\Omega_{\rm F}[0, -1]$, whereas $\Omega'_{\rm TNSHP}[p_1, p_2]$ is the region of support for the FSBM parameters α_{i_1,i_2} and β_{i_1,i_2}. Estimation of $h[n_1, n_2]$ is equivalent to the estimation of the parameters of α_{i_1,i_2} and β_{i_1,i_2}.

8.2.1 Parametric 2-D BSI

Suppose that $y[n_1, n_2]$ is a real stationary random field that can be modeled as (8.1), where the system input $u[n_1, n_2]$ and the noise $w[n_1, n_2]$ satisfy the Assumptions (A8-1) and (A8-3), respectively, and meanwhile the unknown system $h[n_1, n_2]$ satisfies the following assumption

(A8-5) $h[n_1, n_2]$ is a 2-D real FSBM given by (8.30) or (8.33) with p_1 and p_2 known in advance.

With only a given set of measurements $y[n_1, n_2], n_1 = 0, 1, ..., N-1, n_2 = 0, 1, ..., N-1$, the estimation of the 2-D system $h[n_1, n_2]$ includes the estimation of amplitude parameters α_{i_1,i_2} and that of phase parameters β_{i_1,i_2} to be introduced, respectively, next.

Estimation of Amplitude Parameters

The estimation of the amplitude parameters α_{i_1,i_2}, $\forall [i_1, i_2] \in \Omega'_{\rm TNSHP}[p_1, p_2]$, is equivalent to the estimation of the minimum-phase FSBM $H_{\rm MP}(\omega_1, \omega_2)$ given by (8.34). The minimum-phase FSBM $H_{\rm MP}(\omega_1, \omega_2)$ can be estimated using SOS based 2-D minimum-phase LPE filter.

Let $v_{\rm MP}[n_1,n_2]$ be a 2-D IIR filter with the region of support $\Omega_{\rm F}[0,-1]$, and $\varepsilon[n_1,n_2]$ denote the prediction error by processing $y[n_1,n_2]$ with $v_{\rm MP}[n_1,n_2]$ as follows

$$\begin{aligned}\varepsilon[n_1,n_2] &= y[n_1,n_2] \star v_{\rm MP}[n_1,n_2] \\ &= \sum\sum_{[i_1,i_2]\in\Omega_{\rm F}[0,-1]} v_{\rm MP}[i_1,i_2]y[n_1-i_1,n_2-i_2] \\ &= y[n_1,n_2] + \sum\sum_{[i_1,i_2]\in\Omega_{\rm F}[0,0]} v_{\rm MP}[i_1,i_2]y[n_1-i_1,n_2-i_2] \qquad (8.36)\end{aligned}$$

where $v_{\rm MP}[0,0]=1$. The optimum LPE filter $v_{\rm MP}[n_1,n_2]$ by minimizing the MSE $E\{\varepsilon^2[n_1,n_2]\}$ is known to be of minimum phase and infinite length [13, 14]. Therefore, the preceding minimum-phase 2-D FSBM with leading coefficient equal to unity and the region of support $\Omega_{\rm F}[0,-1]$ is suited for the 2-D minimum-phase LPE filter $v_{\rm MP}[n_1,n_2]$ as follows:

$$V_{\rm MP}(\omega_1,\omega_2) = \exp\left\{ \sum\sum_{[i_1,i_2]\in\Omega'_{\rm TNSHP}[p_1,p_2]} a_{i_1,i_2}e^{-j(i_1\omega_1+i_2\omega_2)} \right\}. \qquad (8.37)$$

Let $\boldsymbol{a}$ be a column vector consisting of the LPE filter parameters $a_{i_1,i_2}, \forall[i_1,i_2] \in \Omega'_{\rm TNSHP}[p_1,p_2]$. The optimum $\boldsymbol{a}$ is the one obtained by minimizing

$$J_{\rm MSE}(\boldsymbol{a}) = E\{\varepsilon^2[n_1,n_2]\} \qquad (8.38)$$

and supported by the following theorem which is also the 2-D extension of Theorem 1 reported in [15].

Theorem 8.6. *Suppose that $y[n_1,n_2]$ is a stationary random field given by (8.1) under the Assumptions* (A8-1), (A8-5) and $SNR=\infty$. Then $J_{\rm MSE}(\boldsymbol{a})$ given by (8.38) is minimum if and only if

$$V_{\rm MP}(\omega_1,\omega_2) = 1/H_{\rm MP}(\omega_1,\omega_2) \qquad (8.39)$$

i.e. $a_{i_1,i_2} = -\alpha_{i_1,i_2}, \forall[i_1,i_2]\in\Omega'_{\rm TNSHP}[p_1,p_2]$ and $\min\{J_{\rm MSE}(\boldsymbol{a})\}=\sigma_u^2$.

The proof of Theorem 8.6 is presented in Appendix 8C. Based on Theorem 8.6, the estimation of α_{i_1,i_2} can be performed through finding the optimum a_{i_1,i_2}, denoted as $\widehat{a}_{i_1,i_2}$, by minimizing $J_{\rm MSE}(\boldsymbol{a})$. Then $\widehat{\alpha}_{i_1,i_2}=-\widehat{a}_{i_1,i_2}$, $\forall[i_1,i_2]\in\Omega'_{\rm TNSHP}[p_1,p_2]$, i.e.

$$\widehat{H}_{\rm MP}(\omega_1,\omega_2) = 1/\widehat{V}_{\rm MP}(\omega_1,\omega_2). \qquad (8.40)$$

Because $J_{\rm MSE}(\boldsymbol{a})$ is a highly nonlinear function of $\boldsymbol{a}$, it is almost impossible to find a closed-form solution for the optimum $\boldsymbol{a}$. Again, the iterative BFGS algorithm can be employed to find the optimum $\boldsymbol{a}$. At the ith iteration, $\boldsymbol{a}$ is updated by

$$\boldsymbol{a}^{[i+1]} = \boldsymbol{a}^{[i]} - \mu^{[i]}\mathbf{Q}^{[i]} \cdot \left.\frac{\partial J_{\text{MSE}}(\boldsymbol{a})}{\partial \boldsymbol{a}}\right|_{\boldsymbol{a}=\boldsymbol{a}^{[i]}} \tag{8.41}$$

where $\mu^{[i]}$ is the step size and the real matrix $\mathbf{Q}^{[i+1]}$ (see (2.157)) is updated by

$$\mathbf{Q}^{[i+1]} = \mathbf{Q}^{[i]} + \frac{1}{\mathbf{r}_{i+1}^T\mathbf{s}_{i+1}}\left\{\left(1 + \frac{\mathbf{s}_{i+1}^T\mathbf{Q}^{[i]}\mathbf{s}_{i+1}}{\mathbf{r}_{i+1}^T\mathbf{s}_{i+1}}\right)\mathbf{r}_{i+1}\mathbf{r}_{i+1}^T \right.$$
$$\left. - \mathbf{r}_{i+1}\mathbf{s}_{i+1}^T\mathbf{Q}^{[i]} - \mathbf{Q}^{[i]}\mathbf{s}_{i+1}\mathbf{r}_{i+1}^T\right\} \tag{8.42}$$

in which vectors $\mathbf{r}_i$ and $\mathbf{s}_i$ are updated as follows

$$\mathbf{r}_{i+1} = \boldsymbol{a}^{[i+1]} - \boldsymbol{a}^{[i]} \tag{8.43}$$

and

$$\mathbf{s}_{i+1} = \left.\frac{\partial J_{\text{MSE}}(\boldsymbol{a})}{\partial \boldsymbol{a}}\right|_{\boldsymbol{a}=\boldsymbol{a}^{[i+1]}} - \left.\frac{\partial J_{\text{MSE}}(\boldsymbol{a})}{\partial \boldsymbol{a}}\right|_{\boldsymbol{a}=\boldsymbol{a}^{[i]}}. \tag{8.44}$$

The computation of the gradient $\partial J_{\text{MSE}}(\boldsymbol{a})/\partial \boldsymbol{a}$ in (8.41) needs the computation of $\partial \varepsilon[n_1, n_2]/\partial a_{i_1,i_2}$ that can be easily shown, from (8.36) and (8.37), to be

$$\frac{\partial \varepsilon[n_1, n_2]}{\partial a_{i_1,i_2}} = \varepsilon[n_1 - i_1, n_2 - i_2], \quad [i_1, i_2] \in \Omega'_{\text{TNSHP}}[p_1, p_2]. \tag{8.45}$$

The proof of (8.45) is left as an exercise (Problem 8.2). By (8.38) and (8.45), we obtain

$$\frac{\partial J_{\text{MSE}}(\boldsymbol{a})}{\partial a_{i_1,i_2}} = 2E\left\{\varepsilon[n_1, n_2]\frac{\partial \varepsilon[n_1, n_2]}{\partial a_{i_1,i_2}}\right\}$$
$$= 2E\{\varepsilon[n_1, n_2]\varepsilon[n_1 - i_1, n_2 - i_2]\}, \ [i_1, i_2] \in \Omega'_{\text{TNSHP}}[p_1, p_2] \tag{8.46}$$

which implies that a local minimum $J_{\text{MSE}}(\boldsymbol{a})$ is achieved when

$$E\{\varepsilon[n_1, n_2]\varepsilon[n_1 - i_1, n_2 - i_2]\} = 0, \quad [i_1, i_2] \in \Omega'_{\text{TNSHP}}[p_1, p_2]. \tag{8.47}$$

Moreover, by (8.33) and (8.39), the optimum prediction error $\varepsilon[n_1, n_2]$ is a 2-D white random field in the absence of noise as follows:

$$\varepsilon[n_1, n_2] = y[n_1, n_2] \star \widehat{v}_{\text{MP}}[n_1, n_2] = u[n_1, n_2] \star h_{\text{AP}}[n_1, n_2]. \tag{8.48}$$

In other words, $\widehat{v}_{\text{MP}}[n_1, n_2]$ is a 2-D whitening filter and $\varepsilon[n_1, n_2]$ is an amplitude equalized signal with a flat power spectral density equal to $\sigma_\varepsilon^2 = \sigma_u^2$.

The above BFGS algorithm for finding the optimum $\boldsymbol{\alpha} = -\boldsymbol{a}$ is referred to as the *Amplitude Parameter Estimation Algorithm (APEA)*, which is summarized in Table 8.1. In practice, for a given set of data $y[n_1, n_2]$, the sample mean-square-error (see (8.38)) and sample correlations of $\varepsilon[n_1, n_2]$ (see (8.46)) are used to compute $J_{\text{MSE}}(\boldsymbol{a})$ and $\partial J_{\text{MSE}}(\boldsymbol{a})/\partial a_{i_1,i_2}$. Because the sample mean-square-error and sample correlations of $\varepsilon[n_1, n_2]$ are consistent estimates, the $\widehat{\boldsymbol{\alpha}}$ obtained by the APEA is also a consistent estimate by Slutsky's Theorem (Theorem 3.58).

As a remark, when the 2-D LSI system $h[n_1, n_2]$ is a 2-D FSBM with unknown p_1 and p_2, the $\widehat{H}_{\text{MP}}(\omega_1, \omega_2)$ obtained is merely an approximation to $H_{\text{MP}}(\omega_1, \omega_2)$ if the chosen values for p_1 and p_2 in (8.37) are smaller than the true values of p_1 and p_2. On the other hand, as the 2-D LSI system $h[n_1, n_2]$ is not a 2-D FSBM, the larger the chosen values for p_1 and p_2 in (8.37), the better the approximation $\widehat{H}_{\text{MP}}(\omega_1, \omega_2)$ to the minimum-phase system associated with $h[n_1, n_2]$.

Estimation of Phase Parameters

Estimation of the phase parameters β_{i_1,i_2}, $[i_1, i_2] \in \Omega'_{\text{TNSHP}}[p_1, p_2]$, is equivalent to estimation of the 2-D allpass FSBM $H_{\text{PS}}(\omega_1, \omega_2)$ given by (8.32) from $y[n_1, n_2]$, and also equivalent to the estimation of the 2-D allpass FSBM $H_{\text{AP}}(\omega_1, \omega_2)$ given by (8.35) from the amplitude equalized signal $\varepsilon[n_1, n_2]$ given by (8.48). The estimation of both $H_{\text{PS}}(\omega_1, \omega_2)$ and $H_{\text{AP}}(\omega_1, \omega_2)$ through allpass filtering is based on the following theorem.

Theorem 8.7. *Suppose that $y[n_1, n_2]$ is a real stationary random field given by (8.1) satisfying the Assumptions* (A8-1), (A8-3) *and* (A8-5). *Let $v_{\text{AP}}[n_1, n_2]$ be a 2-D allpass filter with region of support $\Omega_{\text{F}}[-\infty, -\infty]$, $\boldsymbol{v}_{\text{AP}}$ a vector consisting of all the coefficients of $v_{\text{AP}}[n_1, n_2]$, and*

$$\mathsf{z}[n_1, n_2] = y[n_1, n_2] \star v_{\text{AP}}[n_1, n_2]. \tag{8.49}$$

Then

$$\mathcal{J}_p(\boldsymbol{v}_{\text{AP}}) = |C_p\{\mathsf{z}[n_1, n_2]\}|, \ p \geq 3, \tag{8.50}$$

is maximum if and only if

$$\arg[V_{\text{AP}}(\omega_1, \omega_2)] = -\arg[H(\omega_1, \omega_2)] + \tau_1\omega_1 + \tau_2\omega_2 \tag{8.51}$$

where τ_1 and τ_2 are unknown integers.

Note that Theorem 8.7 is actually a special case of Theorem 7.1 with the 2-D deconvolution filter $v[n_1, n_2]$ required to be an allpass deconvolution filter $v_{\text{AP}}[n_1, n_2]$, because $J_p(\boldsymbol{v}_{\text{AP}}) = \mathcal{J}_p(\boldsymbol{v}_{\text{AP}})/\sigma_{\mathsf{z}}^p = \mathcal{J}_p(\boldsymbol{v}_{\text{AP}})/\sigma_y^p$.

Table 8.1 2-D APEA

Parameter setting		Choose filter order (p_1, p_2) for the region of support $\Omega'_{\text{TNSHP}}[p_1, p_2]$, initial conditions $\boldsymbol{a}^{[0]}$ and $\mathbf{Q}^{[0]}$, convergence tolerance $\zeta > 0$, initial step size μ_0 and a positive integer K for the step size search range $[\mu_0/2^0, \mu_0/2^K]$.
Steps	(S1)	Set the iteration number $i = 0$.
	(S2)	Compute the amplitude equalized signal $\varepsilon^{[0]}[n_1, n_2]$ associated with $\boldsymbol{a}^{[0]}$ and obtain the corresponding gradient function $\left.\frac{\partial J_{\text{MSE}}(\boldsymbol{a})}{\partial \boldsymbol{a}}\right\vert_{\boldsymbol{a}=\boldsymbol{a}^{[0]}}$ using (8.36) and (8.46), respectively.
	(S3)	Update $\boldsymbol{a}^{[i+1]}$ by $$\boldsymbol{a}^{[i+1]} = \boldsymbol{a}^{[i]} - \mu^{[i]}\mathbf{Q}^{[i]} \cdot \left.\frac{\partial J_{\text{MSE}}(\boldsymbol{a})}{\partial \boldsymbol{a}}\right\vert_{\boldsymbol{a}=\boldsymbol{a}^{[i]}}$$ where an integer $l \in [0, K]$ is determined such that $\mu^{[i]} = \mu_0/2^l$ is the maximum step size leading to $J_{\text{MSE}}(\boldsymbol{a}^{[i+1]}) < J_{\text{MSE}}(\boldsymbol{a}^{[i]})$.
	(S4)	If $$\frac{\left\vert J_{\text{MSE}}(\boldsymbol{a}^{[i+1]}) - J_{\text{MSE}}(\boldsymbol{a}^{[i]})\right\vert}{J_{\text{MSE}}(\boldsymbol{a}^{[i]})} \geq \zeta,$$ then go to Step (S5); otherwise, obtain $\widehat{\alpha}_{i_1,i_2} = -a^{[i+1]}_{i_1,i_2}$, $\forall [i_1, i_2] \in \Omega'_{\text{TNSHP}}[p_1, p_2]$ and stop.
	(S5)	Update $\mathbf{r}_{i+1}$, $\mathbf{s}_{i+1}$ and $\mathbf{Q}^{[i+1]}$ by (8.43), (8.44) and (8.42), respectively.
	(S6)	Compute the amplitude equalized signal $\varepsilon^{[i+1]}[n_1, n_2]$ associated with $\boldsymbol{a}^{[i+1]}$ and the gradient function $\left.\frac{\partial J_{\text{MSE}}(\boldsymbol{a})}{\partial \boldsymbol{a}}\right\vert_{\boldsymbol{a}=\boldsymbol{a}^{[i+1]}}$ using (8.36) and (8.46), respectively.
	(S7)	Update the iteration number i by $(i + 1)$ and go to Step (S3).

By Theorem 8.7, the phase parameters β_{i_1,i_2} of the unknown FSBM $H(\omega_1, \omega_2)$ can be estimated by finding the optimum allpass filter $v_{\text{AP}}[n_1, n_2]$ also modeled as an FSBM given by

$$V_{\text{AP}}(\omega_1, \omega_2) = \exp\left\{ j \sum\sum_{[i_1,i_2]\in\Omega'_{\text{TNSHP}}[p_1,p_2]} \gamma_{i_1,i_2} \sin(i_1\omega_1 + i_2\omega_2) \right\}. \tag{8.52}$$

Let $\boldsymbol{\gamma}$ be a vector including the allpass filter coefficients γ_{i_1,i_2}, $\forall [i_1, i_2] \in \Omega'_{\text{TNSHP}}[p_1, p_2]$. Let

$$\mathsf{z}_1[n_1, n_2] = \varepsilon[n_1, n_2] \star v_{\text{AP}}[n_1, n_2] \tag{8.53}$$

$$\mathsf{z}_2[n_1, n_2] = y[n_1, n_2] \star v_{\text{AP}}[n_1, n_2] \tag{8.54}$$

where $\varepsilon[n_1, n_2]$ is the amplitude equalized signal (see (8.48)), $y[n_1, n_2]$ is given by (8.1), $v_{\text{AP}}[n_1, n_2]$ is given by (8.52) and

$$\mathcal{J}_p(\boldsymbol{\gamma}) = \mathcal{J}_p(\mathsf{z}[n_1, n_2]) = |C_p\{\mathsf{z}[n_1, n_2]\}| \tag{8.55}$$

where $p \geq 3$ and $\mathsf{z}[n_1, n_2] = \mathsf{z}_1[n_1, n_2]$ or $\mathsf{z}[n_1, n_2] = \mathsf{z}_2[n_1, n_2]$. Then the following two facts needed for the estimation of β_{i_1,i_2} can be shown.

Fact 8.8. *For* $\mathsf{z}[n_1, n_2] = \mathsf{z}_1[n_1, n_2]$, $\mathcal{J}_p(\boldsymbol{\gamma})$ *is maximum if and only if*

$$\arg[V_{\text{AP}}(\omega_1, \omega_2)] = -\arg[H_{\text{AP}}(\omega_1, \omega_2)], \tag{8.56}$$

i.e.

$$\gamma_{i_1,i_2} = -(\alpha_{i_1,i_2} + \beta_{i_1,i_2}), \forall [i_1, i_2] \in \Omega'_{\text{TNSHP}}[p_1, p_2]. \tag{8.57}$$

Fact 8.9. *For* $\mathsf{z}[n_1, n_2] = \mathsf{z}_2[n_1, n_2]$, $\mathcal{J}_p(\boldsymbol{\gamma})$ *is maximum if and only if*

$$\arg[V_{\text{AP}}(\omega_1, \omega_2)] = -\arg[H_{\text{PS}}(\omega_1, \omega_2)], \tag{8.58}$$

i.e.

$$\gamma_{i_1,i_2} = -\beta_{i_1,i_2}, \forall [i_1, i_2] \in \Omega'_{\text{TNSHP}}[p_1, p_2]. \tag{8.59}$$

The proof of Fact 8.9 is presented in Appendix 8D and the proof of Fact 8.8 is left as an exercise (Problem 8.3) since their proofs are similar.

Next, we concentrate on how to obtain the optimum $\boldsymbol{\gamma}$ by maximizing (8.55) that provides the optimum $\boldsymbol{\beta}$, a vector composed of all the phase parameters β_{i_1,i_2} of the FSBM $H(\omega_1, \omega_2)$, by Facts 8.8 and 8.9. Again, the objective function $\mathcal{J}_p(\boldsymbol{\gamma})$ is also a highly nonlinear function of $\boldsymbol{\gamma}$ without a closed-form solution for the optimum $\boldsymbol{\gamma}$. Instead, the iterative BFGS algorithm is considered for finding the optimum $\boldsymbol{\gamma}$.

At the ith iteration, $\boldsymbol{\gamma}$ is updated by

$$\boldsymbol{\gamma}^{[i+1]} = \boldsymbol{\gamma}^{[i]} + \mu^{[i]} \mathbf{Q}^{[i]} \cdot \left. \frac{\partial \mathcal{J}_p(\boldsymbol{\gamma})}{\partial \boldsymbol{\gamma}} \right|_{\boldsymbol{\gamma}=\boldsymbol{\gamma}^{[i]}} \tag{8.60}$$

where $\mu^{[i]}$ is a positive step size and the real matrix $\mathbf{Q}^{[i]}$ is updated by (8.42), in which vectors $\mathbf{r}_i$ and $\mathbf{s}_i$ are updated as follows

$$\mathbf{r}_{i+1} = \boldsymbol{\gamma}^{[i+1]} - \boldsymbol{\gamma}^{[i]} \tag{8.61}$$

and

$$\mathbf{s}_{i+1} = \left.\frac{\partial \mathcal{J}_p(\boldsymbol{\gamma})}{\partial \boldsymbol{\gamma}}\right|_{\boldsymbol{\gamma}=\boldsymbol{\gamma}^{[i+1]}} - \left.\frac{\partial \mathcal{J}_p(\boldsymbol{\gamma})}{\partial \boldsymbol{\gamma}}\right|_{\boldsymbol{\gamma}=\boldsymbol{\gamma}^{[i]}} . \tag{8.62}$$

Each component of the gradient $\partial \mathcal{J}_p(\boldsymbol{\gamma})/\partial \boldsymbol{\gamma}$ needed in (8.60) can be shown to be

$$\begin{aligned}\frac{\partial \mathcal{J}_p(\boldsymbol{\gamma})}{\partial \gamma_{i_1,i_2}} &= \operatorname{sgn}(C_p\{\mathsf{z}[n_1,n_2]\}) \cdot \frac{\partial C_p\{\mathsf{z}[n_1,n_2]\}}{\partial \gamma_{i_1,i_2}} \\ &= \frac{p}{2} \operatorname{sgn}(C_p\{\mathsf{z}[n_1,n_2]\}) \cdot (\operatorname{cum}\{\mathsf{z}[n_1,n_2] : p-1, \mathsf{z}[n_1+i_1,n_2+i_2]\} \\ &\quad - \operatorname{cum}\{\mathsf{z}[n_1,n_2] : p-1, \mathsf{z}[n_1-i_1,n_2-i_2]\})\end{aligned} \tag{8.63}$$

where $\operatorname{sgn}(x)$ denotes the sign of x. The proof of (8.63) is left as an exercise (Problem 8.4).

The above BFGS algorithm for finding the optimum $\boldsymbol{\gamma}$ is referred to as the *2-D Phase Estimation Algorithm (PSEA)(k)*, which is summarized in Table 8.2, where $\mathsf{k} = 1$ or $\mathsf{k} = 2$, denoting that the signal to be processed by the allpass filter $V_{\mathrm{AP}}(\omega_1, \omega_2)$ given by (8.52) is the amplitude equalized signal $\varepsilon[n_1,n_2]$ for $\mathsf{k} = 1$ and the signal $y[n_1,n_2]$ for $\mathsf{k} = 2$. The amplitude estimation algorithm APEA and the phase estimation algorithm PSEA(k) introduced above constitute the 2-D FSBM based BSI algorithm shown in Fig. 8.3, which is summarized as follows.

2-D FSBM Based BSI Algorithm

(S1) Set the order (p_1, p_2) of the 2-D FSBM with the region of support $\Omega'_{\mathrm{TNSHP}}[p_1,p_2]$, convergence tolerance $\zeta > 0$ and a positive integer K for the step size search range $[\mu_0/2^0, \mu_0/2^K]$ for both APEA and PSEA(k). Then set the cumulant order p and parameter k ($= 1$ or 2) needed by the PSEA(k).

(S2) Amplitude Estimation. Find the optimum $\widehat{\boldsymbol{\alpha}}$ and the amplitude equalized signal $\varepsilon[n_1,n_2]$ by processing $y[n_1,n_2]$ using the APEA.

(S3) Phase Estimation. Find the optimum $\widehat{\boldsymbol{\gamma}}$ using the PSEA(k). If $\mathsf{k} = 1$, $\widehat{\boldsymbol{\beta}} = -\widehat{\boldsymbol{\gamma}} - \widehat{\boldsymbol{\alpha}}$ by (8.57). If $\mathsf{k} = 2$, $\widehat{\boldsymbol{\beta}} = -\widehat{\boldsymbol{\gamma}}$ by (8.59).

Notably, the 2-D FSBM based BSI algorithm processes $y[n_1,n_2]$ to obtain $\widehat{\boldsymbol{\alpha}}$ (amplitude parameter estimates) followed by $\widehat{\boldsymbol{\beta}}$ (phase parameter estimates) for $\mathsf{k} = 1$ (serial processing), and $\widehat{\boldsymbol{\alpha}}$ and $\widehat{\boldsymbol{\beta}}$ in parallel for $\mathsf{k} = 2$ (parallel processing) as shown in Fig. 8.3.

Again, for a given set of data $y[n_1,n_2]$, sample cumulants of $\mathsf{z}[n_1,n_2]$ are used to compute $\mathcal{J}_p(\boldsymbol{\gamma})$ (see (8.55)) and $\partial \mathcal{J}_p(\boldsymbol{\gamma})/\partial \gamma_{i_1,i_2}$ (see (8.63)). Because sample cumulants of $\mathsf{z}[n_1,n_2]$ are consistent, and because $\widehat{\boldsymbol{\alpha}}$ obtained using the APEA is consistent, it can easily be inferred that $\widehat{\boldsymbol{\beta}}$ obtained using PSEA(k) is also a consistent estimate by Slutsky's Theorem (Theorem 3.58).

Table 8.2 2-D PSEA(k)

Parameter setting		Choose filter order (p_1, p_2) for the region of support $\Omega'_{\mathrm{TNSHP}}[p_1, p_2]$, cumulant order $p > 2$, initial conditions $\boldsymbol{\gamma}^{[0]}$ and $\mathbf{Q}^{[0]}$, convergence tolerance $\zeta > 0$, initial step size μ_0 and a positive integer K for the step size search range $[\mu_0/2^0, \mu_0/2^K]$. (Note that the signal to be processed by the allpass FSBM is the amplitude equalized signal $\varepsilon[n_1, n_2]$ for $\mathsf{k} = 1$, and the signal $y[n_1, n_2]$ for $\mathsf{k} = 2$.)
Steps	(S1)	Set the iteration number $i = 0$.
	(S2)	Compute the phase equalized signal $\mathsf{z}^{[0]}[n_1, n_2]$ associated with $\boldsymbol{\gamma}^{[0]}$ and obtain the corresponding gradient function $\left.\dfrac{\partial \mathcal{J}_p(\boldsymbol{\gamma})}{\partial \boldsymbol{\gamma}}\right\|_{\boldsymbol{\gamma}=\boldsymbol{\gamma}^{[0]}}$ using (8.63).
	(S3)	Update $\boldsymbol{\gamma}^{[i+1]}$ by $$\boldsymbol{\gamma}^{[i+1]} = \boldsymbol{\gamma}^{[i]} + \mu^{[i]}\mathbf{Q}^{[i]} \cdot \left.\frac{\partial \mathcal{J}_p(\boldsymbol{\gamma})}{\partial \boldsymbol{\gamma}}\right\|_{\boldsymbol{\gamma}=\boldsymbol{\gamma}^{[i]}}$$ where an integer $l \in [0, K]$ is determined such that $\mu^{[i]} = \mu_0/2^l$ is the maximum step size leading to $\mathcal{J}_p(\boldsymbol{\gamma}^{[i+1]}) > \mathcal{J}_p(\boldsymbol{\gamma}^{[i]})$.
	(S4)	If $$\frac{\left\|\mathcal{J}_p(\boldsymbol{\gamma}^{[i+1]}) - \mathcal{J}_p(\boldsymbol{\gamma}^{[i]})\right\|}{\mathcal{J}_p(\boldsymbol{\gamma}^{[i]})} \geq \zeta,$$ then go to Step (S5); otherwise, obtain $\widehat{\gamma}_{i_1,i_2} = \gamma^{[i+1]}_{i_1,i_2}$, $\forall[i_1, i_2] \in \Omega'_{\mathrm{TNSHP}}[p_1, p_2]$ and stop.
	(S5)	Update $\mathbf{r}_{i+1}$, $\mathbf{s}_{i+1}$ and $\mathbf{Q}^{[i+1]}$ by (8.61), (8.62) and (8.42), respectively.
	(S6)	Compute the phase equalized signal $\mathsf{z}^{[i+1]}[n_1, n_2]$ associated with $\boldsymbol{\gamma}^{[i+1]}$ using (8.53) for $\mathsf{k} = 1$ and (8.54) for $\mathsf{k} = 2$, and the gradient function $\left.\dfrac{\partial \mathcal{J}_p(\boldsymbol{\gamma})}{\partial \boldsymbol{\gamma}}\right\|_{\boldsymbol{\gamma}=\boldsymbol{\gamma}^{[i+1]}}$ using (8.63).
	(S7)	Update the iteration number i by $(i + 1)$ and go to Step (S3).

When the LSI system $h[n_1, n_2]$ is not a 2-D FSBM, the unknown linear phase terms $\omega_1\tau_1 + \omega_2\tau_2$ in (8.51) may affect the resultant phase parameter estimates. As p_1 and p_2 are chosen sufficiently large, PSEA(2) may well end up with the optimum

$$\widehat{\gamma}_{i_1,i_2} = -\beta_{i_1,i_2} + d_{i_1,i_2}, \ \forall[i_1, i_2] \in \Omega'_{\mathrm{TNSHP}}[p_1, p_2] \tag{8.64}$$

where d_{i_1,i_2} are coefficients of the 2-D Fourier series expansion of the linear function $\tau_1\omega_1 + \tau_2\omega_2$ given by (8.100) in Appendix 8D, leading to a 2-D space

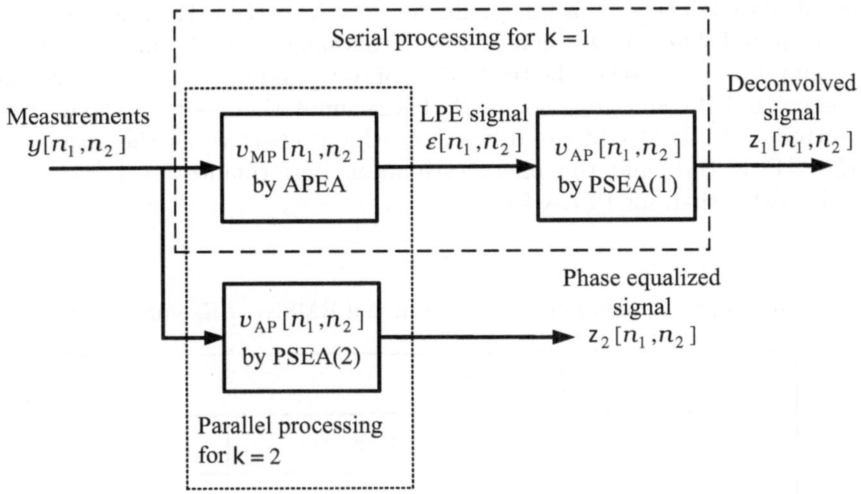

Fig. 8.3 2-D FSBM based BSI algorithm

shift in the resultant estimate $\widehat{h}[n_1 + \tau_1, n_2 + \tau_2]$. This can also happen when PSEA(1) is used.

Simulation Examples

Two simulation examples are presented to test the 2-D FSBM based BSI algorithm. In the two examples, the driving input signal $u[n_1, n_2]$ was a zero-mean, exponentially distributed, i.i.d., random field with variance $\sigma_u^2 = 1$, normalized skewness $\gamma_3 = 2$ and normalized kurtosis $\gamma_4 = 6$ that was convolved with a 2-D LSI system followed by addition of white Gaussian noise to generate the synthetic $N \times N$ data $y[n_1, n_2]$. Then the FSBM based BSI algorithm with $\mathsf{k} = 1$ and $\mathsf{k} = 2$ were used to process $y[n_1, n_2]$. In (S1), $\mu_0 = 1$, $\zeta = 10^{-8}$, $K = 10$ were chosen for the APEA and PSEA(k) and the cumulant order $p = 3$ was chosen for the PSEA(k). In (S2) and (S3), $\mathbf{Q}^{[0]} = \mathbf{I}$ and the initial condition $a_{i_1,i_2}^{[0]} = 0, \forall [i_1, i_2]$ and $\gamma_{i_1,i_2}^{[0]} = 0, \forall [i_1, i_2]$ were used by the APEA and PSEA(k), respectively. Thirty independent runs were performed in each of the two examples.

Example 8.10 (Estimation of a 2-D FSBM)
A 2-D FSBM $H(\omega_1, \omega_2)$ given by (8.30) with orders $p_1 = p_2 = 1$ and parameters

$$\alpha_{0,1} = -0.97 \quad \alpha_{1,-1} = -0.5 \quad \alpha_{1,0} = 1.04 \quad \alpha_{1,1} = 0.52$$
$$\beta_{0,1} = -0.27 \quad \beta_{1,-1} = -0.95 \quad \beta_{1,0} = 0.18 \quad \beta_{1,1} = -0.92$$

(with region of support $\Omega'_{\mathrm{TNSHP}}[1, 1]$) was used in this example. Mean and RMS error (RMSE) of the thirty amplitude and phase parameter estimates were calculated. The simulation results for $N \times N = 128 \times 128$ and SNR$= \infty$ are shown in Table 8.3. One can see, from this table, that the mean values of $\widehat{\alpha}_{i_1,i_2}$ and $\widehat{\beta}_{i_1,i_2}$ are close to the true values of α_{i_1,i_2} and β_{i_1,i_2}, respectively, and the associated RMSEs are also small. These simulation results validate that the introduced 2-D FSBM based BSI algorithm is effective for the estimation of the 2-D FSBM, and the phase estimation performance for PSEA(1) is slightly better than for PSEA(2).

Table 8.3 Simulation results (mean and RMSE) of Example 8.10

α	$\widehat{\alpha}$		β	$\widehat{\beta}$			
	APEA			PSEA(1)		PSEA(2)	
	MEAN	RMSE		MEAN	RMSE	MEAN	RMSE
$\alpha_{0,1}$	− 0.9696	0.0068	$\beta_{0,1}$	− 0.2717	0.0126	− 0.2703	0.0175
$\alpha_{1,-1}$	− 0.4987	0.0085	$\beta_{1,-1}$	−0.9526	0.0103	−0.9496	0.0230
$\alpha_{1,0}$	1.0394	0.0089	$\beta_{1,0}$	0.1784	0.0110	0.1774	0.0356
$\alpha_{1,1}$	0.5199	0.0101	$\beta_{1,1}$	−0.9232	0.0123	−0.9304	0.0231

□

Example 8.11 (Approximation to an MA Model)
In this example, $h[n_1, n_2]$ used was the same 2-D MA model given by (7.111). The true system $h[n_1, n_2]$ and the estimate $\widehat{h}_i[n_1, n_2]$ obtained at the ith run were normalized by unity energy, and the 2-D space shift between $\widehat{h}_i[n_1, n_2]$ and the true $h[n_1, n_2]$ was artificially removed before calculation of the normalized MSE (NMSE) as follows

$$\text{NMSE} = \frac{1}{30}\sum_{i=1}^{30} \sum_{n_1=-15}^{15} \sum_{n_2=-15}^{15} \left(h[n_1, n_2] - \widehat{h}_i[n_1, n_2]\right)^2. \tag{8.65}$$

Table 8.4 shows some simulation results (NMSE) for $N \times N = 128 \times 128$, and $(p_1, p_2) = (1, 1)$, $(2, 2)$, $(3, 3)$ and SNR $= 5$, 10, 15, 20 dB, respectively. One can see, from this table, that the NMSEs are small and decrease with SNR for $p_1(= p_2) \geq 2$, and decrease with $p_1(= p_2)$ for all SNR. Therefore, one can also observe that the 2-D FSBM estimates obtained by the 2-D FSBM based BSI algorithm are good approximations to the true 2-D MA system, as SNR is high and the order of the FSBM model used is sufficient.

□

Table 8.4 Simulation results (NMSE) of Example 8.11

	PSEA(1)				PSEA(2)			
	SNR (dB)							
$p_1(=p_2)$	5	10	15	20	5	10	15	20
1	0.0760	0.0610	0.0603	0.0614	0.0750	0.0594	0.0584	0.0593
2	0.0309	0.0122	0.0083	0.0077	0.0306	0.0122	0.0084	0.0078
3	0.0307	0.0097	0.0052	0.0043	0.0303	0.0098	0.0054	0.0047

8.2.2 Texture Image Classification

Complex cepstra of speech signals with the vocal tract-filter modeled as a minimum-phase AR model have been widely used in speech recognition and speaker identification [12,16–18]. Similarly, the application of amplitude parameters $\boldsymbol{\alpha}$ of the 2-D FSBM to texture image classification [12] is also reasonable simply because $\boldsymbol{\alpha}$ and the complex cepstrum $\mathrm{h}_{\mathrm{MP}}[n_1, n_2]$ given by (7.36) are the same. However, the phase parameters $\boldsymbol{\beta}$ of the 2-D FSBM cannot be used because of unknown 2-D space shift $[\tau_1, \tau_2]$ inherent in the estimate $\widehat{\boldsymbol{\beta}}$ by PSEA(k). Next, let us present this texture image classification method reported in [12].

As introduced in Section 8.1.2, a texture image can be modeled as a 2-D linear nonGaussian processes with nonzero mean (see (8.24)). Suppose that $y[n_1, n_2]$ is a texture image with mean removed and modeled as

$$y[n_1, n_2] = u[n_1, n_2] \star h[n_1, n_2] \tag{8.66}$$

where $u[n_1, n_2]$ is zero-mean nonGaussian and $h[n_1, n_2]$ is a 2-D FSBM satisfying the Assumption (A8-5). Then the 2-D FSBM amplitude parameters $\boldsymbol{\alpha}$ and phase parameters $\boldsymbol{\beta}$ can be obtained by processing $y[n_1, n_2]$ using the 2-D FSBM based BSI algorithm (see Fig. 8.3) introduced in Section 8.2.1. With the obtained $\boldsymbol{\alpha}$ and $\boldsymbol{\beta}$, two feature vectors for texture image classification, denoted by $\boldsymbol{\theta}_1$ and $\boldsymbol{\theta}_2$, are defined as follows

$$\boldsymbol{\theta}_1 = \left[\widehat{\boldsymbol{\alpha}}^T, \sigma_\varepsilon^2/\sigma_x^2\right]^T \tag{8.67}$$

$$\boldsymbol{\theta}_2 = \left[\widehat{\boldsymbol{\alpha}}^T, \gamma_p\{\mathsf{z}[n_1, n_2]\}\right]^T \tag{8.68}$$

where $\varepsilon[n_1, n_2]$ is the amplitude equalized signal given by (8.48), $\gamma_p\{\mathsf{z}[n_1, n_2]\}$ is the normalized pth-order cumulant of the deconvolved signal $\mathsf{z}[n_1, n_2]$

$$\mathsf{z}[n_1, n_2] = y[n_1, n_2] \star \widehat{h}_{\mathrm{INV}}[n_1, n_2] \tag{8.69}$$

where $\widehat{h}_{\rm INV}[n_1,n_2]$ is the inverse system of the system estimate $\widehat{h}[n_1,n_2]$ characterized by $\widehat{\boldsymbol{\alpha}}$ and $\widehat{\boldsymbol{\beta}}$. Note that as PSEA(1) is used, $\mathsf{z}[n_1,n_2]=\mathsf{z}_1[n_1,n_2]$ is available once $\widehat{\boldsymbol{\beta}}$ is obtained (see Fig. 8.3).

A popular criterion for image classification is the *Euclidean distance (ED) criterion* (to be minimized)

$$\mathcal{D}(c)=\|\boldsymbol{\theta}-\boldsymbol{\theta}^{(c)}\|^2 \tag{8.70}$$

where $\boldsymbol{\theta}$ is the feature vector (either $\boldsymbol{\theta}_1$ or $\boldsymbol{\theta}_2$) to be classified, and $\boldsymbol{\theta}^{(c)}$ is the mean vector of all the training feature vectors $\boldsymbol{\theta}$s used for class c.

The leave-one-out strategy [19] was used to perform classification. Suppose that there are $\mathcal{A}$ classes of images with $\mathcal{B}$ images in each class available for classification. To perform classification with the chosen sub-image of a specific class c, the mean feature vector $\boldsymbol{\theta}^{(c)}$ was calculated from the other $\mathcal{B}-1$ training sub-images of the class c, while for the other $\mathcal{A}-1$ classes, $\boldsymbol{\theta}^{(c)}$ was calculated from all the $\mathcal{B}$ training sub-images of each class. The classification procedure was repeated for $\mathcal{A}\times\mathcal{B}$ sub-images. The number of misclassifications out of $\mathcal{A}\times\mathcal{B}$ classification operations is used as the performance index. Next, let us present some experimental results using the feature vectors $\boldsymbol{\theta}_1$ and $\boldsymbol{\theta}_2$, respectively.

Example 8.12 (Texture Image Classification)

Twelve 512×512 texture images shown in Fig. 8.4, taken from the USC-SIPI Image Data Base, were used for classification, including grass, treebark, straw, herringbone, wool, leather, water, wood, raffia, brickwall, plastic and sand. Each image was divided into sixteen 128×128 nonoverlapping sub-images to provide twelve classes of sixteen sub-images each. For each sub-image, the proposed 2-D FSBM based BSI algorithm is employed to obtain $\boldsymbol{\theta}_1$ and $\boldsymbol{\theta}_2$ with the following settings. In (S1), $p_1=p_2=3$ (for the region of support $\Omega'_{\rm TNSHP}[p_1,p_2]$ of the FSBM parameters), $\mu_0=1$, $\zeta=10^{-8}$, $K=10$ and either of $p=3$ and 4 for the PSEA(k); in (S2) $\mathbf{Q}^{[0]}=\mathbf{I}$ and the initial condition $a^{[0]}_{i_1,i_2}=0,\forall[i_1,i_2]$ for the APEA; in (S3), $\mathbf{Q}^{[0]}=\mathbf{I}$ and the initial condition $\gamma^{[0]}_{i_1,i_2}=0,\forall[i_1,i_2]$.

Let $A(B)$ denote B sub-images classified to texture class A over the 16 sub-image classifications for each of 12 texture classes. The classification results of $A(B)$ using $\boldsymbol{\theta}_1$ are shown in Table 8.5 and those using $\boldsymbol{\theta}_2$ for $p=3$ and $p=4$ are shown in Table 8.6 (where the PSEA(1) was used) and Table 8.7 (where the PSEA(2) was used).

The minimum ED classifier using $\boldsymbol{\theta}_1$ has five misclassifications (Table 8.5). The one using $\boldsymbol{\theta}_2$ associated with PSEA(1) has 4 and 6 misclassifications for $p=3$ and $p=4$ (Table 8.6), respectively. The one using $\boldsymbol{\theta}_2$ associated with PSEA(2) has four misclassifications for both of $p=3$ and $p=4$ (Table 8.7). These experimental results show that the feature vectors $\boldsymbol{\theta}_1$ and $\boldsymbol{\theta}_2$ are effective for texture image classification.

□

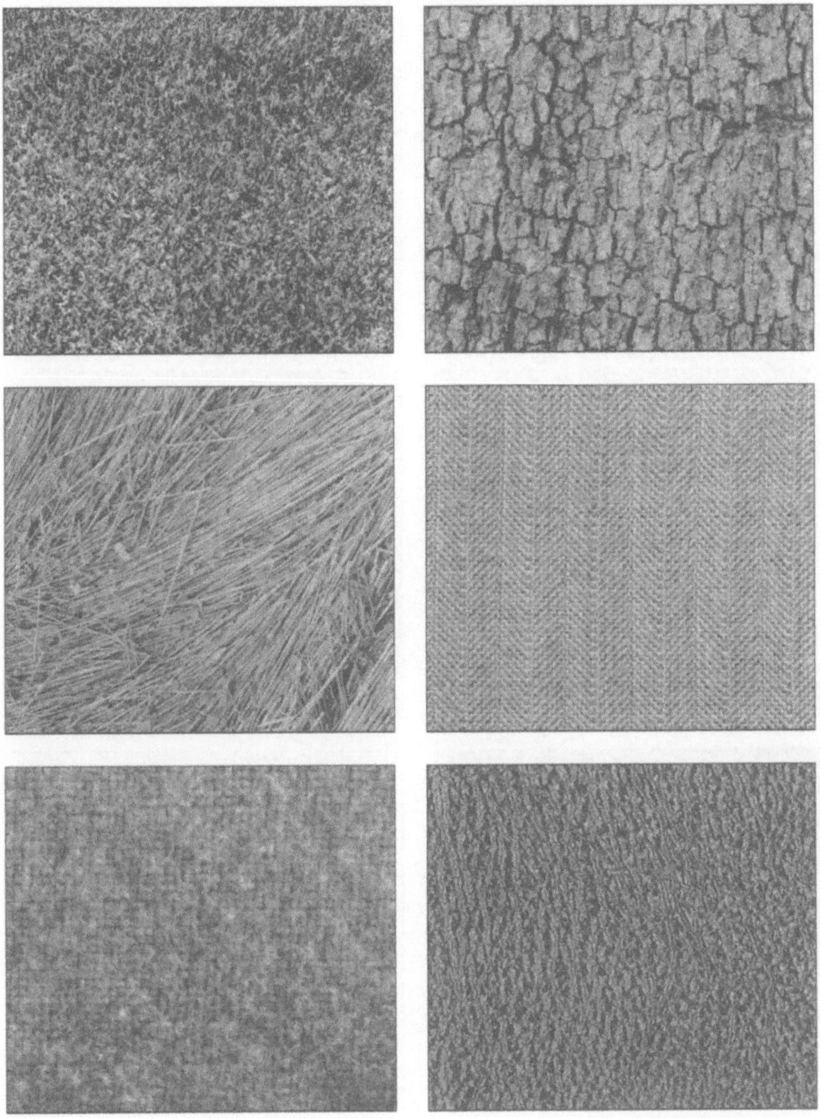

Fig. 8.4 Texture images used for classification: (row 1, column 1): grass; (row 1, column 2): treebark; (row 2, column 1): straw; (row 2, column 2): herringbone; (row 3, column 1): wool; (row 3, column 2): leather

Fig. 8.4 (Continued) (row 1, column 1): water; (row 1, column 2): wood; (row 2, column 1): raffia; (row 2, column 2): brickwall; (row 3, column 1): plastic; (row 3, column 2): sand

Table 8.5 Experimental results using feature vector $\boldsymbol{\theta}_1$ associated with the APEA for $p_1 = p_2 = 3$

Misclassifications: 5 of 192	
Texture	Classification results
1. grass	**1**(16)
2. treebark	**2**(16)
3. straw	**3**(13), **1**(1), **6**(1), **12**(1)
4. herringbone	**4**(16)
5. wool	**5**(16)
6. leather	**6**(16)
7. water	**7**(15), **8**(1)
8. wood	**8**(16)
9. raffia	**9**(16)
10. brickwall	**10**(16)
11. plastic	**11**(15), **1**(1)
12. sand	**12**(16)

Table 8.6 Experimental results using feature vector $\boldsymbol{\theta}_2$ associated with the PSEA(1) for $p_1 = p_2 = 3$

Misclassifications: 4 and 6 of 192 for $p = 3$ and $p = 4$, respectively		
Texture	Classification results	
	$p = 3$	$p = 4$
1. grass	**1**(16)	**1**(15), **12**(1)
2. treebark	**2**(16)	**2**(16)
3. straw	**3**(14), **1**(1), **6**(1)	**3**(13), **1**(1), **6**(1), **12**(1)
4. herringbone	**4**(16)	**4**(16)
5. wool	**5**(16)	**5**(16)
6. leather	**6**(16)	**6**(16)
7. water	**7**(14), **8**(1), **10**(1)	**7**(15), **8**(1)
8. wood	**8**(16)	**8**(16)
9. raffia	**9**(16)	**9**(16)
10. brickwall	**10**(16)	**10**(16)
11. plastic	**11**(16)	**11**(15), **2**(1)
12. sand	**12**(16)	**12**(16)

Table 8.7 Experimental results using feature vector $\boldsymbol{\theta}_2$ associated with the PSEA(2) for $p_1 = p_2 = 3$

Misclassifications: 4 of 192 for $p = 3$ and $p = 4$		
	Classification results	
Texture	$p = 3$	$p = 4$
1. grass	**1**(16)	**1**(16)
2. treebark	**2**(16)	**2**(16)
3. straw	**3**(13), **1**(1), **6**(1), **12**(1)	**3**(13), **1**(1), **6**(1), **12**(1)
4. herringbone	**4**(16)	**4**(16)
5. wool	**5**(16)	**5**(16)
6. leather	**6**(16)	**6**(16)
7. water	**7**(15), **8**(1)	**7**(16)
8. wood	**8**(16)	**8**(16)
9. raffia	**9**(16)	**9**(16)
10. brickwall	**10**(16)	**10**(16)
11. plastic	**11**(16)	**11**(15), **2**(1)
12. sand	**12**(16)	**12**(16)

8.3 Summary and Discussion

In this chapter, a 2-D nonparametric system-MNC filter based BSI algorithm was introduced for estimation of an unknown 2-D LSI system $h[n_1, n_2]$ with only a given set of data $y[n_1, n_2]$ (see (8.1)). This BSI algorithm is based on the key relation (8.2) between the 2-D MNC deconvolution filter $v_{\mathrm{MNC}}[n_1, n_2]$ introduced in Chapter 7 and the unknown system $h[n_1, n_2]$, which holds valid for any SNR, and therefore, its performance is robust against additive Gaussian noise if the data length is sufficient. Some analytical results show that the system estimate obtained by this BSI algorithm is identifiable except for a scale factor and a space shift (see Properties 8.1 and 8.3).

Because texture images can be modeled as 2-D linear nonGaussian random processes, the texture image model can be estimated by the 2-D system-MNC filter based BSI algorithm. A TSM making use of the estimated texture image model and the histogram of the deconvolved signal (an estimate of the model input) was then introduced. Some simulation results and experimental results were presented to verify the efficacy of the 2-D system-MNC filter based BSI algorithm and the TSM, respectively.

A 2-D FSBM based BSI algorithm, which is guaranteed stable and characterized by amplitude parameters $\boldsymbol{\alpha}$ and phase parameters $\boldsymbol{\beta}$, was also introduced for the estimation of an unknown 2-D LSI system. This BSI algorithm estimates the FSBM parameters $\boldsymbol{\alpha}$ and $\boldsymbol{\beta}$ either sequentially or concurrently (see Fig. 8.3). When the unknown 2-D LSI system is an FSBM of known order, the FSBM parameter estimates obtained are consistent; when the unknown 2-D LSI system is not an FSBM, the system estimate obtained is an approximation to the unknown system. The FSBM parameters of texture images obtained by this BSI algorithm, normalized variance and cumulants of the estimated model input (see (8.67) and (8.68)) are used as feature vectors for texture image classification. Some simulation results were presented to validate the good performance of the 2-D FSBM based BSI algorithm, and some experimental results were provided to show its effective application to texture classification.

The 2-D BSI algorithms and their application to texture image synthesis and classification introduced in this chapter are not comprehensive. However, the 2-D MNC deconvolution criterion given by (7.85) is a useful criterion for 2-D deconvolution and system identification, and the parametric 2-D FSBM model may potentially have other applications because its use is free from stability concerns.

Appendix 8A
Proof of Property 8.2

Let us use the notations of $\boldsymbol{\omega}$, $\boldsymbol{\omega}_i$ given by (7.41) and

$$\int_{-\pi}^{\pi} x(\boldsymbol{\omega})d\boldsymbol{\omega} = \int_{-\pi}^{\pi}\int_{-\pi}^{\pi} x(\omega_1,\omega_2)d\omega_1 d\omega_2$$

for ease of later use in the proof.

Since both $v[n_1,n_2]$ and $h[n_1,n_2]$ are stable LSI systems, the overall system $g[n_1,n_2]$ is also stable. The zero-phase assumption of $g[n_1,n_2]$ implies that $G(\omega_1,\omega_2)$ is a continuous function of $\boldsymbol{\omega}$ and

$$G(\omega_1,\omega_2) \geq 0, \ \forall -\pi < \omega_1 \leq \pi, \quad -\pi < \omega_2 \leq \pi. \tag{8.71}$$

The DSFT $G_1(\omega_1,\omega_2)$ of $g_1[n_1,n_2]$ is the $(2p-1)$-fold 2-D periodic convolution given by

$$\begin{aligned} G_1(\omega_1,\omega_2) &= \underbrace{G(\omega_1,\omega_2) \otimes G(\omega_1,\omega_2) \otimes \cdots \otimes G(\omega_1,\omega_2)}_{(2p-1)\text{ terms}} \\ &= \left(\frac{1}{2\pi}\right)^{2(2p-1)} \cdot \int_{-\pi}^{\pi} \cdots \int_{-\pi}^{\pi} G(\boldsymbol{\omega}_1)\cdots G(\boldsymbol{\omega}_{2p-1}) \\ &\quad \cdot G\left(\boldsymbol{\omega} - \sum_{i=1}^{2p-1} \boldsymbol{\omega}_i\right) d\boldsymbol{\omega}_1 \cdots d\boldsymbol{\omega}_{2p-1}. \end{aligned} \tag{8.72}$$

From (8.72), it is easy to see that $G_1(\omega_1,\omega_2) > 0, \forall(\omega_1,\omega_2)$ due to (8.71) and Assumption (A8-4).

Q.E.D.

Appendix 8B
Proof of Property 8.3

Let $\widetilde{h}[n_1, n_2]$ be an arbitrary system satisfying (8.2) and thus

$$G_1(\omega_1, \omega_2)H^*(\omega_1, \omega_2) = \widetilde{G}_1(\omega_1, \omega_2)\widetilde{H}^*(\omega_1, \omega_2) \\ = \beta \cdot V_{\text{MNC}}(\omega_1, \omega_2)S_y(\omega_1, \omega_2) \tag{8.73}$$

where $\widetilde{G}_1(\omega_1, \omega_2)$ is the DSFT of $\widetilde{g}_1[n_1, n_2]$ given by

$$\widetilde{g}_1[n_1, n_2] = \widetilde{g}^{2p-1}[n_1, n_2] \tag{8.74}$$

in which

$$\widetilde{g}[n_1, n_2] = \widetilde{h}[n_1, n_2] \star v_{\text{MNC}}[n_1, n_2]. \tag{8.75}$$

Without loss of generality, assume that both $G(\omega_1, \omega_2)$ and $\widetilde{G}(\omega_1, \omega_2)$ are zero-phase with positive $g[0, 0]$ and $\widetilde{g}[0, 0]$, i.e.

$$g[n_1, n_2] = g[-n_1, -n_2], \text{ and } g[0, 0] > 0, \tag{8.76}$$

$$\widetilde{g}[n_1, n_2] = \widetilde{g}[-n_1, -n_2], \text{ and } \widetilde{g}[0, 0] > 0. \tag{8.77}$$

It can be obtained, from (8.73), that

$$\widetilde{H}(\omega_1, \omega_2) = \Upsilon(\omega_1, \omega_2)H(\omega_1, \omega_2) \tag{8.78}$$

where

$$\Upsilon(\omega_1, \omega_2) = \frac{G_1^*(\omega_1, \omega_2)}{\widetilde{G}_1^*(\omega_1, \omega_2)} = \frac{G_1(\omega_1, \omega_2)}{\widetilde{G}_1(\omega_1, \omega_2)} > 0, \text{ (by Property 8.2)} \tag{8.79}$$

and that

$$G_1(\omega_1, \omega_2)H^*(\omega_1, \omega_2)V_{\text{MNC}}^*(\omega_1, \omega_2) = G_1(\omega_1, \omega_2)G^*(\omega_1, \omega_2) \\ = \widetilde{G}_1(\omega_1, \omega_2)\widetilde{H}^*(\omega_1, \omega_2)V_{\text{MNC}}^*(\omega_1, \omega_2) \\ = \widetilde{G}_1(\omega_1, \omega_2)\widetilde{G}^*(\omega_1, \omega_2) \geq 0 \quad \text{(by Property 8.2).} \tag{8.80}$$

Let $s[n_1, n_2]$ and $\widetilde{s}[n_1, n_2]$ be the inverse DSFT of $G_1(\omega_1, \omega_2)G^*(\omega_1, \omega_2)$ and $\widetilde{G}_1(\omega_1, \omega_2)\widetilde{G}^*(\omega_1, \omega_2)$, respectively. That is,

$$s[n_1, n_2] = g_1[n_1, n_2] \star g[-n_1, -n_2] \\ = \sum_{m_1=-\infty}^{\infty} \sum_{m_2=-\infty}^{\infty} g^{2p-1}[m_1, m_2]g[m_1 - n_1, m_2 - n_2] \tag{8.81}$$

and

$$\widetilde{s}[n_1,n_2] = \widetilde{g}_1[n_1,n_2] \star \widetilde{g}[-n_1,-n_2]$$
$$= \sum_{m_1=-\infty}^{\infty} \sum_{m_2=-\infty}^{\infty} \widetilde{g}^{2p-1}[m_1,m_2]\widetilde{g}[m_1-n_1,m_2-n_2]. \tag{8.82}$$

One can infer from (8.80) that

$$s[n_1,n_2] = \widetilde{s}[n_1,n_2]. \tag{8.83}$$

Let us further assume that $g[n_1,n_2] \neq 0$ for all $[n_1,n_2] \in \Omega[\mathcal{K},\mathcal{K}]$ and $g[n_1,n_2] = 0$ for all $[n_1,n_2] \notin \Omega[\mathcal{K},\mathcal{K}]$ where $\mathcal{K}$ is a positive integer and

$$\Omega[p_1,p_2] = \{[n_1,n_2] : -p_1 \leq n_1 \leq p_1, -p_2 \leq n_2 \leq p_2\}, \tag{8.84}$$

and thus $g_1[n_1,n_2] \neq 0$ only for $[n_1,n_2] \in \Omega[\mathcal{K},\mathcal{K}]$ by (8.3), and $s[n_1,n_2] \neq 0$ only for $[n_1,n_2] \in \Omega[2\mathcal{K},2\mathcal{K}]$. Then, $s[n_1,n_2]$ in (8.81) can be expressed as

$$s[n_1,n_2] = \sum_{m_1=-\mathcal{K}}^{\mathcal{K}} \sum_{m_2=-\mathcal{K}}^{\mathcal{K}} g^{2p-1}[m_1,m_2]g[m_1-n_1,m_2-n_2],$$
$$[n_1,n_2] \in \Omega[2\mathcal{K},2\mathcal{K}] \tag{8.85}$$

and the equality $s[n_1,n_2] = \widetilde{s}[n_1,n_2]$ implies that $g[n_1,n_2] \neq 0$ and $\widetilde{g}[n_1,n_2] \neq 0$ only for $[n_1,n_2] \in \Omega[\mathcal{K},\mathcal{K}]$. Furthermore, from (8.83) and (8.85) that

$$s[2\mathcal{K},2\mathcal{K}] = g^{2p}[\mathcal{K},\mathcal{K}]$$
$$= \widetilde{s}[2\mathcal{K},2\mathcal{K}] = \widetilde{g}^{2p}[\mathcal{K},\mathcal{K}] \tag{8.86}$$

which implies

$$g[\mathcal{K},\mathcal{K}] = \widetilde{g}[\mathcal{K},\mathcal{K}] \text{ or } -\widetilde{g}[\mathcal{K},\mathcal{K}]. \tag{8.87}$$

Again, by (8.85) and $s[2\mathcal{K},2\mathcal{K}-1] = \widetilde{s}[2\mathcal{K},2\mathcal{K}-1]$, we have

$$g[\mathcal{K},\mathcal{K}]g[\mathcal{K},\mathcal{K}-1]\left(g^{2p-2}[\mathcal{K},\mathcal{K}] + g^{2p-2}[\mathcal{K},\mathcal{K}-1]\right)$$
$$= \widetilde{g}[\mathcal{K},\mathcal{K}]\widetilde{g}[\mathcal{K},\mathcal{K}-1]\left(\widetilde{g}^{2p-2}[\mathcal{K},\mathcal{K}] + \widetilde{g}^{2p-2}[\mathcal{K},\mathcal{K}-1]\right), \tag{8.88}$$

which by (8.87), (8.88) and Lemma 6.10 further gives rise to

$$|g[\mathcal{K},\mathcal{K}-1]| = |\widetilde{g}[\mathcal{K},\mathcal{K}-1]|. \tag{8.89}$$

Moreover, one can infer from (8.88) and (8.89) that

$$\frac{\widetilde{g}[\mathcal{K},\mathcal{K}-1]}{g[\mathcal{K},\mathcal{K}-1]} = \frac{g[\mathcal{K},\mathcal{K}]}{\widetilde{g}[\mathcal{K},\mathcal{K}]}. \tag{8.90}$$

Similarly, simplifying $s[n_1,n_2] = \widetilde{s}[n_1,n_2]$ for $[n_1,n_2] = [2\mathcal{K},2\mathcal{K}-2]$, $[2\mathcal{K},2\mathcal{K}-3]$, ..., $[2\mathcal{K},0]$, $[2\mathcal{K}-1,2\mathcal{K}]$, $[2\mathcal{K}-1,2\mathcal{K}-1]$, ..., $[2\mathcal{K}-1,0]$, ..., $[\mathcal{K}+1,2\mathcal{K}]$,

$[\mathcal{K}+1, 2\mathcal{K}-1]$, ..., $[\mathcal{K}+1, 0]$, $[\mathcal{K}, 2\mathcal{K}]$, $[\mathcal{K}, 2\mathcal{K}-1]$, ..., $[\mathcal{K}, \mathcal{K}]$, by (8.85) and Lemma 6.10, we can end up with

$$|g[n_1, n_2]| = |\widetilde{g}[n_1, n_2]|, \ \ \forall [n_1, n_2] \in \Omega_{\mathrm{TNSHP}}[\mathcal{K}, \mathcal{K}] \tag{8.91}$$

and

$$\frac{\widetilde{g}[n_1, n_2]}{g[n_1, n_2]} = \frac{g[\mathcal{K}, \mathcal{K}]}{\widetilde{g}[\mathcal{K}, \mathcal{K}]}, \ \ \forall [n_1, n_2] \in \Omega_{\mathrm{TNSHP}}[\mathcal{K}, \mathcal{K}], \ \ [n_1, n_2] \neq [\mathcal{K}, \mathcal{K}], \tag{8.92}$$

which together with (8.87) lead to

$$g[n_1, n_2] = \widetilde{g}[n_1, n_2], \ \ \forall [n_1, n_2] \in \Omega_{\mathrm{TNSHP}}[\mathcal{K}, \mathcal{K}] \tag{8.93}$$

since $g[0,0] > 0$ and $\widetilde{g}[0,0] > 0$. Moreover, one can infer from (8.76), (8.77) and (8.93) that

$$g[n_1, n_2] = \widetilde{g}[n_1, n_2], \ \forall [n_1, n_2] \in \Omega[\mathcal{K}, \mathcal{K}], \tag{8.94}$$

which implies that $G_1(\omega_1, \omega_2) = \widetilde{G}_1(\omega_1, \omega_2)$ and thus $\Upsilon(\omega_1, \omega_2) = 1, \forall (\omega_1, \omega_2)$ by (8.79). Therefore, we obtain from (8.78) that

$$H(\omega_1, \omega_2) = \widetilde{H}(\omega_1, \omega_2) \tag{8.95}$$

under the zero-phase assumption for both $g[n_1, n_2]$ and $\widetilde{g}[n_1, n_2]$. Furthermore, by Property 8.1 and (8.95), $H'(\omega_1, \omega_2) = \alpha \widetilde{H}(\omega_1, \omega_2) \cdot e^{j(\omega_1 \tau_1 + \omega_2 \tau_2)} = \alpha H(\omega_1, \omega_2) \cdot e^{j(\omega_1 \tau_1 + \omega_2 \tau_2)}$ is also a solution of (8.2). The assumption that $\widetilde{g}[n_1, n_2] \neq 0$ only for $[n_1, n_2] \in \Omega[\mathcal{K}, \mathcal{K}]$ can be relaxed by allowing $\mathcal{K} \to \infty$ for the proof to be true as $g[n_1, n_2]$ is of infinite length. A general proof without the condition $\widetilde{g}[n_1, n_2] \neq 0$ for $[n_1, n_2] \in \Omega[\mathcal{K}, \mathcal{K}]$ is still unknown.

Q.E.D.

Appendix 8C Proof of Theorem 8.6

In the absence of noise, the power spectrum of $\varepsilon[n_1, n_2]$ can easily be seen to be

$$\begin{aligned} S_\varepsilon(\omega_1, \omega_2) &= \sigma_u^2 |H_{\mathrm{MP}}(\omega_1, \omega_2) H_{\mathrm{AP}}(\omega_1, \omega_2) V_{\mathrm{MP}}(\omega_1, \omega_2)|^2 \\ &= \sigma_u^2 |\widetilde{G}(\omega_1, \omega_2)|^2 \end{aligned} \tag{8.96}$$

where

$$\widetilde{G}(\omega_1, \omega_2) \doteq H_{\mathrm{MP}}(\omega_1, \omega_2) V_{\mathrm{MP}}(\omega_1, \omega_2). \tag{8.97}$$

Since both $H_{\mathrm{MP}}(\omega_1, \omega_2)$ and $V_{\mathrm{MP}}(\omega_1, \omega_2)$ are causal minimum-phase filters with the same leading coefficient $h_{\mathrm{MP}}[0,0] = v_{\mathrm{MP}}[0,0] = 1$, the inverse DSFT

$\widetilde{g}[n_1, n_2]$ of $\widetilde{G}(\omega_1, \omega_2)$ is also causal minimum phase with leading coefficient $\widetilde{g}[0, 0] = 1$. Therefore,

$$E\{\varepsilon^2[n_1, n_2]\} = \sigma_u^2 \sum\sum_{[n_1,n_2]\in\Omega_{\text{TNSHP}}[\infty,\infty]} \widetilde{g}^2[n_1, n_2] \geq \sigma_u^2 \quad (8.98)$$

where the equality holds only when $\widetilde{g}[n_1, n_2] = \delta[n_1, n_2]$ or $\widetilde{G}(\omega_1, \omega_2) = 1$. Thus, the optimum minimum-phase LPE filter $\widehat{V}_{\text{MP}}(\omega_1, \omega_2) = 1/H_{\text{MP}}(\omega_1, \omega_2)$ by (8.97) and $\min\{E\{\varepsilon^2[n_1, n_2]\}\} = \sigma_u^2$.

Q.E.D.

Appendix 8D
Proof of Fact 8.9

Since higher-order cumulants of Gaussian processes (due to $w[n_1, n_2]$ in (8.1)) are equal to zero, the Gaussian noise $w[n_1, n_2]$ can be ignored in the following proof. By Theorem 8.7, (8.30) and (8.52), we have

$$\begin{aligned}
&\arg\{V_{\text{AP}}(\omega_1, \omega_2)\} + \arg\{H_{\text{PS}}(\omega_1, \omega_2)\} \\
&= \sum\sum_{[i_1,i_2]\in\Omega'_{\text{TNSHP}}[p_1,p_2]} (\gamma_{i_1,i_2} + \beta_{i_1,i_2}) \sin(i_1\omega_1 + i_2\omega_2), \quad |\omega_1| \leq \pi, \ |\omega_2| \leq \pi \\
&= \tau_1\omega_1 + \tau_2\omega_2, \quad |\omega_1| \leq \pi, \ |\omega_2| \leq \pi \\
&= \sum\sum_{[i_1,i_2]\in\Omega'_{\text{TNSHP}}[\infty,\infty]} d_{i_1,i_2} \sin(i_1\omega_1 + i_2\omega_2), \quad |\omega_1| \leq \pi, \ |\omega_2| \leq \pi
\end{aligned} \quad (8.99)$$

where τ_1 and τ_2 are integers, and

$$d_{i_1,i_2} = \begin{cases} \dfrac{2\tau_1}{i_1}(-1)^{i_1+1}, & i_2 = 0 \text{ and } i_1 > 0 \\ \dfrac{2\tau_2}{i_2}(-1)^{i_2+1}, & i_1 = 0 \text{ and } i_2 > 0 \\ 0, & \text{otherwise} \end{cases} \quad (8.100)$$

are coefficients of the Fourier series expansion of the linear function $\tau_1\omega_1 + \tau_2\omega_2$. From the second line and the fourth line of (8.99), one can see that

$$d_{i_1,i_2} = 0, \ \forall [i_1, i_2] \in \Omega'_{\text{TNSHP}}[\infty, \infty] \text{ but } \notin \Omega'_{\text{TNSHP}}[p_1, p_2], \quad (8.101)$$

which together with (8.100) leads to $\tau_1 = \tau_2 = 0$, i.e.

$$d_{i_1,i_2} = 0, \forall [i_1, i_2] \in \Omega'_{\text{TNSHP}}[\infty, \infty]. \quad (8.102)$$

Therefore, from (8.99) and (8.102), one can obtain $\gamma_{i_1,i_2} = -\beta_{i_1,i_2}$, $\forall [i_1, i_2] \in \Omega'_{\text{TNSHP}}[p_1, p_2]$ for finite p_1 and p_2.

Q.E.D.

Problems

8.1. Prove Property 8.1.

8.2. Prove the gradient $\dfrac{\partial \varepsilon[n_1,n_2]}{\partial a_{i_1,i_2}}$ given by (8.45).

8.3. Prove Fact 8.8.

8.4. Prove the gradient $\dfrac{\partial \mathcal{J}_p(\gamma)}{\partial \gamma_{i_1,i_2}}$ given by (8.63).

Computer Assignments

8.1. 2-D BSI of an MA System
Consider that $y[n_1,n_2] = h[n_1,n_2] \star u[n_1,n_2] + w[n_1,n_2]$ where $h[n_1,n_2]$ is a separable system whose transfer function is given by

$$H(z_1,z_2) = (1 - 0.8z_1^{-1}) \cdot (1 - 0.8z_2^{-1}),$$

$u[n_1,n_2]$ is a zero-mean, i.i.d., exponentially distributed stationary random process with variance $\sigma_u^2 = 1$, normalized skewness $\gamma_3\{u[n_1,n_2]\} = 2$ and normalized kurtosis $\gamma_4\{u[n_1,n_2]\} = 6$ and $w[n_1,n_2]$ is white Gaussian noise. Generate thirty realizations of synthetic data $\{y[n_1,n_2], n_1 = 0,...,127, n_2 = 0,...,127\}$. With the generated synthetic data, estimate the 2-D MA system $h[n_1,n_2]$ using the two 2-D BSI algorithms presented in this chapter.

(a) Write a computer program to implement the 2-D system-MNC filter based BSI algorithm and perform simulations with the following settings. Initial step size $\mu_0 = 1$ and integer $K = 10$ for step size search range in (S2) of (T1) and (S5) of (T2); convergence tolerance $\zeta = 10^{-5}$ in (S2) of (T1) and (S6) of (T2); $p_1 = p_2 = 5$ in (S1) and (S2) of (T1); $\mathbf{Q}^{[0]} = \mathbf{I}$ and the initial condition $v^{[0]}[n_1,n_2] = \delta[n_1-2, n_2-2]$ for the hybrid MNC algorithm with cumulant order set to $2p = 4$ in (S2) of (T1); FFT size $L \times L = 32 \times 32$ in (S3) of (T1) and $H^{[0]}[k_1,k_2] = 1$ for all k_1 and k_2 in (T2). Obtain thirty estimates of $h[n_1,n_2]$ for SNR $= 5$ dB and SNR $= 20$ dB, and exhibit the true system $h[n_1,n_2]$, the true $v_{\rm MS}[n_1,n_2]$, the averages $\bar{h}[n_1,n_2]$ and $\bar{v}_{\rm MS}[n_1,n_2]$ given by (8.20) and (8.21), and the RMSEs $\sigma(\widehat{h}[n_1,n_2])$ and $\sigma(\widehat{v}_{\rm MS}[n_1,n_2])$ given by (8.22) and (8.23), respectively. Explain what you observe from these results.

(b) Write a computer program to implement the 2-D FSBM based BSI algorithm and perform simulations with the following settings. Initial step size $\mu_0 = 1$; integer $K = 10$ for step size search range; convergence tolerance $\zeta = 10^{-5}$; cumulant order $p = 3$, $\mathsf{k} = 1$ and 2 in (S1); $\mathbf{Q}^{[0]} = \mathbf{I}$ and the initial condition $a^{[0]}_{i_1,i_2} = 0, \forall[i_1, i_2]$ in (S2) and $\mathbf{Q}^{[0]} = \mathbf{I}$ and $\gamma^{[0]}_{i_1,i_2} = 0, \forall[i_1, i_2]$ in (S3). Obtain thirty estimates of $h[n_1, n_2]$ for SNR = 5, 10, 15 and 20 dB and $p_1 = p_2 = 1$, 2 and 3 for the region of support $\Omega'_{\text{TNSHP}}[p_1, p_2]$, and exhibit the associated NMSEs given by (8.65). Explain what you observe from these results.

References

1. J. K. Tugnait, "Estimation of linear parametric models of nonGaussian discrete random fields with application to texture synthesis," *IEEE Trans. Image Processing*, vol. 3, no. 2, pp. 109–127, Feb. 1994.
2. T. E. Hall and G. B. Giannakis, "Bispectral analysis and model validation of texture images," *IEEE Trans. Image Processing*, vol. 4, no. 7, pp. 996–1009, July 1995.
3. T. E. Hall and G. B. Giannakis, "Image modeling using inverse filtering criteria with application to textures," *IEEE Trans. Image Processing*, vol. 5, no. 6, pp. 938–949, June 1996.
4. M. K. Tsatsanis and G. B. Giannakis, "Object and texture classification using higher order statistics," *IEEE Trans. Pattern Analysis and Machine Intelligence*, vol. 14, no. 7, pp. 733–750, July 1992.
5. K. B. Eom, "2-D moving average models for texture synthesis and analysis," *IEEE Trans. Image Processing*, vol. 7, no. 12, pp. 1741–1746, Dec. 1998.
6. R. L. Kashyap and R. Chellappa, "Estimation and choice of neighbors in spatial-interaction models of images," *IEEE Trans. Information Theory*, vol. 29, no. 1, pp. 58–72, Jan. 1983.
7. R. L. Kashyap and K. B. Eom, "Robust image modeling techniques with an image restoration application," *IEEE Trans. Acoustics, Speech and Signal Processing*, vol. 36, no. 8, pp. 1313–1325, Aug. 1988.
8. C.-Y. Chi and C.-H. Chen, "Two-dimensional frequency-domain blind system identification using higher order statistics with application to texture synthesis," *IEEE Trans. Signal Processing*, vol. 49, no. 4, pp. 864–877, April 2001.
9. C.-H. Chen, "Blind Multi-channel Equalization and Two-dimensional System Identification Using Higher-order Statistics," Ph.D. Dissertation, Department of Electrical Engineering, National Tsing Hua University, Taiwan, R.O.C., June 2001.
10. R. Chellappa and S. Chatterjee, "Classification of texture using Gaussian Markov random fields," *IEEE Trans. Acoustics, Speech and Signal Processing*, vol. 33, no. 4, pp. 959–963, April 1985.
11. M. H. Hayes, *Statistical Digital Signal Processing and Modeling.* New York: John Wiley & Sons, 1996.
12. C.-H. Chen, C.-Y. Chi, and C.-Y. Chen, "2-D Fourier series based model for nonminimum-phase linear shift-invariant systems and texture image classification," *IEEE Trans. Signal Processing*, vol. 50, no. 4, pp. 945–955, April 2002.

13. S. M. Kay, *Modern Spectral Estimation: Theory and Application.* New Jersey: Prentice-Hall, 1988.
14. D. E. Dudgeon and R. M. Mersereau, *Multidimensional Digital Signal Processing.* New Jersey: Prentice-Hall, 1984.
15. C.-Y. Chi, "Fourier series based nonminimum phase model for statistical signal processing," *IEEE Trans. Signal Processing*, vol. 47, no. 8, pp. 2228–2240, Aug. 1999.
16. A. V. Oppenheim and R. W. Schafer, *Discrete-Time Signal Processing.* New Jersey: Prentice-Hall, 1989.
17. L. R. Rabiner and R. W. Schafer, *Digital Processing of Speech Signals.* New Jersey: Prentice-Hall, 1978.
18. L. R. Rabiner, K. C. Pan, and F. K. Soong, "On the performance of isolated word speech recognizers using vector quantization and temporal energy contours," *AT&T Bell Laboratories Technical Journal*, vol. 63, no. 7, pp. 1245–1260, 1984.
19. R. L. Kashyap, R. Chellappa, and A. Khotanzad, "Texture classification using features derived from random field models," *Pattern Recognition Letters*, vol. 1, no. 1, pp. 43–50, 1982.

Index